T0091724

Numerical Electromagnetics

The FDTD Method

Beginning with the development of Finite Difference Equations, and leading to the complete FDTD algorithm, this is a coherent introduction to the FDTD method (the method of choice for modeling Maxwell's equations). It provides students and professional engineers with everything they need to know to begin writing FDTD simulations from scratch and to develop a thorough understanding of the inner workings of commercial FDTD software. Stability, numerical dispersion, sources, and boundary conditions are all discussed in detail, as are dispersive and anisotropic materials. A comparative introduction of the finite volume and finite element methods is also provided. All concepts are introduced from first principles, so no prior modeling experience is required, and they are made easier to understand through numerous illustrative examples and the inclusion of both intuitive explanations and mathematical derivations.

Umran S. Inan is a Professor in the Department of Electrical Engineering at Stanford University, where he has led pioneering research in plasma physics and electromagnetics for over 30 years. He also currently serves as president of Koç University in Istanbul. As a committed teacher, he has authored two previous books that have become standard textbooks for electromagnetics courses. He has also received numerous awards, including the Tau Beta Pi Excellence in Undergraduate Teaching Award and the Outstanding Service Award from the Electrical Engineering Department for Excellence in Teaching. He is a fellow of the Institute of Electrical and Electronics Engineers (IEEE), the American Geophysical Union (AGU), and the American Physical Society. He is the recipient of the 2008 Appleton Prize of the International Union of Radio Science (URSI) and the Royal Society, and the 2007 Allan V. Cox Medal of Stanford University for faculty excellence in fostering undergraduate research.

Robert A. Marshall is a Research Scientist at Boston University, where he has worked since receiving his Ph.D. in Electrical Engineering from Stanford University in 2009. He has received numerous awards from both Stanford University and the academic community, including the International Radio Science Union (URSI) and the American Geophysical Union (AGU).

FDTD is presently the method of choice for solving most numerical electromagnetics studies. The Inan and Marshall book is a very thorough, yet readable, account of all the details of the method, very valuable for students and professionals alike, with problems included, ready for a course. Even those who use commercial programs would benefit from a better understanding of dispersion, accuracy, media descriptions, basic limitations and so on, all treated in great detail. I am sure it will be a good reference for a number of years. No special background needed since basic material for a good understanding is given, such as electromagnetics, or numerical solution of differential equations. Highly recommended.

Jørgen Bach Andersen, Professor Emeritus,
Aalborg University

This book is what has been needed to teach in a systematic fashion the theoretical developments and numerical implementations of one of the most powerful numerical techniques in computational electromagnetic with vast applications in a variety of disciplines. The authors have elegantly and masterfully presented the FDTD method from the Maxwell's equations with sufficient rigor to make the book very reachable by students and practicing researchers.

Yahya Rahmat-Samii, Distinguished Professor,
University of California at Los Angeles

Numerical Electromagnetics

The FDTD Method

UMRAN S. INAN
Stanford University*

ROBERT A. MARSHALL
Boston University

* Now Serving as President of Koç University, Istanbul

CAMBRIDGE
UNIVERSITY PRESS

CAMBRIDGE UNIVERSITY PRESS
Cambridge, New York, Melbourne, Madrid, Cape Town,
Singapore, São Paulo, Delhi, Tokyo, Mexico City

Cambridge University Press
The Edinburgh Building, Cambridge CB2 8RU, UK

Published in the United States of America by Cambridge University Press, New York

www.cambridge.org
Information on this title: www.cambridge.org/9780521190695

First published 2011

Printed in the United Kingdom at the University Press, Cambridge

A catalog record for this publication is available from the British Library

Library of Congress Cataloging in Publication data
Inan, Umran S.
Numerical electromagnetics : the FDTD method / Umran S. Inan, Robert A. Marshall.
p. cm.
Includes index.
ISBN 978-0-521-19069-5 (hardback)
1. Electromagnetism – Computer simulation.   2. Finite differences.   3. Time-domain analysis.
I. Marshall, Robert A.   II. Title.
QC760.I589   2011
530.14'10113 – dc22   2010048442

ISBN 978-0-521-19069-5 Hardback

Additional resources for this publication at www.cambridge.org/9780521190695

To my parents, my beautiful wife Elif, my dear children
Ayse and Ali, and my very special granddaughter Ayla.
  USI

To my brother Pete.
  RM

Contents

Preface

Our purpose in this text is to provide an introduction to Numerical Electromagnetics, sometimes referred to as Computational Electromagnetics (CEM), a subject much too broad to cover in a single volume. Fifteen years ago, we might have found it difficult to choose between the different techniques to emphasize in our relatively brief coverage. However, due to a number of developments in the late 1980s and the 1990s, partial differential equation (PDE)-based methods, and in particular the so-called Finite Difference Time Domain (FDTD) method, have emerged as the methods with arguably the broadest range of applicability. This is especially true for electromagnetic problems involving complex and dispersive media, photonics applications, and modeling of high-speed circuits and devices. In addition, FDTD modeling of practical problems can now be undertaken with computer resources readily available to individual users. Finally, and quite importantly for our purposes, FDTD methods are relatively straightforward and intuitively follow from a physical understanding of Maxwell's equations, making this topic particularly suitable for both undergraduate and first-year graduate students in the context of a mezzanine-level course. Students with limited or no prior modeling experience will find that the FDTD method is the simplest and most insightful method from which to start their modeling education, and they can write practical and useful simulations in a matter of minutes. With an understanding of the FDTD method under their belts, students can move on to understanding more complicated methods with relative ease.

A number of software packages exist for modeling in the FDTD method, both free, open-source packages and very expensive industrial software suites. Many of these packages are very powerful. In our experience teaching the FDTD method, however, we find that students get far more out of the course by coding their own simple FDTD problems rather than simply pushing buttons in a software package. To this end, this book is targeted toward a comprehensive understanding of the fundamental inner workings and details of the FDTD method. We provide all of the basic background material to understanding how difference equations are formed from differential equations; how instability and numerical dispersion arise; and how to deal with sources, boundary conditions, complex materials, and complicated structures. We provide a number of problems in each chapter, some of which require coding simulations from scratch. These problems are presented in a logical progression, so that codes developed in early chapters can be easily modified to include the new material in later chapters. Furthermore, we encourage the students to modify the basic codes outlined in the problems in order to explore problems of interest to them.

Acknowledgments

The material presented in this book was initially put together as class notes for the EE256 Numerical Electromagnetics course, newly introduced and taught by one of us (USI) at Stanford University in the spring quarter of 1999. The course was then taught regularly every other year, for graduate students from Electrical Engineering, Applied Physics, and Physics. Then Ph.D. student Dr. Michael Chevalier served as the Teaching Assistant (TA) for the course in the spring of 1999, 2001, and 2003, while the other one of us (RAM) served as TA in the spring of 2005 and 2007 and also taught the course in the summer of 2008. Ph.D. student Forrest Foust served as the TA in spring 2009, when the course was taught for the last time by one of us (USI). The original set of course notes from 1999 significantly improved over the years, with the help of the TAs cited above, and by the feedback, comments, and suggestions of the many students who took the course. We would like to extend our sincere gratitude especially to Forrest Foust, who put in countless hours of proofreading on short notice and made many valuable suggestions. Sarah Matthews, Julie Lancashire, and Emma Walker at Cambridge were extraordinarily helpful and patient during the development of this text. Finally, both of us want to acknowledge our families for their patience and encouragement during the process.

On the cover: 2D FDTD simulation of a Luneburg lens. A Luneburg lens, proposed by Rudolf Luneburg in 1944, is a dielectric sphere whose permittivity varies with radius as $\epsilon = \epsilon_0[2 - (r/R)^2]$, where R is the radius of the sphere. When a point source is excited at the surface of the lens, the varying index of refraction creates a plane wave at the output of the lens. The opposite is also true: a plane wave incident on the lens will focus to a point at the opposite side. Recently, an adaptation of the Luneburg lens has been proposed as a compact, high-speed receiver for internet connectivity on trains.

1 Introduction

1.1 Why FDTD?

With the continued growth of computing power, modeling and numerical simulation has grown immensely as a tool for understanding and analyzing just about any problem in science. Where in the mid-twentieth century, detailed analyses were required to get any meaningful insight out of complex problems, today we can simply plug the governing differential equations into a computer, the results of which can provide an immense amount of information, which is of course complementary to theoretical analyses. The growth of computing power has brought with it a smorgasbord of modeling methods, applicable in any number of fields. The problem, then, is knowing when to use which method.

In electromagnetic problems, which are of interest to us in this book, there are quite a number of useful numerical methods, including the Method of Moments, Finite Volume methods, Finite Element methods, and Spectral methods, just to name a few. The FDTD method, however, grew to become the method of choice in the 1990s, for a number of reasons. First, it has always had the advantage of being a very simple method; we shall see in Chapter 3 that the derivation of difference equations is very straightforward. However, before the 1990s, the FDTD method was hindered by the need to discretize the simulation space on sub-wavelength scales, with time steps commensurately small. Hence, any reasonable problem would require a large amount of computer memory and time. Since the 1990s, however, with the growth of computing power, the FDTD method has taken off.

As an example, a typical 3D problem would require, at minimum, 100 grid cells in each dimension, or 10^6 grid cells total. With a minimum of six fields to compute (three components each of the electric field $\mathcal{E}$ and magnetic field $\mathcal{H}$), and 2 bytes per value (for 16-bit resolution), we require 12 MB of memory. As for computation time, our simulation might require 1,000 time steps. Each of six equations will have four additions and two multiplications (at minimum, for the free-space algorithm in Chapter 4) at each of the 1 million grid cells, for $\sim$36 billion operations over the time of our simulation. In 1990, 12 MB of memory and 36 billion operations was a significant calculation; today, you could quite easily run this simulation on your mobile phone.[1]

---

[1] More accurately, this simulation used 48 MB on a desktop PC running Matlab, since it stored the arrays at 8-byte doubles; and it ran 1,000 time steps in under 5 minutes.

The advantages of the FDTD method over other methods include:

- *Short development time.* Thanks to the simple discretization process, a realistic 2D or 3D simulation can be written in only a few minutes in less than 100 lines of code. In other methods, such as the Finite Element method, creating the numerical grid alone can require entire software packages, and understanding of the discretization procedure can be quite convoluted.
- *Ease of understanding.* Again, thanks to the simple discretization procedure, the FDTD method is easily understandable and directly follows from the differential form of Maxwell's equations. The stability and dispersion characteristics of the method also follow from a simple, intuitive understanding of the updating procedure.
- *Explicit nature.* In the traditional explicit FDTD method, no linear algebra or matrix inversions are required, and as such there is no inherent limit to the size of a simulation; computer time is the only limitation.

However, there are also a number of disadvantages of the FDTD method:

- *Stair-stepping edges.* The orthogonal grid structure of the FDTD method implies that edges of structures within the simulation have edges that follow the grid structure. This can become a problem for curved surfaces, for which greater accuracy is sought. Some methods for FDTD have been developed to overcome this limitation, including the subcell structures discussed in Chapter 12, but other methods are generally better suited to these complex geometries.
- *Computational time.* In the FDTD method, the time step at which we advance the solution is limited by the spatial size, and cannot be larger than a certain maximum size, as we will derive in Chapter 5. For simulations with large spaces or multiple scales (largely varying wavelengths), this means the simulation must be run for a very long time. Other methods can often be better at dealing with multiscale problems.

1.2   Other methods

This book covers the FDTD method in detail, with little coverage of other methods. As mentioned above, this is due to the increasing prevalence of the FDTD method in electromagnetic problems. However, any good engineer or scientist should have a good understanding of other available methods, and should develop knowledge of the appropriate conditions under which different methods are used. Here we provide a brief mention of some other methods commonly used in electromagnetic problems.

1.2.1   Finite volume time domain

The Finite Volume Time Domain (FVTD) method became popular in modeling electromagnetic problems due to its flexibility in modeling irregular structures [1, 2]. As we will see throughout this book, the FDTD method is somewhat restricted to regular, structured grids, and any curved surfaces become "staircased" when the discretized grid is formulated. We will see in Chapter 15 that FDTD methods can be developed around irregular grids; however, the FVTD method is another way of working around irregular structures.

In short, the FVTD method defines the fields $\overline{\mathscr{E}}$ and $\overline{\mathscr{H}}$ in small volumes of space, rather than at the nodes of grid cells as in the FDTD method. These small volumes can be arbitrarily defined, but are typically taken to be tetrahedra in 3D or triangles in 2D. These shapes simplify the resulting equations and can be designed around curved and complex structures quite well. The FVTD method then uses the integral forms of Maxwell's equations to conserve the field quantities. For example, in a small volume V_e with surface area A_e, any change in the electric or magnetic flux inside the volume from one time step to the next must be balanced by the flux moving across the boundary area A_e, which moves into (or out of) the adjacent cells.

An introduction to the Finite Volume method for electromagnetics is provided in the book by S. Rao [3], and we will introduce it in some detail in Chapter 15. The method has the obvious advantage that irregular structures can be modeled quite easily. The simulation time is typically very similar to the FDTD method. Disadvantages include the need to create and define an irregular grid of tetrahedral cells, which is quite cumbersome outside of commercial software.

1.2.2 Finite difference frequency domain

Time domain methods such as FDTD are extremely useful when a transient or broadband analysis is required. For example, we may be interested in the scattering pattern of a broadband pulse of energy from a particular scatterer. However, in cases where a steady-state solution is sought only at a single frequency, the FDTD method is rather inefficient. Instead, frequency domain methods can be much more efficient, since they avoid the need to step in time.

The Finite Difference Frequency Domain (FDFD) method is highly applicable, since it maintains the spatial features of the FDTD method, but removes time stepping. Rather, the steady-state solution is found at a single frequency through a matrix inversion process. We will briefly introduce the FDFD method in Chapter 14.

The FDFD method has the additional advantage that dispersive materials become trivial to implement. As we shall see in Chapter 10, dispersive materials in the FDTD method require either convolution terms or auxiliary equations. In the FDFD method, one simply uses the scalar (or vector, in the case of anisotropic dispersive materials) values of ϵ, μ, and σ at the frequency of interest.

Particular problems that are better suited to FDFD are those where the solution is required at only a single frequency. For example, mobile communications typically operate in a narrow bandwidth around a carrier frequency, so that the bandwidth can be approximated by a single frequency. However, the FDFD method can also be used for broadband simulations, by running multiple simulations, one at each frequency of interest. In this way the spectral response of a problem can be determined, with the frequency resolution limited only by the number of simulations one is willing to run. This can be useful for problems involving dispersive media, whose material parameters vary with frequency in a way that cannot be easily modeled in FDTD. In Chapter 10 we will discuss modeling of dispersive materials in the FDTD method, where methods have been derived for some typical dispersion characteristics.

1.2.3   Finite element methods

The finite element method has become prominent in electromagnetic problems in the past decade or so, but has been around much longer than that, having originated in the 1940s with the work of A. Hrennikoff and R. Courant[2] [4]. The finite element method (often known as Finite Element Analysis or FEA) was developed in the 1950s for airframe and structural analysis.

Like the finite volume method, the finite element method divides the simulation space into small areas or volumes (in 2D and 3D, respectively) which can be arbitrarily shaped and oriented; for this reason, the finite element method is well suited to problems with complex geometry. Also similar to the finite volume method, while the small "subdomains" can have arbitrary shapes, triangles and tetrahedra are most commonly used for their simplicity.

Now, in each of the other methods we have discussed so far, Maxwell's equations are discretized and values of the fields are found which satisfy these equations. In the finite element method, however, the *solution* to Maxwell's equations is approximated over each subdomain with some functional form, usually a low-order polynomial, which is known as a basis function. The solutions in each subdomain are then made to be continuous across their boundaries, and the solution must be made to fit with the global boundary conditions enforced by any scattering structures.

The finite element method in electromagnetics has the disadvantage of being rather complicated, and as such we will provide only a brief introduction to the method in Chapter 15. However, Finite Element Time Domain (FETD) is the state-of-the-art for time domain solutions of electromagnetic problems. Many books have been written on the method, and we refer the reader to those books in Chapter 15.

Discontinuous Galerkin methods

One of the drawbacks of the finite element method in electromagnetic problems is that it requires some level of *global* knowledge of the simulation space. The basis functions used are local, defined in each grid element, but to enforce continuity at element boundaries, a large, sparse matrix must be solved, which can heavily increase the computational cost.

More recently, discontinuous Galerkin methods have moved to the forefront of electromagnetic simulation. These methods enforce strict locality by relaxing the requirement of continuity between elements. Discontinuous Galerkin methods borrow ideas from finite volume methods to connect elements together at their boundaries, and result in explicit, local, and highly accurate algorithms. We will provide a brief introduction to discontinous Galerkin methods in Section 15.4.

---

[2] The same Courant for whom the Courant-Friedrichs-Lewy (CFL) condition in FDTD is named.

1.2.4   Spectral methods

In each of the methods described above, the discretization of space requires on the order of 10 or more grid cells per wavelength, for the smallest wavelength in the simulation, in order to achieve reasonable accuracy in the results. As we have mentioned, for large problems or multiscale problems, this restriction becomes cumbersome and leads to long simulation times. Spectral methods take advantage of the Nyquist theorem, which states that only two points are needed per wavelength to perfectly reconstruct a wave; indeed, spectral methods have been shown to require only two grid cells per wavelength.

In spectral methods, the simulation space is broken into grid cells as usual, but the solution at a given time step is approximated by a function covering the entire simulation space; the methods are "spectral" because the functional form is usually a Fourier decomposition. This type of spectral method is very similar, in fact, to the finite element method; the primary difference is that the finite element method is *local*: the functional forms are assumed to be piecewise continuous over small subdomains; whereas the spectral methods are *global*, where the functional forms cover the entire simulation space. The largest frequency used in the Fourier representation of the solution defines the smallest wavelength, and in turn the grid cell size.

The spectral method of choice in recent years for time domain simulations of Maxwell's equations has been the Pseudospectral Time Domain (PSTD) method. In this method, the spatial derivatives are approximated by taking the Fast Fourier Transform (FFT) of the spatial distribution of fields along an axis; multiplying by jk to achieve the spatial derivative; then taking the inverse FFT to get back to the spatial domain. These spatial derivatives are then used directly in the update equations, and time marching proceeds as in the FDTD method.

These methods provide the stated advantage that a far more coarse grid is required; similar accuracy can be achieved compared to the FDTD method for considerably fewer grid cells. Liu [5, 6] reports a reduction in computer storage and time of a factor of 8^D, where D is the dimensionality, compared to the FDTD algorithm. The costs of the PSTD methods are a slightly stricter stability criterion (by a factor of $\pi/2$) and slightly increased numerical dispersion.

In the interest of brevity, we will not provide a detailed overview of spectral methods or the PSTD method in this book. The interested reader is referred to Chapter 17 of [7] for a good overview of PSTD methods, and the book by Hesthaven et al. [8] for an introduction to spectral methods for time-domain problems.

1.3   Organization

This book is intended to be used as a teaching tool, and has thus been written in the order that we feel is most appropriate for a one-semester course on the FDTD method. The book can be thought of as loosely organized into three sections.

Chapters 2 to 6 introduce the basics required to create, and understand, a simple FDTD problem. Chapter 2 provides a review of Maxwell's equations and the elements of electromagnetic theory that are essential to understanding the FDTD method, and numerical electromagnetics in general. Chapter 3 describes the methods by which partial differential equations (PDEs) are discretized and transformed into finite difference equations (FDEs). The Yee cell and the FDTD algorithm are introduced in Chapter 4, in one, two, and three dimensions, as well as in other coordinate systems and in lossy materials. The stability and accuracy of the FDTD method are discussed in Chapters 5 and 6, respectively. An understanding of the accuracy of an FDTD simulation is extremely crucial; too often modelers simply run a simulation for stability, and ignore the loss of accuracy that comes with many of the inherent assumptions.

Chapters 7 to 11 provide the next level of understanding required for FDTD simulations. Chapter 7 describes methods by which sources are introduced into the simulation, including the total-field / scattered-field formulation. Chapter 8 introduces some analytical boundary conditions, used to absorb fields at the edge of the simulation space. While these methods are introduced partially for historical and mathematical interest, in many cases they are still the best choice in certain scenarios. The perfectly matched layer (PML) boundary condition is discussed in Chapter 9; the PML is the state of the art in absorbing boundary conditions. Chapter 10 describes methods for simulating wave propagation in dispersive (frequency-dependent) materials, and Chapter 11 describes the FDTD method in anisotropic materials, including materials that are both dispersive and anisotropic.

Chapters 12 to 15 introduce some more advanced topics. We have chosen topics that should be of interest to the general audience, rather than choosing particular applications. Chapter 12 describes a variety of topics, including modeling periodic structures; modeling structures that are smaller than the grid cell size; the bodies of revolution (BOR) method for modeling cylindrical structures; and the near-to-far field transformation for calculating the far-field pattern of a scattering or radiation problem. Chapter 13 introduces implicit FDTD methods, which circumvent the stability restriction of the classic explicit method. The finite difference frequency domain method, mentioned briefly above, is introduced in some detail in Chapter 14. Finally, Chapter 15 provides an overview of nonuniform, nonorthogonal, and irregular grids, which can be used to improve accuracy and efficiency near complex structures, as well as brief introductions to the finite volume, finite element, and discontinuous Galerkin methods for electromagnetics.

References

[1] V. Shankar and W. F. Hall, "A time domain differential solver for electromagnetic scattering," in *URSI National Meeting, Boulder, CO*, 1988.

[2] N. K. Madsen and R. W. Ziolkowski, "Numerical solutions of Maxwell's equations in the time domain using irregular nonorthogonal grids," *Waves Motion*, vol. 10, pp. 538–596, 1988.

[3] S. M. Rao, ed., *Time Domain Electromagnetics*. Academic Press, 1999.

[4] G. Pelosi, "The finite-element method, part i: R. I. Courant," *IEEE Antennas and Propagation Magazine*, vol. 49, pp. 180–182, 2007.

[5] Q. H. Liu, "The PSTD algorithm: A time-domain method requiring only two cells per wavelength," *Microwave Opt. Technol. Lett.*, vol. 15, pp. 158–165, 1997.

[6] O. H. Liu, "Large-scale simulations of electromagnetic and acoustic measurements using the pseudospectral time-domain (PSTD) algorithm," *IEEE Micro. Guided Wave Lett.*, vol. 37, pp. 917–926, 1999.

[7] A. Taflove and S. Hagness, *Computational Electrodynamics: The Finite-Difference Time-Domain Method*, 3rd edn. Artech House, 2005.

[8] J. S. Hesthaven, S. Gottlieb, and D. Gottlieb, *Spectral Methods for Time-Dependent Problems*. Cambridge University Press, 2007.

2 Review of electromagnetic theory

A study of Numerical Electromagnetics must rely on a firm base of knowledge in the foundations of electromagnetics as stated in Maxwell's equations. Accordingly, we undertake in this chapter a review of Maxwell's equations and associated boundary conditions.

All classical electromagnetic phenomena are governed by a compact and elegant set of fundamental rules known as *Maxwell's equations*. This set of four coupled partial differential equations was put forth as the complete classical theory of electromagnetics in a series of brilliant papers[1] written by James Clerk Maxwell between 1856 and 1865, culminating in his classic paper [2]. In this work, Maxwell provided a mathematical framework for Faraday's primarily experimental results, clearly elucidated the different behavior of conductors and insulators under the influence of fields, imagined and introduced the concept of displacement current [3, Sec. 7.4], and inferred the electromagnetic nature of light. A most fundamental prediction of this theoretical framework is the existence of electromagnetic waves, a conclusion to which Maxwell arrived in the absence of experimental evidence that such waves can exist and propagate through empty space. His bold hypotheses were to be confirmed 23 years later (in 1887) in the experiments of Heinrich Hertz [4].[2]

When most of classical physics was fundamentally revised as a result of Einstein's introduction [6][3] of the special theory of relativity, Maxwell's equations remained intact.[4] To this day, they stand as the most general mathematical statements of fundamental natural laws which govern all of classical electrodynamics. The basic justification and validity of Maxwell's equations lie in their consistency with physical experiments over the entire range of the experimentally observed electromagnetic spectrum, extending from cosmic rays at frequencies greater than 10^{22} Hz to the so-called micropulsations at frequencies of about 10^{-3} Hz. The associated practical applications cover an equally wide range, from the use of gamma rays ($10^{18} - 10^{22}$ Hz) for cancer therapy to use of waves at frequencies

---

[1] For an excellent account with passages quoted from Maxwell's papers, see [1, Ch. 5].
[2] For a collected English translation of this and other papers by H. Hertz, see [5].
[3] The English translation of [6] is remarkably readable and is available in a collection of original papers [7].
[4] Maxwell's formulation was in fact one of the major motivating factors which led to the development of the theory of special relativity. The fact that Galilean relativity was consistent with classical mechanics but inconsistent with electromagnetic theory suggested either that Maxwell's equations were incorrect or that the laws of mechanics needed to be modified. For discussions of the relationship between electromagnetism and the special theory of relativity, see [8, Sec. 15]; [9, Ch. 10]; [1, Ch. 2]; [10, Ch. 11].

of a few Hz and below for geophysical prospecting. Electromagnetic wave theory as embodied in Maxwell's equations has provided the underpinning for the development of many vital practical tools of our technological society, including broadcast radio, radar, television, cellular phones, optical communications, Global Positioning Systems (GPS), microwave heating and processing, X-ray imaging, and numerous others.

We now continue with a brief review of Maxwell's equations [11, pp. 247–262] and their underlying foundations. Maxwell's equations are based on experimentally established facts, namely Coulomb's law, which states that electric charges attract or repel one another in a manner inversely proportional to the square of the distance between them [12, p. 569]; Ampère's law, which states that current-carrying wires create magnetic fields and exert forces on one another, with the amplitude of the magnetic field (and thus force) depending on the inverse square of the distance [13]; Faraday's law, which states that magnetic fields which vary with time induce electromotive force or electric field [14, pp. 1–109]; and the principle of conservation of electric charge. Discussion of the experimental bases of Maxwell's equations is available elsewhere.[5] The validity of Maxwell's equations is based on their consistency with all of our experimental knowledge to date concerning electromagnetic phenomena. The physical meaning of the equations is better perceived in the context of their integral forms, which are listed below together with their differential counterparts:

1. Faraday's law is based on the experimental fact that time-changing magnetic flux induces electromotive force:

$$\oint_C \overline{\mathcal{E}} \cdot d\mathbf{l} = -\int_S \frac{\partial \overline{\mathcal{B}}}{\partial t} \cdot d\mathbf{s} \qquad \nabla \times \overline{\mathcal{E}} = -\frac{\partial \overline{\mathcal{B}}}{\partial t}, \qquad (2.1)$$

where the contour C is that which encloses the surface S, and where the sense of the line integration over the contour C (i.e., direction of $d\mathbf{l}$) must be consistent with the direction of the surface vector $d\mathbf{s}$ in accordance with the right-hand rule.

2. Gauss's law is a mathematical expression of the experimental fact that electric charges attract or repel one another with a force inversely proportional to the square of the distance between them (i.e., Coulomb's law):

$$\oint_S \overline{\mathcal{D}} \cdot d\mathbf{s} = \int_V \tilde{\rho} \, dv \qquad \nabla \cdot \overline{\mathcal{D}} = \tilde{\rho}, \qquad (2.2)$$

where the surface S encloses the volume V. The volume charge density is represented with $\tilde{\rho}$ to distinguish it from its phasor form ρ used in the time-harmonic version of Maxwell's equations.

3. Maxwell's third equation is a generalization of Ampère's law, which states that the line integral of the magnetic field over any closed contour must equal the total current

---

[5] Numerous books on fundamental electromagnetics have extensive discussion of Coulomb's law, Ampère's law, Faraday's law, and Maxwell's equations. For a recent reference that provides a physical and experimentally based point of view, see [3, Ch. 4–7].

enclosed by that contour:

$$\oint_C \overline{\mathcal{H}} \cdot d\mathbf{l} = \int_S \overline{\mathcal{J}} \cdot d\mathbf{s} + \int_S \frac{\partial \overline{\mathcal{D}}}{\partial t} \cdot d\mathbf{s} \qquad \nabla \times \overline{\mathcal{H}} = \overline{\mathcal{J}} + \frac{\partial \overline{\mathcal{D}}}{\partial t}, \qquad (2.3)$$

where the contour C is that which encloses the surface S, and $\overline{\mathcal{J}}$ is the electrical current density (see below). Maxwell's third equation expresses the fact that time-varying electric fields produce magnetic fields. This equation with only the first term on the right-hand side (also referred to as the conduction-current term) is Ampère's law, which is a mathematical statement of the experimental findings of Oersted, whereas the second term, known as the displacement-current term, was introduced theoretically by Maxwell in 1862 and verified experimentally many years later (1888) in Hertz's experiments [4].

4. Maxwell's fourth equation is based on the fact that there are no magnetic charges (i.e., magnetic monopoles) and that, therefore, magnetic field lines always close on themselves:

$$\oint_S \overline{\mathcal{B}} \cdot d\mathbf{s} = 0 \qquad \nabla \cdot \overline{\mathcal{B}} = 0, \qquad (2.4)$$

where the surface S encloses the volume V. This equation can actually be derived [3, Sec. 6.5–6.7] from the Biot-Savart law, so it is not completely independent.[6]

The continuity equation, which expresses the principle of conservation of charge in differential form, is contained in Maxwell's Equations and in fact can be readily derived by taking the divergence of Equation (2.3) and using Equation (2.2). For the sake of completeness, we give the integral and differential forms of the continuity equation:

$$- \oint_S \overline{\mathcal{J}} \cdot d\mathbf{s} = \frac{\partial}{\partial t} \int_V \tilde{\rho} \, dv \qquad \nabla \cdot \overline{\mathcal{J}} = -\frac{\partial \tilde{\rho}}{\partial t}, \qquad (2.5)$$

where the surface S encloses the volume V. The fact that the continuity equation can be derived from Equations (2.2) and (2.3) indicates that Maxwell's Equations (2.2) and (2.3) are not entirely independent, if we accept conservation of electric charge as a fact; i.e., using Equations (2.3) and (2.5), one can derive Equation (2.2).

Note that for all of the Equations (2.1) through (2.5), the differential forms can be derived from the integral forms (or vice versa) by using either Stokes's or the divergence

---

[6] Note that Equation (2.4) can be derived from Equation (2.1) by taking the divergence of the latter and using the vector identity of $\nabla \cdot (\nabla \times \overline{\mathcal{G}}) \equiv 0$, which is true for any vector $\overline{\mathcal{G}}$. We find

$$\nabla \cdot (\nabla \times \overline{\mathcal{E}}) = -\nabla \cdot \left(\frac{\partial \overline{\mathcal{B}}}{\partial t} \right) \qquad \rightarrow \qquad 0 = -\frac{\partial (\nabla \cdot \overline{\mathcal{B}})}{\partial t} \qquad \rightarrow \qquad \text{const.} = \nabla \cdot \overline{\mathcal{B}}$$

The constant can then be shown to be zero by the following argument. If we suppose that the $\overline{\mathcal{B}}$ field was produced a finite time ago, i.e., it has not always existed, then, if we go back far enough in time, we have $\overline{\mathcal{B}} = 0$ and therefore $\nabla \cdot \overline{\mathcal{B}} = 0$. Hence it would appear that

$$\nabla \cdot \overline{\mathcal{B}} = 0 \quad \text{and} \quad \oint_S \overline{\mathcal{B}} \cdot d\mathbf{s} = 0$$

theorem, both of which are valid for any arbitrary vector field $\overline{\mathcal{G}}$. Stokes's theorem states that the flux of the curl of a field $\overline{\mathcal{G}}$ through the surface S is equal to the line integral of the field along the closed contour C enclosing S:

$$\oint_C \overline{\mathcal{G}} \cdot d\mathbf{l} = \int_S (\nabla \times \overline{\mathcal{G}}) \cdot d\mathbf{s} \qquad \text{(Stokes's theorem)}, \qquad (2.6)$$

where, as stated, the contour C encloses the surface S. The divergence theorem states that the divergence of a field $\overline{\mathcal{G}}$, over the volume V, is equal to the flux of that field across the surface S:

$$\oint_S \overline{\mathcal{G}} \cdot d\mathbf{s} = \int_V (\nabla \cdot \overline{\mathcal{G}}) \, dv \qquad \text{(Divergence theorem)}, \qquad (2.7)$$

where, again, the surface S encloses volume V.

2.1 Constitutive relations and material properties

The "constitutive relations" relate the electric field intensity $\overline{\mathcal{E}}$ to the electric flux density $\overline{\mathcal{D}}$, and similarly the magnetic field intensity $\overline{\mathcal{B}}$ to the magnetic flux density $\overline{\mathcal{H}}$:

$$\overline{\mathcal{D}} = \tilde{\epsilon}\,\overline{\mathcal{E}}$$

$$\overline{\mathcal{B}} = \tilde{\mu}\,\overline{\mathcal{H}}.$$

These two constitutive relations (the latter of which is more properly expressed as $\overline{\mathcal{H}} = \tilde{\mu}^{-1}\overline{\mathcal{B}}$ since, in the magnetostatic case, $\overline{\mathcal{H}}$ is the medium-independent quantity [3, Sec. 6.2.3, 6.8.2, footnote 71]) govern the manner by which the electric and magnetic fields, $\overline{\mathcal{E}}$ and $\overline{\mathcal{B}}$, are related to the medium-independent quantities, $\overline{\mathcal{D}}$ and $\overline{\mathcal{H}}$, in material media, where, in general, the permittivity $\tilde{\epsilon} \neq \epsilon_0$ and the permeability $\tilde{\mu} \neq \mu_0$. The electrical current density $\overline{\mathcal{J}}$ is in general given by $\overline{\mathcal{J}} = \overline{\mathcal{J}}_i + \overline{\mathcal{J}}_c$, where $\overline{\mathcal{J}}_i$ represents the externally impressed source currents, while $\overline{\mathcal{J}}_c = \tilde{\sigma}\overline{\mathcal{E}}$ represents the conduction current which flows in electrically conducting media (with conductivity $\tilde{\sigma} \neq 0$) whenever there is an electric field present in the material. The volume charge density $\tilde{\rho}$ represents the sources from which electric fields originate.

Note that the permittivity $\tilde{\epsilon}$, the permeability $\tilde{\mu}$, and the conductivity $\tilde{\sigma}$ are macroscopic parameters that describe the relationships among macroscopic field quantities, but they are based on the microscopic behavior of the atoms and molecules in response to the fields. These parameters are simple constants only for *simple* material media, which are linear, homogeneous, time-invariant, and isotropic. Otherwise, for complex material media that are nonlinear, inhomogeneous, time-variant, and/or anisotropic, $\tilde{\epsilon}$, $\tilde{\mu}$, and $\tilde{\sigma}$ may depend on the magnitudes of $\overline{\mathcal{E}}$ and $\overline{\mathcal{B}}$ (nonlinear), on spatial coordinates (x, y, z) (inhomogeneous), on frequency (dispersive), on time (time-variant), or on the orientations of $\overline{\mathcal{E}}$ and $\overline{\mathcal{H}}$ (anisotropic). For anisotropic media, $\tilde{\epsilon}$, $\tilde{\mu}$, and $\tilde{\sigma}$ are generally expressed as matrices (i.e., as *tensors*) whose entries relate each component (e.g., the x, y, or z components) of $\overline{\mathcal{E}}$ (or $\overline{\mathcal{H}}$) to the other three components (e.g., x, y, and z components) of $\overline{\mathcal{D}}$ or $\overline{\mathcal{J}}$ (or $\overline{\mathcal{B}}$); for example, while in free space $\mathcal{D}_x = \epsilon_0 \mathcal{E}_x$, in an isotropic medium there may be cross terms such as $\mathscr{D}_x = \tilde{\epsilon}_{x,z}\mathscr{E}_z$. In ferromagnetic materials, the magnetic field $\overline{\mathscr{B}}$ is determined by the past history of the field $\overline{\mathscr{H}}$ rather than by its instantaneous value; as such the medium is time-variant. Such substances are said to exhibit *hysteresis*. Some hysteresis effects can also be seen in certain dielectric materials.

It is possible to eliminate the vectors $\overline{\mathscr{D}}$ and $\overline{\mathscr{H}}$ from Maxwell's equations by substituting for them as follows:

$$\overline{\mathscr{D}} = \tilde{\epsilon}\overline{\mathscr{E}} + \overline{\mathscr{P}} \tag{2.8}$$

$$\overline{\mathscr{H}} = \frac{\overline{\mathscr{B}}}{\mu_0} - \overline{\mathscr{M}}, \tag{2.9}$$

where $\mathscr{P}$ is the polarization vector in a dielectric in units of Coulombs-m^{-2}, and $\overline{\mathscr{M}}$ is the magnetization vector in a magnetic medium in units of Ampères-m^{-1}. These two quantities account for the presence of matter at the points considered. Maxwell's equations then take the following form:

$$\nabla \times \overline{\mathscr{E}} = -\frac{\partial \overline{\mathscr{B}}}{\partial t} \tag{2.10a}$$

$$\nabla \cdot \overline{\mathscr{E}} = \frac{1}{\tilde{\epsilon}}(\tilde{\rho} - \nabla \cdot \overline{\mathscr{P}}) \tag{2.10b}$$

$$\nabla \times \overline{\mathscr{B}} = \tilde{\epsilon}\tilde{\mu}\frac{\partial \overline{\mathscr{E}}}{\partial t} + \tilde{\mu}\left(\overline{\mathscr{J}} + \frac{\partial \overline{\mathscr{P}}}{\partial t} + \nabla \times \overline{\mathscr{M}}\right) \tag{2.10c}$$

$$\nabla \cdot \overline{\mathscr{B}} = 0. \tag{2.10d}$$

The preceding equations are completely general, but are expressed in a way that stresses the contributions of the medium. Note that the presence of matter has the effect of adding the bound volume charge density $-\nabla \cdot \overline{\mathscr{P}}$, the polarization current density $\partial \overline{\mathscr{P}}/\partial t$, and the equivalent volume magnetization current density $\nabla \times \overline{\mathscr{M}}$. Note that $\tilde{\rho}$ is free charge density, while the electrical current density $\overline{\mathscr{J}}$ could be a source current $\overline{\mathscr{J}}_i$ and/or an electrical conduction current given by $\overline{\mathscr{J}}_c = \tilde{\sigma}\overline{\mathscr{E}}$.

One of the important features of the FDTD method (as compared to other techniques) that we shall study in this book is the ease with which it can be applied to numerical solution of electromagnetic problems involving propagation or scattering in media that are not simple, i.e., that may be *inhomogeneous, dispersive, anisotropic,* and/or *nonlinear* media. For *inhomogeneous* materials, the constitutive relations are valid at each point and at all times, i.e.,

$$\overline{\mathscr{D}}(\mathbf{r}, t) = \tilde{\epsilon}(\mathbf{r}, t)\overline{\mathscr{E}}(\mathbf{r}, t) \tag{2.11a}$$

$$\overline{\mathscr{B}}(\mathbf{r}, t) = \tilde{\mu}(\mathbf{r}, t)\overline{\mathscr{H}}(\mathbf{r}, t) \tag{2.11b}$$

$$\overline{\mathscr{J}}(\mathbf{r}, t) = \tilde{\sigma}(\mathbf{r}, t)\overline{\mathscr{E}}(\mathbf{r}, t). \tag{2.11c}$$

In this case, it is simply necessary to specify the medium parameters $\tilde{\epsilon}(\mathbf{r}, t)$, $\tilde{\mu}(\mathbf{r}, t)$, and $\tilde{\sigma}(\mathbf{r}, t)$ separately at each of the grid points in the FDTD solution space, so that the only computational "penalty" we pay is in increased storage requirements. FDTD modeling of *anisotropic* materials is also generally straightforward, since the corresponding multiplying terms in the FDTD update equations are separately calculated and stored for each point in the FDTD space, before the time-stepping solution is carried out.

For *dispersive* materials, for which medium parameters are functions of frequency, the constitutive relations need to be expressed in terms of a convolution integral. For example, consider a material for which the frequency domain permittivity is $\epsilon(\omega) = \epsilon_0[1 + \chi_e(\omega)]$, where χ_e is the Fourier transform of the electric susceptibility of the material. We then have:

$$\overline{\mathscr{D}}(\mathbf{r}, t) = \epsilon_0 \, \epsilon_\infty \, \overline{\mathscr{E}}(\mathbf{r}, t) + \underbrace{\epsilon_0 \int_0^t \overline{\mathscr{E}}[\mathbf{r}, (t - \tau)] \, \tilde{\chi}_e(\mathbf{r}, \tau) \, d\tau}_{\overline{\mathscr{P}}(\mathbf{r}, t)} \tag{2.12}$$

where $\tilde{\chi}_e(\mathbf{r}, t)$ is the electric susceptibility, i.e., the inverse Fourier transform of $\chi_e(\omega)$. The quantity ϵ_∞ is the relative permittivity at the upper end of the frequency band considered in the particular application. Note that the particular frequency dependence of χ_e (and thus of ϵ) is dependent upon the particular material properties. We will explore the details of FDTD simulations of anisotropic and dispersive materials in Chapters 10 and 11; we will not deal explicitly with nonlinear materials in this book.

2.2 Time-harmonic Maxwell's equations

While the FDTD algorithm uses Maxwell's equations in their time domain form, as shown above, there are occasions when understanding of the time-harmonic (i.e., frequency domain) versions of these equations is useful. For example, numerous practical applications (e.g., broadcast radio and TV, radar, optical, and microwave applications) involve transmitting sources that operate in such a narrow band of frequencies that the behavior of all the field components is very similar to that of the central single-frequency sinusoid (i.e., the carrier).[7] Most generators also produce sinusoidal voltages and currents, and hence electric and magnetic fields that vary sinusoidally with time. In many applications, the transients involved at the time the signal is switched on (or off) are not of concern, so the steady-state sinusoidal approximation is most suitable. For example, for an AM broadcast station operating at a carrier frequency of 1 MHz, any turn-on transients would last only a few µs and are of little consequence to the practical application. For all practical purposes, the signal propagating from the transmitting antenna to the receivers can be treated as a sinusoid, with its amplitude modulated within a narrow bandwidth (e.g., ±5 kHz) around the carrier frequency. Since the characteristics of the propagation medium do not vary significantly over this bandwidth, we can describe the propagation behavior of the AM broadcast signal by studying a single sinusoidal carrier at a frequency of 1 MHz.

---

[7] In many of these cases, the astute engineer would find it more practical to use finite difference frequency domain (FDFD) approaches, which we will introduce in Chapter 14.

The time-harmonic (sinusoidal steady-state) forms of Maxwell's equations[8] are listed here with their more general versions:

$$\nabla \times \overline{\mathscr{E}} = -\frac{\partial \overline{\mathscr{B}}}{\partial t} \qquad\qquad \nabla \times \mathbf{E} = -j\omega\mathbf{B} \qquad\qquad (2.13a)$$

$$\nabla \cdot \overline{\mathscr{D}} = \tilde{\rho} \qquad\qquad \nabla \cdot \mathbf{D} = \rho \qquad\qquad (2.13b)$$

$$\nabla \times \overline{\mathscr{H}} = \overline{\mathscr{J}} + \frac{\partial \overline{\mathscr{D}}}{\partial t} \qquad\qquad \nabla \times \mathbf{H} = \mathbf{J} + j\omega\mathbf{D} \qquad\qquad (2.13c)$$

$$\nabla \cdot \overline{\mathscr{B}} = 0 \qquad\qquad \nabla \cdot \mathbf{B} = 0. \qquad\qquad (2.13d)$$

Note that in Equations (2.13a)–(2.13d), the field vectors $\overline{\mathscr{E}}$, $\overline{\mathscr{D}}$, $\overline{\mathscr{H}}$, and $\overline{\mathscr{B}}$ are real (measurable) quantities that can vary with time, whereas the vectors $\mathbf{E}$, $\mathbf{D}$, $\mathbf{H}$, and $\mathbf{B}$ are complex phasors that do not vary with time. In general, we can obtain the former from the latter by multiplying by $e^{j\omega t}$ and taking the real part. For example,

$$\mathscr{E}(x, y, z, t) = \mathscr{R}e\{\mathbf{E}(x, y, z)e^{j\omega t}\}.$$

Note that the same is true for all of the other phasor quantities. For example,

$$\tilde{\rho}(x, y, z, t) = \mathscr{R}e\{\rho(x, y, z)e^{j\omega t}\}.$$

Note that for dispersive materials, for which ϵ, μ, and σ are functions of frequency (i.e., frequency-dependent), the phasor forms of the constitutive relations are valid at each frequency, so that we have:

$$\mathbf{D}(\omega) = \epsilon(\omega)\,\mathbf{E}(\omega) \qquad\qquad (2.14a)$$

$$\mathbf{B}(\omega) = \mu(\omega)\,\mathbf{H}(\omega) \qquad\qquad (2.14b)$$

$$\mathbf{J}(\omega) = \sigma(\omega)\,\mathbf{E}(\omega). \qquad\qquad (2.14c)$$

Note that the convolution Equation (2.12) was obtained by taking the inverse Fourier transform of Equation (2.14a). It is evident that similar convolution relations would result from inverse transformation of Equations (2.14b) and (2.14c), as would be required for dispersive magnetic materials and materials for which the electrical conductivity is a function of frequency. The latter is the case for metals in the optical frequency range, often known as *Drude* materials.

---

[8] The actual derivation of the time-harmonic form, for example, for Equation (2.13a) is as follows:

$$\nabla \times \overline{\mathscr{E}} = -\frac{\partial \overline{\mathscr{B}}}{\partial t}$$

$$\rightarrow \quad \nabla \times \underbrace{[\mathscr{R}e\{\mathbf{E}(x, y, z)e^{j\omega t}\}]}_{\overline{\mathscr{E}}} = -\frac{\partial}{\partial t}\underbrace{[\mathscr{R}e\{\mathbf{B}(x, y, z)e^{j\omega t}\}]}_{\overline{\mathscr{B}}}$$

$$\rightarrow \quad \mathscr{R}e\{e^{j\omega t}\,\nabla \times \mathbf{E}\} = \mathscr{R}e\{-j\omega e^{j\omega t}\mathbf{B}\} \quad \rightarrow \quad \nabla \times \mathbf{E} = -j\omega\mathbf{B}.$$

2.3   Complex permittivity: dielectric losses

In most dielectrics that are good insulators, the direct conduction current (which is due to finite conductivity) is usually negligible. However, at high frequencies, an alternating current that is in phase with the applied field is present because the rapidly varying applied electric field has to do work against molecular forces in alternately polarizing the bound electrons. As a result, materials that are good insulators at low frequencies can consume considerable energy when they are subjected to high-frequency fields. The heat generated as a result of such radio-frequency heating is used in molding plastics; in microwave cooking; and in microwave drying of paper, printing ink, glued products, leather, textile fibers, wood, foundry materials, rubbers, and plastics [15].

The microphysical bases of such effects are different for solids, liquids, and gases and are too complex to be summarized here. When an external time-varying field is applied to a material, displacement of bound charges occurs, giving rise to volume polarization density $\mathbf{P}$. At sinusoidal steady state, the polarization $\mathbf{P}$ varies at the same frequency as the applied field $\mathbf{E}$. At low frequencies, $\mathbf{P}$ is also in phase with $\mathbf{E}$, both quantities reaching their maxima and minima at the same points in the radio-frequency cycle. As the frequency is increased, however, the inertia of the charged particles (not just because of their mass but also because of the elastic and frictional forces that keep them attached to their molecules) tends to prevent the polarization $\mathbf{P}$ from keeping in phase with the applied field. The work that must be done against the frictional damping forces causes the applied field to lose power, and this power is deposited in the medium as heat. This condition of out-of-phase polarization that occurs at higher frequencies can be characterized by a complex electric susceptibility χ_e, and hence a complex permittivity ϵ_c. In such cases, both the frictional damping and any other ohmic losses (e.g., due to nonzero electrical conductivity σ) are included in the imaginary part of the complex dielectric constant ϵ_c:

$$\epsilon_c = \epsilon' - j\epsilon''$$

We can analyze the resultant effects by substituting the preceding ϵ_c into Equation (2.13c):

$$\nabla \times \mathbf{H} = +j\omega\epsilon_c\mathbf{E} = (\omega\epsilon'')\mathbf{E} + j\omega\epsilon'\mathbf{E}. \tag{2.15}$$

Note that we have assumed $\mathbf{J} = \sigma\mathbf{E} = 0$ since the effects of any nonzero σ are included in ϵ''. It thus appears that the imaginary part of ϵ_c (namely, ϵ'') leads to a volume current density term that is in phase with the electric field, as if the material had an effective conductivity $\sigma_{\text{eff}} = \omega\epsilon''$. At low frequencies, $\omega\epsilon''$ is small due to ω being small and due to the fact that ϵ'' is itself small so that the losses are largely negligible. However, at high frequencies, $\omega\epsilon''$ increases and produces the same macroscopic effect as if the dielectric had effective conductivity $\sigma_{\text{eff}} = \omega\epsilon''$. When a steady current $\mathbf{J} = \sigma\mathbf{E}$ flows in a conducting material in response to a constant applied field $\mathbf{E}$, the electrical power per unit volume dissipated in the material is given by $\mathbf{E} \cdot \mathbf{J} = \sigma|\mathbf{E}|^2$ in units of W-m^{-3}. Similarly, when a dielectric is excited at frequencies high enough for $\omega\epsilon''$ to be appreciable, an alternating current density of $\omega\epsilon''\mathbf{E}$ flows (in addition to the displacement

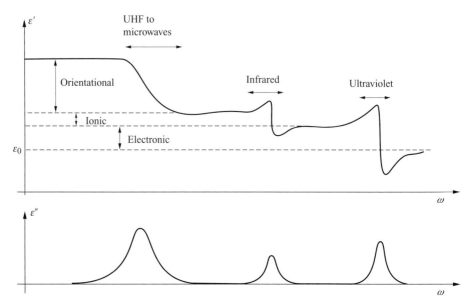

Figure 2.1 Dielectric constant as a function of frequency. The real and imaginary parts of the dielectric constant go through resonances, where the real part settles to a new value and the imaginary part increases, leading to absorption. Adapted from [16].

current density given by $j\omega\epsilon'\mathbf{E}$) in response to an applied alternating field $\mathbf{E}$, leading to an instantaneous power dissipation of $(\omega\epsilon''\mathbf{E}) \cdot \mathbf{E} = \omega\epsilon''|\mathbf{E}|^2$ in units of W-m^{-3}. When the electric field is time-harmonic, i.e., $\mathscr{E}(t) = |\mathbf{E}| \cos(\omega t \pm \beta z)$, the time-average power dissipated in the material per unit volume at any arbitrary point (e.g., $z = 0$) is given by

$$P_{\mathrm{av}} = T_p^{-1} \int_0^{T_p} \omega\epsilon''|\mathscr{E}(t)|^2 dt = T_p^{-1} \int_0^{T_p} \omega\epsilon''|\mathbf{E}|^2 \cos^2(\omega t) dt = \frac{1}{2}\omega\epsilon''|\mathbf{E}|^2, \quad (2.16)$$

where T_p is the period of oscillation. This dissipated power density is the basis for microwave heating of dielectric materials. Since the values of ϵ'' are often determined by measurement, it is not necessary to distinguish between losses due to nonzero conductivity σ and the dielectric losses discussed here.

In general, both ϵ' and ϵ'' can depend on frequency in complicated ways, exhibiting several resonances over wide frequency ranges. The typical behavior in the vicinity of resonances is enhanced losses (i.e., large values of ϵ'') and a reduction of ϵ' to a new level. As an example, Figure 2.1 shows the real and imaginary parts of the dielectric constant (ϵ' and ϵ'' respectively) for a hypothetical dielectric. This example exhibits sharp resonances, at each of which the real part changes dramatically before settling to a new constant value, and the imaginary part exhibits a peak, demonstrating higher absorption. For example, strong resonances in the atmosphere at high altitude, due primarily to ozone and nitrogen, cause high absorption of ultraviolet (UV) energy from the sun, protecting us from harmful UV radiation.

2.4 Complex permeability: magnetic losses

Although our discussion above dealt exclusively with lossy dielectrics, an analogous effect also occurs in magnetic materials, although it is of less importance for electromagnetic wave applications. If a magnetic field is suddenly applied to a paramagnetic[9] material, the magnetization exhibits some inertia and does not immediately reach its static value, but instead approaches it gradually. Similar inertial delay also occurs when the field is turned off. The inertia exhibited is attributable to the energy exchanges between the spinning electrons and the lattice vibrations as well as energy exchanges between neighboring spins. For further details, see [1, Sec. 7.16] and [19, Ch. 16]. In analogy with dielectric relaxation, such effects can be described by introducing a complex permeability μ_c such that

$$\mu_c = \mu' - j\mu''$$

To represent the fact that this effect would lead to power dissipation in the medium, consider the time rate of change of magnetic energy density, or

$$\frac{\partial w_m}{\partial t} = \frac{\partial}{\partial t}\left(\frac{1}{2}\overline{\mathcal{H}} \cdot \overline{\mathcal{B}}\right) \qquad \rightarrow \qquad \frac{1}{2}j\omega(\mathbf{H} \cdot \mathbf{B}).$$

With $\mathbf{B} = (\mu' - j\mu'')\mathbf{H}$, we have

$$\frac{1}{2}j\omega(\mathbf{H} \cdot \mathbf{B}) = \frac{1}{2}(j\omega\mu')H^2 + \frac{1}{2}\omega\mu''H^2,$$

where $H = |\mathbf{H}|$. For a time-harmonic magnetic field, the real part of the preceding quantity represents the time-average power dissipated in a medium. We note that the real part of the first term is zero; that is, this term simply represents the volume density of energy stored in the magnetic field per unit time. The second term, on the other hand, is purely real and represents the time-average power density of the magnetization loss.

In principle, diamagnetic materials also exhibit time-varying relaxation behavior. Although the resultant effects are so small that one might think such phenomena are of little practical interest, resonant absorption in diamagnetic liquids is the cause of the phenomenon of nuclear magnetic resonance,[10] which in turn is the basis for magnetic resonance imaging (MRI) technology [see, e.g., 20].

2.5 Equivalent magnetic currents

In solving electromagnetic problems, it is often useful to view the fields $\overline{\mathcal{E}}$ and $\overline{\mathcal{H}}$, as given by Equations (2.13a) and (2.13c), as being specified by the "current" terms on the

---

[9] Paramagnetic materials are those for which the permeability μ is slightly above μ_0; see [17, Ch. 9] and [18, Ch. 14].

[10] Resonant absorption occurs because of the highly frequency-dependent nature of the permeability μ, similar to the resonances for permittivity in Figure 2.1. For further discussion and relevant references, see [19, Ch. 16].

left-hand side. In this connection, we can view the curl of the magnetic field intensity $\overline{\mathcal{H}}$ as being determined by the sum of three different electric current sources, namely, an externally impressed electrical source current $\mathbf{J}_i$, the electrical conduction current $\mathbf{J}_c$, and a displacement current $\mathbf{J}_d$:

$$\nabla \times \mathbf{H} = \mathbf{J}_i + \mathbf{J}_c + \mathbf{J}_d$$
$$= \mathbf{J}_i + \sigma\,\mathbf{E} + j\omega\epsilon\,\mathbf{E}.$$

We also note that, in general, for materials with complex permittivity $\epsilon_c = \epsilon' - j\epsilon''$, the displacement current term $\mathbf{J}_d$ will contain a dissipative component that is in phase with the electric field:

$$\mathbf{J}_d = j\omega\epsilon_c\,\mathbf{E} = j\omega\epsilon'\mathbf{E} + \omega\epsilon''\mathbf{E}$$

To the best of our knowledge at this time, magnetic "charges" do not exist in nature, thus there cannot be any free magnetic charge or magnetic conduction current, equivalent respectively to ρ in Equation (2.13b) and $\mathbf{J}_c$. However, there are many electromagnetic problems[11] for which it is useful to use fictitious magnetic currents and charges. In such problems, the electric and magnetic fields actually produced by distributions of electric charges and currents can be more easily computed from an "equivalent" distribution of fictitious magnetic currents and charges. To allow for the possible use of magnetic currents, we can rewrite Equations (2.13a) and (2.13b) as follows:

$$\nabla \times \mathbf{E} = -j\omega\mu\,\mathbf{H} - \mathbf{M}_i = -\mathbf{M}_d - \mathbf{M}_i \tag{2.17a}$$
$$\nabla \times \mathbf{H} = \quad j\omega\epsilon\,\mathbf{E} + \mathbf{J}_i, \tag{2.17b}$$

where we have assumed a nonconducting medium ($\sigma = 0$) to emphasize the symmetry between the two curl equations.

Note that for lossy magnetic materials with complex permeability $\mu_c = \mu' - j\mu''$, the magnetic displacement current term contains a dissipative component:

$$\mathbf{M}_d = j\omega\mu_c\,\mathbf{H} = j\omega\mu'\mathbf{H} + \omega\mu''\mathbf{H}.$$

Note that the quantity $\omega\mu''\mathbf{H}$ arises from the physical effects which occur in some materials due to magnetic hysteresis or high-frequency magnetic relaxation.[12] In the context of our discussions of numerical electromagnetics in the FDTD method, recognition of the possible presence of magnetic as well as electric losses is particularly important. We shall see in Chapter 9 that such losses can be introduced artificially at the boundaries of the FDTD solution space in order to absorb the incident fields and eliminate unwanted reflections.

---

[11] Examples of such problems are those which involve radiation from "aperture antennas," or from distributions of electric and magnetic fields over a given surface.

[12] For a brief note on magnetic relaxation see [3, pp. 616–617]; for further details, see [1, Sec. 7.16] and [19, Ch. 16].

2.6   Electromagnetic potentials

In some classes of electromagnetic problems, especially involving antenna applications, it is easier to find electromagnetic fields by first evaluating the scalar and/or vector potentials ($\tilde{\Phi}$, $\overline{\mathcal{A}}$) as an intermediate step, and subsequently deriving the electric and magnetic fields from them. Discussion of electromagnetic potentials in general is not necessary for our coverage of FDTD techniques, since the electromagnetic fields are directly evaluated by time domain solutions of Maxwell's equations. However, other numerical electromagnetics methods, in particular the Method of Moments commonly used in antenna problems, do rely on potentials. Accordingly, we provide a brief review of the relationships between potential functions and source currents and charges.

By way of notation, we denote scalar electric and vector magnetic potentials as $\tilde{\Phi}(\mathbf{r}, t)$ and $\overline{\mathcal{A}}(\mathbf{r}, t)$, respectively. The vector magnetic potential is defined so that $\overline{\mathcal{B}} = \nabla \times \overline{\mathcal{A}}$, but that for $\overline{\mathcal{A}}$ to be uniquely specified, we must also define its divergence. For time-varying electromagnetic fields, the sources ($\overline{\mathcal{J}}$, $\tilde{\rho}$) vary with time, and the electric and magnetic potentials are defined in terms of sources as

$$\overline{\mathcal{A}}(\mathbf{r}, t) = \frac{\mu}{4\pi} \int_{V'} \frac{\overline{\mathcal{J}}_s(\mathbf{r}', t - R/v_p)}{R} dv' \tag{2.18a}$$

$$\tilde{\Phi}(\mathbf{r}, t) = \frac{1}{4\pi\epsilon} \int_{V'} \frac{\tilde{\rho}_s(\mathbf{r}', t - R/v_p)}{R} dv', \tag{2.18b}$$

where $R = |\mathbf{r} - \mathbf{r}'|$ is the distance between sources located at $\mathbf{r}'$ and the observation point at $\mathbf{r}$, and R/v_p is the propagation time, with v_p being the phase velocity of uniform plane waves in the medium in which the sources are embedded. Note that since electromagnetic waves have finite propagation times, there is a time delay (retardation) between the sources and the potentials (and therefore the fields) at a distance from the sources. In other words, for an electromagnetic wave propagating with a finite phase velocity, the variation of the source current at position $\mathbf{r}'$ is conveyed to the observation point at a slightly later time because of the finite velocity. We note that the time t in the above is the time at the point of observation, while $t' = t - R/v_p$ is the time at the source point. Thus, the above relations (2.18) indicate that the sources which had the configurations of $\overline{\mathcal{J}}$ and $\tilde{\rho}$ at t' produced the potentials $\overline{\mathcal{A}}$ and $\tilde{\Phi}$ at a later time t. Because of this time-delayed aspect of the solutions, the potentials $\overline{\mathcal{A}}(\mathbf{r}, t)$ and $\tilde{\Phi}(\mathbf{r}, t)$ are referred to as the *retarded potentials*.

The integral forms of the retarded potential expressions (2.18) can be rigorously derived.[13] Differential equations describing the relationship between $\tilde{\Phi}$ and $\overline{\mathcal{A}}$ and electric and magnetic fields can be derived from Maxwell's equations. Using $\nabla \times \overline{\mathcal{A}} = \overline{\mathcal{B}}$ we have:

$$\nabla \times \overline{\mathcal{E}} = -\frac{\partial \overline{\mathcal{B}}}{\partial t} \qquad \rightarrow \qquad \nabla \times \left(\overline{\mathcal{E}} + \frac{\partial \overline{\mathcal{A}}}{\partial t} \right) = 0.$$

---

[13] The derivation is given in many advanced electromagnetics texts, including [21, Ch. 8].

Noting that the scalar potential for the electrostatic case was defined on the basis of the fact that $\nabla \times \overline{\mathscr{E}} = 0$ (in the absence of time-varying fields) as $\overline{\mathscr{E}} = -\nabla \tilde{\Phi}$, we can now define the *scalar potential* $\tilde{\Phi}$ for time-varying fields as:

$$\overline{\mathscr{E}} + \frac{\partial \overline{\mathscr{A}}}{\partial t} = -\nabla \tilde{\Phi} \quad \rightarrow \quad \overline{\mathscr{E}} = -\nabla \tilde{\Phi} - \frac{\partial \overline{\mathscr{A}}}{\partial t}.$$

Note that the above expression for $\overline{\mathscr{E}}$ recognizes two possible sources for the electric field, namely, charges (represented by the $\tilde{\Phi}$ term) and time-varying magnetic fields (represented by the $\partial \overline{\mathscr{A}}/\partial t$ term).

In order for the vector potential $\overline{\mathscr{A}}$ to be uniquely specified, we must specify its divergence (in addition to its curl, which is already specified as $\nabla \times \overline{\mathscr{A}} = \overline{\mathscr{B}}$). The choice that leads to the simplest differential equations for $\overline{\mathscr{A}}$ and $\tilde{\Phi}$, and to solutions of $\overline{\mathscr{A}}$ and $\tilde{\Phi}$ in the form of retarded potentials as given in Equation (2.18), is

$$\nabla \cdot \overline{\mathscr{A}} = -\mu\epsilon \frac{\partial \tilde{\Phi}}{\partial t}. \tag{2.19}$$

This condition, known as the *Lorentz condition* or the *Lorentz gauge*, is also consistent with the source charge and current distributions being related through the continuity equation. Using Equation (2.19), and Maxwell's equations, the wave equations can be written in terms of $\overline{\mathscr{A}}$ and $\tilde{\Phi}$ as:

$$\nabla^2 \overline{\mathscr{A}} - \mu\epsilon \frac{\partial^2 \overline{\mathscr{A}}}{\partial t^2} = -\mu \overline{\mathscr{J}} \tag{2.20a}$$

$$\nabla^2 \tilde{\Phi} - \mu\epsilon \frac{\partial^2 \tilde{\Phi}}{\partial t^2} = -\frac{\tilde{\rho}}{\epsilon}. \tag{2.20b}$$

The fact that the retarded potentials as given in Equation (2.18) are indeed the solutions of Equation (2.20) can be verified by direct substitution[14] of Equation (2.18) into the respective differential equations given by Equation (2.20), although this process is rather involved mathematically.

The time-harmonic forms of Equations (2.20) are given as

$$\nabla^2 \mathbf{A} + \omega^2 \mu\epsilon \, \mathbf{A} = -\mu \, \mathbf{J} \tag{2.21a}$$

$$\nabla^2 \Phi + \omega^2 \mu\epsilon \, \Phi = -\frac{\rho}{\epsilon}. \tag{2.21b}$$

Note that the magnetic vector potential $\mathbf{A}$ which originates from electrical currents completely specifies the time-harmonic electric and magnetic fields:

$$\mathbf{B} = \nabla \times \mathbf{A}$$

$$\mathbf{E} = -j\omega\mathbf{A} + \frac{\nabla(\nabla \cdot \mathbf{A})}{j\omega\epsilon\mu},$$

where we have used Equation (2.19).

---

[14] Such substitution must be done with considerable care; see [22, p. 164, footnote]. Expressions for A and Φ which satisfy Equation (2.18) can also be written with $(t + R/v_p)$ instead of $(t - R/v_p)$ in Equation (2.18), and are called *advanced potentials*. However, their use would clearly violate causality, implying that the potentials at a given time and place depend on later events elsewhere.

In some types of electromagnetic problems, it is convenient to introduce an analogous electric vector potential $\mathbf{F}$ which has magnetic current distributions as its source. In a region of no impressed sources, i.e., free of charges ($\tilde{\rho}_s = 0$) and source currents ($\overline{\mathcal{J}}_s = 0$), we have $\nabla \cdot \mathbf{D} = 0$, so that the vector $\mathbf{D}$ can be expressed as the curl of another vector, since $\nabla \cdot (-\nabla \times \mathbf{F}) \equiv 0$, for an arbitrary vector $\mathbf{F}$. The electric vector potential $\mathbf{F}$ is thus defined as

$$\mathbf{D} = \epsilon \mathbf{E} = -\nabla \times \mathbf{F} \qquad \rightarrow \qquad \mathbf{E} = -\frac{1}{\epsilon} \nabla \times \mathbf{F}.$$

An analysis similar to that provided above for the magnetic vector potential $\mathbf{A}$ indicates that the electric vector potential satisfies the Helmholtz equation:

$$\nabla^2 \mathbf{F} + \omega^2 \mu \epsilon \mathbf{F} = -\epsilon \mathbf{M}$$

and that it is therefore related to the magnetic current sources via the following equation analogous to Equation (2.18a):

$$\mathbf{F}(\mathbf{r}) = \frac{\epsilon}{4\pi} \int_{V'} \frac{\mathbf{M}(\mathbf{r}')e^{-j\beta R}}{R} dv'$$

$$\rightarrow \qquad \overline{\mathcal{F}}(\mathbf{r}, t) = \frac{\epsilon}{4\pi} \int_{V'} \frac{\overline{\mathcal{M}}(\mathbf{r}', t - R/v_p)}{R} dv'.$$

The electric and magnetic fields originating from magnetic current distributions can be written in terms of the electric vector potential as:

$$\mathbf{E} = -\frac{1}{\epsilon} \nabla \times \mathbf{F}$$

$$\mathbf{H} = -j\omega \mathbf{F} + \frac{\nabla(\nabla \cdot \mathbf{F})}{j\omega\epsilon\mu}.$$

2.7   Electromagnetic boundary conditions

The interfaces between materials are where the interesting interactions of electromagnetic waves occur. It is at these interfaces, of course, that reflection and refraction occur, the physical laws for which are described in many electromagnetic texts. In FDTD simulations, we are often interested in the interactions of electromagnetic waves with surfaces that cannot be easily described analytically. However, the basic rules for the electric and magnetic field vectors at these interfaces are well defined, and are summarized here.

The integral forms of Equations (2.1) through (2.4) can be used to derive the relationships between electric- and magnetic field components on both sides of interfaces between two different materials (i.e., different μ, ϵ, and/or σ).

The electromagnetic boundary conditions can be summarized as follows:

1. It follows from applying the integral form of Equation (2.1) to a contour C, as shown in Figure 2.2b, and letting Δh tend toward zero (so the area in the right-hand side of the integral goes to zero), that the tangential component of the electric field $\overline{\mathcal{E}}$ is

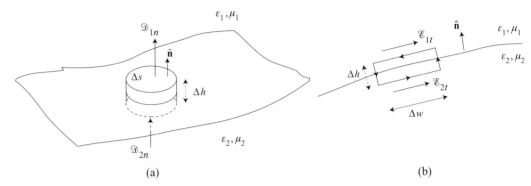

Figure 2.2 Interfaces between two different materials. The boundary conditions for the electromagnetic fields are derived by applying the surface integrals to the cylindrical surface as shown in (a) and the line integrals to the rectangular contour in (b). The normal and tangential components of a general vector **A** are shown for reference.

continuous across any interface:

$$\hat{\mathbf{n}} \times [\overline{\mathscr{E}}_1 - \overline{\mathscr{E}}_2] = 0 \qquad \rightarrow \qquad \mathscr{E}_{1t} = \mathscr{E}_{2t}, \tag{2.22}$$

where $\hat{\mathbf{n}}$ is the unit vector perpendicular to the interface and outward from medium 2 (shown as $\hat{\mathbf{n}}$ in Figure 2.2b).

2. It can be shown by similarly applying the integral form of Equation (2.3) to the same contour C in Figure 2.2b that the tangential component of the magnetic field $\mathscr{H}$ is continuous across any interface:

$$\hat{\mathbf{n}} \times [\overline{\mathscr{H}}_1 - \overline{\mathscr{H}}_2] = 0 \qquad \rightarrow \qquad \mathscr{H}_{1t} = \mathscr{H}_{2t}, \tag{2.23}$$

except where surface currents $(\overline{\mathscr{J}}_s)$ may exist, such as at the surface of a perfect conductor (i.e., $\sigma = \infty$):

$$\hat{\mathbf{n}} \times \overline{\mathscr{H}}_1 = \overline{\mathscr{J}}_s, \tag{2.24}$$

Noting that the field $\mathscr{H}_2$ inside the perfect conductor is zero.[15]

3. It can be shown by applying the integral form of Equation (2.2) to the surface of the cylinder shown in Figure 2.2a, and again letting Δh tend toward zero, that the normal component of electric flux density $\mathscr{D}$ is continuous across interfaces, except where surface charge $(\tilde{\rho}_s)$ may exist, such as at the surface of a metallic conductor or at the interface between two lossy dielectrics ($\sigma_1 \neq 0$, $\sigma_2 \neq 0$):

$$\hat{\mathbf{n}} \cdot [\overline{\mathscr{D}}_1 - \overline{\mathscr{D}}_2] = \tilde{\rho}_s \qquad \rightarrow \qquad \mathscr{D}_{1n} - \mathscr{D}_{2n} = \tilde{\rho}_s. \tag{2.25}$$

4. A consequence of similarly applying the integral form of Equation (2.4) to the surface of the cylinder in Figure 2.2a is that the normal component of the magnetic field $\overline{\mathscr{B}}$

---

[15] The interior of a solid perfect conductor is void of both static and dynamic magnetic fields; for a discussion at an appropriate level see [3, Sec. 6.8.4].

is continuous across interfaces:

$$\hat{\mathbf{n}} \cdot [\overline{\mathscr{B}}_1 - \overline{\mathscr{B}}_2] = 0 \qquad \rightarrow \qquad \mathscr{B}_{1n} = \mathscr{B}_{2n}. \tag{2.26}$$

5. It follows from applying the integral form of Equation (2.5) to the surface of the cylinder in Figure 2.2a that, at the interface between two lossy media (i.e., $\sigma_1 \neq 0$, $\sigma_2 \neq 0$), the normal component of the electric current density $\overline{\mathscr{J}}$ is continuous, except where time-varying surface charge may exist, such as at the surface of a perfect conductor or at the interface between lossy dielectrics ($\epsilon_1 \neq \epsilon_2$ and $\sigma_1 \neq \sigma_2$):

$$\hat{\mathbf{n}} \cdot [\overline{\mathscr{J}}_1 - \overline{\mathscr{J}}_2] = -\frac{\partial \tilde{\rho}_s}{\partial t} \qquad \rightarrow \qquad \mathscr{J}_{1n} - \mathscr{J}_{2n} = -\frac{\partial \tilde{\rho}_s}{\partial t}. \tag{2.27}$$

Note that Equation (2.27) is not completely independent of Equations (2.22) through (2.26), since Equation (2.5) is contained in Equations (2.1) through (2.4), as mentioned previously. For the stationary case ($\partial/\partial t = 0$), Equation (2.27) implies $\mathscr{J}_{1n} = \mathscr{J}_{2n}$, or $\sigma_1 \mathscr{E}_{1n} = \sigma_2 \mathscr{E}_{2n}$, which means that at the interface between lossy dielectrics (i.e., $\epsilon_1 \neq \epsilon_2$ and $\sigma_1 \neq \sigma_2$) there must in general be finite surface charge (i.e., $\tilde{\rho}_s \neq 0$) because otherwise Equation (2.25) demands that $\epsilon_1 \mathscr{E}_{1n} = \epsilon_2 \mathscr{E}_{2n}$.

2.8 Electromagnetic waves

Maxwell's equations embody all of the essential aspects of electromagnetics, including the idea that light is electromagnetic in nature, that electric fields which change in time create magnetic fields in the same way as time-varying voltages induce electric currents in wires, and that the source of electric and magnetic energy does not only reside on the body which is electrified or magnetized, but also, and to a far greater extent, in the surrounding medium. However, arguably the most important and far-reaching implication of Maxwell's equations is the idea that electric and magnetic effects can be transmitted from one point to another through the intervening space, whether that be empty or filled with matter.

Electromagnetic energy propagates, or travels from one point to another, as *waves*. The propagation of electromagnetic waves results in the phenomenon of *delayed action at a distance*; in other words, electromagnetic fields can exert forces, and hence can do work, at distances far away from the places where they are generated and at later times. Electromagnetic radiation is thus a means of transporting energy and momentum from one set of electric charges and currents (those at the source end) to another (those at the receiving end). Since whatever can carry energy can also convey information, *electromagnetic waves* thus provide the means of transmitting energy and information over a distance.

To appreciate the concept of propagation of electromagnetic waves in empty space, it is useful to think of other wave phenomena which we may observe in nature. When a pebble is dropped into a body of water, the water particles in the vicinity of the pebble are immediately displaced from their equilibrium positions. The motion of these particles disturbs adjacent particles, causing them to move, and the process continues,

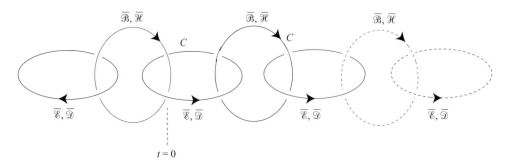

Figure 2.3 Plausibility of electromagnetic wave propagation as dictated by Equations (2.1) and (2.3). Starting with a time-varying magnetic field at the location of the dashed line, electric and magnetic fields are successively generated at surrounding regions.

creating a wave. Because of the finite velocity of the wave, a finite time elapses before the disturbance causes the water particles at distant points to move. Thus, the initial disturbance produces, at distant points, effects that are *retarded* in time. The water wave consists of ripples that move along the surface away from the disturbance. Although the motion of any particular water particle is essentially a small up-and-down movement (i.e., a *transverse* wave, where the motion is perpendicular to the direction of the wave propagation), the cumulative effects of all the particles produce the wave which moves radially outward from the point at which the pebble is dropped. Another excellent example of wave propagation is the motion of sound through a medium. In air, this motion occurs through the to-and-fro movement of the air molecules (i.e., a *longitudinal* wave, where the motion is parallel to the direction of the wave propagation), but these molecules do not actually move along with the wave.

Electromagnetic waves consist of time-varying electric and magnetic fields. Suppose an electrical disturbance, such as a change in the current through a conductor, occurs at some point in space. The time-changing electric field resulting from the disturbance generates a time-changing magnetic field. The time-changing magnetic field, in turn, produces an electric field. These time-varying fields continue to generate one another in an ever-expanding sphere, and the resulting wave propagates away from the source. When electromagnetic waves propagate in a pure vacuum, there is no vibration of physical particles, as in the case of water and sound waves. The velocity of electromagnetic wave propagation in free space is that of the speed of light, so that the fields produced at distant points are *retarded* in time with respect to those near the source.

Careful examination of Equations (2.1) to (2.4) provides a qualitative understanding of the plausibility of electromagnetic wave propagation in space. Consider, for example, Equations (2.1) and (2.3) in a nonconducting medium (i.e., $\sigma = 0$) and in the absence of impressed sources (i.e., $\overline{\mathcal{J}}_i$, $\tilde{\rho} = 0$). According to Equation (2.1), any magnetic field $\overline{\mathcal{B}}$ that varies with time generates an electric field along a contour C surrounding it, as shown in Figure 2.3. On the other hand, according to Equation (2.3), this electric field $\overline{\mathcal{E}}$, which typically varies in time because $\overline{\mathcal{B}}$ is taken to be varying with time, in turn generates a magnetic field along a contour C surrounding itself. This process continues indefinitely, as shown in Figure 2.3. It thus appears that if we start with a magnetic field at

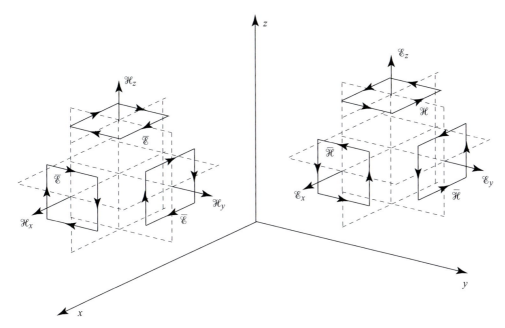

Figure 2.4 Geometrical relationship between $\overline{\mathcal{E}}$ and $\overline{\mathcal{H}}$ fields. Note that electric fields are encircled by magnetic fields and vice versa.

one point in space and vary it with time, Maxwell's equations dictate that magnetic fields and electric fields are created at surrounding points — i.e., that the disturbance initiated by the changing magnetic field propagates away from its point of origin, indicated by the dashed line in Figure 2.3.

Note that although Figure 2.3 shows propagation toward the left and right, the same sequence of events takes place in all directions. In fact, all components of the electric field are encircled by loops of magnetic fields and vice versa, as depicted in Figure 2.4. This geometrical relationship between the $\overline{\mathcal{E}}$ and $\overline{\mathcal{B}}$ fields is the key to the FDTD algorithm, as we shall see in later chapters.

2.8.1 The wave equation

In the vast majority of this book, we will deal with discretized versions of Maxwell's equations (2.1)–(2.4), rather than the wave equation. However, we introduce it here to demonstrate the concept of a propagating wave, and to introduce the concepts of phase velocity and group velocity.

Consider Maxwell's equations in a source-free medium, i.e., $\tilde{\rho} = 0$ and $\overline{\mathcal{J}} = 0$. If the material parameters ϵ and μ are not functions of time we can write:

$$\nabla \times \overline{\mathcal{E}} = -\mu \frac{\partial \overline{\mathcal{H}}}{\partial t} \qquad \text{and} \qquad \nabla \times \overline{\mathcal{H}} = \epsilon \frac{\partial \overline{\mathcal{E}}}{\partial t}.$$

Taking the curl of the first equation, and inserting the second, we find:

$$\nabla \times \nabla \times \overline{\mathcal{E}} = -\mu \frac{\partial}{\partial t} \nabla \times \overline{\mathcal{H}} = -\mu\epsilon \frac{\partial^2 \overline{\mathcal{E}}}{\partial t^2},$$

where we have taken advantage of the fact that spatial and temporal derivative operators can be applied in any order. Finally, using the vector identity

$$\nabla \times \nabla \times \overline{\mathscr{A}} = \nabla(\nabla \cdot \overline{\mathscr{A}}) - \nabla^2 \overline{\mathscr{A}}$$

and noting from Equation (2.2) that when $\tilde{\rho} = 0$, $\nabla \cdot \overline{\mathscr{E}} = 0$, we find:

$$\nabla^2 \overline{\mathscr{E}} - \mu\epsilon \frac{\partial^2 \overline{\mathscr{E}}}{\partial t^2} = 0. \tag{2.28}$$

Equation (2.28) is the simplest form of the *Wave Equation*. It can become significantly more complicated in complex media, whether they are dispersive, anisotropic, nonlinear, time-dependent, or include sources. In such cases, for example, we may not be able to move the constants μ and ϵ outside of the time derivatives, and we may find $\nabla \cdot \overline{\mathscr{E}} \neq 0$.

Consider the 1D case, with $\overline{\mathscr{E}} = \mathscr{E}_x$ and $\partial/\partial y = \partial/\partial x = 0$. The general solution of the wave equation is given by the simple relation:

$$\mathscr{E}_x(z, t) = f(z - v_p t) + g(z + v_p t). \tag{2.29}$$

This equation describes two fluctuations of the electric field, given by the general functions f and g, which are propagating in the $+z$ and $-z$ directions, respectively, with a *phase velocity* given by $v_p = 1/\sqrt{\mu\epsilon}$. Simple substitution of Equation (2.29) into Equation (2.28) demonstrates that this is a solution of the wave equation. For the time-harmonic Maxwell's equations, where $\partial/\partial t \to j\omega$ and we have assumed sinusoidally varying electric and magnetic fields, the 1D wave equation has the form:

$$\frac{\partial^2 \mathscr{E}_x}{\partial z^2} + k^2 \mathscr{E}_x = 0, \tag{2.30}$$

where $k^2 = \omega^2 \mu\epsilon$. The general solution of this equation has the form:

$$\mathscr{E}_x(z, t) = \mathfrak{Re}\left\{(C_1 e^{-jkz} + C_2 e^{jkz})e^{j\omega t}\right\}$$
$$= C_1 \cos(kz - \omega t) + C_2 \cos(kz + \omega t), \tag{2.31}$$

where $k = 2\pi/\lambda$ is the *wavenumber* and $\omega = 2\pi f$ is the angular frequency of the wave. This equation describes two sinusoids traveling in the $\pm z$ directions. The phase velocity of the wave can be found by following a point on the wave in time, or equivalently, following the argument of the cosine:

$$\omega t \pm kz = \text{constant} \quad \rightarrow \quad v_p = \frac{dz}{dt} = \frac{\omega}{k}. \tag{2.32}$$

The relationship $v_p = \omega/k$ is often used as the definition of the phase velocity, and is the relationship used to derive the phase velocity in more complex media. For example, in dispersive media the frequency ω can be a complicated function of k, and when the ratio ω/k is evaluated, we find that the phase velocity is also a function of k.

2.8.2 Group velocity

The phase velocity v_p describes the speed of propagation of the primary wave at frequency ω. However, in most problems of interest (especially those of interest in the FDTD method), the wave contains a spectrum of frequencies, rather than a single frequency. Each frequency propagates at a different phase velocity v_p, but the wave packet propagates at an altogether different velocity, known as the *group velocity*.

Consider a superposition of two waves at very similar frequencies, ω and $\omega + \Delta\omega$, and correspondingly similar wavenumbers, k and $k + \Delta k$. For this illustration we can consider only the wave propagating in the $+z$ direction, given by:

$$\mathscr{E}_{x,1} = C\cos(\omega t - kz)$$

$$\mathscr{E}_{x,2} = C\cos[(\omega + \Delta\omega)t - (k + \Delta k)z].$$

Using the trigonometric identity for the sum of two cosines,[16] we find the superposition of these two waves:

$$\mathscr{E}_x = \mathscr{E}_{x,1} + \mathscr{E}_{x,2} = 2C\cos\left(\frac{\Delta\omega\, t - \Delta k\, z}{2}\right)\cos\left[\left(\omega + \frac{\Delta\omega}{2}\right)t - \left(k + \frac{\Delta k}{2}\right)z\right].$$

$$(2.33)$$

This is equivalent to the propagation of a wave at frequency $\omega + \Delta\omega/2$, which for small Δw is negligibly different from ω. The amplitude of the wave, however, has an envelope which varies in time and space, given by:

$$A_{\text{env}}(z, t) = 2C\cos\left(\frac{\Delta\omega\, t - \Delta k\, z}{2}\right).$$

As we did for the phase velocity, we can track the envelope as it moves by following the argument of the cosine:

$$\Delta\omega\, t - \Delta k\, z = \text{constant} \qquad \rightarrow \qquad v_g = \frac{dz}{dt} = \frac{\Delta\omega}{\Delta k}.$$

Taking the limit as $\Delta\omega \to 0$,

$$v_g = \frac{d\omega}{dk}. \qquad (2.34)$$

This equation is the definition of the *group velocity*. It is the velocity of the envelope, which contains the different frequencies propagating at their own phase velocities v_p. Quite often, the group velocity is also the speed at which energy or information is carried by the wave; it is for this reason that in complex media, the phase velocity can exceed the "speed of light" of $c = 3 \times 10^8$ m/s, but the group velocity can never exceed this physical limit.[17]

---

[16]
$$\cos A + \cos B = 2\cos\left(\frac{A + B}{2}\right)\cos\left(\frac{A - B}{2}\right).$$

[17] While this statement is most often accurate and convenient for illustrative purposes, the group velocity is not always the speed of energy propagation. In some lasers, for example, the group velocity can be made to exceed c; however, in those cases the phase velocity is less than c and the signal travels at this speed [e.g., 23].

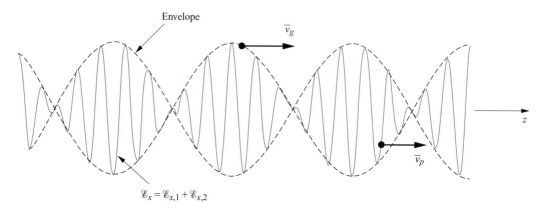

Figure 2.5 Concept of phase and group velocities. The figure shows the sum of two waves with slightly different frequencies, from which arises the envelope amplitude. This envelope propagates at the group velocity v_g.

The phase and group velocities are illustrated in Figure 2.5. In the simple medium where there is no dispersion and $\omega = \pm v_p k$, we find that $v_g = d\omega/dk = \pm v_p$, and the phase and group velocities are the same, and independent of frequency. In the more general case, however, the phase and group velocities are both functions of frequency. Furthermore, in the case of discretized versions of Maxwell's equations, such as in the FDTD method, we often find that the *numerical* phase and group velocities are not identical to the continuous cases, in which case we have *numerical dispersion*. This numerical dispersion is the subject of Chapter 6.

2.9   The electromagnetic spectrum

The properties of uniform plane waves are identical over all of the electromagnetic spectrum that has been investigated experimentally, frequencies ranging from milli-hertz to 10^{24} Hz. Regardless of their frequency, all uniform plane electromagnetic waves propagate in unbounded free space with the same velocity, namely, $v_p = c \simeq 3 \times 10^8$ m/s, but with different wavelengths, as determined by $\lambda = c/f$. Table 2.1, adapted from [16], lists the various designated frequency and wavelength ranges of the electro-magnetic spectrum and selected applications for each range. Maxwell's equations and the results derived from them thus encompass a truly amazing range of physical phe-nomena and applications that affect nearly every aspect of human life and our physical environment.

At frequencies in the ultraviolet range and higher, physicists are more accustomed to thinking in terms of the associated energy level of the photon (a quantum of radiation), which is given by hf, where $h \simeq 6.63 \times 10^{-34}$ J-s is *Planck's constant*. Gamma rays, which are photons at energies of 1 MeV or greater, are constantly present in our universe; they ionize the highest reaches of Earth's atmosphere and help maintain the ionosphere

Table 2.1 The electromagnetic spectrum and related applications; adapted from [16]

Frequency	Designation	Selected applications	Wavelength (in free space)
10^{18}–10^{22} Hz	γ-rays	Cancer therapy, astrophysics	
10^{16}–10^{21} Hz	X-rays	Medical diagnosis	
10^{15}–10^{18} Hz	Ultraviolet	Sterilization	0.3–300 nm
3.95×10^{14}– 7.7×10^{14} Hz	Visible light	Vision, astronomy, optical communications	390–760 nm
		Violet	390–455
		Blue	455–492
		Green	492–577
		Yellow	577–600
		Orange	600–625
		Red	625–760
10^{12}–10^{14} Hz	Infrared	Heating, night vision, optical communications	3–300 μm
0.3–1 THz	Millimeter	Astronomy, meteorology	0.3–1 mm
30–300 GHz	EHF	Radar, remote sensing	0.1–1 cm
80–100		W-band	
60–80		V-band	
40–60		U-band	
27–40		K_a-band	
3–30 GHz	SHF	Radar, satellite comm.	1–10 cm
18–27		K-band	
12–18		K_u-band	
8–12		X-band	
4–8		C-band	
0.3–3 GHz	UHF	Radar, TV, GPS, cellular phone	10–100 cm
2–4		S-band	
2.45		Microwave ovens	
1–2		L-band, GPS system	
470–890 MHz		TV Channels 14–83	
30–300 MHz	VHF	TV, FM, police	1–10 m
3–30 MHz	HF	Short-wave, citizens' band	10–100 m
0.3–3 MHz	MF	AM broadcasting	0.1–1 km
30–300 kHz	LF	Navigation, radio beacons	1–10 km
3–30 kHz	VLF	Navigation, positioning, naval communications	10–100 km
0.3–3 kHz	ULF	Telephone, audio	0.1–1 Mm
30–300 Hz	SLF	Power transmission, submarine communications	1–10 Mm
3–30 Hz	ELF	Detection of buried metals	10–100 Mm
<3 Hz		Geophysical prospecting	>100 Mm

at night, in the absence of solar radiation. Short-duration bursts of γ-rays, which bathe our solar system (and our galaxy) about three times a day, are believed to be produced in the most powerful explosions in the universe, releasing (within a few seconds or minutes) energies of 10^{51} ergs — more energy than our sun will produce in its entire 10 billion years of existence [24].

It is also interesting to note that only a very narrow portion of the electromagnetic spectrum is perceptible to human vision, namely, the visible range. We can also "feel" infrared as heat, and our bodies can be damaged by excessive amounts of UV, microwave radiation, X-rays, and γ-rays. Applications of X-rays, UV, visible, and infrared light are far too numerous to be commented on here and include vision, lasers, optical fiber communications, and astronomy.

At present, a relatively unexplored part of the electromagnetic spectrum is the transition region between the low-frequency end of infrared and the high-frequency end of the millimeter range. The frequency in this region is measured in terahertz (THz), or 10^{12} Hz, with the corresponding wavelengths being in the submillimeter range. Fabrication of antennas, transmission lines, and other receiver components usable in these frequencies requires the use of novel quasi-optical techniques. At present, the terahertz region is primarily used in astronomy and remote sensing; however, industrial, medical, and other scientific applications are currently under investigation.

Each decade of the electromagnetic spectrum below the millimeter range of frequencies is divided[18] into designated ranges, with the acronyms indicated in Table 2.1: extremely high frequency (EHF); super high frequency (SHF); ultra high frequency (UHF); very high frequency (VHF); high frequency (HF); medium frequency (MF); low frequency (LF); very low frequency (VLF); ultra low frequency (ULF); super low frequency (SLF); extremely low frequency (ELF).

The microwave band of frequencies is vaguely defined as the range from 300 MHz up to 1 THz, including the millimeter range. It is extensively utilized for radar, remote sensing, and a host of other applications too numerous to cite here. In radar work, the microwave band is further subdivided into bands with alphabetical designations, which are listed in Table 2.1. The VLF range of frequencies is generally used for global navigation and naval communications, using transmitters that utilize huge radiating structures. Although VLF transmissions are used for global communications with surface ships and submarines near the water surface, even lower frequencies are required for communication with deeply submerged submarines.

The lowest frequencies of the experimentally investigated electromagnetic spectrum are commonly used to observe so-called micropulsations, which are electromagnetic waves at frequencies 0.001 to 10 Hz generated as a result of large-scale currents flowing in the Earth's auroral regions and by the interaction between the Earth's magnetic field and the energetic particles that stream out of the sun in the form of the solar wind.

2.10   Summary

This chapter introduced Maxwell's equations, electric and magnetic losses, boundary conditions, and the wave equation. The Finite Difference Time Domain (FDTD) method

---

[18] There is a certain arbitrariness in the designations of these frequency ranges. In geophysics and solar terrestrial physics, the designation ELF is used for the range 3 Hz to 3 kHz, while ULF is used to describe frequencies typically below 3 Hz.

relies primarily on the differential forms of Maxwell's equations:

$$\nabla \times \overline{\mathscr{E}} = -\frac{\partial \overline{\mathscr{B}}}{\partial t} \qquad \text{(Faraday's law)}$$

$$\nabla \cdot \overline{\mathscr{D}} = \tilde{\rho} \qquad \text{(Gauss's law)}$$

$$\nabla \times \overline{\mathscr{H}} = \overline{\mathscr{J}} + \frac{\partial \overline{\mathscr{D}}}{\partial t} \qquad \text{(Ampère's law)}$$

$$\nabla \cdot \overline{\mathscr{B}} = 0 \qquad \text{(Gauss's magnetic law)}$$

$$\nabla \cdot \overline{\mathscr{J}} = -\frac{\partial \tilde{\rho}}{\partial t} \qquad \text{(Continuity equation)}$$

where each of the fields is a function of time and space, i.e., $\overline{\mathscr{E}} = \overline{\mathscr{E}}(x, y, z, t)$. The equations above hold for a general medium, which may be inhomogeneous, dispersive, anisotropic, and/or nonlinear; the fields are related through the constitutive relations:

$$\overline{\mathscr{D}}(\mathbf{r}, t) = \tilde{\epsilon}(\mathbf{r}, t)\overline{\mathscr{E}}(\mathbf{r}, t)$$

$$\overline{\mathscr{B}}(\mathbf{r}, t) = \tilde{\mu}(\mathbf{r}, t)\overline{\mathscr{H}}(\mathbf{r}, t) \qquad \text{(Constitutive relations)}$$

$$\overline{\mathscr{J}}(\mathbf{r}, t) = \tilde{\sigma}(\mathbf{r}, t)\overline{\mathscr{E}}(\mathbf{r}, t).$$

Lossy media can be characterized as having complex ϵ and/or μ:

$$\epsilon_c = \epsilon' - j\epsilon'' \qquad \text{and/or} \qquad \mu_c = \mu' - j\mu'' \qquad \text{(Lossy media)}$$

leading to electric and magnetic losses, respectively:

$$P_{\text{av},e} = \frac{1}{2}\omega\epsilon''|\mathbf{E}|^2 \qquad \text{and} \qquad P_{\text{av},h} = \frac{1}{2}\omega\mu''|\mathbf{H}|^2 \qquad \text{(Material losses)}$$

where $\mathbf{E}$ and $\mathbf{H}$ represent the time-harmonic electric and magnetic fields, and ω is the time-harmonic wave frequency.

In the FDTD method, the enforcement of boundary conditions is of considerable importance for accurate modeling of wave interactions with scattering objects. At the interface between any two media, the following boundary conditions must hold:

$$\mathscr{E}_{1t} = \mathscr{E}_{2t} \qquad\qquad \mathscr{D}_{1n} - \mathscr{D}_{2n} = \tilde{\rho}_s$$

$$\mathscr{B}_{1n} = \mathscr{B}_{2n} \qquad\qquad \mathscr{H}_{1t} - \mathscr{H}_{2t} = \overline{\mathscr{J}}_s \qquad \text{(Boundary conditions)}$$

$$\mathscr{J}_{1n} - \mathscr{J}_{2n} = -\frac{\partial \tilde{\rho}_s}{\partial t}$$

where the subscripts n and t refer to the normal and tangential components of each field, and the 1 and 2 subscripts refer to regions 1 and 2 on either side of the interface.

Finally, the combination of Maxwell's equations leads to the equation for a propagating wave, given below for a simple medium:

$$\nabla^2\overline{\mathscr{E}} - \mu\epsilon\frac{\partial^2\overline{\mathscr{E}}}{\partial t^2} = 0. \qquad \text{(Wave equation)}$$

The phase velocity is the speed at which wavefronts propagate through the medium, and is given by

$$v_p = \frac{\omega}{k} \qquad \text{(Phase velocity)}$$

where $\omega = 2\pi f$ is the wave angular frequency and $k = 2\pi/\lambda$ is the wavenumber. When the wavenumber is a function of frequency, so is the phase velocity. The group velocity of a wave is the speed at which energy propagates through the medium, and it is given by

$$v_g = \frac{\partial \omega}{\partial k}. \qquad \text{(Group velocity)}$$

In a simple medium in which the wavenumber is not a function of frequency, we find $v_g = \partial\omega/\partial k = \partial(v_p k)/\partial k = v_p$.

2.11   Problems

2.1. **Electromagnetic potentials.** Show by directly differentiating Equations (2.18) that they satisfy the Lorentz condition in Equation (2.19).

2.2. **Group velocity.** Rederive the expression for group velocity, Equation (2.34), by analyzing two waves closely spaced in *direction*, rather than frequency; i.e., in 2D, one of the waves has the k-vector $\mathbf{k}_1 = k_x + k_y$ and the other $\mathbf{k}_2 = k_x - k_y$, with $k_y \ll k_x$.

2.3. **Boundary conditions at perfect conductors.** Write the expressions for the incident and reflected waves incident normally on a perfect conductor wall, for (a) a perfect electric conductor (PEC, $\sigma = \infty$) and (b) a perfect magnetic conductor (PMC, $\sigma_m = \infty$). What happens to the polarization of the wave upon reflection?

2.4. **Direction of energy flow.** Both the group velocity and phase velocity are vector velocities, giving not only the speed but the direction of propagation. In an isotropic material, the direction of energy propagation is given by the time-averaged Poynting vector, show that this is also the direction of the phase velocity.

2.5. **Poynting vector and energy loss.** In a lossy material with conductivity σ, the energy dissipated per unit volume is given by Equation (2.16), where $\omega\epsilon'' = \sigma$. Show that this energy dissipation is related to the divergence of the time-averaged Poynting vector. Explain what this means physically. Hint: consider the divergence theorem.

2.6. **Impedance matching.** The impedance of any material is given by $\eta = \sqrt{\mu/\epsilon}$. Consider a material with $\epsilon = \epsilon_0$ and $\mu = \mu_0$, but with nonzero conductivities σ and σ_m (where the latter is not physical). Derive the relationship between σ and σ_m for this material to have the same impedance as that of free space, $\eta_0 = \sqrt{\mu_0/\epsilon_0}$. Hint: start by showing the relationship between σ_m and μ'' in Section 2.4.

2.7. Dispersion relation. Starting with the frequency domain Maxwell's equations, derive a dispersion relation (i.e., a relationship between k and ω) for a lossy medium with frequency-dependent conductivity $\sigma(\omega)$. Write this relationship as $k = k_0 f(\omega)$, where k_0 is the wavenumber in free space. Plot the real and imaginary parts of the resulting relationship for $0 \leq (\sigma/\omega\epsilon) \leq 1$ assuming σ is constant with frequency.

References

[1] R. S. Elliott, *Electromagnetics*. IEEE Press, 1993.

[2] J. C. Maxwell, "A dynamical theory of the electromagnetic field," *Phil. Trans. Royal Soc.*, vol. 155, p. 450, 1865.

[3] U. S. Inan and A. S. Inan, *Engineering Electromagnetics*. Addison-Wesley, 1999.

[4] H. Hertz, "On the finite velocity of propagation of electromagnetic actions," *Sitzb. d. Berl. Akad. d. Wiss*, February 1888.

[5] H. Hertz, *Electric Waves*. London: MacMillan, 1893.

[6] A. Einstein, "Über die von der molekularkinetischen theorie der wrme geforderte bewegung von in ruhenden flüssigkeiten suspendierten teilchen," *Annalen der Physik*, vol. 322, pp. 549–650, 1905.

[7] A. Einstein, H. A. Lorentz, H. Minkowski, and H. Weyl, *The Principle of Relativity, Collected Papers*. New York: Dover, 1952.

[8] D. M. Cook, *The Theory of the Electromagnetic Field*. Prentice-Hall, 1975.

[9] D. J. Griffiths, *Introduction to Electrodynamics*, 2nd edn. Prentice-Hall, 1989.

[10] J. D. Jackson, *Classical Electrodynamics*, 2nd edn. Wiley, 1975.

[11] J. C. Maxwell, *A Treatise in Electricity and Magnetism*. Clarendon Press, 1862, vol. 2.

[12] C. A. de Coulomb, *Première Mémoire sur l'Électricité et Magnétisme (First Memoir on Electricity and Magnetism)*. Historie de l'Académie Royale des Sciences, 1785.

[13] A. M. Ampère, *Recueil d'Observation Électrodynamiques*. Crochard, 1822.

[14] M. Faraday, *Experimental Researches in Electricity*. Taylor, 1839, vol. 1.

[15] J. Thuery, *Microwaves: Industrial, Scientific and Medical Applications*. Artech House, 1992.

[16] U. S. Inan and A. S. Inan, *Electromagnetic Waves*. Prentice-Hall, 2000.

[17] M. A. Plonus, *Applied Electromagnetism*. McGraw-Hill, 1978.

[18] P. S. Neelakanta, *Handbook of Electromagnetic Materials*. CRC Press, 1995.

[19] C. Kittel, *Introduction to Solid State Physics*, 5th edn. Wiley, 1976.

[20] M. A. Brown and R. C. Semelka, *MRI Basic Principles and Applications*. Wiley-Liss, 1995.

[21] J. A. Stratton, *Electromagnetic Theory*. McGraw-Hill, 1941.

[22] H. H. Skilling, *Fundamentals of Electric Waves*. R. E. Krieger Publishing Co., 1974.

[23] M. S. Bigelow, N. N. Lepeshkin, H. Shin, and R. W. Boyd, "Propagation of a smooth and discontinuous pulses through materials with very large or very small group velocities," *J. Phys: Cond. Matt.*, vol. 18, pp. 3117–3126, 2006.

[24] G. J. Fishman and D. H. Hartmann, "Gamma-ray bursts," *Scientific American*, vol. July, pp. 46–51, 1997.

3 Partial differential equations and physical systems

Most physical systems are described by one or more partial differential equations, which can be derived by direct application of the governing physical laws. For electromagnetic phenomena, the governing physical laws are experimentally based and include Faraday's law, Ampère's law, Coulomb's law, and the conservation of electric charge. Application of these laws leads to Maxwell's equations, as described in the previous chapter. In this chapter, we will look at different types of differential equations; Maxwell's equations are an example of a set of coupled partial differential equations. We will also look at how these equations are discretized in order to be solved in a discrete space (i.e., on a computer).

In these introductory chapters, we will occasionally revert to a description of waves on transmission lines, as discussed in detail in [1, Ch. 2–3]. As shown in Figure 3.1, voltage and current signals propagate along transmission lines as waves, similar to electromagnetic waves. In this case, however, the waves are confined to one spatial dimension, and so provide a simple but realistic 1D example.

For voltage and current waves on transmission lines, application of Kirchoff's voltage and current laws (which are contained in Maxwell's equations) leads to the *telegrapher's equations*:[1]

$$\frac{\partial \mathcal{V}}{\partial x} = -R\,\mathcal{I} - L\,\frac{\partial \mathcal{I}}{\partial t} \tag{3.1a}$$

$$\frac{\partial \mathcal{I}}{\partial x} = -G\,\mathcal{V} - C\,\frac{\partial \mathcal{V}}{\partial t} \tag{3.1b}$$

where $\mathcal{V}(x, t)$ and $\mathcal{I}(x, t)$ are respectively the line voltage and current, while R, L, G, and C are respectively the distributed resistance, inductance, conductance, and capacitance of the line, in units of Ω/m, H/m, S/m, and F/m, respectively. A segment of a transmission line is shown in Figure 3.1, showing the equivalent circuit diagram for a short piece of the line. Manipulation of Equations (3.1a) and (3.1b) leads to the general *wave equation* for a transmission line:

$$\frac{\partial^2 \mathcal{V}}{\partial x^2} = RG\,\mathcal{V} + (RC + LG)\,\frac{\partial \mathcal{V}}{\partial t} + LC\,\frac{\partial^2 \mathcal{V}}{\partial t^2}. \tag{3.2}$$

---

[1] We will use x as the spatial variable in this and future chapters when discussing the telegrapher's equations, in order to be consistent with the 1D versions of Maxwell's equations, which will be addressed in the next chapter.

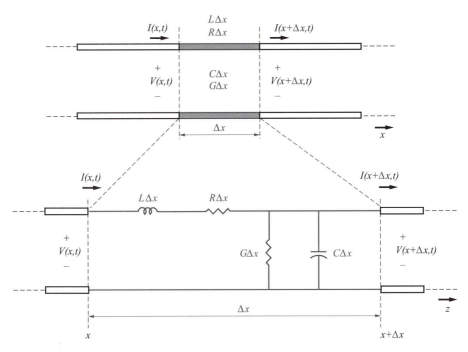

Figure 3.1 Voltage and current waves on a transmission line. Voltage and current propagate along the transmission line as waves, as though through a medium with parameters R, G, C, and L. Each small segment of the transmission line Δx can be treated as a small lumped-circuit element.

Note that making the substitution $\mathscr{V} \leftrightarrow \mathscr{I}$ in (3.2) gives the wave equation in terms of current $\mathscr{I}(x, t)$.

A similar wave equation can be derived in terms of the electric field $\overline{\mathscr{E}}$ by manipulating Equations (2.1) and (2.3), for simple media for which ϵ, μ, σ, and σ_m are simple constants, and Equation (2.9) is used with $M = \sigma_m H$:

$$\nabla^2 \overline{\mathscr{E}} = \sigma \sigma_m \overline{\mathscr{E}} + (\mu \sigma + \epsilon \sigma_m) \frac{\partial \overline{\mathscr{E}}}{\partial t} + \mu \epsilon \frac{\partial^2 \overline{\mathscr{E}}}{\partial t^2}. \tag{3.3}$$

The corresponding equation for the magnetic field $\overline{\mathscr{H}}$ can be obtained by making the substitution $\overline{\mathscr{E}} \leftrightarrow \overline{\mathscr{H}}$. Note that while Equation (3.2) is a scalar wave equation, Equation (3.3) for the $\overline{\mathscr{E}}$ field is a vector wave equation.

General solutions of Equations (3.2) and (3.3) are in general quite complex, except in special cases. Solutions for Equation (3.2) take a simple and analytically tractable form when we assume that the variations in time are sinusoidal (time-harmonic), or when we assume that the loss terms (i.e., those involving R, G in Equation (3.2)) are zero. Solutions for Equation (3.3) also become tractable for the time-harmonic and lossless (assuming that terms involving σ, σ_m in Equation (3.3) are zero) cases, if we make additional simplifying assumptions about the direction and spatial dependency of $\overline{\mathscr{E}}(x, y, z, t)$.

Partial differential equations arise in a variety of ways and in the context of a wide range of different physical problems, remarkably having forms quite similar to Equations (3.1) and (2.1) through (2.5). The basic reason for the commonality of the form of governing equations for different physical systems is the fact that most of classical physics is based on principles of conservation, whether it be principle of conservation of mass, energy, momentum, or electric charge.

For example, application of the principle of conservation of momentum to an incremental volume of a compressible fluid (e.g., a gas such as air) gives the "Telegrapher's equations" for acoustic waves:

$$\text{Transmission-line waves} \qquad \frac{\partial \mathcal{V}}{\partial x} = -L\,\frac{\partial \mathcal{I}}{\partial t} \qquad \frac{\partial \mathcal{I}}{\partial x} = -C\,\frac{\partial \mathcal{V}}{\partial t}$$

$$\text{Acoustic waves} \qquad \frac{\partial u_x}{\partial x} = -\frac{1}{\gamma_g p_a}\frac{\partial p}{\partial t} \qquad \frac{\partial p}{\partial x} = -\rho_v\,\frac{\partial u_x}{\partial t} \qquad (3.4)$$

where u_x (in m/s) is the velocity in the x direction, p (in kg/m/s^2) is the variation in pressure (above an ambient level), ρ_v is the mass per unit volume, γ_g (unitless) is the ratio of specific heats for the gas at constant pressure and constant volume, and p_a is the ambient pressure. The similarity between Equation (3.4) and the analogous transmission line equations, which are the lossless forms (i.e., $R = 0$ and $G = 0$) of Equation (3.1), is obvious. The two Equations (3.4) can be combined to find the wave equation for acoustic waves, written below in both 1D and 3D form:

$$\text{Trans. line / Electromagnetic} \qquad \frac{\partial^2 \mathcal{V}}{\partial x^2} = LC\,\frac{\partial^2 \mathcal{V}}{\partial t^2} \qquad \nabla^2 \overline{\mathscr{E}} = \mu\epsilon\,\frac{\partial^2 \overline{\mathscr{E}}}{\partial t^2}$$

$$\text{Acoustic wave equation} \qquad \frac{\partial^2 p}{\partial x^2} = \frac{\rho_v}{\gamma_g p_a}\frac{\partial^2 p}{\partial t^2} \qquad \nabla^2 p = \frac{\rho_v}{\gamma_g p_a}\frac{\partial^2 p}{\partial t^2}. \qquad (3.5)$$

The fact that the equations for electromagnetic and voltage waves are identical to that for acoustic waves indicates that wave-like solutions will result in both cases.

Similarly, application of the principles of conservation of mass and the continuity of mass to problems involving diffusion of one substance into another, leads to the telegrapher's equations for the diffusion process:

$$\text{Transmission line} \qquad \frac{\partial \mathcal{V}}{\partial x} = -R\,\mathcal{I} \qquad \frac{\partial \mathcal{I}}{\partial x} = -C\,\frac{\partial \mathcal{V}}{\partial t}$$

$$\text{Diffusion process} \qquad \frac{\partial c}{\partial x} = -\frac{1}{D}\,m_{vx} \qquad \frac{\partial m_{vx}}{\partial x} = -\frac{\partial c}{\partial t} \qquad (3.6)$$

where c is the mass density in kg/m^3, m_{vx} is the mass transfer per unit area in the x direction (in units of kg/m^2/s), and D is a constant known as the diffusivity or diffusion constant, which has units of m^2/s. Combining Equations (3.6) gives the equation for diffusion processes, written below in both 1D and 3D form:

$$\text{Trans. line / Electromagnetic} \qquad \frac{\partial^2 \mathcal{V}}{\partial x^2} = RC \frac{\partial \mathcal{V}}{\partial t} \qquad \nabla^2 \overline{\mathcal{E}} = \mu\sigma \frac{\partial \overline{\mathcal{E}}}{\partial t} \qquad (3.7a)$$

$$\text{Diffusion equation} \qquad \frac{\partial^2 c}{\partial x^2} = \frac{1}{D}\frac{\partial c}{\partial t} \qquad \nabla^2 c = \frac{1}{D}\frac{\partial c}{\partial t}. \qquad (3.7b)$$

Equations (3.7b) also describe the conduction of heat, if we take $c(x)$ (or $c(\mathbf{r})$ in the 3D case) to be the temperature distribution, and the "diffusion" coefficient D to be the product of the thermal capacity and thermal resistance of the material (in which heat is conducted). In the transmission-line context, Equation (3.7a) arises in cases when the line inductance is negligible, i.e., $L = 0$, and the line conductance G is also zero (the leakage conductance is usually small in most cases). Such lines, sometimes called RC lines, occur commonly in practice; for example, most on-chip interconnect structures and some thin film package wires exhibit negligible values of L, and have significant resistance, often much larger than the characteristic impedance. Some telephone lines, particularly closely packed lines and twisted-pair lines, can also be treated as RC lines. When an RC line is excited at its input, the line voltage does not exhibit oscillatory behavior (as expected due to the absence of inductance); rather, the voltage builds up along the line in a manner entirely analogous to the diffusion which occurs in heat-flow problems.[2] The underlying process is the same as that of diffusion of electrons and holes in semiconductors. The solutions of equations such as (3.6) and (3.7) thus exhibit an entirely different character than those of Equations (3.4) and (3.5).

3.1 Classification of partial differential equations

The general form of a quasi-linear (i.e., linear in its highest-order derivative) second-order partial differential equation (PDE) is:

$$A\frac{\partial^2 f}{\partial x^2} + B\frac{\partial^2 f}{\partial x \partial y} + C\frac{\partial^2 f}{\partial y^2} + D\frac{\partial f}{\partial x} + E\frac{\partial f}{\partial y} + Ff = G \qquad (3.8)$$

where the coefficients A, B, and C can be functions of x, y, f, $\partial f/\partial x$, and $\partial f/\partial y$, the coefficients D and E may be functions of x, y, and f, and the coefficients F and G may be functions of x and y.

The classification of PDEs is based on the analogy between the form of the discriminant $B^2 - 4AC$ for a PDE as given in Equation (3.8) and the form of the discriminant $B^2 - 4AC$ that classifies conic sections, described by the second-order algebraic equation:

$$Ax^2 + Bxy + Cy^2 + Dx + Ey + F = 0 \qquad (3.9)$$

---

[2] For example, when a solid heated object is dropped into cold water, heat slowly diffuses out to the surrounding body of water. For a detailed discussion and a formulation of 1D heat flow in terms of diffusion, see [2, Ch. 18].

where the relationship between Equations (3.8) and (3.9) should be obvious, by substituting $x \leftrightarrow \partial f/\partial x$ and $y \leftrightarrow \partial f/\partial y$. The type of conic section described by Equation (3.9) depends on the sign of the discriminant:

$B^2 - 4AC$	Conic type
Negative	Ellipse
Zero	Parabola
Positive	Hyperbola

Simply on the basis of the analogy between Equations (3.8) and (3.9), partial differential equations are classified as *elliptic*, *parabolic*, and *hyperbolic*. We now briefly discuss examples of the different type of PDEs.

3.1.1 Elliptic PDEs

Elliptic PDEs generally arise in physical problems involving equilibrium or static conditions, i.e., where there is no time dependence. A good example of an elliptic PDE is the electrostatic Laplace's equation in two dimensions:

$$\nabla^2 V = 0 \qquad \rightarrow \qquad \frac{\partial^2 V}{\partial x^2} + \frac{\partial^2 V}{\partial y^2} = 0.$$

In this case, comparing with Equation (3.8) with V substituted for f, we have $A = 1$, $B = 0$, and $C = 1$, so that $B^2 - 4AC < 0$, and Laplace's equation is therefore an *elliptic* PDE. We shall not deal with elliptic equations in this book, since they do not lend themselves to numerical solutions using time-marching methods, such as the FDTD method.

3.1.2 Parabolic PDEs

Parabolic PDEs generally arise in propagation problems involving infinite propagation "speed," so that the physical quantity of interest (e.g., line voltage) at each point in the solution domain is influenced by the present and past values of the quantity across the entire solution domain. A good example is the diffusion equation for an RC line, as in Equation (3.7a):

$$\frac{\partial V}{\partial t} - \frac{1}{RC} \frac{\partial^2 V}{\partial x^2} = 0. \tag{3.10}$$

Once again we let $f = V$ and $y = t$, in which case we have $A = -(RC)^{-1}$, $B = 0$, and $C = 0$,[3] so that $B^2 - 4AC = 0$, and the diffusion equation is therefore a *parabolic* PDE. Time-marching techniques such as FDTD are the preferred numerical method for the solution of parabolic PDEs. While we shall not explicitly deal with parabolic PDEs in this book, some of the applications we shall consider will involve highly lossy materials and slowly varying systems, in which case the displacement current term in

---

[3] The $\partial V/\partial t$ term implies $E = 1$ in Equation (3.8); however, it does not affect the type of PDE.

Equation (2.3) is negligibly small, and the electromagnetic wave Equation (3.3) effectively reduces to the diffusion wave Equation (3.7).

3.1.3 Hyperbolic PDEs

Maxwell's two curl Equations (2.1) and (2.3) are primary examples of hyperbolic PDEs, and together lead to the electromagnetic wave Equation (3.3), which is also a hyperbolic PDE.[4] These equations describe harmonically varying and oscillatory electromagnetic waves, and will be the focus of our attention in this book. Consider the wave equation for a transmission line:

$$\frac{\partial^2 \mathcal{V}}{\partial t^2} - v_p^2 \frac{\partial^2 \mathcal{V}}{\partial x^2} = 0 \qquad (3.11)$$

where $v_p = (LC)^{-1/2}$. Equation (3.11) corresponds to Equation (3.8) if we let $f = \mathcal{V}$ and $y = t$, in which case we have $A = -v_p^2$, $B = 0$, and $C = 1$, so that $B^2 - 4AC > 0$, and the wave Equation (3.11) is therefore a *hyperbolic* PDE.

It is well known (and can be easily shown by substitution) that the exact general solution of the wave Equation (3.11) is:

$$\mathcal{V}(x, t) = f(x - v_p t) + g(x + v_p t) \qquad (3.12)$$

where $f(\cdot)$ and $g(\cdot)$ are two different arbitrary functions. The two terms in (3.12) respectively represent waves traveling in the $+x$ and $-x$ directions. The solution of this second-order wave equation requires the specification of two initial conditions:

$$\mathcal{V}(x, 0) = \phi(x) \qquad (3.13a)$$

$$\left[\frac{\partial \mathcal{V}(x, t)}{\partial t} \right]_{t=0} = \theta(x). \qquad (3.13b)$$

Note that Equation (3.13a) amounts to specifying the voltage at $t = 0$ at all points x along the line, while Equation (3.13b) specifies the initial time derivative of the voltage at all points x.[5] Substituting Equation (3.13) into (3.12) gives:

$$\phi(x) = f(x) + g(x) \qquad (3.14a)$$

$$\theta(x) = -v_p\, f'(x) + v_p\, g'(x). \qquad (3.14b)$$

where the primes denote differentiation of $f(\cdot)$ and $g(\cdot)$ with respect to their arguments. We can now integrate Equation (3.14b) to find:

$$- f(x) + g(x) = \frac{1}{v_p} \int_{-\infty}^{x} \theta(\zeta) d\zeta. \qquad (3.15)$$

---

[4] In the interest of completeness, note that in most practical cases, equations are neither strictly parabolic nor hyperbolic; when loss terms are added, equations are often referred to as "strongly hyperbolic" or "weakly hyperbolic."

[5] Or, as we shall see shortly, when we consider coupled hyperbolic equations, this second initial condition is actually the *current* at $t = 0$.

Combination of Equations (3.15) and (3.14a) yields:

$$f(x) = \frac{1}{2}\left[\phi(x) - \frac{1}{v_p}\int_{-\infty}^{x}\theta(\varsigma)d\varsigma\right] \tag{3.16a}$$

$$g(x) = \frac{1}{2}\left[\phi(x) + \frac{1}{v_p}\int_{-\infty}^{x}\theta(\varsigma)d\varsigma\right]. \tag{3.16b}$$

Substituting Equation (3.16) into (3.12) we find:

$$\mathcal{V}(x,t) = \frac{1}{2}\left[\phi(x - v_pt) + \phi(x + v_pt) + \frac{1}{v_p}\int_{x-v_pt}^{x+v_pt}\theta(\varsigma)d\varsigma\right]. \tag{3.17}$$

This elegant and exact solution of the 1D wave equation is known as *D'Alembert's solution*[6] and is particularly applicable to problems involving vibrating systems such as a stretched string.

It is useful to note that the wave Equation (3.11) can be factored and expressed as a product of two first-order hyperbolic equations:

$$\frac{\partial^2\mathcal{V}}{\partial t^2} - v_p^2\frac{\partial^2\mathcal{V}}{\partial x^2} = 0 \tag{3.18a}$$

$$\underbrace{\left(\frac{\partial}{\partial t} + v_p\frac{\partial}{\partial x}\right)}_{\mathcal{L}^+}\underbrace{\left(\frac{\partial}{\partial t} - v_p\frac{\partial}{\partial x}\right)}_{\mathcal{L}^-}\mathcal{V} = 0 \tag{3.18b}$$

where the two operators $\mathcal{L}^-$ and $\mathcal{L}^+$ are successively applied to $\mathcal{V}(x,t)$. It thus appears that the wave equation is in fact a combination of two equations:

$$\frac{\partial\mathcal{V}}{\partial t} + v_p\frac{\partial\mathcal{V}}{\partial x} = 0 \qquad \rightarrow \qquad \mathcal{V}(x,t) = f(x - v_pt) \tag{3.19a}$$

$$\frac{\partial\mathcal{V}}{\partial t} - v_p\frac{\partial\mathcal{V}}{\partial x} = 0 \qquad \rightarrow \qquad \mathcal{V}(x,t) = g(x + v_pt). \tag{3.19b}$$

While the wave Equation (3.11) supports waves propagating in either $+x$ or $-x$ directions, Equations (3.19) are known as the *one-way wave equations* and only allow for waves traveling respectively in the $+x$ and $-x$ directions. Note that since the operators $\mathcal{L}^+$ and $\mathcal{L}^-$ may be applied in any order, any solution that is a linear combination of positive and negative going waves, i.e., $\mathcal{V}(x,t) = af(x - v_pt) + bg(x - v_pt)$, is also a solution of the wave Equation (3.18a).

Equations of the type (3.19a) and (3.19b) are also hyperbolic PDEs, but under a somewhat different classification criterion[7] than that which was described above in

[6] Named for the French mathematician Jean le Rond D'Alembert [1717–1783].
[7] A general first-order PDE of the form:

$$\frac{\partial f}{\partial t} + A\frac{\partial f}{\partial x} + Bf = G$$

is said to be *hyperbolic* if the matrix A is diagonalizable with real eigenvalues. This is certainly the case for Equations (3.19), since $A = \pm v_p$ are simple real numbers.

connection with Equation (3.8). In fact, one-way wave equations of the form:

$$\frac{\partial f}{\partial t} + v_p \frac{\partial f}{\partial x} = 0 \tag{3.20}$$

are the prototype for all hyperbolic equations. Equation (3.20) is known as the *convection* equation since it arises when the property of a fluid (i.e., whatever the quantity f represents) is convected in the $+x$ direction at a speed v_p.

In our applications of Finite Difference Time Domain methods to electromagnetic problems, we shall typically find it more convenient to work with the original Maxwell's Equations (2.1) and (2.3), rather than the wave Equation (3.3). Written out in their 1D form for TEM[8] waves, Maxwell's equations in free space consist of two coupled first-order convection equations:

$$\frac{\partial \mathcal{E}_y}{\partial t} + \frac{1}{\epsilon} \frac{\partial \mathcal{H}_z}{\partial x} = 0 \tag{3.21a}$$

$$\frac{\partial \mathcal{H}_z}{\partial t} + \frac{1}{\mu} \frac{\partial \mathcal{E}_y}{\partial x} = 0. \tag{3.21b}$$

Note that Equations (3.21) are identical to the coupled voltage-current equations for a lossless transmission line (i.e., Equation (3.1) with $R = 0$ and $G = 0$), when we make the substitutions $\mathcal{V} \to \mathcal{E}_y$, $\mathcal{I} \to \mathcal{H}_z$, $\epsilon \to C$, and $\mu \to L$.

3.2 Numerical integration of ordinary differential equations

Before we embark on the formulation of Partial Differential Equation (PDE) solutions using finite difference methods, it is useful to consider briefly numerical integration of Ordinary Differential Equations (ODEs). In this context, we note that there is a wide range of different techniques for integration of ODEs which we by no means attempt to cover completely. Our purpose here is simply to look at a few different finite difference methods, in particular the so-called *first-order Euler*, the *leapfrog*, and the *Runge-Kutta* methods, in order to develop an initial feel for evaluating the accuracy and stability of such methods.

We shall consider the simple ordinary differential equation:

$$\frac{dy}{dt} + f(y, t) = 0 \tag{3.22}$$

where $y = y(t)$ and the initial condition (which must be known for solution of Equation 3.22) is $y(0) = y_{\text{init}}$. In order to integrate Equation (3.22) in time using small steps, we first establish a *time-mesh* as shown in Figure 3.2. We will adopt here the commonly used notation of representing the values of the function at particular time points with a superscript, so that we have:

$$y(n\Delta t) = y^n \tag{3.23}$$

---

[8] The nomenclature TEM, TE, and TM are defined differently in FDTD compared to traditional waveguide modes; these will be discussed in detail in later chapters.

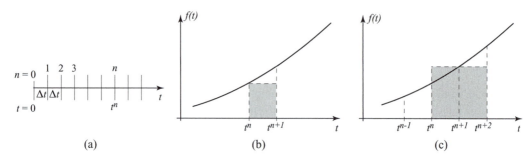

Figure 3.2 Finite difference integration of an ODE. (a) The time-mesh. (b) First-order Euler method. The shaded region is the approximation to the integral under the curve between t^n and t^{n+1}. (c) Leapfrog method. The shaded area is the approximation to the integral under the curve between t^n and t^{n+2}.

where $\Delta t = t^{n+1} - t^n$ is the time increment on our mesh. Integrating Equation (3.22) on the time-mesh between the time points t^n and t^{n+1} allows us to relate two successive values of $y(t)$ as:

$$y^{n+1} = y^n - \int_{t^n}^{t^{n+1}} f(y, t) \, dt. \qquad (3.24)$$

The methods considered below involve different approximations of the integral in Equation (3.24) above.

3.2.1   First-order Euler method

The simplest way in which we can approximate the integral in Equation (3.24) is the so-called first-order Euler method:

$$y^{n+1} = y^n - f(y^n, t^n)\Delta t \qquad n = 0, 1, 2, \ldots . \qquad (3.25)$$

In this method, we are simply approximating the value of the function $f(y, t)$ as $f(y^n, t^n)$ over the interval $[t^n, t^{n+1}]$, as shown in Figure 3.2. With the initial condition specified as $y^0 = y_{\text{init}}$, Equation (3.25) can be straightforwardly used to determine all the subsequent values of $y(t)$ at our time-mesh points t^n. Clearly, we must make sure that Δt is sufficiently small, so that errors do not build up, i.e., that the method is stable. To determine the conditions under which the first-order Euler method is stable, we assume that the value of y^n at time t^n contains an error of ϵ_n and determine whether the error ϵ^{n+1} in y^{n+1} is larger, i.e., whether the finite differencing procedure amplifies the error. We define this amplification or growth factor q as follows:

$$\epsilon^{n+1} = q \, \epsilon^n. \qquad (3.26)$$

Using (3.25) we have

$$y^{n+1} + \epsilon^{n+1} = y^n + \epsilon^n - f[(y^n + \epsilon^n), t^n]\Delta t. \qquad (3.27)$$

For small errors, i.e., $|\epsilon^n| \ll |y^n|$, we can expand the function f into a Taylor series[9] around y^n:

$$f(y^n + \epsilon^n, t^n) = f(y^n, t^n) + (y^n + \epsilon^n - y^n) \left. \frac{\partial f}{\partial y} \right|_n + O(\epsilon^n) \qquad (3.28)$$

where the notation $O(\epsilon^n)$, to be read as "order ϵ^n", denotes the remainder term and indicates that it depends linearly (in this case) on ϵ^n, i.e., that it approaches zero as the first power of ϵ^n. Using Equation (3.28) together with (3.25), we find an equation describing the amplification of a small error:

$$\epsilon^{n+1} = \epsilon^n - \Delta t \, \epsilon^n \left. \frac{\partial f}{\partial y} \right|_n + O(\epsilon^n) \qquad \rightarrow \qquad q = \frac{\epsilon^{n+1}}{\epsilon^n} \simeq 1 - \Delta t \left. \frac{\partial f}{\partial y} \right|_n. \qquad (3.29)$$

In order for the method to be stable, errors must not grow in time, i.e., we must have $|q| \leq 1$. Assuming that $[\partial f / \partial y]_n$ is positive, we then have:

$$|q| \leq 1 \qquad \rightarrow \qquad \Delta t \left. \frac{\partial f}{\partial y} \right|_n \leq 2 \qquad \rightarrow \qquad \Delta t \leq \frac{2}{[\partial f / \partial y]_n}. \qquad (3.30)$$

As an example, consider the case when $f(y, t) = \tau^{-1} y$, i.e., the simple ODE describing a first-order system exhibiting exponential decay (e.g., a lumped electrical circuit consisting of a resistance and capacitance):

$$\frac{dy}{dt} + \frac{y}{\tau} = 0 \qquad (3.31)$$

with an initial condition $y(0) = 1$. We know that the analytical solution of Equation (3.31) is:

$$y(t) = e^{-t/\tau}.$$

For this case, we have $[\partial f / \partial y]_n = \tau^{-1}$, so that the criterion in Equation (3.30) reduces to:

$$\Delta t \leq 2\tau, \qquad (3.32)$$

revealing the physically sensible result that the integration time increment must be smaller than the characteristic time described by the ODE. Experimentation with different values of Δt to determine the degree of stability is the topic of Problem 3 at the end of this chapter.

The first-order Euler method is stable as long as Equation (3.30) is satisfied; what's more, the accuracy of the numerical solution can be arbitrarily increased by using smaller and smaller values of the time increment Δt. Such is not the case, however, for ODEs that have an oscillatory type of solution, for which the first-order Euler method is always

[9] Taylor's series expansion of a function $g(x)$ around $x = x_0$ is given by:

$$g(x) = g(x_0) + (x - x_0) \left. \frac{\partial g}{\partial x} \right|_{x=x_0} + \frac{1}{2}(x - x_0)^2 \left. \frac{\partial^2 g}{\partial x^2} \right|_{x=x_0} + \cdots + \frac{1}{n!}(x - x_0)^n \left. \frac{\partial^n g}{\partial x^n} \right|_{x=x_0} + \cdots .$$

(i.e., for any value of Δt) unstable. Consider the second-order differential equation for a lumped LC circuit:

$$\frac{d^2 \mathcal{V}}{dt^2} + \omega_0^2 \mathcal{V} = 0 \tag{3.33}$$

where $\omega_0 = (LC)^{-1/2}$ is the frequency of oscillation. We can factor this equation in the same manner as was done in Equation (3.18):

$$\left(\frac{d}{dt} + j\omega_0\right)\left(\frac{d}{dt} - j\omega_0\right)\mathcal{V} = 0. \tag{3.34}$$

The general solution of Equation (3.33) consists of a linear superposition of the solutions of the two first-order Equations contained in Equation (3.34). Consider the first one:

$$\frac{d\mathcal{V}}{dt} + j\omega_0 \mathcal{V} = 0. \tag{3.35}$$

Note that (3.35) is same as the general first-order ODE given in Equation (3.22), when we note that $y \leftrightarrow \mathcal{V}$ and $f(\mathcal{V}, t) \leftrightarrow j\omega_0\mathcal{V}$. Noting that we then have $[\partial f/\partial y]_n = j\omega_0$, the error amplification factor from Equation (3.29) is:

$$q = 1 - j\omega_0 \,\Delta t \qquad \rightarrow \qquad |q|^2 = q\,q^* = 1 + \omega_0^2(\Delta t)^2,$$

which is clearly always larger than unity for any time increment Δt, indicating that the first-order Euler method is *unconditionally unstable* and is thus not appropriate for purely oscillatory solutions.

3.2.2   The leapfrog method

The basic reason for the unsuitability of the first-order Euler method for oscillatory equations is the fact that the integral in Equation (3.24) is evaluated by using only the value of the integrand f at time t^n, resulting in only first-order accuracy. Better accuracy can be obtained by evaluating the area under the integrand with a time-centered method. One such commonly used method is the so-called leapfrog method.[10] In this method, the time derivative is determined across a double time step, using the intermediate time step to determine the integral of the function:

$$y^{n+1} = y^{n-1} - f(y^n, t^n)\,2\Delta t \tag{3.36a}$$
$$y^{n+2} = y^n - f(y^{n+1}, t^{n+1})\,2\Delta t. \tag{3.36b}$$

It is clear that y^{n+1} and y^{n+2} can in turn be used to determine y^{n+3}, proceeding in this "leapfrog" fashion.

---

[10] The leapfrog method was first introduced apparently by [3] for the solution of parabolic PDEs. Ironically, it turns out that the leapfrog method is unstable for parabolic PDEs, but is well suited for the solution of hyperbolic PDEs such as (3.20) and in fact forms the basis of most FDTD solutions of electromagnetic problems.

One difficulty with the leapfrog method is that it requires the specification of not just the initial condition $y^0 = y(0)$, but also the value at the second time-mesh point, namely y^1. It is possible simply to use a less accurate method to determine y^1 and then proceed using Equation (3.36) for the rest of the time increments. However, the overall accuracy of the leapfrog method is actually sensitively dependent on the value of y^1, so that in practice one needs to determine y^1 with an even higher-order method.

To analyze the stability of the leapfrog scheme, we can apply the same type of error propagation analysis used for the first-order Euler method to one of the Equations (3.36). In this case, the errors at the three time-mesh points t^{n-1}, t^n, and t^{n+1} are related so that we have:

$$\epsilon^{n+1} = \epsilon^{n-1} - 2\Delta t \, \epsilon^n \frac{\partial f}{\partial y}\bigg|_n \tag{3.37}$$

$$\rightarrow \quad \underbrace{\frac{\epsilon^{n+1}}{\epsilon^n}}_{q} = \underbrace{\frac{\epsilon^{n-1}}{\epsilon^n}}_{q^{-1}} - 2\Delta t \frac{\partial f}{\partial y}\bigg|_n$$

$$\rightarrow \quad q^2 \simeq 1 - 2q \, \Delta t \frac{\partial f}{\partial y}\bigg|_n \tag{3.38}$$

$$\rightarrow \quad q = -\zeta \pm \sqrt{\zeta^2 + 1} \qquad \text{where} \qquad \zeta = \Delta t \frac{\partial f}{\partial y}\bigg|_n.$$

It is interesting to note that for equations with non-oscillatory solutions, for example the exponential decay Equation (3.31), we have $\zeta = \tau^{-1}\Delta t$, which is purely real, so that one of the roots of Equation (3.38) is always larger than unity. Thus, it seems that the leapfrog scheme may not be appropriate for equations with decay-type solutions. However, for equations with oscillatory solutions, such as Equation (3.33), we have $\zeta = j\omega_o \Delta t$, so that we have:

$$q = -j(\omega_0 \Delta t) \pm \sqrt{-(\omega_0 \Delta t)^2 + 1}$$

$$\rightarrow \quad |q|^2 = qq^* = 1 \quad \text{for} \quad (\omega_0 \Delta t) \leq 1$$

$$\rightarrow \quad \Delta t \leq \frac{1}{\omega_0}.$$

Thus, the leapfrog method is stable for oscillatory solutions, as long as the time step we use is less than the characteristic time of the oscillation, i.e., ω_0^{-1}.

3.2.3 Runge-Kutta methods

Of the two methods described above, the Euler method is only first-order accurate and is rarely sufficient for real problems; the leapfrog method is second-order accurate, and is the most frequently used time-stepping algorithm in finite difference methods. In certain circumstances, however, higher-order accuracy is required, and for time discretization, the Runge-Kutta methods are a popular choice.

The Runge-Kutta or RK methods rely on intermediate values of the function at fractional time steps. Consider the general ODE given by Equation (3.22). The *second-order* Runge-Kutta, or RK2, solution is given by [4]:

$$y^{n+1} = y^n + \left(1 - \frac{1}{2\alpha}\right) k_1 + \left(\frac{1}{2\alpha}\right) k_2 \tag{3.39}$$

where the functions k_1 and k_2 are defined as:

$$k_1 = \Delta t \, f(y^n, t^n) \tag{3.40a}$$

$$k_2 = \Delta t \, f(y^n + \alpha k_1, t^n + \alpha \Delta t). \tag{3.40b}$$

and the constant α can be varied, but is frequently taken to be $\alpha = 1/2$. Now, like the Euler method, the RK2 method is *unconditionally unstable* for oscillatory solutions; however, the RK2 method is far less unstable: for 100 time steps and $\omega \Delta t = 0.2$, the solution is amplified only 2%.

The most commonly used RK method is RK4, which is fourth-order accurate. The complete algorithm is given by:

$$y^{n+1} = y^n + \frac{1}{6}k_1 + \frac{1}{3}k_2 + \frac{1}{3}k_3 + \frac{1}{6}k_4 \tag{3.41a}$$

where the k's are given by:

$$k_1 = \Delta t \, f(y^n, t^n) \tag{3.41b}$$

$$k_2 = \Delta t \, f(y^n + k_1/2, t^n + \Delta t/2) \tag{3.41c}$$

$$k_3 = \Delta t \, f(y^n + k_2/2, t^n + \Delta t/2) \tag{3.41d}$$

$$k_4 = \Delta t \, f(y^n + k_3, t^n + \Delta t). \tag{3.41e}$$

Figure 3.3 shows the stability diagrams for the RK2 and RK4 methods. Notice that RK2 does not enclose the imaginary axis, which is why it is unstable for oscillatory solutions (complex λ). The RK4 method, however, encloses part of the imaginary axis, and so it is stable for oscillatory solutions, albeit for $|\lambda \Delta t| \leq 2.83$.

3.3   Finite difference approximations of partial differential equations

We are now ready to tackle the discretization of partial differential equations, which will ultimately lead us to discretization of Maxwell's equations. The main goal of a finite difference method applied for the solution of a partial differential equation (or set of equations) is to transform a calculus problem into an algebra problem by discretizing the continuous physical domain and time, and by approximating the exact partial derivatives with finite difference approximations. In other words, we aim to transform the equation(s) from the *continuous* domain, where one deals with *differential* equations, to the *discrete* domain, whereupon it becomes a *difference* equation. This procedure approximates a

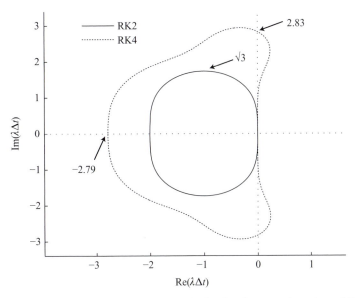

Figure 3.3 Stability of Runge-Kutta methods. The diagrams show the stability regions for the general ODE given by $dy/dt = \lambda y$. When λ is complex, oscillatory solutions result; only the RK4 method is stable under these circumstances.

partial differential equation (PDE) with a finite difference equation (FDE), which can then be readily implemented on a computer.

The simplest example of a 1D finite difference grid is shown in Figure 3.4. Note that when we refer to it as being "one-dimensional" we are referring to its one spatial dimension; since these equations are also functions of time, this results in a 2D grid. The intersections of the gridlines shown are the grid points. It is customary to use subscript i to denote the spatial location of the grid lines in the x dimension, i.e., $x_i = (i - 1)\Delta x$. The gridlines in time are indicated by a superscript n, i.e., $t^n = n\Delta t$. Thus, the grid point (i, n) corresponds to location (x_i, t^n) on the finite difference grid. In this simplest case, the spatial grid lines are equally spaced and perpendicular to a principal axis, i.e., the x axis. In some problems, it is necessary to use nonuniform grids in which Δx varies with i. It is also possible to use various transformations and work with grids that are not perpendicular to a principal coordinate axis; these advanced techniques will be briefly introduced in Chapter 15.

The solution to the finite difference equation is obtained by time marching, i.e., using past, present, and sometimes future values of the physical quantity (e.g., voltage or electric field) to determine its values at grid points corresponding to increasingly higher n. Although the discretization of the time axis is usually uniform, nonuniform time-stepping may be necessary in cases when the decay rates of the physical quantity are exceptionally rapid [5, Sec. 3.6.10].

Note that extension of the finite difference grid to 2D or 3D physical spaces is straightforward; for example, in 3D physical space, the grid point (i, j, k, n) corresponds to location (x_i, y_j, z_k, t^n) on the finite difference grid.

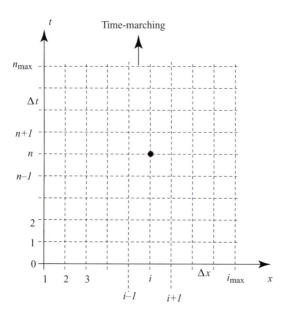

Figure 3.4 The finite difference grid. Note that we have indexed in i starting with 1, and in time starting with zero; $t = 0$ (the lower edge) represents the initial condition at all x.

The first step in any finite-difference problem is the establishment of a grid as described above; we will have much more to say on this process in later chapters. Once the finite difference grid is established, the next step is to approximate partial derivatives with finite differences; i.e., transform from PDEs to FDEs. This is typically achieved by using Taylor series expansions of the function around different grid points in time and space. The next sections illustrate this procedure.

3.3.1 Derivatives in time

Before tackling the partial differential equations directly, we must derive the finite difference versions of the partial derivatives themselves. We first consider time derivatives. The general form of the Taylor series expansion of a function $f(x, t)$ of two variables around a given time t_0 is

$$f(x, t) = f(x, t_0) + (t - t_0) \left.\frac{\partial f}{\partial t}\right|_x^{t_0} + \frac{1}{2}(t - t_0)^2 \left.\frac{\partial^2 f}{\partial t^2}\right|_x^{t_0} + \cdots + \frac{1}{n!}(t - t_0)^n \left.\frac{\partial^n f}{\partial t^n}\right|_x^{t_0} + \cdots .$$

$$(3.42)$$

Consider the Taylor series expansion in time (i.e., in time index n) of f_i^{n+1} around the grid point (i, n):

$$f_i^{n+1} = f_i^n + \underbrace{(t^{n+1} - t^n)}_{\Delta t} \left.\frac{\partial f}{\partial t}\right|_i^n + \frac{1}{2} \underbrace{(t^{n+1} - t^n)^2}_{(\Delta t)^2} \left.\frac{\partial^2 f}{\partial t^2}\right|_i^n + \cdots . \qquad (3.43)$$

Equation (3.43) follows directly from (3.42), where the time point n corresponds to t_0 and the time point $n + 1$ corresponds to time t. We can now solve (3.43) for $[\partial f/\partial t]_i^n$ to obtain

$$\left.\frac{\partial f}{\partial t}\right|_i^n = \frac{f_i^{n+1} - f_i^n}{\Delta t} - \underbrace{\frac{1}{2}\Delta t \left.\frac{\partial^2 f}{\partial t^2}\right|_i^n}_{O(\Delta t)} - \cdots . \tag{3.44}$$

It is clear that all the other terms are multiplied by higher powers of Δt, so that the "order" of the remaining term is first-order in Δt; i.e., the remainder (or error) increases linearly with Δt. A finite difference approximation to the first derivative in time is obtained by dropping the error terms; i.e.,

First-order forward difference

$$\boxed{\left.\frac{\partial f}{\partial t}\right|_i^n \simeq \frac{f_i^{n+1} - f_i^n}{\Delta t}} \tag{3.45}$$

Equation (3.45) is a first-order forward difference approximation of $[\partial f/\partial t]_i^n$ at the grid point (i, n). It is first order because it was obtained by dropping terms of $O(\Delta t)$. We refer to this approximation as "forward difference" because the derivative at time step n is approximated using the value of the function f at time step n and the next step forward in time, $n + 1$.

There are many other ways in which we can approximate the first derivative of the function at the grid point (i, n). For example, by expanding (in Taylor series) f_i^{n+1} and f_i^{n-1} around the grid point (i, n) and subtracting them, we find a more accurate second-order time-centered approximation for $[\partial f/\partial t]_i^n$, given by:

$$\left.\frac{\partial f}{\partial t}\right|_i^n = \frac{f_i^{n+1} - f_i^{n-1}}{2\Delta t} - \underbrace{\frac{1}{6}(\Delta t)^2 \left.\frac{\partial^3 f}{\partial t^3}\right|_i^n}_{O[(\Delta t)^2]} - \cdots . \tag{3.46}$$

Dropping the error terms gives:

Second-order centered difference

$$\boxed{\left.\frac{\partial f}{\partial t}\right|_i^n \simeq \frac{f_i^{n+1} - f_i^{n-1}}{2\Delta t}} \tag{3.47}$$

This approximation is known as "centered difference" because it is centered in time between the field values used to make the approximation. The derivation of the above formulation is the subject of Problem 1 at the end of this chapter. Note that this equation requires that we know the value of the field f at time steps $n + 1$ and at $n - 1$, two steps apart. In some finite difference schemes, it is prudent (or even necessary) to define temporary or intermediate values of the time derivative at half-time steps.[11] A second-order centered difference approximation of $[\partial f/\partial t]_i^{n+\frac{1}{2}}$ can be obtained by subtracting the Taylor series expansions of f_i^{n+1} and f_i^n around the grid point $(i, n + \frac{1}{2})$. We find

$$\frac{\partial f}{\partial t}\bigg|_i^{n+\frac{1}{2}} = \frac{f_i^{n+1} - f_i^n}{\Delta t} - \underbrace{\frac{1}{24}(\Delta t)^2 \frac{\partial^3 f}{\partial t^3}\bigg|_i^n}_{O[(\Delta t)^2]} - \cdots . \tag{3.48}$$

Dropping the error terms gives:

Second-order forward difference

$$\boxed{\frac{\partial f}{\partial t}\bigg|_i^{n+\frac{1}{2}} \simeq \frac{f_i^{n+1} - f_i^n}{\Delta t}} . \tag{3.49}$$

The reader should notice two things about this result. First, the right-hand side of the equation is identical to that of Equation (3.44); however, Equation (3.48) is second-order accurate, while Equation (3.44) is only first-order accurate. Hence, the fraction on the right-hand side is a better approximation of the derivative at time step $n + \frac{1}{2}$ than at time step n. Second, Equations (3.47) and (3.48) are essentially the same if one considers that the time step Δt is simply halved in the latter case.

Finite difference approximations for the second derivatives in time can be obtained in a similar manner, using the various Taylor series expansions. Consider Equation (3.43), rewritten below for convenience:

$$f_i^{n+1} = f_i^n + \underbrace{(t^{n+1} - t^n)}_{\Delta t} \frac{\partial f}{\partial t}\bigg|_i^n + \frac{1}{2}\underbrace{(t^{n+1} - t^n)^2}_{(\Delta t)^2} \frac{\partial^2 f}{\partial t^2}\bigg|_i^n + \frac{1}{6}\underbrace{(t^{n+1} - t^n)^3}_{(\Delta t)^3} \frac{\partial^3 f}{\partial t^3}\bigg|_i^n + \cdots$$

$$f_i^{n+1} = f_i^n + \Delta t \frac{\partial f}{\partial t}\bigg|_i^n + \frac{1}{2}(\Delta t)^2 \frac{\partial^2 f}{\partial t^2}\bigg|_i^n + \frac{1}{6}(\Delta t)^3 \frac{\partial^3 f}{\partial t^3}\bigg|_i^n + \cdots . \tag{3.50}$$

Now consider the Taylor series expansion of f_i^{n-1} around the grid point (i, n):

$$f_i^{n-1} = f_i^n + \underbrace{(t^{n-1} - t^n)}_{-\Delta t} \frac{\partial f}{\partial t}\bigg|_i^n + \frac{1}{2}\underbrace{(t^{n-1} - t^n)^2}_{(-\Delta t)^2} \frac{\partial^2 f}{\partial t^2}\bigg|_i^n + \frac{1}{6}\underbrace{(t^{n-1} - t^n)^3}_{(-\Delta t)^3} \frac{\partial^3 f}{\partial t^3}\bigg|_i^n + \cdots$$

$$f_i^{n-1} = f_i^n - \Delta t \frac{\partial f}{\partial t}\bigg|_i^n + \frac{1}{2}(\Delta t)^2 \frac{\partial^2 f}{\partial t^2}\bigg|_i^n - \frac{1}{6}(\Delta t)^3 \frac{\partial^3 f}{\partial t^3}\bigg|_i^n + \cdots . \tag{3.51}$$

Adding Equations (3.50) and (3.51) allows us to cancel the second terms on the right-hand side of each of these equations, the terms proportional to Δt; rearranging terms

---

[11] This is particularly the case for the solution of Maxwell's equations using FDTD methods, since the electric and magnetic field components are interleaved spatially and temporally and advanced in time in a leapfrog manner using intermediate half-time steps.

we find:

$$\frac{\partial^2 f}{\partial t^2}\bigg|_i^n = \frac{f_i^{n+1} - 2f_i^n + f_i^{n-1}}{(\Delta t)^2} + O[(\Delta t)^2]. \tag{3.52}$$

Dropping the error terms gives us a second-order centered difference approximation to the second derivative in time:

Second-order centered difference

$$\boxed{\frac{\partial^2 f}{\partial t^2}\bigg|_i^n \simeq \frac{f_i^{n+1} - 2f_i^n + f_i^{n-1}}{(\Delta t)^2}}. \tag{3.53}$$

Other variations of these second derivatives follow similar derivations; some of these will be discussed in later chapters. Similarly, higher-order derivatives can be found along similar lines.

3.3.2 Derivatives in space

The finite difference approximations for the derivatives in space can be obtained exactly in the same manner as the time derivatives, by expanding the function into Taylor series (in the space variable x) around various grid points. As an example, consider the Taylor series expansion (in terms of the space variable x) of f_{i+1}^n around the grid point (i, n):

$$f_{i+1}^n = f_i^n + \underbrace{(x_{i+1} - x_n)}_{\Delta x}\frac{\partial f}{\partial x}\bigg|_i^n + \frac{1}{2}\underbrace{(x_{i+1} - x_n)^2}_{(\Delta x)^2}\frac{\partial^2 f}{\partial x^2}\bigg|_i^n + \frac{1}{6}\underbrace{(x_{i+1} - x_n)^3}_{(\Delta x)^3}\frac{\partial^3 f}{\partial x^3}\bigg|_i^n + \cdots \tag{3.54}$$

and the expansion of f_{i-1}^n around the grid point (i, n):

$$f_{i-1}^n = f_i^n + \underbrace{(x_{i-1} - x_n)}_{-\Delta x}\frac{\partial f}{\partial x}\bigg|_i^n + \frac{1}{2}\underbrace{(x_{i-1} - x_n)^2}_{(-\Delta x)^2}\frac{\partial^2 f}{\partial x^2}\bigg|_i^n + \frac{1}{6}\underbrace{(x_{i-1} - x_n)^3}_{(-\Delta x)^3}\frac{\partial^3 f}{\partial x^3}\bigg|_i^n + \cdots. \tag{3.55}$$

Subtracting Equation (3.55) from (3.54) and rearranging terms we find:

$$\frac{\partial f}{\partial x}\bigg|_i^n = \frac{f_{i+1}^n - f_{i-1}^n}{2\Delta x} - \underbrace{\frac{1}{6}(\Delta x)^2\frac{\partial^3 f}{\partial x^3}\bigg|_i^n - \cdots}_{O[(\Delta x)^2]}. \tag{3.56}$$

Dropping the error term from the above expression gives the second-order centered difference approximation for $[\partial f/\partial x]_i^n$:

Second-order centered difference

$$\boxed{\frac{\partial f}{\partial x}\bigg|_i^n \simeq \frac{f_{i+1}^n - f_{i-1}^n}{2\Delta x}}. \tag{3.57}$$

Other approximations for the first spatial derivative can be obtained by manipulations of different Taylor series expansions. One approximation that gives the first derivate at an intermediate half-time step is:

Second-order half-time step

$$\frac{\partial f}{\partial x}\bigg|_i^{n+\frac{1}{2}} \simeq \frac{1}{2}\left(\frac{f_{i+1}^{n+1} - f_{i-1}^{n+1}}{2\Delta x} + \frac{f_{i+1}^n - f_{i-1}^n}{2\Delta x}\right) \tag{3.58}$$

Note that this is simply the average of the second-order centered difference approximations at time steps n and $n + 1$.

The finite difference approximations for the second spatial derivative can be obtained using a procedure similar to that followed for the second derivative in time. We find:

Second-order centered difference

$$\frac{\partial^2 f}{\partial x^2}\bigg|_i^n \simeq \frac{f_{i+1}^n - 2f_i^n + f_{i-1}^n}{(\Delta x)^2} \tag{3.59}$$

3.3.3   Finite difference versions of PDEs

Using the various finite difference approximations for the space and time derivatives, we can now write the corresponding finite difference equations (FDEs) for a few important PDEs. For this purpose, we use the second-order centered difference approximations in Equations (3.47), (3.53), (3.57), and (3.59).

The convection equation is given as:

$$\frac{\partial \mathcal{V}}{\partial t} + v_p \frac{\partial \mathcal{V}}{\partial x} = 0.$$

By substituting the second-order centered difference approximations for first derivatives in time and space from above, we find the finite difference equation:

$$\frac{\mathcal{V}_i^{n+1} - \mathcal{V}_i^{n-1}}{2\Delta t} + v_p \frac{\mathcal{V}_{i+1}^n - \mathcal{V}_{i-1}^n}{2\Delta x} = 0. \tag{3.60}$$

Similarly for the diffusion equation on a dissipative RC transmission line:

$$\frac{\partial \mathcal{V}}{\partial t} - \frac{1}{RC}\frac{\partial^2 \mathcal{V}}{\partial x^2} = 0$$

which after the appropriate substitutions becomes:

$$\frac{\mathcal{V}_i^{n+1} - \mathcal{V}_i^{n-1}}{2\Delta t} - \frac{1}{RC}\frac{\mathcal{V}_{i+1}^n - 2\mathcal{V}_i^n + \mathcal{V}_{i-1}^n}{(\Delta x)^2} = 0. \tag{3.61}$$

Finally, the wave equation on a transmission line is given by:

$$\frac{\partial^2 \mathcal{V}}{\partial t^2} - v_p^2 \frac{\partial^2 \mathcal{V}}{\partial x^2} = 0.$$

After substituting the second-order centered difference equations for second derivatives in time and space, we find:

$$\frac{\mathcal{V}_i^{n+1} - 2\mathcal{V}_i^n + \mathcal{V}_i^{n-1}}{(\Delta t)^2} - v_p \frac{\mathcal{V}_{i+1}^n - 2\mathcal{V}_i^n + \mathcal{V}_{i-1}^n}{(\Delta x)^2} = 0. \qquad (3.62)$$

Note that these are not the only finite difference approximations to the PDEs considered; different FDE approximations result from the use of different approximations for the time and space derivatives. We will consider different FDE approximations for the convection equation in the next section.

It will be convenient at this point to introduce the concept of a "semidiscrete" form of a PDE. Owing to causality, we can separately treat the time and spatial derivatives in hyperbolic PDEs, and for that reason they can also be discretized separately, and with different types of discretization. For example, we can write the convection equation, discretized in space with the second-order centered difference, as:

$$\frac{\partial \mathcal{V}}{\partial t}\Big|_i^n = -v_p \frac{\mathcal{V}_{i+1}^n - \mathcal{V}_{i-1}^n}{2\Delta x}.$$

This is the semidiscrete form; we can then proceed to discretize the time derivative at spatial location i and time step n with whatever method suits the problem and the required accuracy. For instance, rather than using the centered difference as in Equation (3.60), we could use the fourth-order Runge-Kutta discretization in time, as described in Section 3.2.3, to achieve higher accuracy in time.

3.4 Finite difference solutions of the convection equation

As the simplest example of a PDE, we consider finite difference solutions of the convection Equation (3.60). As mentioned before, this equation constitutes the simplest prototype for hyperbolic equations, and is also highly relevant for our FDTD analyses of Maxwell's equations, since it has the same form as the component curl equations that we shall solve in FDTD solutions of electromagnetic problems. Consider the convection equation written in terms of voltage on a transmission line:

$$\frac{\partial \mathcal{V}}{\partial t} + v_p \frac{\partial \mathcal{V}}{\partial x} = 0 \qquad (3.63)$$

where $v_p \simeq 1$ ft-(ns)$^{-1}$, i.e., the speed of light in free space.[12] As mentioned in Section 3.1.3, the exact solution of this convection equation is:

$$\mathcal{V}(x, t) = \mathcal{V}^+(x - v_p t) \qquad (3.64)$$

---

[12] Note that the well-known speed of light in a vacuum is 2.9979×10^8 m-s^{-1}, which is equivalent to 0.984 ft-(ns)$^{-1}$; as such 1 ft-(ns)$^{-1}$ makes a very accurate and convenient approximation.

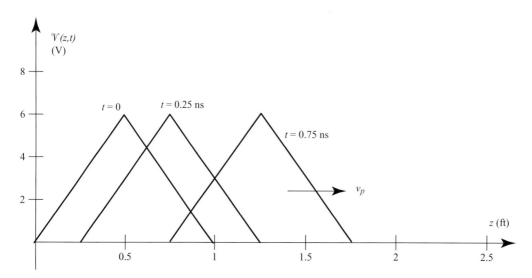

Figure 3.5 Exact solution of the convection equation for a triangular initial voltage distribution. The triangular distribution simply moves to the right at the speed $v_p \simeq 1$ ft-(ns)$^{-1}$.

where $\mathcal{V}^+(\cdot)$ is an arbitrary function. A particular solution is determined by means of the initial condition at time $t = 0$, i.e., $\mathcal{V}^+(x, 0) = \Phi(x)$, where $\Phi(x)$ is the initial voltage distribution on the line. The solution at any later time is then given by:

$$\mathcal{V}(x, t) = \Phi(x - v_p t). \qquad (3.65)$$

In other words, the initial voltage distribution is simply convected (or propagated) in the positive x-direction (i.e., to the right) at the velocity v_p, preserving its magnitude and shape.

We shall now consider finite difference solutions of Equation (3.63), using $v_p \simeq 1$ ft-(ns)$^{-1}$, for a particular initial voltage distribution:

$$\mathcal{V}(x, 0) = \Phi(x) = \begin{cases} 12\,x & V \quad 0 \le x \le 0.5 \text{ ft,} \\ -12\,(1 - x) & V \quad 0.5 \text{ ft} \le x \le 1.0 \text{ ft,} \\ 0 & V \quad \text{elsewhere.} \end{cases} \qquad (3.66)$$

The exact solution is simply the triangular pulse with a peak of 6 V, moving to the right at the speed of $v_p \simeq 1$ ft-(ns)$^{-1}$, as shown in Figure 3.5.

We now proceed to numerically solve for $\mathcal{V}(x, t)$, using two different methods, namely (i) the forward-time centered space method, and (ii) the leapfrog method.

3.4.1   The forward-time centered space method

As our first approximation of the PDE given in (3.63), we use the first-order forward difference approximation (i.e., Equation 3.45) for the time derivative and the second-order centered difference approximation (i.e., Equation 3.57) for the space derivative.

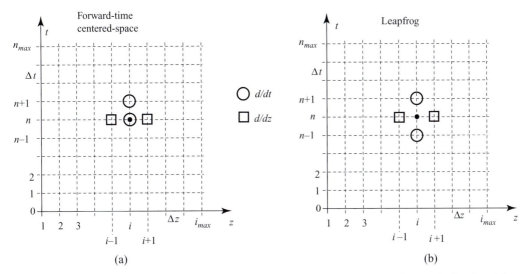

Figure 3.6 Mesh diagram for the forward-time centered space and leapfrog methods. The grid points which are involved in the evaluation of the time derivative are indicated as open circles, whereas those used in calculating the space derivatives are indicated by open squares.

We thus have:

$$\frac{\partial \mathcal{V}}{\partial t} + v_p \frac{\partial \mathcal{V}}{\partial x} = 0 \qquad \rightarrow \qquad \frac{\mathcal{V}_i^{n+1} - \mathcal{V}_i^n}{\Delta t} + v_p \frac{\mathcal{V}_{i+1}^n - \mathcal{V}_{i-1}^n}{2\Delta x} = 0. \qquad (3.67)$$

We now solve (3.67) for $\mathcal{V}_i^{n+1}$ to find:

Forward-time centered space

$$\boxed{\mathcal{V}_i^{n+1} = \mathcal{V}_i^n - \left(\frac{v_p \Delta t}{2\Delta x}\right)\left[\mathcal{V}_{i+1}^n - \mathcal{V}_{i-1}^n\right].} \qquad (3.68)$$

This result is known as the "update equation" for the field or parameter $\mathcal{V}$; the field is "updated" in time by stepping forward in increments of Δt. Note that each of the voltage components on the right side of the equation is at time step n, and the "updated" voltage on the right is at time step $n + 1$. In comparing different finite difference methods, it is useful to graphically depict the grid points involved in the evaluation of the time and space derivatives. Such a mesh diagram is given in Figure 3.6, for both the methods used in this section.

For our first attempt at a solution of Equation (3.68), we choose $\Delta x = 0.05$ ft, and $\Delta t = 0.025$ ns. Given that our triangular "wave" is about 1 ft long, this gives us 20 sample points in one "wavelength". The resulting solution is given in the left-hand column of Figure 3.7, depicting the $\mathcal{V}(x, t)$ across the spatial grid at time instants $t = 0.25, 0.5, 0.75$, and 1.0 ns. The exact solution (i.e., the triangular pulse of Figure 3.5, appropriately shifted to the right at a speed of 1 ft-ns^{-1}) is also shown for comparison.

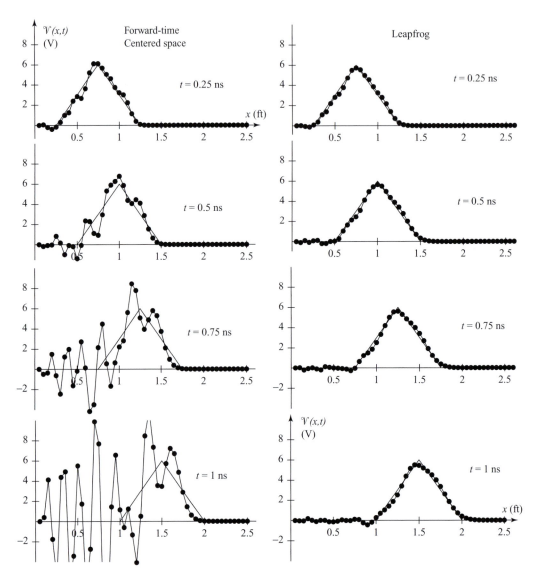

Figure 3.7 FDTD solutions of the one-dimensional convection equation. Time snapshots of $\mathcal{V}(x, t)$ determined with the forward-time centered space (left) and leapfrog (right) methods are shown. With $v_p \simeq 1$ ft-(ns)$^{-1}$, the solutions shown correspond to $\Delta x = 0.05$ ft, and $\Delta t = 0.025$ ns. The exact solution (i.e., the triangular pulse) is also shown for comparison.

The fact that the amplitude of the solution increases as the wave propagates to the right indicates that the method is *unstable*. We will examine stability of these methods in great detail in Chapter 5. In fact, as we shall see later, the forward-time centered space method is *unconditionally unstable* when applied to the convection equation, regardless of the value of the spatial grid size (i.e., Δx).

3.4.2   The leapfrog method

For our second attempt at an approximation of the PDE given in (3.21), we use the second-order centered difference approximation (i.e., Equation 3.47) for the time derivative and the second-order centered difference approximation (i.e., Equation 3.57) for the space derivative. We thus have:

$$\frac{\partial \mathcal{V}}{\partial t} + v_p \frac{\partial \mathcal{V}}{\partial x} = 0 \qquad \rightarrow \qquad \frac{\mathcal{V}_i^{n+1} - \mathcal{V}_i^{n-1}}{2\Delta t} + v_p \frac{\mathcal{V}_{i+1}^n - \mathcal{V}_{i-1}^n}{2\Delta x} = 0. \qquad (3.69)$$

We now solve (3.27) for $\mathcal{V}_i^{n+1}$ to find the update equation for $\mathcal{V}$:

Leapfrog

$$\mathcal{V}_i^{n+1} = \mathcal{V}_i^{n-1} - \left(\frac{v_p \Delta t}{\Delta x} \right) \left[\mathcal{V}_{i+1}^n - \mathcal{V}_{i-1}^n \right]. \qquad (3.70)$$

The mesh diagram corresponding to the leapfrog method is given in Figure 3.6. Note in this case that we use the previous time point (i.e., $\mathcal{V}^{n-1}$) as well as the next time point (i.e., $\mathcal{V}^{n+1}$) to evaluate the derivative at the current time point (n). Thus, while the equation involves derivatives about n, we actually solve for the field at time step $n + 1$; however, this requires knowledge of $\mathcal{V}$ two time steps back, and thus extra storage in computer memory.

The solution of (3.70), for the same parameters as those used for the forward-time centered space method (i.e., $\Delta x = 0.05$ ft, and $\Delta t = 0.025$ ns), is given in the right-hand column of Figure 3.7. As before, $\mathcal{V}(x, t)$ is depicted across the spatial grid at time instants $t = 0.25, 0.5, 0.75,$ and 1.0 ns, and the exact solution (i.e., the triangular pulse of Figure 3.5) is also shown for comparison. It is clear that unlike the forward-time centered space method, the leapfrog technique is *stable*, at least for the chosen values of Δx and Δt. Because of its stability, this method is also far more *accurate* than the forward-time centered space method; the accuracy is a direct result of the higher-order terms that are dropped from the Taylor series expansions in Equations (3.46) and (3.56). We will have much more to say on the accuracy of finite difference methods in Chapter 6.

It is instructive at this point to investigate the dependence of the solution on the chosen values of Δt and Δx. Plots of the line voltage distribution at $t \simeq 0.25$ ns are shown in Figure 3.8, calculated with the leapfrog method using different values of Δt, with $\Delta x = 0.05$ ft. It is clear that the leapfrog method remains stable and reasonably accurate as the time step is increased from $\Delta t = 0.025$ ns to Δt slightly less than 0.05 ns. The result for $\Delta t = 0.05$ ns appears to be stable, but is substantially less accurate than that of $\Delta t = 0.0475$ ns, and when Δt is increased to 0.0525 ns, the solution becomes very unstable. In fact, the result for $\Delta t = 0.06$ ns is not shown since none of its values lie in the range of the plots! Thus, it appears that for this set of parameters (i.e., our choice of $\Delta x = 0.05$ ft and the known value of v_p), a condition for stability of the numerical method is $\Delta t \leq 0.05$ ns. It is interesting to note from (3.28) that with $v_p = 1$ ft-(ns)$^{-1}$

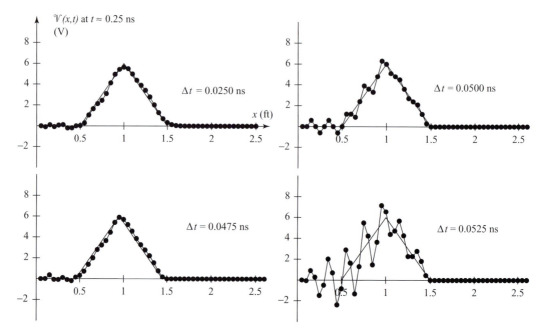

Figure 3.8 Dependence of the leapfrog method solution on Δt. The voltage distribution $\mathcal{V}(x, t \simeq 0.25$ ns$)$ is shown for different values of Δt. The spatial grid size Δz was kept constant at $\Delta x = 0.05$ ft.

and $\Delta x = 0.05$ ft, this stability criterion can also be written as

$$\Delta t \le 0.05 \text{ ns} \quad \rightarrow \quad \frac{v_p \Delta t}{\Delta x} \le \frac{(1)(0.05)}{0.05} = 1. \tag{3.71}$$

This condition, that $v_p \Delta t / \Delta x$ must be less than 1, is known as the *Courant-Friedrichs-Lewy* (CFL) stability condition, and will be derived more formally in Chapter 5 by using the von Neumann method for analyzing the stability of finite difference schemes.

Note that since the only dependence of the finite difference algorithm (3.28) on Δt and Δx is in terms of the ratio $v_p \Delta t / \Delta x$, one might at first think that the results given in Figure 3.8 for fixed Δx and varying Δt also represent corresponding results for fixed Δt and different values of Δx; in other words, one might suppose that we can simply halve (or double) both Δt and Δx together and get the same results. That this is not the case is illustrated in Figure 3.9, where we illustrate snapshots of $\mathcal{V}(x, t)$ at $t = 1$ ns for triangular pulses of different lengths and also for different values of Δx.

The left column shows solutions using $\Delta x = 0.05$ ft as before, and the first three also use $\Delta t = 0.025$ ns, similar to the top left solution in Figure 3.8. However, the pulse width has been shortened from (a) 1 ns to (b) 0.4 ns and finally to (c) 0.2 ns. In the fourth example (d), the time step Δt has been cut in half to 0.0125 ns. In the first three solutions in the right column, the spatial grid size has been halved to $\Delta x = 0.025$ ft and the time step Δt has been progressively increased, (f) close to, and (g) beyond, the CFL condition. The last solution on the right (h) should be compared to the bottom left

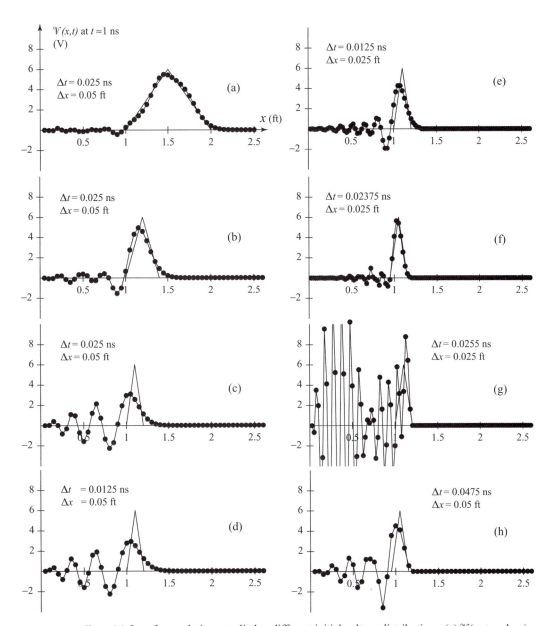

Figure 3.9 Leapfrog technique applied to different initial voltage distributions. (a) $\mathcal{V}(x, t = 1 \text{ ns})$ for triangular pulse of beginning-to-end length of 1 ns. (b) Same for a 0.4 ns pulse. (c) Same for a 0.2 ns pulse, with $(\Delta t / \Delta x) = 0.5$. (d) Same as (c) but with Δt halved, $(\Delta t / \Delta x) = 0.25$. (e) Same as (c) but with both Δx and Δt halved; $(\Delta t / \Delta x) = 0.5$. (f) Same as (c) but with Δx reduced by a factor of two; $(\Delta t / \Delta x) = 0.95$. (g) Same as (f) but $(\Delta t / \Delta x) = 1.02$. (h) Same as (c) but with Δt approximately doubled; $(\Delta t / \Delta x) = 0.95$.

solution of Figure 3.8; the parameters are the same, but the pulse has been reduced to 0.2 ns.

We can make a number of important observations from Figure 3.9:

1. For any given set of values of $(\Delta t, \Delta x)$, the accuracy of the leapfrog algorithm deteriorates as the duration of the input pulse becomes shorter.
2. For a short pulse, e.g., the 0.2 ns pulse in Figure 3.9c, better results are obtained by decreasing the mesh size Δx (i.e., Figure 3.9f) rather than decreasing Δt (i.e., Figure 3.9d).
3. For any given mesh size Δx, better results are obtained when the ratio $(\Delta t/\Delta x)$ is closer to unity, although the method becomes unstable when $(\Delta t/\Delta x)$ exceeds unity by even a small amount. This can be seen by comparing panels (e) and (f): although (f) has a larger time step, its accuracy seems better than (e).
4. For a fixed ratio $(\Delta t/\Delta x)$, better results are obtained if both Δt and Δx are reduced by the same factor, as can be seen by comparing panels (c) and (e).

The fundamental reason for the behavior of the finite difference solution as illustrated in Figure 3.9 pertains to our *sampling* of the initial waveform. This is properly interpreted as the phenomenon of *numerical dispersion*, which occurs because of the fact that components of the signal at different frequencies (and thus different wavelengths) propagate at different numerical velocities. The triangular pulse contains energy in a range of frequencies, though most of its energy is around the frequency corresponding to the wavelength of 1 ft (i.e., 1 GHz). In simple terms, although the finite differencing scheme used may well be stable, we still have to ensure that the spatial mesh size used is fine enough to properly resolve the full spectrum of wavelengths contained in the signal.

The original triangular input voltage distribution (Figure 3.9a) can either be thought of in terms of its beginning-to-end spatial extent, which is 1 ft, or in terms of its temporal duration, as would be observed at a fixed point. In this case, the voltage distribution would pass by an observer in 1 ns, since the propagation speed is $v_p \simeq 1$ ft-(ns)$^{-1}$. The frequency components of such a 1 ns pulse are contained in a narrow band centered around $\sim (1 \text{ ns})^{-1} = 1$ GHz, corresponding to a wavelength (with $v_p \simeq 1$ ft-(ns)$^{-1}$) of $\lambda = v_p/(1 \text{ GHz}) \simeq 1$ ft. In practice, it is typically necessary to select a mesh size so that the principal frequency components are resolved with at least 10 mesh points per wavelength.

The numerical dispersion properties of different finite difference methods will be more formally studied in Chapter 6.

3.4.3   The Lax-Wendroff methods

Many other finite difference methods exist, some more accurate than the leapfrog method, that are suitable for the solution of the convection equation. As particularly popular examples, we briefly describe the Lax-Wendroff methods [6] in this section. There are basically two categories of Lax-Wendroff methods, namely one-step and two-step methods.

The Lax-Wendroff one-step method

To derive the algorithm for the so-called Lax-Wendroff one-step method applied to the convection equation, we start with the Taylor series expansion of the solution $\mathcal{V}(x, t)$ in time around the grid point (i, n), as before:

$$\mathcal{V}_i^{n+1} = \mathcal{V}_i^n + \Delta t \left.\frac{\partial \mathcal{V}}{\partial t}\right|_i^n + \frac{1}{2}(\Delta t)^2 \left.\frac{\partial^2 \mathcal{V}}{\partial t^2}\right|_i^n + O[(\Delta t)^3] + \cdots . \tag{3.72}$$

We now isolate the time derivative $[\partial \mathcal{V}/\partial t]_i^n$ directly from the convection equation:

$$\frac{\partial \mathcal{V}}{\partial t} + v_p \frac{\partial \mathcal{V}}{\partial x} = 0 \qquad \rightarrow \qquad \frac{\partial \mathcal{V}}{\partial t} = -v_p \frac{\partial \mathcal{V}}{\partial x}. \tag{3.73}$$

The second derivative $[\partial^2 \mathcal{V}/\partial t^2]_i^n$ can be found by differentiating (3.73) with respect to time:

$$\frac{\partial^2 \mathcal{V}}{\partial t^2} = \frac{\partial}{\partial t}\left(-v_p \frac{\partial \mathcal{V}}{\partial x}\right) = -v_p \frac{\partial}{\partial x}\frac{\partial \mathcal{V}}{\partial t} = -v_p \frac{\partial}{\partial x}\left(-v_p \frac{\partial \mathcal{V}}{\partial x}\right) = v_p^2 \frac{\partial^2 \mathcal{V}}{\partial x^2}. \tag{3.74}$$

Note that this simply gives us the wave equation. Substituting (3.73) and (3.74) into the Taylor series expansion in (3.72) we find:

$$\mathcal{V}_i^{n+1} = \mathcal{V}_i^n - v_p \Delta t \left.\frac{\partial \mathcal{V}}{\partial x}\right|_i^n + \frac{1}{2} v_p (\Delta t)^2 \left.\frac{\partial^2 \mathcal{V}}{\partial x^2}\right|_i^n + O[(\Delta t)^3] + \cdots . \tag{3.75}$$

We now use second-order centered difference approximations for the first and second spatial derivatives to find:

$$\mathcal{V}_i^{n+1} = \mathcal{V}_i^n - v_p \Delta t \left[\frac{\mathcal{V}_{i+1}^n - \mathcal{V}_{i-1}^n}{2\Delta x}\right] + \frac{1}{2} v_p^2 (\Delta t)^2 \left[\frac{\mathcal{V}_{i+1}^n - 2\mathcal{V}_i^n + \mathcal{V}_{i-1}^n}{(\Delta x)^2}\right]. \tag{3.76}$$

The resultant Lax-Wendroff algorithm is then:

Lax-Wendroff one-step

$$\mathcal{V}_i^{n+1} = \mathcal{V}_i^n - \frac{1}{2}\left(\frac{v_p \Delta t}{\Delta x}\right)\left[\mathcal{V}_{i+1}^n - \mathcal{V}_{i-1}^n\right] + \frac{1}{2}\left(\frac{v_p \Delta t}{\Delta x}\right)^2 \left[\mathcal{V}_{i+1}^n - 2\mathcal{V}_i^n + \mathcal{V}_{i-1}^n\right]$$

$$\tag{3.77}$$

The mesh diagram showing the various derivatives used in the Lax-Wendroff one-step method is shown in Figure 3.10. Note that the time derivative is essentially evaluated in a forward manner, whereas the space derivative is evaluated in a centered manner using three points. Now, one might be tempted to think that this method is third-order accurate, since the error term that is dropped from (3.33) is $O[(\Delta t)^3]$; however, since both of the spatial derivatives used in (3.34) are second-order accurate, the resulting algorithm is also second-order accurate.

The two-step Lax-Wendroff method

The Lax-Wendroff one-step method is an efficient and accurate method for solving the linear convection Equation (3.21). For nonlinear equations or systems of equations (like

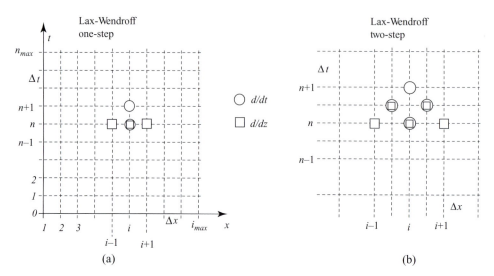

Figure 3.10 Mesh diagram for the one- and two-step Lax-Wendroff methods. The grid points which are involved in the evaluation of the time derivative are indicated as open circles, whereas those used in calculating the space derivatives are indicated by open squares.

Maxwell's equations), the one-step method becomes too complicated, and different two-step methods have been proposed which are simpler and more accurate. Since Maxwell's equations are clearly linear, we shall generally not need to resort to such methods in this course. However, we list one such method here for illustration, as it may be useful in modeling electromagnetic problems involving nonlinear media. In fact, this two-step Lax-Wendroff algorithm has a number of other desirable properties, including minimal numerical dissipation and dispersion.

The two-step Lax-Wendroff method, first proposed by Richtmeyer [7] and then modified by Burstein [8] who applied it over half-mesh spacings, is evaluated by first solving for the field $\mathcal{V}$ at time step $n + 1/2$, at both spatial points $i + 1/2$ and $i - 1/2$, using the forward-time centered space method:

$$\mathcal{V}_{i+1/2}^{n+1/2} = \frac{1}{2}[\mathcal{V}_{i+1}^{n} + \mathcal{V}_{i}^{n}] - \frac{1}{2}\left(\frac{v_p \Delta t}{\Delta x}\right)[\mathcal{V}_{i+1}^{n} - \mathcal{V}_{i}^{n}] \tag{3.78}$$

$$\mathcal{V}_{i-1/2}^{n+1/2} = \frac{1}{2}[\mathcal{V}_{i}^{n} + \mathcal{V}_{i-1}^{n}] - \frac{1}{2}\left(\frac{v_p \Delta t}{\Delta x}\right)[\mathcal{V}_{i}^{n} - \mathcal{V}_{i-1}^{n}]. \tag{3.79}$$

Note that the first term on the right-hand side of both equations is effectively the previous value of the field at the same spatial point, i.e., $\mathcal{V}_{i+1/2}^{n}$ in Equation (3.78) and $\mathcal{V}_{i-1/2}^{n}$ in Equation (3.79); both are found by spatial averaging of the field at known, integer grid points. Next, these two equations are used in the leapfrog method (3.70), using half-steps, to advance from time step $n + 1/2$ to time step $n + 1$:

Lax-Wendroff two-step

$$\mathcal{V}_{i}^{n+1} = \mathcal{V}_{i}^{n} - \left(\frac{v_p \Delta t}{\Delta x}\right)\left[\mathcal{V}_{i+1/2}^{n+1/2} - \mathcal{V}_{i-1/2}^{n+1/2}\right] \tag{3.80}$$

Written in terms of only the known, integer-location fields, we can write the two-step Lax-Wendroff method as:

$$\mathcal{V}_i^{n+1} = \mathcal{V}_i^n - \frac{1}{2} \left(\frac{v_p \Delta t}{\Delta x} \right) [\mathcal{V}_{i+1}^n - \mathcal{V}_{i-1}^n] + \frac{1}{2} \left(\frac{v_p \Delta t}{\Delta x} \right)^2 [\mathcal{V}_{i+1}^n - 2\mathcal{V}_i^n + \mathcal{V}_{i-1}^n].$$

(3.81)

Hence, note that while the method is inherently a two-step method, the two "half-steps" can be combined into a single equation, leaving what is essentially a one-step method – and, we should note, is exactly the same as the one-step method! Thus, we see that for the simple case of linear PDEs such as the convection equation, the two-step method collapses to the one-step method; however, this is not the case for nonlinear PDEs or systems of PDEs.

The mesh diagram showing the various derivatives used in the Lax-Wendroff two-step method is shown in Figure 3.10. Note that the time derivative is essentially evaluated in a forward manner, whereas the space derivative is evaluated in a centered manner using three points.

3.5 Finite difference methods for two coupled first-order convection equations

We have seen in previous sections that single PDEs can be converted to FDEs by straightforward replacement of time and space derivatives with appropriate finite difference approximations. In the same manner, we can convert two coupled PDEs into their equivalent two coupled FDEs and then algebraically solve them to determine an algorithm to use for time-marching type of solutions. As an example of two coupled first-order convection equations, we consider the telegrapher's Equations (3.4) for a lossless transmission line, rewritten in a slightly different form:

$$\frac{\partial \mathcal{V}}{\partial t} + \frac{1}{C} \frac{\partial \mathcal{I}}{\partial x} = 0$$

(3.82a)

$$\frac{\partial \mathcal{I}}{\partial t} + \frac{1}{L} \frac{\partial \mathcal{V}}{\partial x} = 0.$$

(3.82b)

Directly applying Equation (3.47) for the time derivatives and Equation (3.57) for the space derivatives and manipulating, we find:

Forward-time centered space

$$\boxed{\begin{aligned} \mathcal{V}_i^{n+1} &= \mathcal{V}_i^n - \left(\frac{\Delta t}{2C \Delta x} \right) [\mathcal{I}_{i+1}^n - \mathcal{I}_{i-1}^n] \\ \mathcal{I}_i^{n+1} &= \mathcal{I}_i^n - \left(\frac{\Delta t}{2L \Delta x} \right) [\mathcal{V}_{i+1}^n - \mathcal{V}_{i-1}^n] \end{aligned}}.$$

(3.83)

Whereas using (3.48) for the time derivatives and (3.58) for the space derivatives yields the leapfrog algorithm:

Leapfrog

$$
\boxed{
\begin{aligned}
\mathcal{V}_i^{n+1} &= \mathcal{V}_i^{n-1} - \left(\frac{\Delta t}{C\,\Delta x}\right)\left[\mathcal{I}_{i+1}^n - \mathcal{I}_{i-1}^n\right] \\[2mm]
\mathcal{I}_i^{n+1} &= \mathcal{I}_i^{n-1} - \left(\frac{\Delta t}{L\,\Delta x}\right)\left[\mathcal{V}_{i+1}^n - \mathcal{V}_{i-1}^n\right]
\end{aligned}
}
\tag{3.84}
$$

In practice, when advancing (in time) two different variables $\mathcal{V}$ and $\mathcal{I}$, which are related to one another through their derivatives in time and space, interleaving the finite difference meshes for the two variables often leads to better results, as one might intuitively expect. In this case, the first-order coupled Equations (3.82) are converted to finite difference approximations at different grid points; the voltage $\mathcal{V}$ is discretized at integer grid points in both time and space (i, n), while the current $\mathcal{I}$ is discretized at half grid points $(i + 1/2, n + 1/2)$:

$$
\frac{\mathcal{V}_i^{n+1} - \mathcal{V}_i^n}{\Delta t} + \frac{1}{C}\,\frac{\mathcal{I}_{i+1/2}^{n+1/2} - \mathcal{I}_{i-1/2}^{n+1/2}}{\Delta x} = 0
\tag{3.85a}
$$

$$
\frac{\mathcal{I}_{i+1/2}^{n+1/2} - \mathcal{I}_{i+1/2}^{n-1/2}}{\Delta t} + \frac{1}{L}\,\frac{\mathcal{V}_{i+1}^n - \mathcal{V}_i^n}{\Delta x} = 0 .
\tag{3.85b}
$$

Note that the first equation is discretized at time step $n + 1/2$ and spatial step i, while the second equations is discretized at time step n and spatial step $i + 1/2$. Rearranging, this leads to the interleaved leapfrog algorithm:

Interleaved leapfrog

$$
\boxed{
\begin{aligned}
\mathcal{V}_i^{n+1} &= \mathcal{V}_i^n - \left(\frac{\Delta t}{C\,\Delta x}\right)\left[\mathcal{I}_{i+1/2}^{n+1/2} - \mathcal{I}_{i-1/2}^{n+1/2}\right] \\[2mm]
\mathcal{I}_{i+1/2}^{n+1/2} &= \mathcal{I}_{i+1/2}^{n-1/2} - \left(\frac{\Delta t}{L\,\Delta x}\right)\left[\mathcal{V}_{i+1}^n - \mathcal{V}_i^n\right]
\end{aligned}
}
\tag{3.86}
$$

Note, as mentioned above, that the values of voltage are only available at points (i, n) whereas those of the current are kept at mesh points $(i + 1/2, n + 1/2)$. This spatial arrangement and the leapfrog time-marching are indicated in Figure 3.11. Note that the two quantities (i.e., $\mathcal{V}$ and $\mathcal{I}$) are advanced in time by using the value of the other quantity at an intermediate time point.

This type of interleaved placement of the physical quantities and the leapfrog time-marching forms the basis for the FDTD method for solving Maxwell's equations, the so-called *Yee algorithm*, which we shall study in the next chapter.

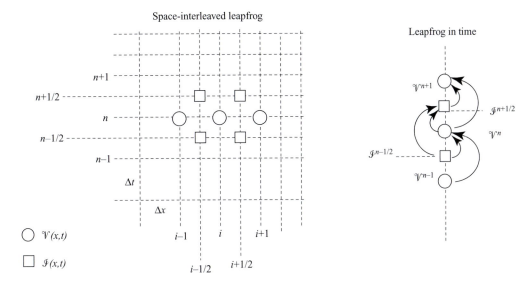

Figure 3.11 Mesh diagram for the interleaved leapfrog method. The grid points for the voltage $\mathcal{V}(x, t)$ are marked as open circles, whereas those used for the current $\mathcal{I}(x, t)$ are indicated by open squares. The leapfrog time-marching algorithm is also depicted.

3.6 Higher-order differencing schemes

Each of the differencing schemes presented above was *second-order accurate*, meaning the difference between the original PDE and the FDE (i.e., the error term) goes as $(\Delta t)^2$ or $(\Delta x)^2$. This means that if we halve Δx, for example, the spatial error is reduced by a factor of 4.

The FDTD algorithm, which will be presented in the next chapter, is also second-order accurate, and we will restrict our discussion in this book to second-order accurate methods. However, it should not be assumed that this is the limiting accuracy of the FDTD method, but rather it is the conventional choice. One can easily apply higher-order algorithms when greater accuracy is required (the accuracy of each method will be discussed in greater detail in Chapter 6). This higher-order accuracy, however, comes at the expense of computational speed, as well as complexity of the algorithm and associated boundary conditions.

Higher-order algorithms can be derived in the same fashion as the second-order accurate algorithms above. As an example, let us consider a fourth-order accurate scheme for coupled convection Equations (3.82). We will first consider Equation (3.82a),

$$\frac{\partial \mathcal{V}}{\partial t} + \frac{1}{C} \frac{\partial \mathcal{I}}{\partial x} = 0.$$

Consider the spatial derivative $\partial \mathcal{I}/\partial x$ on the right-hand side. We would like to discretize this equation at $(n + 1/2, i)$ as in the interleaved leapfrog algorithm given by Equation (3.86), so we need a finite difference approximation for $\partial \mathcal{I}/\partial x$ at this location in time and space. For fourth-order accuracy, the astute reader might guess that we need an equation spanning four spatial grid locations. The finite difference equations at four half-cell locations surrounding i can be written using Fourier series expansions as:

$$\mathcal{I}_{i+1/2} = \mathcal{I}_i + \frac{\Delta x}{2} \mathcal{I}'\big|_i + \frac{1}{2!} \left(\frac{\Delta x}{2}\right)^2 \mathcal{I}''\big|_i + \frac{1}{3!} \left(\frac{\Delta x}{2}\right)^3 \mathcal{I}'''\big|_i + \cdots \quad (3.87a)$$

$$\mathcal{I}_{i-1/2} = \mathcal{I}_i - \frac{\Delta x}{2} \mathcal{I}'\big|_i + \frac{1}{2!} \left(\frac{\Delta x}{2}\right)^2 \mathcal{I}''\big|_i - \frac{1}{3!} \left(\frac{\Delta x}{2}\right)^3 \mathcal{I}'''\big|_i + \cdots \quad (3.87b)$$

$$\mathcal{I}_{i+3/2} = \mathcal{I}_i + \frac{3\Delta x}{2} \mathcal{I}'\big|_i + \frac{1}{2!} \left(\frac{3\Delta x}{2}\right)^2 \mathcal{I}''\big|_i + \frac{1}{3!} \left(\frac{3\Delta x}{2}\right)^3 \mathcal{I}'''\big|_i + \cdots \quad (3.87c)$$

$$\mathcal{I}_{i-3/2} = \mathcal{I}_i - \frac{3\Delta x}{2} \mathcal{I}'\big|_i + \frac{1}{2!} \left(\frac{3\Delta x}{2}\right)^2 \mathcal{I}''\big|_i - \frac{1}{3!} \left(\frac{3\Delta x}{2}\right)^3 \mathcal{I}'''\big|_i + \cdots \quad (3.87d)$$

where $\mathcal{I}' = \partial\mathcal{I}/\partial x$, $\mathcal{I}'' = \partial^2\mathcal{I}/\partial x^2$, and so forth, and we have dropped the time step superscript $n + 1/2$ for now. While we have kept only the terms up to $\mathcal{I}'''$ above, we will actually need terms up to $\partial^5\mathcal{I}/\partial x^5$. The next step is to subtract Equation (3.87b) from (3.87a), and similarly Equation (3.87d) from (3.87c), which eliminates the even-order spatial derivatives:

$$\left(\mathcal{I}_{i+1/2} - \mathcal{I}_{i-1/2}\right) = \Delta x\, \mathcal{I}'\big|_i + \frac{2}{3!} \left(\frac{\Delta x}{2}\right)^3 \mathcal{I}'''\big|_i + \frac{2}{5!} \left(\frac{\Delta x}{2}\right)^5 \mathcal{I}^v\big|_i + \cdots \quad (3.88a)$$

$$\left(\mathcal{I}_{i+3/2} - \mathcal{I}_{i+3/2}\right) = 3\Delta x\, \mathcal{I}'\big|_i + \frac{2}{3!} \left(\frac{3\Delta x}{2}\right)^3 \mathcal{I}'''\big|_i + \frac{2}{5!} \left(\frac{3\Delta x}{2}\right)^5 \mathcal{I}^v\big|_i + \cdots$$

$$\quad (3.88b)$$

where, as above, $\mathcal{I}^v$ is shorthand for $\partial^5\mathcal{I}/\partial x^5$. We are looking for a finite difference equation for $\mathcal{I}'$ that will eliminate the third-order term, so next we subtract Equation (3.88b) from 27 times Equation (3.88a):

$$\left(-\mathcal{I}_{i+3/2} + 27\,\mathcal{I}_{i+1/2} - 27\,\mathcal{I}_{i-1/2} + \mathcal{I}_{i-3/2}\right) = 24\Delta x\, \frac{\partial\mathcal{I}}{\partial x}\bigg|_i - \frac{9}{80}(\Delta x)^5 \frac{\partial^5\mathcal{I}}{\partial x^5}\bigg|_i + \cdots$$

$$\quad (3.89)$$

where the coefficient $(-9/80) = (2/5!)(27 - 3^5)/2^5$. Finally, solving for $\partial\mathcal{I}/\partial x$, we find:

$$\frac{\partial\mathcal{I}}{\partial x}\bigg|_i = \left[\frac{-\mathcal{I}_{i+3/2} + 27\,\mathcal{I}_{i+1/2} - 27\,\mathcal{I}_{i-1/2} + \mathcal{I}_{i-3/2}}{24\Delta x}\right] + \frac{3}{640}(\Delta x)^4 \frac{\partial^5\mathcal{I}}{\partial x^5}\bigg|_i + \cdots . \quad (3.90)$$

We can then use this fourth-order spatial derivative with the standard second-order time stepping to find the update equation analogous to Equation (3.86):

$$\mathcal{V}_i^{n+1} = \mathcal{V}_i^n - \left(\frac{\Delta t}{24 C \Delta x}\right)\left[-\mathcal{I}_{i+3/2}^{n+1/2} + 27\,\mathcal{I}_{i+1/2}^{n+1/2} - 27\,\mathcal{I}_{i-1/2}^{n+1/2} + \mathcal{I}_{i-3/2}^{n+1/2}\right]. \quad (3.91)$$

Note that not only is this algorithm fourth-order accurate in space, but the coefficient of the error term is $3/640 = 0.0047$, whereas the coefficient for the second-order spatial difference from (3.56) was $1/6 = 0.1667$. This improved accuracy comes from the $1/n!$ term in the Taylor expansions. Notice, however, that while we have gained fourth-order accuracy in space, we are still second-order accurate in time, due to the choice of difference scheme in time. We can, of course, implement a similar fourth-order differencing scheme in time to improve accuracy; however, the update equation at time step $n + 1$ will then involve the values of the field at time steps n, $n - 1$, and $n - 2$, requiring that we store extra fields in memory.

The algorithm derived in Equation (3.91) is the most commonly used fourth-order scheme for the FDTD method. However, we will not deal with higher-order algorithms in this book, instead sticking to the second-order accurate FDTD method; the derivation above shows how one would go about deriving a higher-order algorithm when greater accuracy is required.

However, it should be noted that these higher-order finite difference methods suffer from problems at discontinuities in the simulation space, for example in the presence of perfectly conducting surfaces. In fact, it can be shown that in the presence of such discontinuities, *any* finite difference method degrades to first-order accuracy, in which case the fourth-order discretization increases computational time without improving the accuracy at all.

3.7   Summary

This chapter presented and classified partial differential equations (PDEs), and showed how they are discretized to form finite difference equations (FDEs). Most of the chapter focused on the discretization of components of the telegrapher's equations, which relate the voltage and current on a transmission line:

$$\frac{\partial \mathcal{V}}{\partial x} = -R\,\mathcal{I} - L\,\frac{\partial \mathcal{I}}{\partial t}$$

$$\frac{\partial \mathcal{I}}{\partial x} = -G\,\mathcal{V} - C\,\frac{\partial \mathcal{V}}{\partial t}.$$

(telegrapher's equations)

PDEs are classified in this chapter as either elliptic, parabolic, or hyperbolic, according to criteria analogous to the definitions of the ellipse, parabola, and hyperbola. Maxwell's Equations (2.1) and (2.3) are hyperbolic PDEs, as are the wave Equation and the one-way wave Equations (3.19).

Following a brief introduction to the numerical integration of ordinary differential equations in Section 3.2, the derivation of the FDE versions of PDEs was presented, beginning with the discretization of individual partial derivatives. Of greatest importance to the FDTD method is the second-order centered difference approximation, which for the time derivative is given by:

$$\left.\frac{\partial f}{\partial t}\right|_i^n \simeq \frac{f_i^{n+1} - f_i^{n-1}}{2\Delta t}$$

(second-order centered difference)

and similarly for the spatial derivative:

$$\frac{\partial f}{\partial x}\Big|_i^n \simeq \frac{f_{i+1}^n - f_{i-1}^n}{2\Delta x}. \qquad \text{(second-order centered difference)}$$

Using these two equations, the convection equation could be discretized as (3.60):

$$\frac{\mathcal{V}_i^{n+1} - \mathcal{V}_i^{n-1}}{2\Delta t} + v_p \frac{\mathcal{V}_{i+1}^n - \mathcal{V}_{i-1}^n}{2\Delta x} = 0.$$

Practical methods for creating FDEs were presented in Section 3.4; in particular, the leapfrog method and the Lax-Wendroff methods are practical and stable for the convection equation. The leapfrog method leads to an "update equation" for the FDE version of the convection equation:

$$\mathcal{V}_i^{n+1} = \mathcal{V}_i^{n-1} - \left(\frac{v_p \Delta t}{\Delta x}\right)\left[\mathcal{V}_{i+1}^n - \mathcal{V}_{i-1}^n\right]$$

which can be derived directly from the equation above. It was found that this equation is stable under the condition that $v_p \Delta t / \Delta x \leq 1$, known as the *Courant-Friedrichs-Lewy* or CFL condition.

Finally, methods were presented for the finite difference approximations of two coupled PDEs, such as the telegrapher's equations or Maxwell's equations. This led to the introduction of the interleaved leapfrog method, which for the telegrapher's equations is:

$$\mathcal{V}_i^{n+1} = \mathcal{V}_i^n - \left(\frac{\Delta t}{C\,\Delta x}\right)\left[\mathcal{I}_{i+1/2}^{n+1/2} - \mathcal{I}_{i-1/2}^{n+1/2}\right]$$

$$\mathcal{I}_{i+1/2}^{n+1/2} = \mathcal{I}_{i+1/2}^{n-1/2} - \left(\frac{\Delta t}{L\,\Delta x}\right)\left[\mathcal{V}_{i+1}^n - \mathcal{V}_i^n\right].$$

This interleaved leapfrog method, as we shall see in the next chapter, is the basis for the Yee algorithm used in the FDTD method.

3.8   Problems

3.1. Centered difference update equations.
   (a) Derive Equation (3.46), the second-order centered difference approximation of $\partial f / \partial t$ around grid point (i, n).
   (b) In a similar fashion, derive Equation (3.48) by taking the difference of the Taylor series expansions of f_i^{n+1} and f_i^n around the grid point $(i, n + 1/2)$.

3.2. First-order Euler method for an ODE. Consider using the first-order Euler method for the solution of Equation (3.31). Find and plot solutions for $\Delta t = 0.1\tau$, 2.0τ, and 2.5τ and compare your numerical results with the exact solution. Your plots should cover the time range $0 < t < 10\tau$. Comment on your results. How small does Δt have to be in order for the normalized mean squared error (averaged over $0 < t < 10\tau$) between your solution and the exact one to be less than 1%?

3.3. **Comparison of Euler, leapfrog, and RK4.** Plot the analytical solution to the simple ODE given by $dy/dt = (\alpha + j\omega)y$, where $\omega = 2$, for $0 < t < 10$ seconds, with the initial condition $y(0) = 1$. Consider the two cases of $\alpha = 0$ and $\alpha = \pm 0.1$.

Now, solve this equation numerically using the Euler, Leapfrog, and RK4 methods. Plot the difference between the exact solution and real part of each of the numerical solutions, for $\Delta t = 0.01$, 0.1, and 0.5 seconds. For the RK4 method, vary Δt further to verify the stability diagram in Figure 3.3.

3.4. **Pulse propagation on a lossless microstrip line.** Consider the transmission line Equations (3.4) and the voltage wave Equation (3.11) for a lossless transmission line. Write a computer program to implement the following algorithm to solve the voltage wave Equation (3.11):

$$\mathcal{V}_i^{n+1} = \left(\frac{v_p \Delta t}{\Delta x}\right)^2 \left[\mathcal{V}_{i+1}^n - 2\mathcal{V}_i^n + \mathcal{V}_{i-1}^n\right] + 2\mathcal{V}_i^n - \mathcal{V}_i^{n-1}$$

where $v_p = (LC)^{-\frac{1}{2}}$. Your program should be parameterized in Δt and Δx so that you can easily change the values of these quantities. Consider a 20-cm long uniform lossless microstrip line having per-unit-length parameters of $L = 4$ nH-$(\text{cm})^{-1}$ and $C = 1.6$ pF-$(\text{cm})^{-1}$. The initial voltage distribution on the line is given as:

$$\mathcal{V}(x, 0) = \begin{cases} x - 8 & \text{Volts} & 8 \text{ cm} \leq x \leq 10 \text{ cm}, \\ -x + 12 & \text{Volts} & 10 \text{ cm} \leq x \leq 12 \text{ cm}, \\ 0 & \text{Volts} & \text{elsewhere}, \end{cases}$$

whereas the initial current is zero everywhere. Find the exact solution (see Equation 3.17) for the line voltage $\mathcal{V}_{\text{exact}}(x, t)$ to compare your numerical solutions against.

(a) To terminate the ends of the transmission line at $x = 0$ and $x = 20$ cm, try the following two simple boundary conditions that we shall learn about in later chapters:

Boundary Condition A: $x = 0 \qquad \mathcal{V}_1^{n+1} = \mathcal{V}_2^n$

$$x = 20 \text{ cm} \quad \mathcal{V}_{i_m}^{n+1} = \mathcal{V}_{i_m-1}^n$$

Boundary Condition B: $x = 0 \qquad \mathcal{V}_1^{n+1} = \left(1 - \frac{v_p \Delta t}{\Delta x}\right)\mathcal{V}_1^n + \frac{v_p \Delta t}{\Delta x}\mathcal{V}_2^n$

$$x = 20 \text{ cm} \quad \mathcal{V}_{i_m}^{n+1} = \left(1 - \frac{v_p \Delta t}{\Delta x}\right)\mathcal{V}_{i_m}^n + \frac{v_p \Delta t}{\Delta x}\mathcal{V}_{i_m-1}^n$$

where i_m is the index corresponding to the spatial FDTD grid point at $x = 20$ cm.

(b) Use your program to numerically determine $\mathcal{V}(x^n, t_i)$. Verify the numerical dispersion and stability properties of the algorithm by varying the spatial resolution and the relationship between Δt and Δx. To properly observe the wave phenomena, plot representative snapshots of your solutions over the spatial grid.

3.5. **The leapfrog method.** Repeat Problem 3.4 but this time solve for the line voltage by means of a FDTD solution of the coupled transmission line Equations (3.4) using the following leapfrog algorithm:

$$\mathcal{V}_i^{n+1} = \mathcal{V}_i^{n-1} - \left(\frac{\Delta t}{C\,\Delta x}\right)\left[\mathcal{I}_{i+1}^n - \mathcal{I}_{i-1}^n\right]$$

$$\mathcal{I}_i^{n+1} = \mathcal{I}_i^{n-1} - \left(\frac{\Delta t}{L\,\Delta x}\right)\left[\mathcal{V}_{i+1}^n - \mathcal{V}_{i-1}^n\right].$$

Comment on the relative accuracy, stability, and speed of the two algorithms used in this and in the previous problem. Assume that the current distribution is initially zero everywhere.

3.6. **Low-loss coaxial line.** The RG17A/U is a low-loss radio frequency coaxial line with line capacitance $C \simeq 96.8$ pF-m^{-1} and $L \simeq 0.242$ µH-m^{-1}. In this problem, we will use the interleaved leapfrog method to measure the standing wave ratio on this line when it is terminated by a short circuit. Theoretically, we know that the standing wave ratio on a short-circuited lossless line is infinite; our goal in this problem is to use this knowledge to assess the accuracy of our numerical solution.

(a) Launch a sinusoidal signal as given below from the input end of the transmission line:

$$\mathcal{V}_{i=1}^n = \sin[\omega_0 \, n \Delta t]$$

where $f_0 = \omega_0/2\pi = 100$ MHz. A short circuit is placed at a distance of $x = 20$ m from the input end. Using $\Delta x = \lambda_0/20$, and $\Delta t = \Delta x/v_p$, write a program to numerically determine $\mathcal{V}(x^i, t^n)$, and plot it by using appropriate graphical displays, both before the reflection of the wave at the short circuit. Devise a scheme to measure the standing wave ratio $S = \max|\mathcal{V}|/\min|\mathcal{V}|$. Theoretically, S should be infinity; how large is your numerical S and why? Comment on your results.

(b) Repeat (a) for an open-circuited transmission line of the same length.

3.7. **Fourth-order leapfrog method.** Consider applying the interleaved leapfrog method, similar to Equations (3.89), but with fourth-order discretization in space, using Equation (3.91). Use second-order discretization in time as usual.

(a) Derive the fourth-order update equation for Equation (3.85b), i.e., the update equation for $\mathcal{I}$.

(b) Repeat Problem 3.6 and calculate the standing wave ratios for the open- and short-circuit cases. Compare the results to the previous problem.

Note that with this method, you will need to determine $\mathcal{I}$ and $\mathcal{V}$ at the grid points adjacent to the boundaries using a different method, since the fourth-order method requires two points on either side of i. These "second" points can be updated using the standard second-order centered difference in space, or you can derive a left- or right-sided fourth-order finite difference approximation on the boundaries.

3.8. **Fourth-order method for convection equation.** Consider solving the convection Equation (3.66) with fourth-order accuracy in both time and space. Derive an update equation for the convection equation that uses the RK4 integration in time, and a fourth-order centered difference in space. What is $f(y, t)$ in this case?

References

[1] U. S. Inan and A. S. Inan, *Engineering Electromagnetics*. Addison-Wesley, 1999.

[2] R. N. Bracewell, *The Fourier Transform and its Applications*, 3rd edn. McGraw-Hill, 2000.

[3] L. F. Richardson, "The approximate arithmetical solution by finite differences of physical problems involving differential equations, with an application to the stresses in a masonry dam," *Phil. Trans. Roy. Soc. London*, vol. 210, Series A, pp. 305–357, 1910.

[4] P. Moin, *Fundamentals of Engineering Numerical Analysis*. Cambridge University Press, 2001.

[5] A. Taflove and S. Hagness, *Computational Electrodynamics: The Finite-Difference Time-Domain Method*, 3rd edn. Artech House, 2005.

[6] P. D. Lax and B. Wendroff, "Systems of conservation laws," *Comm. Pure Appl. Math.*, vol. 13, pp. 217–237, 1960.

[7] R. D. Richtmeyer, "A survey of difference methods for nonsteady fluid dynamics; NCAR Technical Note 63-2," National Center for Atmospheric Research, Tech. Rep., 1963.

[8] S. Z. Burstein, "Finite difference calculations for hydrodynamic flows containing discontinuities," *J. Comput. Phys.*, vol. 2, pp. 198–222, 1967.

4 The FDTD grid and the Yee algorithm

After a brief exposure to different finite difference algorithms and methods, we now focus our attention on the so-called FDTD algorithm, or alternatively the Yee algorithm [1], for time domain solutions of Maxwell's equations. In this algorithm, the continuous derivatives in space and time are approximated by second-order accurate, two-point centered difference forms; a staggered spatial mesh is used for interleaved placement of the electric and magnetic fields; and leapfrog integration in time is used to update the fields. This yields an algorithm very similar to the interleaved leapfrog described in Section 3.5.

The cell locations are defined so that the grid lines pass through the electric field components and coincide with their vector directions, as shown in Figure 4.1. As a practical note, the choice here of the electric field rather than the magnetic field is somewhat arbitrary. However, in practice, boundary conditions imposed on the electric field are more commonly encountered than those for the magnetic field, so that placing the mesh boundaries so that they pass through the electric field vectors is more advantageous. We will have more to say about the locations of field components later in this chapter. Note that the Yee cell depicted in [2, Fig. 3.1] has the cell boundaries to be aligned with the magnetic field components, rather than with the electric field components as in Figure 4.1. This choice of the particular way of associating the spatial indices i, j, and k with the field quantities is obviously arbitrary and should not matter in the final analysis, as long as one is consistent.

The fundamental unit of our 3D grid, known as the Yee cell, is shown in Figure 4.1. With the Yee cell so defined, the spatial derivatives of various quantities are evaluated using the simple two-point centered difference method. For example, the z derivative of any given field component $\mathcal{G}$ evaluated at time $n\Delta t$ and at the mesh point (i, j, k) is given as

$$\left.\frac{\partial \mathcal{G}}{\partial z}\right|_{i,j,k}^{n} = \frac{\mathcal{G}_{i,j,k+1/2}^{n} - \mathcal{G}_{i,j,k-1/2}^{n}}{\Delta z} + O[(\Delta z)^2]. \tag{4.1}$$

This implies that the field $\mathcal{G}$ is defined at integer grid points in x and y (i.e., i and j) and at half grid points in z (i.e., $z + 1/2$). Now consider a real field, for example, the z component of the magnetic field, $\mathcal{H}_z$, which is defined at mesh points $(i \pm 1/2, j \pm 1/2, k)$ as shown in Figure 4.1. The derivative of $\mathcal{H}_z$ with respect to x at

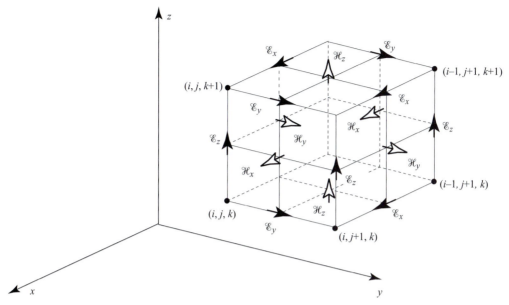

Figure 4.1 Placement of electric and magnetic field components in a 3D staggered mesh, known as the Yee cell. The small vectors with thick arrows are placed at the point in the mesh at which they are defined and stored. For example, $\mathcal{E}_y$ is defined/stored at mesh points $(i + m_1, j + 1/2m_2, k + m_3)$, where $m_{1,2,3} = 0, \pm1, \pm2, \ldots$, while $\mathcal{H}_y$ is defined/stored at mesh points $(i - 1/2 + m_1, j + m_2, k + 1/2 + m_3)$, where $m_{1,2,3} = 0, \pm1, \pm2, \ldots$.

the point $(i,\ j + 1/2,\ k)$ (at which, by the way, $\mathcal{E}_y$ is defined) and at time step n is given by:

$$\left.\frac{\partial \mathcal{H}_z}{\partial x}\right|^n_{i,j+1/2,k} = \frac{\mathcal{H}_z\big|^n_{i+1/2,j+1/2,k} - \mathcal{H}_z\big|^n_{i-1/2,j+1/2,k}}{\Delta x} + O[(\Delta x)^2]. \tag{4.2}$$

The derivatives in time are also discretized with second-order centered difference approximations such as that given in Equation (3.49), but the updating of $\overline{\mathcal{E}}$ and $\overline{\mathcal{H}}$ are staggered in time by one half time step; in fact, the components of $\overline{\mathcal{H}}$ are defined at half time steps, $n + 1/2$. In vector form we can write the discretized Maxwell's equations as:

$$\left.\frac{\partial \overline{\mathcal{E}}}{\partial t}\right|^{n+1/2} \simeq \frac{\overline{\mathcal{E}}^{n+1} - \overline{\mathcal{E}}^n}{\Delta t} = \frac{1}{\epsilon}[\nabla \times \overline{\mathcal{H}}]^{n+1/2}$$

$$\left.\frac{\partial \overline{\mathcal{H}}}{\partial t}\right|^{n+1} \simeq \frac{\overline{\mathcal{H}}^{n+3/2} - \overline{\mathcal{H}}^{n+1/2}}{\Delta t} = -\frac{1}{\mu}[\nabla \times \overline{\mathcal{E}}]^{n+1}.$$

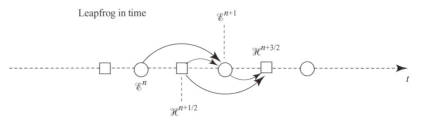

Figure 4.2 Leapfrog time marching of electric and magnetic field vectors.

Rearranging terms, we have the time-marching procedure:

$$\overline{\mathscr{E}}^{n+1} = \overline{\mathscr{E}}^n + \frac{\Delta t}{\epsilon} [\nabla \times \overline{\mathscr{H}}]^{n+1/2} \tag{4.3a}$$

$$\overline{\mathscr{H}}^{n+3/2} = \overline{\mathscr{H}}^{n+1/2} - \frac{\Delta t}{\mu} [\nabla \times \overline{\mathscr{E}}]^{n+1}. \tag{4.3b}$$

This *leapfrog* updating in time is identical to that given in Figure 3.11 for voltage $\mathscr{V}$ instead of $\overline{\mathscr{E}}$ and current $\mathscr{I}$ instead of $\overline{\mathscr{H}}$. This is depicted in Figure 4.2.

The next step in the derivation of the FDTD algorithm is to discretize the spatial components on the right-hand side of Equations (4.3). In Section 4.6, we will look at cylindrical and spherical coordinate systems; for simplicity, however, we begin with Cartesian coordinates. Maxwell's equations in a source-free, simple medium in Cartesian coordinates are given by:

$$\frac{\partial \mathscr{E}_x}{\partial t} = \frac{1}{\epsilon} \left(\frac{\partial \mathscr{H}_z}{\partial y} - \frac{\partial \mathscr{H}_y}{\partial z} \right) \qquad \frac{\partial \mathscr{H}_x}{\partial t} = \frac{1}{\mu} \left(\frac{\partial \mathscr{E}_y}{\partial z} - \frac{\partial \mathscr{E}_z}{\partial y} \right)$$

$$\frac{\partial \mathscr{E}_y}{\partial t} = \frac{1}{\epsilon} \left(\frac{\partial \mathscr{H}_x}{\partial z} - \frac{\partial \mathscr{H}_z}{\partial x} \right) \qquad \frac{\partial \mathscr{H}_y}{\partial t} = \frac{1}{\mu} \left(\frac{\partial \mathscr{E}_z}{\partial x} - \frac{\partial \mathscr{E}_x}{\partial z} \right) \tag{4.4}$$

$$\frac{\partial \mathscr{E}_z}{\partial t} = \frac{1}{\epsilon} \left(\frac{\partial \mathscr{H}_y}{\partial x} - \frac{\partial \mathscr{H}_x}{\partial y} \right) \qquad \frac{\partial \mathscr{H}_z}{\partial t} = \frac{1}{\mu} \left(\frac{\partial \mathscr{E}_x}{\partial y} - \frac{\partial \mathscr{E}_y}{\partial x} \right).$$

Next, we will look at the discretization of these equations, starting with 1D and building up to the 2D and 3D cases.

4.1 Maxwell's equations in one dimension

The simplest expression of Maxwell's equations in the FDTD algorithm can be found by limiting ourselves to the consideration of electromagnetic fields and systems with no variations in two dimensions, namely, both z and y. We thus drop all of the y and z derivatives in Equations (4.4) above. For now, we neglect magnetic or electric losses (i.e., let $\sigma = 0$ and $\sigma_m = 0$) and assume simple and source-free media, so that $\tilde{\rho} = 0$, $\overline{\mathscr{J}} = 0$, and ϵ and μ are simple constants, independent of position (i.e., the medium is homogeneous), direction (isotropic), or time. With these simplifications, Maxwell's Equations (2.1) and (2.3) simplify to:

$$\frac{\partial \mathscr{E}_x}{\partial t} = 0 \qquad\qquad \frac{\partial \mathscr{H}_x}{\partial t} = 0$$

$$\frac{\partial \mathscr{E}_y}{\partial t} = -\frac{1}{\epsilon}\frac{\partial \mathscr{H}_z}{\partial x} \qquad\qquad \frac{\partial \mathscr{H}_y}{\partial t} = \frac{1}{\mu}\frac{\partial \mathscr{E}_z}{\partial x} \qquad (4.5)$$

$$\frac{\partial \mathscr{E}_z}{\partial t} = \frac{1}{\epsilon}\frac{\partial \mathscr{H}_y}{\partial x} \qquad\qquad \frac{\partial \mathscr{H}_z}{\partial t} = -\frac{1}{\mu}\frac{\partial \mathscr{E}_y}{\partial x}.$$

Note that two of the equations have disappeared completely; the remaining equations involve only derivatives with respect to x and with respect to t. We can now organize the remaining four equations into pairs that are completely independent of each other:

$$\text{1D TM} \qquad\qquad \frac{\partial \mathscr{H}_y}{\partial t} = \frac{1}{\mu}\frac{\partial \mathscr{E}_z}{\partial x} \qquad (4.6a)$$

$$\frac{\partial \mathscr{E}_z}{\partial t} = \frac{1}{\epsilon}\frac{\partial \mathscr{H}_y}{\partial x} \qquad (4.6b)$$

and

$$\text{1D TE} \qquad\qquad \frac{\partial \mathscr{E}_y}{\partial t} = -\frac{1}{\epsilon}\frac{\partial \mathscr{H}_z}{\partial x} \qquad (4.7a)$$

$$\frac{\partial \mathscr{H}_z}{\partial t} = -\frac{1}{\mu}\frac{\partial \mathscr{E}_y}{\partial x}. \qquad (4.7b)$$

For example, note that Equations (4.6) involve x- and t-derivatives of $\mathscr{E}_z$ and $\mathscr{H}_y$, but do not involve $\mathscr{H}_z$ or $\mathscr{E}_y$; the opposite is true of Equations (4.7). The first pair are referred to as the TM mode, and the second comprise the TE mode.[1] The placement of the electric and magnetic field vectors on a one-dimensional Yee grid is shown in Figure 4.3 for both TM and TE cases.

The corresponding finite difference equations are found by applying second-order centered differences to both the time and space derivatives in Equations (4.6) and (4.7). As shown in Figure 4.3, the components of $\overline{\mathscr{E}}$ are located at integer grid points in x, and integer grid points in time t as well; the components of $\mathscr{H}$ are located at half grid points.

---

[1] Note that these TM and TE mode definitions are different from those in classic waveguide analysis. The meaning of the nomenclature will become clear when we discuss the 2D FDTD algorithm.

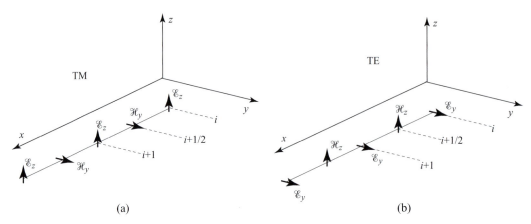

Figure 4.3 The placement of electric and magnetic field vectors for Maxwell's equations in one dimension. Both the TM and TE cases are shown.

After rearranging, we find:

1D TE mode

$$
\mathcal{E}_y\Big|_i^{n+1} = \mathcal{E}_y\Big|_i^{n} - \frac{\Delta t}{\epsilon_i\,\Delta x}\left[\mathcal{H}_z\Big|_{i+1/2}^{n+1/2} - \mathcal{H}_z\Big|_{i-1/2}^{n+1/2}\right]
$$

$$
\mathcal{H}_z\Big|_{i+1/2}^{n+1/2} = \mathcal{H}_z\Big|_{i+1/2}^{n-1/2} - \frac{\Delta t}{\mu_{i+1/2}\,\Delta x}\left[\mathcal{E}_y\Big|_{i+1}^{n} - \mathcal{E}_y\Big|_{i}^{n}\right]
$$

(4.8)

and

1D TM mode

$$
\mathcal{H}_y\Big|_{i+1/2}^{n+1/2} = \mathcal{H}_y\Big|_{i+1/2}^{n-1/2} + \frac{\Delta t}{\mu_{i+1/2}\,\Delta x}\left[\mathcal{E}_z\Big|_{i+1}^{n} - \mathcal{E}_z\Big|_{i}^{n}\right]
$$

$$
\mathcal{E}_z\Big|_i^{n+1} = \mathcal{E}_z\Big|_i^{n} + \frac{\Delta t}{\epsilon_i\,\Delta x}\left[\mathcal{H}_y\Big|_{i+1/2}^{n+1/2} - \mathcal{H}_y\Big|_{i-1/2}^{n+1/2}\right]
$$

(4.9)

Note that Equations (4.8) and (4.9) are identical to the interleaved leapfrog algorithm given in Equation (3.86) in terms of voltage $\mathcal{V}$ and current $\mathcal{I}$ except for the substitution of ϵ for C and μ for L.

A comment is necessary about *starting* the FDTD simulation. If we begin with an initial electric field distribution $\mathcal{E}_z^0$ over the simulation space, then Equation (4.9) suggests that we need $\mathcal{H}_y^{-1/2}$ in order to proceed with updating. Usually, one simply estimates $\mathcal{H}_y^{1/2}$ instead, and proceeds to the update equation for $\mathcal{E}_z^n$. However, the accuracy of the simulation is highly dependent on the first time step $\mathcal{H}_y^{1/2}$, so higher-order methods are required to make that estimate.

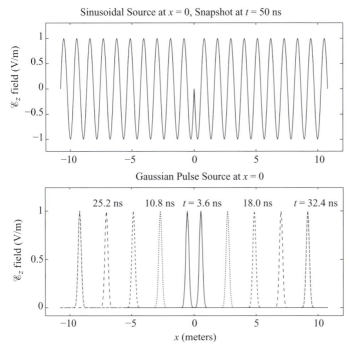

Figure 4.4 1D FDTD simulation examples. The top panel shows a snapshot of $\mathcal{E}_z$ at $t = 50$ ns of a simulation using a sinusoidal source at the center of the grid. The lower panel shows successive snapshots of $\mathcal{E}_z$ where the source is a Gaussian pulse.

This problem goes away in most circumstances, when the simulation space is assumed to have zero fields initially; then both $\mathcal{E}_z^0$ and $\mathcal{H}_y^{-1/2}$ are zero everywhere.

4.1.1   Example 1D simulations

Figure 4.4 shows example simulations in a simple, lossless 1D space for the TM mode. In both simulations, $\Delta t = 0.18$ ns, $\Delta x = v_p \Delta t = 5.4$ cm, and $\epsilon = \epsilon_0$, $\mu = \mu_0$. The grid is 401 cells in length, giving just under 22 meters in total length.

In the topmost panel, the space is excited by a sinusoidal source at the center point given by $\mathcal{E}_{z,\text{in}}(n) = \mathcal{E}_0 \sin(2\pi f_0 n \Delta t)$, with a center frequency $f_0 = 278$ MHz, corresponding to exactly 20 grid cells per wavelength. In this way we resolve the wave sufficiently to avoid numerical dispersion, which will be discussed in Chapter 6. The amplitude of the input wave is $\mathcal{E}_0 = 1$ V/m. In this example, the wave has propagated beyond the simulation space boundaries, but we have used an absorbing boundary condition (ABC) to truncate the simulation space; these will be described in Chapters 8 and 9.

In the lower panel, the space is excited by a Gaussian pulse source at the center point given by $\mathcal{E}_{z,\text{in}}(n) = \mathcal{E}_0 e^{-(n-8)^2/4^2}$, i.e., the pulse has a $1/e$ half-width of four time steps, and ramps up smoothly from zero. The amplitude is given as $\mathcal{E}_0 = 1$ V/m. The figure shows that our Gaussian pulse propagates to the left and right at the speed of light: specifically, at $t = 32.4$ ns the pulse has traveled exactly 9.72 meters, giving a numerical phase velocity of 3×10^8 m/s.

4.2   Maxwell's equations in two dimensions

We now proceed with discretization of Maxwell's equations based on the staggered mesh placement of the fields as given in Figure 4.1 for the 2D case. For this purpose, we assume that there are no variations of either the fields or the excitation in one of the directions, say the z direction. Thus, all derivatives with respect to z drop out from the two curl equations. Again neglecting magnetic or electric losses (i.e., let $\sigma = 0$ and $\sigma_m = 0$) and assuming simple and source-free media, so that $\tilde{\rho} = 0$, $\overline{\mathcal{J}} = 0$, and ϵ and μ are simple constants, independent of position, direction, or time, Maxwell's Equations (2.1) and (2.3) can be written as:

$$-\mu \frac{\partial \overline{\mathcal{H}}}{\partial t} = \nabla \times \overline{\mathcal{E}} \qquad \rightarrow$$

$$\frac{\partial \mathcal{H}_x}{\partial t} = -\frac{1}{\mu} \frac{\partial \mathcal{E}_z}{\partial y} \tag{4.10a}$$

$$\frac{\partial \mathcal{H}_y}{\partial t} = \frac{1}{\mu} \frac{\partial \mathcal{E}_z}{\partial x} \tag{4.10b}$$

$$\frac{\partial \mathcal{H}_z}{\partial t} = \frac{1}{\mu} \left(\frac{\partial \mathcal{E}_x}{\partial y} - \frac{\partial \mathcal{E}_y}{\partial x} \right) \tag{4.10c}$$

and

$$\epsilon \frac{\partial \overline{\mathcal{E}}}{\partial t} = \nabla \times \overline{\mathcal{H}} \qquad \rightarrow$$

$$\frac{\partial \mathcal{E}_x}{\partial t} = \frac{1}{\epsilon} \frac{\partial \mathcal{H}_z}{\partial y} \tag{4.11a}$$

$$\frac{\partial \mathcal{E}_y}{\partial t} = -\frac{1}{\epsilon} \frac{\partial \mathcal{H}_z}{\partial x} \tag{4.11b}$$

$$\frac{\partial \mathcal{E}_z}{\partial t} = \frac{1}{\epsilon} \left(\frac{\partial \mathcal{H}_y}{\partial x} - \frac{\partial \mathcal{H}_x}{\partial y} \right). \tag{4.11c}$$

As we did with the 1D case, we can group Equations (4.10) and (4.11) according to field vector components, one set involving only $\mathcal{E}_x$, $\mathcal{E}_y$, and $\mathcal{H}_z$, and another set involving $\mathcal{H}_x$, $\mathcal{H}_y$, and $\mathcal{E}_z$, referred to respectively as the TE and TM modes. The resulting sets of equations are given as:

$$\frac{\partial \mathcal{H}_x}{\partial t} = -\frac{1}{\mu} \frac{\partial \mathcal{E}_z}{\partial y} \tag{4.12a}$$

TM mode
$$\frac{\partial \mathcal{H}_y}{\partial t} = \frac{1}{\mu} \frac{\partial \mathcal{E}_z}{\partial x} \tag{4.12b}$$

$$\frac{\partial \mathcal{E}_z}{\partial t} = \frac{1}{\epsilon} \left(\frac{\partial \mathcal{H}_y}{\partial x} - \frac{\partial \mathcal{H}_x}{\partial y} \right) \tag{4.12c}$$

and

$$\frac{\partial \mathscr{E}_x}{\partial t} = \frac{1}{\epsilon} \frac{\partial \mathscr{H}_z}{\partial y} \tag{4.13a}$$

TE mode
$$\frac{\partial \mathscr{E}_y}{\partial t} = -\frac{1}{\epsilon} \frac{\partial \mathscr{H}_z}{\partial x} \tag{4.13b}$$

$$\frac{\partial \mathscr{H}_z}{\partial t} = \frac{1}{\mu} \left(\frac{\partial \mathscr{E}_x}{\partial y} - \frac{\partial \mathscr{E}_y}{\partial x} \right). \tag{4.13c}$$

It is important to note that the TE and TM modes above are completely uncoupled from one another; in other words, they contain no common vector field components and can therefore exist independently from one another.

This particular TE and TM designation of electromagnetic wave types is quite unrelated to the separation into TE and TM modes of classic guided electromagnetic waves. For time-harmonic guided waves, the z direction is generally taken to be the direction of propagation, with all field components thus exhibiting variations in z in the form of $e^{-j\bar{\beta}z}$, where $\bar{\beta}$ is the phase constant for the particular waveguide, as determined by the boundary conditions. Thus, all field components do in fact vary with z, and the TE (or TM) modes are identified as those waves which have a magnetic (or electric) field component in the axial (i.e., z) direction. For the 2D electromagnetic waves in the FDTD grid, none of the field components varies with z so that there certainly is no propagation in the z direction; instead, the TE or TM modes are identified on the basis of whether the wave has a nonzero $\mathscr{H}$ or $\mathscr{E}$ component in the third dimension, which we decided had no variation – in this case the z dimension. Hence, Equations (4.12) are identified as the TM mode on account of the $\mathscr{E}_z$ component.

Another way of looking at the TM and TE modes is as follows: since we have decided that there is no variation in the z direction (i.e., all derivatives with respect to z have been set to zero), there can be no propagation in the z direction; but, in general, propagation in either x- or y-directions (or both) is possible. In the TM mode, only $\mathscr{H}_x$ and $\mathscr{H}_y$ components are nonzero, and are *in the plane of propagation*. The reverse is true in the TE mode.

Although Equations (4.12) and (4.13) are entirely symmetric, the TE and TM modes are physically quite different, due to orientation of electric and magnetic field lines with respect to the infinitely long axis (i.e., the z axis) of the system. Note, for example, that the electric fields for the TE mode lie in a plane perpendicular to the long axis, so that when modeling infinitely long metallic structures (aligned along the long axis), large electric fields can be supported near the metallic surface. For the TM mode, however, the electric field is along the z direction and thus must be nearly zero at infinitely long metallic surfaces (which must necessarily be aligned with the infinite dimension in the system) being modeled. As such, when running 2D simulations, the choice of TE or TM modes can depend on the types of structures in the simulation.

We now separately consider the discretization of Equations (4.13) and (4.12), respectively, for TE and TM modes.

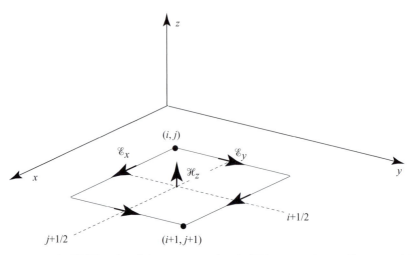

Figure 4.5 An FDTD unit cell for transverse electric (TE) waves. The small vectors with thick arrows are placed at the point in the mesh at which they are defined and stored. For example, $\mathscr{E}_y$ is defined/stored at mesh point $(i, j + 1/2)$, while $\mathscr{H}_z$ is defined/stored at mesh points $(i + 1/2, j + 1/2)$.

4.2.1   Transverse electric (TE) mode

A portion of the Yee cell constituting a unit cell for the TE case is depicted in Figure 4.5. Note the spatial positions of the two electric field components and the single magnetic field component. The $\mathscr{E}_x$ component is located at half x and integer y grid points, i.e., $(i + 1/2, j)$, while the $\mathscr{E}_y$ component is located at integer x and half y grid points, i.e., $(i, j + 1/2)$. The magnetic field component $\mathscr{H}_z$ is located at half x and half y grid points, i.e., $(i + 1/2, j + 1/2)$. These grid points are chosen to accommodate the interleaved leapfrog algorithm that makes up the FDTD method; however, one should be able to identify the trend that makes it easy to remember. This will be spelled out in the next section, when we look at the 3D FDTD method.

The spatially discretized versions of the component Maxwell's Equations (4.13) are:

$$\left. \frac{\partial \mathscr{E}_x}{\partial t} \right|_{i+1/2,j} = \frac{1}{\epsilon} \left[\frac{\left. \mathscr{H}_z \right|_{i+1/2,j+1/2} - \left. \mathscr{H}_z \right|_{i+1/2,j-1/2}}{\Delta y} \right] \tag{4.14a}$$

$$\left. \frac{\partial \mathscr{E}_y}{\partial t} \right|_{i,j+1/2} = -\frac{1}{\epsilon} \left[\frac{\left. \mathscr{H}_z \right|_{i+1/2,j+1/2} - \left. \mathscr{H}_z \right|_{i-1/2,j+1/2}}{\Delta x} \right] \tag{4.14b}$$

$$\left. \frac{\partial \mathscr{H}_z}{\partial t} \right|_{i+1/2,j+1/2} = \frac{1}{\mu} \left[\frac{\left. \mathscr{E}_x \right|_{i+1/2,j+1} - \left. \mathscr{E}_x \right|_{i+1/2,j}}{\Delta y} - \frac{\left. \mathscr{E}_y \right|_{i+1,j+1/2} - \left. \mathscr{E}_y \right|_{i,j+1/2}}{\Delta x} \right].$$

$$\tag{4.14c}$$

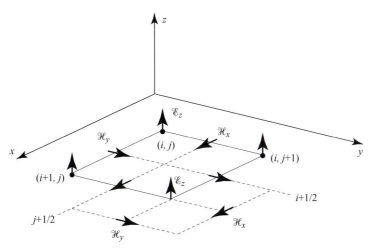

Figure 4.6 An FDTD unit cell for transverse magnetic (TM) waves. The small vectors with thick arrows are placed at the point in the mesh at which they are defined and stored. For example, $\mathscr{E}_z$ is defined/stored at mesh point (i, j), while $\mathscr{H}_y$ is defined/stored at mesh points $(i + 1/2, j)$.

Using (4.3) for the time derivatives, we arrive at the FDTD algorithm for TE waves:

2D TE mode

$$
\mathscr{H}_z \Big|_{i+1/2,j+1/2}^{n+1/2} = \mathscr{H}_z \Big|_{i+1/2,j+1/2}^{n-1/2} + \frac{\Delta t}{\mu_{i+1/2,j+1/2}}
$$

$$
\times \left[\frac{\mathscr{E}_x \Big|_{i+1/2,j+1}^{n} - \mathscr{E}_x \Big|_{i+1/2,j}^{n}}{\Delta y} - \frac{\mathscr{E}_y \Big|_{i+1,j+1/2}^{n} - \mathscr{E}_y \Big|_{i,j+1/2}^{n}}{\Delta x} \right]
$$

$$
\mathscr{E}_x \Big|_{i+1/2,j}^{n+1} = \mathscr{E}_x \Big|_{i+1/2,j}^{n} + \frac{\Delta t}{\epsilon_{i+1/2,j}\,\Delta y} \left[\mathscr{H}_z \Big|_{i+1/2,j+1/2}^{n+1/2} - \mathscr{H}_z \Big|_{i+1/2,j-1/2}^{n+1/2} \right]
$$

$$
\mathscr{E}_y \Big|_{i,j+1/2}^{n+1} = \mathscr{E}_y \Big|_{i,j+1/2}^{n} - \frac{\Delta t}{\epsilon_{i,j+1/2}\,\Delta x} \left[\mathscr{H}_z \Big|_{i+1/2,j+1/2}^{n+1/2} - \mathscr{H}_z \Big|_{i-1/2,j+1/2}^{n+1/2} \right]
$$

(4.15)

where we have explicitly acknowledged that the medium parameter values (ϵ, μ) to be used are those at the spatial point at which the quantity is being updated. In this case, the medium does not need to be homogeneous: the values of ϵ and μ can change throughout the simulation space. However, this algorithm still requires that the medium is *isotropic*, i.e., the parameters ϵ and μ do not vary with direction.

4.2.2   Transverse magnetic (TM) mode

A portion of the Yee cell constituting a unit cell for the TM case is depicted in Figure 4.6. The spatially discretized equations for the TM mode can be written in an analogous fashion to (4.14), by noting that the discrete electric field variables are located at the cell edges, and including an appropriate half-cell shift. Or, one can use the pattern identified for the TE mode (spelled out in the next section) to show that the $\mathcal{E}_z$ component is located at integer grid points (i, j); the $\mathcal{H}_x$ component is located at integer x and half y grid points, i.e., $(i, j + 1/2)$; and the $\mathcal{H}_y$ component is located at half x and integer y components, i.e., $(i + 1/2, j)$.

The spatially discretized versions of Equations (4.12) are:

$$\left. \frac{\partial \mathcal{H}_x}{\partial t} \right|_{i,j+1/2} = -\frac{1}{\mu} \left[\frac{\left. \mathcal{E}_z \right|_{i,j+1} - \left. \mathcal{E}_z \right|_{i,j}}{\Delta y} \right] \tag{4.16a}$$

$$\left. \frac{\partial \mathcal{H}_y}{\partial t} \right|_{i+1/2,j} = \frac{1}{\mu} \left[\frac{\left. \mathcal{E}_z \right|_{i+1,j} - \left. \mathcal{E}_z \right|_{i,j}}{\Delta x} \right] \tag{4.16b}$$

$$\left. \frac{\partial \mathcal{E}_z}{\partial t} \right|_{i,j} = \frac{1}{\epsilon} \left[\frac{\left. \mathcal{H}_y \right|_{i+1/2,j} - \left. \mathcal{H}_y \right|_{i-1/2,j}}{\Delta x} - \frac{\left. \mathcal{H}_x \right|_{i,j+1/2} - \left. \mathcal{H}_x \right|_{i,j-1/2}}{\Delta y} \right]. \tag{4.16c}$$

Again using analogies to Equations (4.3) for the time derivatives, we arrive at the FDTD algorithm for TM waves:

2D TM mode

$$\left. \mathcal{H}_x \right|_{i,j+1/2}^{n+1/2} = \left. \mathcal{H}_x \right|_{i,j+1/2}^{n-1/2} - \frac{\Delta t}{\mu_{i,j+1/2} \Delta y} \left[\left. \mathcal{E}_z \right|_{i,j+1}^{n} - \left. \mathcal{E}_z \right|_{i,j}^{n} \right]$$

$$\left. \mathcal{H}_y \right|_{i+1/2,j}^{n+1/2} = \left. \mathcal{H}_y \right|_{i+1/2,j}^{n-1/2} + \frac{\Delta t}{\mu_{i+1/2,j} \Delta x} \left[\left. \mathcal{E}_z \right|_{i+1,j}^{n} - \left. \mathcal{E}_z \right|_{i,j}^{n} \right]$$

$$\left. \mathcal{E}_z \right|_{i,j}^{n+1} = \left. \mathcal{E}_z \right|_{i,j}^{n} + \frac{\Delta t}{\epsilon_{i,j}} \left[\frac{\left. \mathcal{H}_y \right|_{i+1/2,j}^{n+1/2} - \left. \mathcal{H}_y \right|_{i-1/2,j}^{n+1/2}}{\Delta x} - \frac{\left. \mathcal{H}_x \right|_{i,j+1/2}^{n+1/2} - \left. \mathcal{H}_x \right|_{i,j-1/2}^{n+1/2}}{\Delta y} \right]$$

$$\tag{4.17}$$

where we have again explicitly denoted the medium parameter values (ϵ, μ) as those at the spatial point at which the quantity is being updated, removing the restriction of homogeneity.

4.2.3   Example 2D simulations

Figure 4.7 shows example simulations in a simple, lossless 2D space. In both simulations, $\Delta x = 5.4$ cm, $\Delta t = \Delta x / v_p / \sqrt{2} = 0.13$ ns,[2] and $\epsilon = \epsilon_0, \mu = \mu_0$. The grid is 201×201

---

[2] Note the $\sqrt{2}$ in this formula for Δt; this is a result of operating in 2D. We will derive this stability restriction formally in Chapter 5.

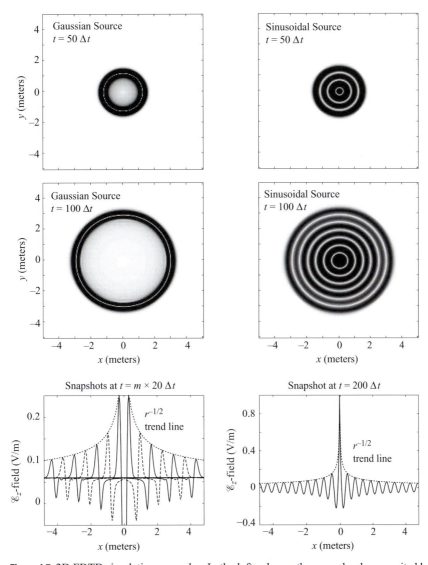

Figure 4.7 2D FDTD simulation examples. In the left column, the space has been excited by the derivative of Gaussian source at the center of the space; in the right column, a sinusoidal source. The lower panels show a slice through the center, demonstrating the $1/\sqrt{r}$ decay of the fields with distance.

cells, giving ~ 11 meters in each dimension; the figure zooms in to the range from -5 to 5 meters.

In the left column, the space is excited by the derivative of a Gaussian pulse at the center point given by $\mathscr{E}_{z,\text{in}} = \mathscr{E}_0\, e^{-(n-8)^2/4^2}$, i.e., the pulse has a $1/e$ half-width of four time steps, and ramps up smoothly from zero. Note that excitation with a Gaussian

Table 4.1 Field component locations in the Yee cell

Field	x location	y location	z location
$\mathcal{E}_x$	$i + 1/2$	j	k
$\mathcal{E}_y$	i	$j + 1/2$	k
$\mathcal{E}_z$	i	j	$k + 1/2$
$\mathcal{H}_x$	i	$j + 1/2$	$k + 1/2$
$\mathcal{H}_y$	$i + 1/2$	j	$k + 1/2$
$\mathcal{H}_z$	$i + 1/2$	$j + 1/2$	k

leads to unphysical fields left in the space due to "grid capacitance" [2]. For this reason it is prudent to use the first derivative of a Gaussian as an excitation source for broadband simulations.

In the right column, the center point is excited by a sinusoidal source at 278 MHz, ensuring exactly 20 grid cells per wavelength. The figures show the magnitude of the $\mathcal{E}_z$ field (i.e., the absolute value), so each concentric ring is half a wavelength. The bottom panels show slices through the center of the expanding concentric rings; in the case of the Gaussian source, we show snapshots of the expanding field at successive times, while for the sinusoidal source we show a single snapshot. In both cases, we can see that the fields decay with radius at exactly $1/\sqrt{r}$: recall that for cylindrical symmetry the *power* decays as $1/r$, so each of the fields $\overline{\mathcal{E}}$ and $\overline{\mathcal{H}}$ decay as $1/\sqrt{r}$.

4.3   FDTD expressions in three dimensions

We have hinted up to this point that there is a pattern in the location of field values in the Yee cell. Looking at Figure 4.1, and writing down the locations of each of the six field components, we arrive at Table 4.1. The pattern is as follows:

- For the $\mathcal{E}_a$ components, for $a = x, y, z$, the a-location is half-integer, while the other two are integer.
- For the $\mathcal{H}_a$ components, it is exactly the opposite: the a-location is integer, while the other two are half-integer.

This is further illustrated in Figure 4.8, which expands Figure 4.1 and shows two slices in the x-plane, at i and $i - 1/2$. These two plane slices are the same at $i \pm m$ and $i \pm m - 1/2$, for any integer m. Note that these two slices are, in fact, the TE and TM mode patterns in the y-z plane.

For completeness, we present here the full FDTD expressions for Maxwell's equations in simple media in three dimensions, derived by discretizing Maxwell's Cartesian

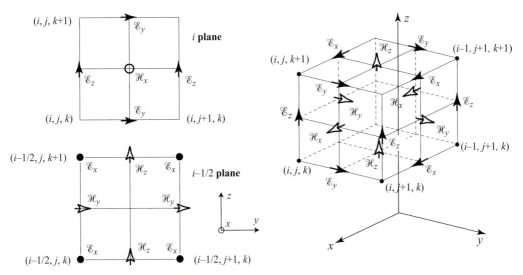

Figure 4.8 Yee cell in 3D. The cube at right is a repeat of Figure 4.1; at left, slices of the cube are shown at i and $i - 1/2$, showing the interleaved field placement in each plane.

component Equations (4.4) with second-order centered differences:

$$\mathcal{E}_x\Big|_{i+1/2,j,k}^{n+1} = \mathcal{E}_x\Big|_{i+1/2,j,k}^{n} \qquad (4.18a)$$

$$+ \frac{\Delta t}{\epsilon_{i+1/2,j,k}} \left[\frac{\mathcal{H}_z\Big|_{i+1/2,j+1/2,k}^{n+1/2} - \mathcal{H}_z\Big|_{i+1/2,j-1/2,k}^{n+1/2}}{\Delta y} \right.$$

$$\left. - \frac{\mathcal{H}_y\Big|_{i+1/2,j,k+1/2}^{n+1/2} - \mathcal{H}_y\Big|_{i+1/2,j,k-1/2}^{n+1/2}}{\Delta z} \right]$$

$$\mathcal{E}_y\Big|_{i,j+1/2,k}^{n+1} = \mathcal{E}_y\Big|_{i,j+1/2,k}^{n} \qquad (4.18b)$$

$$+ \frac{\Delta t}{\epsilon_{i,j+1/2,k}} \left[\frac{\mathcal{H}_x\Big|_{i,j+1/2,k+1/2}^{n+1/2} - \mathcal{H}_x\Big|_{i,j+1/2,k-1/2}^{n+1/2}}{\Delta z} \right.$$

$$\left. - \frac{\mathcal{H}_z\Big|_{i+1/2,j+1/2,k}^{n+1/2} - \mathcal{H}_z\Big|_{i-1/2,j+1/2,k}^{n+1/2}}{\Delta x} \right]$$

$$\mathscr{E}_z\Big|_{i,j,k+1/2}^{n+1} = \mathscr{E}_z\Big|_{i,j,k+1/2}^{n} \tag{4.18c}$$

$$+ \frac{\Delta t}{\epsilon_{i,j,k+1/2}} \left[\frac{\mathscr{H}_y\Big|_{i+1/2,j,k+1/2}^{n+1/2} - \mathscr{H}_y\Big|_{i-1/2,j,k+1/2}^{n+1/2}}{\Delta x} \right.$$

$$\left. - \frac{\mathscr{H}_x\Big|_{i,j+1/2,k+1/2}^{n+1/2} - \mathscr{H}_x\Big|_{i,j-1/2,k+1/2}^{n+1/2}}{\Delta y} \right]$$

$$\mathscr{H}_x\Big|_{i,j+1/2,k+1/2}^{n+1/2} = \mathscr{H}_x\Big|_{i,j+1/2,k+1/2}^{n-1/2} \tag{4.19a}$$

$$+ \frac{\Delta t}{\mu_{i,j+1/2,k+1/2}} \left[\frac{\mathscr{E}_y\Big|_{i,j+1/2,k+1}^{n} - \mathscr{E}_y\Big|_{i,j+1/2,k}^{n}}{\Delta z} \right.$$

$$\left. - \frac{\mathscr{E}_z\Big|_{i,j+1,k+1/2}^{n} - \mathscr{E}_z\Big|_{i,j,k+1/2}^{n}}{\Delta y} \right]$$

$$\mathscr{H}_y\Big|_{i+1/2,j,k+1/2}^{n+1/2} = \mathscr{H}_y\Big|_{i+1/2,j,k+1/2}^{n-1/2} \tag{4.19b}$$

$$+ \frac{\Delta t}{\mu_{i+1/2,j,k+1/2}} \left[\frac{\mathscr{E}_z\Big|_{i+1,j,k+1/2}^{n} - \mathscr{E}_z\Big|_{i,j,k+1/2}^{n}}{\Delta x} \right.$$

$$\left. - \frac{\mathscr{E}_x\Big|_{i+1/2,j,k+1}^{n} - \mathscr{E}_x\Big|_{i+1/2,j,k}^{n}}{\Delta z} \right]$$

$$\mathscr{H}_z\Big|_{i+1/2,j+1/2,k}^{n+1/2} = \mathscr{H}_z\Big|_{i+1/2,j+1/2,k}^{n-1/2} \tag{4.19c}$$

$$+ \frac{\Delta t}{\mu_{i+1/2,j+1/2,k}} \left[\frac{\mathscr{E}_x\Big|_{i+1/2,j+1,k}^{n} - \mathscr{E}_x\Big|_{i+1/2,j,k}^{n}}{\Delta y} \right.$$

$$\left. - \frac{\mathscr{E}_y\Big|_{i+1,j+1/2,k}^{n} - \mathscr{E}_y\Big|_{i,j+1/2,k}^{n}}{\Delta x} \right].$$

Note that the above expressions are for the case where we have no electric or magnetic current sources, $\overline{\mathscr{J}}_i$ or $\overline{\mathscr{M}}_i$. Inclusion of the sources is straightforward and will be discussed in later chapters. Together with the discretization pattern outlined in Table 4.1, the above

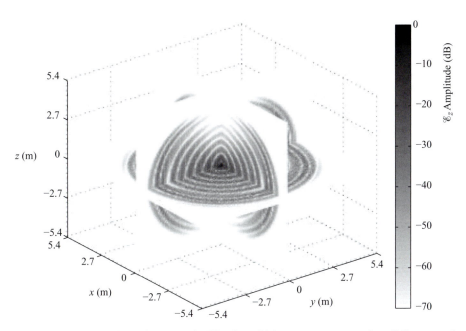

Figure 4.9 3D FDTD simulation example. The sinusoidal source wave expands radially outward from the center of the space, and the field amplitude decays as $1/r$.

Equations (4.18) and (4.19) can be used to make the following observations about the discretization process:

- The update equations for $\mathcal{E}$ components are discretized in time at $n + 1/2$, so that they use the time difference of the $\mathcal{E}$ fields at times $n + 1$ and n.
- The update equations for $\mathcal{H}$ components are discretized at n, so that they use the time difference of the $\mathcal{H}$ fields at times $n + 1/2$ and $n - 1/2$.
- In each equation, the values of μ and/or ϵ are used at the same location as the field component being updated. This arises because the true Maxwell's equations relate $\epsilon \overline{\overline{\mathcal{E}}}$ to $\mathcal{H}$ and $\mu \overline{\mathcal{H}}$ to $\overline{\mathcal{E}}$; as such, ϵ and μ must be discretized along with the fields.

4.3.1   Example 3D simulation

Figure 4.9 shows an example simulation in 3D. This example uses a grid of $100 \times 100 \times 100$ grid cells, yielding 10^6 locations total. One can see from this relatively small example that in 3D the spaces can quickly become prohibitively large. We excite a sinusoidal $\mathcal{E}_z$ source at the center of the space with frequency 278 MHz just as in the 2D example earlier; however, in this case we have only 10 grid cells per wavelength, so $\Delta x = \Delta y = \Delta z = 10.8$ cm.

The figure shows slices of the 3D $\mathcal{E}_z$ field amplitude through the center in each dimension; this plot is created with the Matlab function `slice`. Because of the $\mathcal{E}_z$ source at the center of the simulation space, the characteristic dipole radiation pattern is evident, with very little field intensity radiated in the z direction above and below the source. In the radial direction in the $x-y$ plane, the field decays as $1/r$ as expected, since the power decays as $1/r^2$.

4.4   FDTD algorithm for lossy media

We have so far derived the 1D, 2D, and 3D FDTD algorithms for simple media, by neglecting the loss terms in the two curl equations. Incorporation of these losses into the FDTD algorithm is straightforward. These losses are denoted by electric conductivity σ, and sometimes magnetic conductivity σ_m, which are parameters of the medium just like ϵ and μ. The magnetic conductivity σ_m is, strictly speaking, not physically realistic, but is often used for artificial absorption at simulation space boundaries, as will be discussed in Chapter 9; as such, we include it here. Note that like ϵ and μ in Equations (4.18) and (4.19) above, the conductivities σ and σ_m may vary throughout the space.

4.4.1   1D waves in lossy media: waves on lossy transmission lines

We start with the 1D case, which is posed in terms of uniform transmission line equations; again, one should note that the algorithms we derive for $\mathcal{V}$ and $\mathcal{I}$ are directly applicable to the 1D Maxwell's equations by simple substitution of variables. In this case, the electric and magnetic conductivities are replaced by electrical resistance R and conductance G.

Derivation begins with the lossy transmission line Equations (3.1), repeated here for convenience:

$$\frac{\partial \mathcal{V}}{\partial x} = -R\,\mathcal{I} - L\,\frac{\partial \mathcal{I}}{\partial t} \tag{4.20a}$$

$$\frac{\partial \mathcal{I}}{\partial x} = -G\,\mathcal{V} - C\,\frac{\partial \mathcal{V}}{\partial t}. \tag{4.20b}$$

The discretization of these equations follows as before. The only difference is in the first term on the right-hand sides of both Equations (4.20) above; these terms are not derivatives, which means they can be discretized directly. The finite difference approximations of these equations using the interleaved leapfrog algorithm are given by modified versions of Equations (3.86):

Discretized at $(i + 1/2, n)$:

$$\frac{\mathcal{V}_{i+1}^n - \mathcal{V}_i^n}{\Delta x} = -R\,\mathcal{I}_{i+1/2}^n - L\,\frac{\mathcal{I}_{i+1/2}^{n+1/2} - \mathcal{I}_{i+1/2}^{n-1/2}}{\Delta t}. \tag{4.21a}$$

Discretized at $(i, n + 1/2)$:

$$\frac{\mathcal{I}_{i+1/2}^{n+1/2} - \mathcal{I}_{i-1/2}^{n+1/2}}{\Delta x} = -G\,\mathcal{V}_i^{n+1/2} - C\,\frac{\mathcal{V}_i^{n+1} - \mathcal{V}_i^n}{\Delta t}. \tag{4.21b}$$

Note that Equation (4.21a) is discretized at time step n and spatial step $i + 1/2$, while Equation (4.21b) is discretized at time step $n + 1/2$ and spatial step i. Noting that the current and voltage are advanced in time in a leapfrog manner, there is a problem in using Equation (4.21a) directly, since the current $\mathcal{I}$ is needed at time step n, but is only available at half time steps $n + 1/2$. Similarly, Equation (4.21b) requires the voltage $\mathcal{V}$ at time step $n + 1/2$, but it is only available at integer time steps n. To estimate these quantities with second-order accuracy, it is common to use two point averaging in time, i.e.,

$$\mathcal{I}_{i+1/2}^n \simeq \frac{\mathcal{I}_{i+1/2}^{n+1/2} + \mathcal{I}_{i+1/2}^{n-1/2}}{2} \tag{4.22a}$$

$$\mathcal{V}_i^{n+1/2} \simeq \frac{\mathcal{V}_i^{n+1} + \mathcal{V}_i^n}{2}. \tag{4.22b}$$

Substituting (4.22) in (4.21) and manipulating, we find the update equations for lossy media:

Interleaved leapfrog for lossy line

$$\mathcal{I}_{i+1/2}^{n+1/2} = \left[\frac{2L - R\,\Delta t}{2L + R\,\Delta t}\right]\mathcal{I}_{i+1/2}^{n-1/2} - \left[\frac{2\Delta t}{(2L + R\,\Delta t)\,\Delta x}\right]\left[\mathcal{V}_{i+1}^n - \mathcal{V}_i^n\right]$$

$$\mathcal{V}_i^{n+1} = \left[\frac{2C - G\,\Delta t}{2C + G\,\Delta t}\right]\mathcal{V}_i^n - \left[\frac{2\Delta t}{(2C + G\,\Delta t)\,\Delta x}\right]\left[\mathcal{I}_{i+1/2}^{n+1/2} - \mathcal{I}_{i-1/2}^{n+1/2}\right] \tag{4.23}$$

The corresponding FDTD algorithms for 1D TE and 1D TM electromagnetic waves in lossy media can be obtained from Equation (4.23) by making the appropriate substitutions. In particular, one would substitute $\epsilon, \mu, \sigma, \sigma_m$ for C, L, G, R, respectively; and either $\mathcal{E}_y, \mathcal{H}_z$ (TE Mode) or $\mathcal{E}_z, \mathcal{H}_y$ (TM Mode) in place of $\mathcal{V}$ and $\mathcal{I}$.

It should be apparent, and one can easily show by substitution, that when the resistance and conductance R and G above are set to zero, the update equations in (4.23) collapse to the same equations for lossless media, given in Equations (3.86).

Figure 4.10 shows an example of a lossy 1D simulation, using the same code as in Figure 4.4. The results of Figure 4.4 are repeated in the top panel for comparison; below, we have introduced a conductivity $\sigma = 0.001$ S/m into the FDTD algorithm.

Recall from electromagnetic theory that in general the decay rate of a wave is given by [3]:

$$\alpha = \omega\sqrt{\frac{\mu\epsilon}{2}}\left[\sqrt{1 + \left(\frac{\sigma}{\omega\epsilon}\right)^2} - 1\right]^{1/2} \quad \text{nepers/m}$$

In our example in Figure 4.10, our Gaussian pulse has a width of about 0.6 m, for which we can approximate a frequency of $\sim$500 MHz. Substituting this frequency into the equation above, we find a decay rate of $\sim \alpha = 0.189$ np/m. The smallest pulse at the far right of Figure 4.10 has traveled 9.2 m, so its theoretical amplitude should be $e^{-0.189 \cdot 9.2} = 0.176$ V/m — in fact, measurement shows that the amplitude in the

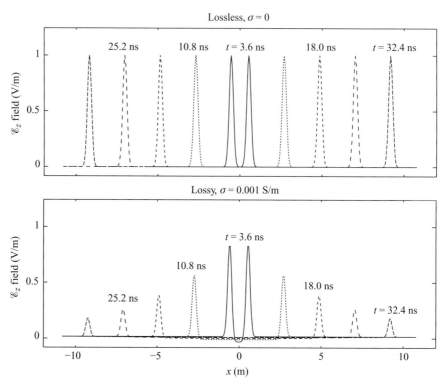

Figure 4.10 Lossy 1D FDTD simulation example. The case with $\sigma = 0.001$ S/m and $\sigma_m = 0$ is compared to the lossless case above it.

simulation shown is ~ 0.178 V/m, so our lossy FDTD method has very accurately simulated the decay of this pulse.

4.4.2   2D and 3D waves in lossy media

The procedure for incorporating the loss terms into the 2D FDTD algorithms follows along very similar lines. For example, the corresponding versions of (4.15) ($\mathcal{H}_z$ component) and (4.17) ($\mathcal{E}_z$ component) for lossy media are:

$$
\mathcal{H}_z\Big|_{i+1/2,j+1/2,k}^{n+1/2} = \left[\frac{2\mu - \sigma_m \Delta t}{2\mu + \sigma_m \Delta t}\right] \mathcal{H}_z\Big|_{i+1/2,j+1/2,k}^{n-1/2} \tag{4.24a}
$$

$$
+ \left[\frac{2\Delta t}{2\mu + \sigma_m \Delta t}\right] \left[\frac{\mathcal{E}_x\Big|_{i+1/2,j+1,k}^{n} - \mathcal{E}_x\Big|_{i+1/2,j,k}^{n}}{\Delta y}\right.
$$

$$
\left. - \frac{\mathcal{E}_y\Big|_{i+1,j+1/2,k}^{n} - \mathcal{E}_y\Big|_{i,j+1/2,k}^{n}}{\Delta x}\right].
$$

Table 4.2 Coefficients for Lossy Media

Update Equation	Lossless C_1	Lossless C_2	Lossy C_1	Lossy C_2
$\mathcal{E}_{x,y,z}$ (4.18)	1	$\dfrac{\Delta t}{\epsilon}$	$\dfrac{2\epsilon - \sigma\,\Delta t}{2\epsilon + \sigma\,\Delta t}$	$\dfrac{2\Delta t}{2\epsilon + \sigma\,\Delta t}$
$\mathcal{H}_{x,y,z}$ (4.19)	1	$\dfrac{\Delta t}{\mu}$	$\dfrac{2\mu - \sigma_m\,\Delta t}{2\mu + \sigma_m\,\Delta t}$	$\dfrac{2\Delta t}{2\mu + \sigma_m\,\Delta t}$

$$
\mathcal{E}_z\Big|_{i,j,k+1/2}^{n+1} = \left[\frac{2\epsilon - \sigma\,\Delta t}{2\epsilon + \sigma\,\Delta t}\right] \mathcal{E}_z\Big|_{i,j,k+1/2}^{n} \tag{4.24b}
$$

$$
+ \left[\frac{2\Delta t}{2\epsilon + \sigma\,\Delta t}\right] \left[\frac{\mathcal{H}_y\Big|_{i+1/2,j,k+1/2}^{n+1/2} - \mathcal{H}_y\Big|_{i-1/2,j,k+1/2}^{n+1/2}}{\Delta x} \right.
$$

$$
\left. - \frac{\mathcal{H}_x\Big|_{i,j+1/2,k+1/2}^{n+1/2} - \mathcal{H}_x\Big|_{i,j-1/2,k+1/2}^{n+1/2}}{\Delta y} \right].
$$

The similarity of the multiplying terms with the 1D version given in Equation (4.23) is evident by making the substitutions $\epsilon \leftrightarrow C$, $\mu \leftrightarrow L$, $\sigma \leftrightarrow G$, and $\sigma_m \leftrightarrow R$.

From the 1D lossy Equations (4.23) and the two component equations above of the 3D algorithm, one should be able to distinguish a pattern in the coefficients of these equations. If we write any particular component equation of the 3D algorithm in the following form:

$$
\mathcal{G}^{n+1} = C_1\,\mathcal{G}^n + C_2\,\nabla_{\mathrm{FD}}\mathcal{F}^{n+1/2}
$$

where we have used the notation ∇_{FD} to denote the spatial differences on the right-hand sides of Equations (4.18) and (4.19), noting that this includes the Δx, Δy, or Δz in the denominator. The fields $\mathcal{G}$ and $\mathcal{F}$ refer to either $\mathcal{E}$ or $\mathcal{H}$. We can then write the coefficients C_1 and C_2 for each equation, as in Table 4.2. From these coefficients one can transform from the lossless 3D update Equations (4.18) and (4.19) to their lossy versions; in the same way one can find the lossy update equations for the 1D and 2D FDTD algorithms for either TE or TM modes.

Figure 4.11 shows a 1D slice through the 3D simulation of Figure 4.9 along the x-axis. The lossless case of Figure 4.9 is shown, as well as a lossy simulation in which we have included $\sigma = 10^{-4}$ S/m. One can easily observe that the field amplitudes decay with distance faster than the lossless case, which approximately follows the $1/r$ trend.[3]

---

[3] This trend line does not fit perfectly because we have not run the simulation long enough to smooth out any start-up transients.

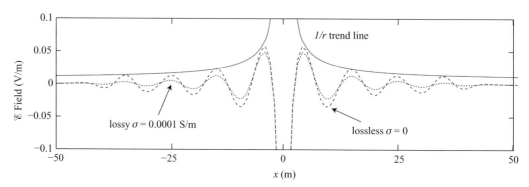

Figure 4.11 Lossy 3D FDTD simulation example. The same simulation as Figure 4.9 is used, but this time with $\sigma = 10^{-4}$ S/m. This figure shows a 1D slice through the 3D simulation.

4.5   Divergence-free nature of the FDTD algorithm

Before proceeding further in our discussion, we must point out an important feature of the FDTD algorithm as presented above. The integral and differential forms of Maxwell's curl equations (Faraday's law and Ampère's law, Equations 2.1 and 2.3) have been shown to contain Gauss's laws (Equations 2.2 and 2.4) implicitly; this implies that the fields are *divergence-free*. However, after discretizing these equations and creating the difference equations above, there is no guarantee that the divergence-free nature of Maxwell's equations is maintained. The conservation of the divergence-free nature of the fields is an important property of the FDTD algorithm and the Yee grid, as we now demonstrate here.

Consider the integral form of Equation (2.2) in a charge-free region (i.e., $\tilde{\rho} = 0$):

$$\nabla \cdot \mathscr{D} = 0 \qquad \rightarrow \qquad \oint_S \mathscr{D} \cdot d\mathbf{s} = 0. \tag{4.25}$$

We apply this integral to the closed rectangular box surface (with sides Δx, Δy, and Δz) shown in Figure 4.12. Noting that $d\mathbf{s}$ is by definition always outward from the surface, we have

$$\oint_S \overline{\mathscr{D}} \cdot d\mathbf{s} = - \left(\epsilon \mathscr{E}_x \Big|_{i+1/2,j} \Delta y \Delta z \right) + \left(\epsilon \mathscr{E}_x \Big|_{i+3/2,j} \Delta y \Delta z \right)$$

$$- \left(\epsilon \mathscr{E}_y \Big|_{i+1,j-1/2} \Delta x \Delta z \right) + \left(\epsilon \mathscr{E}_y \Big|_{i+1,j+1/2} \Delta x \Delta z \right). \tag{4.26}$$

Taking the derivative of (4.26) with respect to time, and applying (4.15) to the $\partial \mathscr{E}_x / \partial t$ and $\partial \mathscr{E}_y / \partial t$ terms, we find:

$$\frac{\partial}{\partial t} \left[\oint_S \overline{\mathscr{D}} \cdot d\mathbf{s} \right] = - \left[\frac{\mathscr{H}_z^B - \mathscr{H}_z^C}{\Delta y} \right] \Delta y \Delta z + \left[\frac{\mathscr{H}_z^A - \mathscr{H}_z^D}{\Delta y} \right] \Delta y \Delta z \tag{4.27}$$

$$- \left[\frac{\mathscr{H}_z^C - \mathscr{H}_z^D}{\Delta x} \right] \Delta x \Delta z + \left[\frac{\mathscr{H}_z^B - \mathscr{H}_z^A}{\Delta x} \right] \Delta x \Delta z = 0,$$

where, for example, $\mathscr{H}_z^B$ simply denotes the z component of the magnetic field at the point marked B in Figure 4.12, as a shorthand for $\mathscr{H}_z \big|_{i+1/2,j+1/2}$. This result indicates

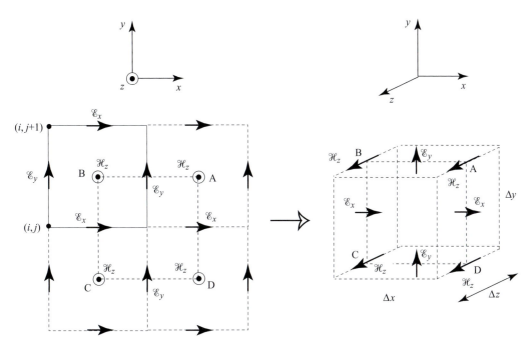

Figure 4.12 FDTD evaluation of the divergence of $\overline{\mathscr{D}} = \epsilon\overline{\mathscr{E}}$. By integrating Gauss's law over a unit cell enclosed by ABCD, we demonstrate that the divergence-free nature of Maxwell's equations holds for the FDTD algorithm.

that if the initial conditions were such that the field was divergence free, then Gauss's law is automatically satisfied throughout any FDTD calculation, since $\partial/\partial t[\oint_S \overline{\mathscr{D}} \cdot d\mathbf{s}] = 0$, so that $\nabla \cdot \overline{\mathscr{D}} = 0$.

It is worth noting, however, that this divergence-free nature of the FDTD algorithm is only valid in source-free media and for our second-order centered differencing. If we were to use a higher-order method for spatial discretization and then apply the same algorithm as above, in general we would find that the method is not divergence-free.

4.6 The FDTD method in other coordinate systems

It goes without saying that the FDTD method can easily be implemented in other orthogonal curvilinear coordinate systems, such as cylindrical or spherical coordinates. For this purpose, we can imagine a Yee cell drawn in such a coordinate system, although it might well be a challenge to accurately draw one, and basically follow the same procedure as before. This is particularly easy for the 2D systems with cylindrical symmetry, since decoupled TE and TM modes can be identified in the same manner as was done above for Cartesian coordinates. In this section we will derive the update equations for 2D cylindrical coordinates, as well as for 3D cylindrical and spherical coordinates.

4.6.1   2D polar coordinates

In this section we will examine the FDTD method for 2D coordinate systems based on the cylindrical coordinates r, ϕ, and z. In 2D, we can implement a polar coordinate system $(r - \phi)$, in which the system is symmetric along the z-axis; or a cylindrical system $(r - z)$, in which the system is azimuthally symmetric. To illustrate these two methods, we begin with Maxwell's equations in 3D cylindrical coordinates, given by:

$$\frac{\partial \mathcal{E}_r}{\partial t} = \frac{1}{\epsilon}\left[\frac{1}{r}\frac{\partial \mathcal{H}_z}{\partial \phi} - \frac{\partial \mathcal{H}_\phi}{\partial z}\right] \qquad \frac{\partial \mathcal{H}_r}{\partial t} = \frac{1}{\mu}\left[\frac{\partial \mathcal{E}_\phi}{\partial z} - \frac{1}{r}\frac{\partial \mathcal{E}_z}{\partial \phi}\right]$$

$$\frac{\partial \mathcal{E}_\phi}{\partial t} = \frac{1}{\epsilon}\left[\frac{\partial \mathcal{H}_r}{\partial z} - \frac{\partial \mathcal{H}_z}{\partial r}\right] \qquad \frac{\partial \mathcal{H}_\phi}{\partial t} = \frac{1}{\mu}\left[\frac{\partial \mathcal{E}_z}{\partial r} - \frac{\partial \mathcal{E}_r}{\partial z}\right] \qquad (4.28)$$

$$\frac{\partial \mathcal{E}_z}{\partial t} = \frac{1}{\epsilon}\frac{1}{r}\left[\frac{\partial}{\partial r}(r\mathcal{H}_\phi) - \frac{\partial \mathcal{H}_r}{\partial \phi}\right] \qquad \frac{\partial \mathcal{H}_z}{\partial t} = \frac{1}{\mu}\frac{1}{r}\left[\frac{\partial \mathcal{E}_r}{\partial \phi} - \frac{\partial}{\partial r}(r\mathcal{E}_\phi)\right]$$

First, let us consider the case where there is no variation in the z direction, i.e., the two-dimensional polar coordinate case. This $r - \phi$ coordinate system is useful when simulating structures that lend themselves to circular cross-sections, such as circular waveguides or coaxial cables [4, 5]. Setting all derivatives with respect to z to zero and grouping the equations, we have, just as in the Cartesian system, two modes; the TE mode:

$$\frac{\partial \mathcal{E}_r}{\partial t} = \frac{1}{\epsilon}\left[\frac{1}{r}\frac{\partial \mathcal{H}_z}{\partial \phi}\right] \qquad (4.29a)$$

TE mode, $\partial/\partial z = 0$
$$\frac{\partial \mathcal{E}_\phi}{\partial t} = -\frac{1}{\epsilon}\left[\frac{\partial \mathcal{H}_z}{\partial r}\right] \qquad (4.29b)$$

$$\frac{\partial \mathcal{H}_z}{\partial t} = \frac{1}{\mu}\left[\frac{1}{r}\frac{\partial \mathcal{E}_r}{\partial \phi} - \frac{1}{r}\frac{\partial(r\mathcal{E}_\phi)}{\partial r}\right] \qquad (4.29c)$$

and the TM mode:

$$\frac{\partial \mathcal{H}_r}{\partial t} = -\frac{1}{\mu}\left[\frac{1}{r}\frac{\partial \mathcal{E}_z}{\partial \phi}\right] \qquad (4.30a)$$

TM mode, $\partial/\partial z = 0$
$$\frac{\partial \mathcal{H}_\phi}{\partial t} = \frac{1}{\mu}\left[\frac{\partial \mathcal{E}_z}{\partial r}\right] \qquad (4.30b)$$

$$\frac{\partial \mathcal{E}_z}{\partial t} = \frac{1}{\epsilon}\left[\frac{1}{r}\frac{\partial(r\mathcal{H}_\phi)}{\partial r} - \frac{1}{r}\frac{\partial \mathcal{H}_r}{\partial \phi}\right]. \qquad (4.30c)$$

As before, the TE designation implies that the singular $\overline{\overline{\mathcal{H}}}$ component ($\mathcal{H}_z$ in this case) in normal to the plane of the 2D coordinate system; or rather, that the $\overline{\overline{\mathcal{E}}}$ components are in the plane. Using the unit cells shown in Figure 4.13, we can discretize Equations (4.29) and (4.30) in a straightforward manner, following a procedure exactly similar to the Cartesian case, except for the $(1/r)$ factors. While these might seem at first glance to be a potential difficulty, it turns out that the $(1/r)$ terms are quite simple: one merely takes the value of r at the current location in the grid wherever r appears in the equations. Similarly, the partial derivatives that include r terms, such as $\partial(r\mathcal{H}_\phi)/\partial r$ in Equation

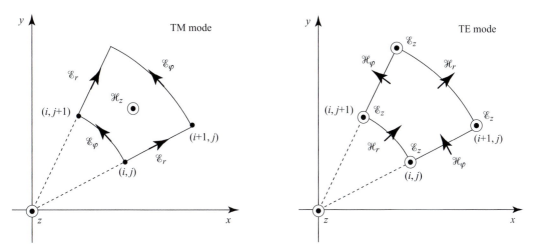

Figure 4.13 Two-dimensional FDTD cell in polar coordinates. As in Cartesian coordinates, fields are located using the pattern in Table 4.1; for example, $\mathcal{H}_\phi$ is located at an integer location in ϕ, but at half-integer locations in r and z.

(4.30c), must be treated as if the "field" $r\mathcal{H}_\phi$ is to be discretized; hence, our second-order approximation of this partial derivative is:

$$\left.\frac{\partial(r\mathcal{H}_\phi)}{\partial r}\right|_{i+1/2,j} \simeq \frac{r_{i+1}\mathcal{H}_\phi|_{i+1,j} - r_{i,j}\mathcal{H}_\phi|_{i,j}}{\Delta r} + O(\Delta r^2). \tag{4.31}$$

Thus, including these partial derivatives, we find the update equations for the TE mode by discretizing Equations (4.29):

$$\mathcal{E}_r\big|_{i+1/2,j}^{n+1} = \mathcal{E}_r\big|_{i+1/2,j}^{n} + \frac{\Delta t}{\epsilon_{i+1/2,j}} \left[\frac{\mathcal{H}_z\big|_{i+1/2,j+1/2}^{n+1/2} - \mathcal{H}_z\big|_{i+1/2,j-1/2}^{n+1/2}}{r_{i+1/2}\Delta\phi}\right] \tag{4.32a}$$

$$\mathcal{E}_\phi\big|_{i,j+1/2}^{n+1} = \mathcal{E}_\phi\big|_{i,j+1/2}^{n} - \frac{\Delta t}{\epsilon_{i,j+1/2}} \left[\frac{\mathcal{H}_z\big|_{i+1/2,j+1/2}^{n+1/2} - \mathcal{H}_z\big|_{i-1/2,j+1/2}^{n+1/2}}{\Delta r}\right] \tag{4.32b}$$

$$\mathcal{H}_z\big|_{i+1/2,j+1/2}^{n+1/2} = \mathcal{H}_z\big|_{i+1/2,j+1/2}^{n-1/2} \tag{4.32c}$$

$$+ \frac{\Delta t}{\mu_{i+1/2,j+1/2}} \left[\frac{\mathcal{E}_r\big|_{i+1/2,j+1}^{n} - \mathcal{E}_r\big|_{i+1/2,j}^{n}}{r_{i+1/2}\Delta\phi}\right.$$

$$\left. - \frac{r_{i+1}\mathcal{E}_\phi\big|_{i+1,j+1/2}^{n} - r_i\mathcal{E}_\phi\big|_{i,j+1/2}^{n}}{r_{i+1/2}\Delta r}\right]$$

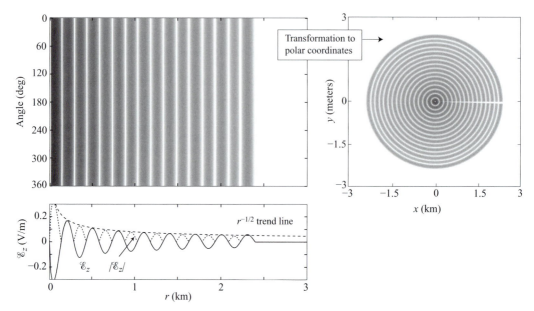

Figure 4.14 2D FDTD simulation in polar coordinates, TM mode. The calculations yield fields in a 2D $r - \phi$ grid, as shown in the top left; these are then transformed to $x - y$ to display in a polar plot at right. The 1D slice shows how the fields decay as $1/\sqrt{r}$.

where $r_i = i\Delta r$ and $r_{i+1/2} = (i + 1/2)\Delta r$, and i and j are respectively the indices in the r and ϕ coordinates; later, in 3D cylindrical coordinates, we will use k as the index in the z direction as usual. Similarly, for the TM mode we find:

$$\mathcal{H}_r \Big|_{i,j+1/2}^{n+1/2} = \mathcal{H}_r \Big|_{i,j+1/2}^{n-1/2} - \frac{\Delta t}{\mu_{i,j+1/2}} \left[\frac{\mathcal{E}_z \Big|_{i,j+1}^{n} - \mathcal{E}_z \Big|_{i,j}^{n}}{r_i \Delta \phi} \right] \tag{4.33a}$$

$$\mathcal{H}_\phi \Big|_{i+1/2,j}^{n+1/2} = \mathcal{H}_\phi \Big|_{i+1/2,j}^{n-1/2} + \frac{\Delta t}{\mu_{i+1/2,j}} \left[\frac{\mathcal{E}_z \Big|_{i,j}^{n} - \mathcal{E}_z \Big|_{i+1,j}^{n}}{\Delta r} \right] \tag{4.33b}$$

$$\mathcal{E}_z \Big|_{i,j}^{n+1} = \mathcal{E}_z \Big|_{i,j}^{n}$$
$$+ \frac{\Delta t}{\epsilon_{i,j}} \left[\frac{r_{i+1/2} \mathcal{H}_\phi \Big|_{i+1/2,j}^{n+1/2} - r_{i-1/2} \mathcal{H}_\phi \Big|_{i-1/2,j}^{n+1/2}}{r_i \Delta r} - \frac{\mathcal{H}_r \Big|_{i,j+1/2}^{n+1/2} - \mathcal{H}_r \Big|_{i,j-1/2}^{n+1/2}}{r_i \Delta \phi} \right]. \tag{4.33c}$$

Example polar simulation

Figure 4.14 shows an example simulation in 2D polar coordinates for the TM mode, in which we have directly discretized Equations (4.33). For this example, we excite an $\mathcal{E}_z$ source at the 'center' of a circular "grid" with center frequency $f_0 = 1$ MHz. We choose $\Delta r = 15$ m, so that we have 20 cells per wavelength in r, and we break the azimuthal grid into 200 cells, so that $\Delta\phi = 1.8$ degrees or 0.0314 radians. The $\mathcal{H}_r$, $\mathcal{H}_\phi$, and $\mathcal{E}_z$ fields are stored in 2D arrays of 200×200 grid points, as they were for Cartesian coordinates; what this means is that the spatial distance in the azimuthal direction grows with r. For example, at the first grid point from the center where $r = \Delta r$, the steps in ϕ have length $\Delta r \Delta\phi$ (in radians), which in our case is only 0.47 m. At the outer edge of the space where $r = 200\Delta r$, the steps in ϕ are 94.2 m! Hence, while this grid is *orthogonal*, since all of the coordinate vectors are orthogonal to each other, it is *non-uniform*.

Recall that for Cartesian coordinates, we found that stability was ensured in 1D when, for a given spatial grid size Δx, Δt was chosen such that $v_p \Delta t/\Delta x \leq 1$. In 2D, this requirement becomes more strict; if we choose $\Delta x \leq \Delta y$, we require $v_p \Delta t/\Delta x \leq 1/\sqrt{2}$; i.e., the stability criterion is set by the smallest spatial grid size. Here, our smallest grid cells are those closest to the center; with $\Delta r = 15$ m and $r\Delta\phi \geq 0.47$ m, we require approximately $v_p \Delta t/r\Delta\phi \leq 1/\sqrt{2}$. The strict 2D stability criterion will be derived in detail in Chapter 5.

The resulting $\mathcal{E}_z$ field in $r - \phi$ coordinates is shown in Figure 4.14, showing a snapshot at time $t = 8$ μs. By transforming the coordinates into $x - y$ pairs, we plot these fields in their correct polar pattern on the right. As one can imagine by comparing the left (uniform) grid with the right (nonuniform) grid, there is a much higher density of points at the center of the grid compared to the edges! As such, this scheme is a simple example of a *nonuniform* grid. The lower panel shows the 1D slice through a radius of the field pattern, showing the $r^{-1/2}$ falloff of the fields with distance as expected.

Finally, note the discontinuity between the first and last array indices in ϕ. In our polar coordinate system in Figure 4.13, for the TM mode, the $\mathcal{E}_z$ component on the x-axis (i.e., at $\phi = 0$) requires the $\mathcal{H}_r$ component *below* (i.e., at $\phi = -\Delta\phi/2$) in its update equation. This minor correction can be implemented quite easily in the update equations; similar corrections are required for the TE mode where the $\mathcal{E}_r$ component is on the $\phi = 0$ axis.

4.6.2   2D cylindrical coordinates

A second situation that arises in many problems is that of cylindrical symmetry, in which there is no variation in the azimuthal (i.e., ϕ) direction, but fields vary with r and z. This scenario might occur, for example, with a vertically oriented antenna source, whose radiation pattern is azimuthally symmetric; however, the medium and scattering objects must also be azimuthally symmetric about the same axis. In this scenario, we set derivatives with respect to ϕ to zero in Equations (4.33), and again find two separable modes:

$$\frac{\partial\mathcal{H}_r}{\partial t} = \frac{1}{\mu}\left[\frac{\partial\mathcal{E}_\phi}{\partial z}\right] \tag{4.34a}$$

$$\text{TM mode}, \partial/\partial\phi = 0 \qquad \frac{\partial\mathcal{E}_\phi}{\partial t} = \frac{1}{\epsilon}\left[\frac{\partial\mathcal{H}_r}{\partial z} - \frac{\partial\mathcal{H}_z}{\partial r}\right] \tag{4.34b}$$

$$\frac{\partial\mathcal{H}_z}{\partial t} = -\frac{1}{\mu}\frac{1}{r}\left[\frac{\partial}{\partial r}(r\mathcal{E}_\phi)\right] \tag{4.34c}$$

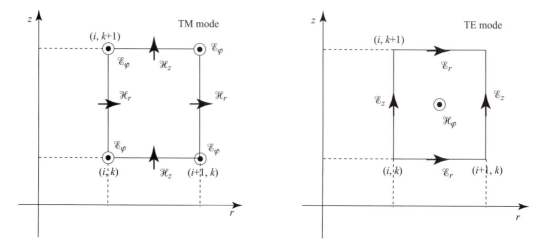

Figure 4.15 Two-dimensional FDTD cell in 2D cylindrical coordinates, with $\partial/\partial\phi = 0$. The same pattern of field placement holds as described in Figure 4.13.

and

$$\frac{\partial \mathscr{E}_r}{\partial t} = -\frac{1}{\epsilon}\left[\frac{\partial \mathscr{H}_\phi}{\partial z}\right] \qquad (4.35a)$$

TE mode, $\partial/\partial\phi = 0$

$$\frac{\partial \mathscr{H}_\phi}{\partial t} = \frac{1}{\mu}\left[\frac{\partial \mathscr{E}_z}{\partial r} - \frac{\partial \mathscr{E}_r}{\partial z}\right] \qquad (4.35b)$$

$$\frac{\partial \mathscr{E}_z}{\partial t} = \frac{1}{\epsilon}\frac{1}{r}\left[\frac{\partial}{\partial r}(r\mathscr{H}_\phi)\right]. \qquad (4.35c)$$

Again, this latter set of equations is designated the TE mode because the components of $\overline{\mathscr{E}}$ are in the $r - z$ plane of the simulation. These equations can be discretized in the same way as the previous case, using the unit cells shown in Figure 4.15. This procedure yields the update equations, first for the TM mode:

$$\mathscr{H}_r\Big|_{i,k+1/2}^{n+1/2} = \mathscr{H}_r\Big|_{i,k+1/2}^{n-1/2} - \frac{\Delta t}{\mu_{i,k+1/2}}\left[\frac{\mathscr{E}_\phi\Big|_{i,k+1}^{n} - \mathscr{E}_\phi\Big|_{i,k}^{n}}{\Delta z}\right] \qquad (4.36a)$$

$$\mathscr{E}_\phi\Big|_{i,k}^{n+1} = \mathscr{E}_\phi\Big|_{i,k}^{n} + \frac{\Delta t}{\epsilon_{i,k}}\left[\frac{\mathscr{H}_r\Big|_{i,k+1/2}^{n+1/2} - \mathscr{H}_r\Big|_{i,k-1/2}^{n+1/2}}{\Delta z} - \frac{\mathscr{H}_z\Big|_{i+1/2,k}^{n+1/2} - \mathscr{H}_z\Big|_{i-1/2,k}^{n+1/2}}{\Delta r}\right] \qquad (4.36b)$$

$$\mathscr{H}_z\Big|_{i+1/2,k}^{n+1/2} = \mathscr{H}_z\Big|_{i+1/2,k}^{n-1/2} - \frac{\Delta t}{\mu_{i+1/2,k}}\left[\frac{r_{i+1}\mathscr{E}_\phi\Big|_{i+1,k}^{n} - r_i\mathscr{E}_\phi\Big|_{i,k}^{n}}{r_{i+1/2}\,\Delta r}\right], \qquad (4.36c)$$

and for the TE mode:

$$
\left.\mathcal{E}_r\right|_{i+1/2,k}^{n+1} = \left.\mathcal{E}_r\right|_{i+1/2,k}^{n} + \frac{\Delta t}{\epsilon_{i+1/2,k}} \left[\frac{\left.\mathcal{H}_\phi\right|_{i+1/2,k+1/2}^{n+1/2} - \left.\mathcal{H}_\phi\right|_{i+1/2,k-1/2}^{n+1/2}}{\Delta z} \right] \tag{4.37a}
$$

$$
\left.\mathcal{H}_\phi\right|_{i+1/2,k+1/2}^{n+1/2} = \left.\mathcal{H}_\phi\right|_{i+1/2,k+1/2}^{n-1/2} \tag{4.37b}
$$

$$
+ \frac{\Delta t}{\mu_{i+1/2,k+1/2}} \left[\frac{\left.\mathcal{E}_z\right|_{i+1,k+1/2}^{n} - \left.\mathcal{E}_z\right|_{i,k+1/2}^{n}}{\Delta r} - \frac{\left.\mathcal{E}_r\right|_{i+1/2,k+1}^{n} - \left.\mathcal{E}_r\right|_{i+1/2,k}^{n}}{\Delta z} \right]
$$

$$
\left.\mathcal{E}_z\right|_{i,k+1/2}^{n+1} = \left.\mathcal{E}_z\right|_{i,k+1/2}^{n} + \frac{\Delta t}{\epsilon_{i,k+1/2}} \left[\frac{r_{i+1/2} \left.\mathcal{H}_\phi\right|_{i+1/2,k+1/2}^{n+1/2} - r_{i-1/2} \left.\mathcal{H}_\phi\right|_{i-1/2,k+1/2}^{n+1/2}}{r_i \, \Delta r} \right]. \tag{4.37c}
$$

A problem with this algorithm arises when one considers the on-axis $\mathcal{E}_z$ (in the TE mode) or $\mathcal{H}_z$ (in the TM mode) fields, i.e. at $r = 0$. Clearly, these update equations will not work due to the $1/r$ terms, leading to division by zero. A correction to the on-axis fields, however, is very straightforward. In the TE mode, for example, consider Ampère's law, Equation (2.3), applied in its integral form to a small circular region enclosing the axis at a radius $r = \Delta r/2$, at the location of $\left.\mathcal{H}_\phi\right|_{i=1/2,k+1/2}$, as shown in the left panel of Figure 4.16. In the absence of any source current, we have:

$$
\oint_C \overline{\mathcal{H}} \cdot d\mathbf{l} = \int_S \frac{\partial \overline{\mathcal{D}}}{\partial t} \cdot d\mathbf{s}.
$$

For our purposes here, we assume that $\mathcal{H}_\phi$ does not vary along the curve C, which is a small circle in ϕ at $r = \Delta r/2$ (i.e., our pre-assumed azimuthal symmetry); and that $\mathcal{E}_z$ does not vary inside the surface S, which is the area enclosed by C. Since $\mathcal{E}_z$ is azimuthally symmetric, this latter assumption implies that $\mathcal{E}_z$ also does not vary with r between $r = 0$ and $r = \Delta r/2$. Thus we have:

$$
\left.\mathcal{H}_\phi\right|_{1/2,k+1/2}^{n+1/2} (2\pi r_{i=1/2}) = \epsilon \left.\frac{\partial \mathcal{E}_z}{\partial t}\right|_{0,k+1/2}^{n+1/2} (\pi r_{i=1/2}^2). \tag{4.38}
$$

Noting that $r_{i=1/2} = \Delta r/2$, we find the update equation for $\mathcal{E}_z$ at $r = 0$:

$$
\left.\mathcal{E}_z\right|_{0,k+1/2}^{n+1} = \left.\mathcal{E}_z\right|_{0,k+1/2}^{n} + \frac{4\Delta t}{\epsilon_0 \Delta r} \left.\mathcal{H}_\phi\right|_{1/2,k+1/2}^{n+1/2}. \tag{4.39}
$$

While this equation appears to suggest that the on-axis field depends only on one value of $\mathcal{H}_\phi$, the application of the integral form of Ampère's law shows that it depends on the value of $\mathcal{H}_\phi$ at the $r = \Delta r/2$ grid location, but at *all* locations in azimuth, over

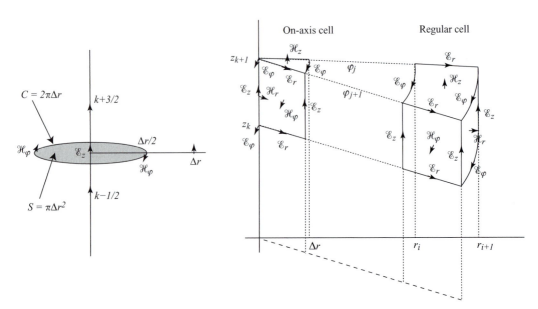

Figure 4.16 Cylindrical coordinate system used in FDTD. The full 3D cylindrical coordinate system is shown at right. At left, the method for determining the on-axis $\mathcal{E}_z$ component for the 2D TE mode, by directly applying Ampère's law.

which the field does not vary. We will see a similar analysis in the 3D cylindrical case next, in which $\mathcal{H}_\phi$ is not azimuthally symmetric.

Example cylindrical simulation

Figure 4.17 shows an example simulation using this TE mode cylindrical coordinate algorithm. This example uses grid cells of $\lambda/20$ with $\lambda = 3$ m (a frequency of 100 MHz). We excite $\mathcal{E}_z$ at $r = 0, z = 0$, and set $\mathcal{E}_z$ at the next five grid cells both above and below the source to zero; this emulates a "half-wave" dipole antenna, 10 grid cells long, with the source excited at the center as a voltage $\mathcal{V} = \mathcal{E}_{z,\text{in}} \Delta z$. Note that attempting to implement a small dipole in this scenario will not work as one would expect from antenna theory; the "small dipole" may be only $\lambda/20$ or 15 cm in length, but it is also 15 cm wide! In Chapter 7 we will introduce a method to circumvent this problem known as the thin-wire approximation.

The right panel in Figure 4.17 plots the simulated dipole radiation pattern, measured by sampling the $\mathcal{E}_z$ field amplitude along a circle of constant radius, shown by the dashed line, and then normalizing. The ideal half-wave dipole radiation pattern is also shown for comparison, given by [e.g., 6, p. 61]:

$$F(\theta) = \frac{\cos[(\pi/2)\cos\theta]}{\sin^2\theta}.$$

It is thus evident that our half-wave dipole simulation approximately emulates the behavior of a real half-wave dipole.

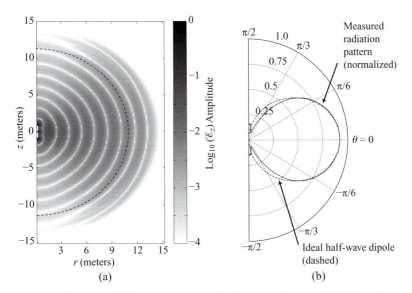

Figure 4.17 Example FDTD simulation in cylindrical coordinates. (a) 2D $\mathscr{E}_z$ field pattern in $r - z$ space at a snapshot in time. (b) The resulting radiation pattern as measured along the dashed line in the left panel. The dashed line on the right is the ideal half-wave dipole radiation pattern.

Since we have chosen a dipole radiator at the center of what is effectively a 3D space with azimuthal symmetry, our simulation is nearly identical to the 3D case in Figure 4.9, except that we have reduced the simulation space from $100 \times 100 \times 100$ grid cells to 100×200 grid cells, and the number of field components from six to three. Hence, for problems which can be approximated with azimuthal symmetry, this 2D formulation is invaluable.

4.6.3   3D cylindrical coordinates

Implementation of the 3D FDTD method in cylindrical coordinates simply requires discretization of the full 3D cylindrical Maxwell's equations, given in Equation (4.28) above. As an example, we will show the discretization of the $\mathscr{E}_z$ component equation, repeated below:

$$\frac{\partial \mathscr{E}_z}{\partial t} = \frac{1}{\epsilon} \frac{1}{r} \left[\frac{\partial}{\partial r} (r \mathscr{H}_\phi) - \frac{\partial \mathscr{H}_r}{\partial \phi} \right].$$

The grid cell for the 3D cylindrical case is shown in Figure 4.16. As in Cartesian coordinates, the locations of field components follow the same pattern as shown in Table 4.1, where now (i, j, k) refer to indices in (r, ϕ, z). Applying the same discretization rules as for the 2D cylindrical case, we find the update equation:

$$
\left. \mathcal{E}_z \right|^{n+1}_{i,j,k+1/2} = \left. \mathcal{E}_z \right|^{n}_{i,j,k+1/2}
\tag{4.40}
$$

$$
+ \frac{\Delta t}{\epsilon} \left[\frac{r_{i+1/2}\left.\mathcal{H}_\phi\right|^{n+1/2}_{i+1/2,j,k+1/2} - r_{i-1/2}\left.\mathcal{H}_\phi\right|^{n+1/2}_{i-1/2,j,k+1/2}}{r_i\,\Delta r} \right.
$$

$$
\left. - \frac{\left.\mathcal{H}_r\right|^{n+1/2}_{i,j+1/2,k+1/2} - \left.\mathcal{H}_r\right|^{n+1/2}_{i,j-1/2,k+1/2}}{r_i\,\Delta\phi} \right].
$$

Similar discretization procedures lead to update equations for the other components in Equation (4.28); the remaining five update equations are given by:

$$
\left. \mathcal{E}_r \right|^{n+1}_{i+1/2,j,k} = \left. \mathcal{E}_r \right|^{n}_{i+1/2,j,k}
\tag{4.41a}
$$

$$
+ \frac{\Delta t}{\epsilon} \left[\frac{\left.\mathcal{H}_z\right|^{n+1/2}_{i+1/2,j+1/2,k} - \left.\mathcal{H}_z\right|^{n+1/2}_{i+1/2,j-1/2,k}}{r_{i+1/2}\,\Delta\phi} \right.
$$

$$
\left. - \frac{\left.\mathcal{H}_\phi\right|^{n+1/2}_{i+1/2,j,k+1/2} - \left.\mathcal{H}_\phi\right|^{n+1/2}_{i+1/2,j,k-1/2}}{\Delta z} \right]
$$

$$
\left. \mathcal{E}_\phi \right|^{n+1}_{i,j+1/2,k} = \left. \mathcal{E}_\phi \right|^{n}_{i,j+1/2,k}
\tag{4.41b}
$$

$$
+ \frac{\Delta t}{\epsilon} \left[\frac{\left.\mathcal{H}_r\right|^{n+1/2}_{i,j+1/2,k+1/2} - \left.\mathcal{H}_r\right|^{n+1/2}_{i,j+1/2,k-1/2}}{\Delta z} \right.
$$

$$
\left. - \frac{\left.\mathcal{H}_z\right|^{n+1/2}_{i+1/2,j+1/2,k} - \left.\mathcal{H}_z\right|^{n+1/2}_{i-1/2,j+1/2,k}}{\Delta r} \right]
$$

$$
\left. \mathcal{H}_r \right|^{n+1/2}_{i,j+1/2,k+1/2} = \left. \mathcal{H}_r \right|^{n-1/2}_{i,j+1/2,k+1/2}
\tag{4.41c}
$$

$$
+ \frac{\Delta t}{\mu} \left[\frac{\left.\mathcal{E}_\phi\right|^{n}_{i,j+1/2,k+1} - \left.\mathcal{E}_\phi\right|^{n}_{i,j+1/2,k}}{\Delta z} \right.
$$

$$
\left. - \frac{\left.\mathcal{E}_z\right|^{n}_{i,j+1,k+1/2} - \left.\mathcal{E}_z\right|^{n}_{i,j,k+1/2}}{r_i\,\Delta\phi} \right]
$$

$$
\mathcal{H}_\phi\Big|_{i+1/2,j,k+1/2}^{n+1/2} = \mathcal{H}_\phi\Big|_{i+1/2,j,k+1/2}^{n-1/2} \tag{4.41d}
$$

$$
+ \frac{\Delta t}{\mu} \left[\frac{\mathcal{E}_z\Big|_{i+1,j,k+1/2}^{n} - \mathcal{E}_z\Big|_{i,j,k+1/2}^{n}}{\Delta r} \right.
$$

$$
\left. - \frac{\mathcal{E}_r\Big|_{i+1/2,j,k+1}^{n} - \mathcal{E}_r\Big|_{i+1/2,j,k}^{n}}{\Delta z} \right]
$$

$$
\mathcal{H}_z\Big|_{i+1/2,j+1/2,k}^{n+1/2} = \mathcal{H}_z\Big|_{i+1/2,j+1/2,k}^{n-1/2} \tag{4.41e}
$$

$$
+ \frac{\Delta t}{\mu} \left[\frac{\mathcal{E}_r\Big|_{i+1/2,j+1,k}^{n} - \mathcal{E}_r\Big|_{i+1/2,j,k}^{n}}{r_{i+1/2}\Delta\phi} \right.
$$

$$
\left. - \frac{r_{i+1}\mathcal{E}_\phi\Big|_{i+1,j+1/2,k}^{n} - r_i\mathcal{E}_\phi\Big|_{i,j+1/2,k}^{n}}{r_{i+1/2}\Delta r} \right].
$$

Note that we have assumed a constant ϵ here, but an inhomogeneous medium can easily be implemented as we have shown earlier, by using the values of ϵ and μ at the grid location at which the equation is discretized ($i, j, k + 1/2$ in Equation 4.40).

As we can see from Figure 4.16, however, there is a problem on the axis, as there was in the 2D cylindrical case. In 3D, if we choose the axis to align with $\mathcal{E}_z$, then we have $\mathcal{E}_z$, $\mathcal{H}_r$, and $\mathcal{E}_\phi$ components on the axis to deal with, as shown in Figure 4.16.

To update $\mathcal{E}_z$ on the axis, we apply Ampère's law directly, similar to the 2D case. In this case, however, we must integrate $\mathcal{H}_\phi$ along the ϕ direction, at $r = \Delta r/2$, rather than simply multiplying by $2\pi(\Delta r/2)$. Integrating in continous space is simply a sum in discrete space, so we find [7]:

$$
\mathcal{E}_z\Big|_{0,j,k+1/2}^{n+1} = \mathcal{E}_z\Big|_{0,j,k+1/2}^{n} + \frac{4\Delta t}{N_\phi\epsilon_0\Delta r} \sum_{p=1}^{N_\phi} \mathcal{H}_\phi\Big|_{i=1/2,p,k+1/2}^{n+1/2}. \tag{4.42}
$$

Note that the summation above, when divided by N_ϕ (the number of grid points in ϕ), is actually the *average* value of $\mathcal{H}_\phi$ around the contour; as such, Equation (4.42) is nearly identical to Equation (4.39).

Now, despite the fact that $\mathcal{E}_\phi$ and $\mathcal{H}_r$ have components on the axis, these axial components need not be computed for the remainder of the space to be updated normally. To see this, consider the on-axis cell in Figure 4.16. The fields that could require knowledge of the on-axis fields are $\mathcal{E}_r$, $\mathcal{H}_z$, and $\mathcal{H}_\phi$ at $r = \Delta r/2$. Now consider Equations (4.41): in Equation (4.41a), $\mathcal{E}_r$ does not require any of the on-axis field components; in Equation (4.41d), $\mathcal{H}_\phi$ requires only $\mathcal{E}_z$ on the axis, which we have updated above in Equation (4.42). Finally, in Equation (4.41e), $\mathcal{H}_z$ requires $\mathcal{E}_\phi$ on the axis in the last term of the equation, but this component is multiplied by r_i which is equal to zero on the axis, so it is not needed. In summary, the full set of update equations can be used for the half-step away from the $r = 0$ axis, and the $\mathscr{E}_\phi$ and $\mathscr{H}_r$ components on-axis are simply ignored.

It is interesting to note that these axial $\mathscr{H}_r$ and $\mathscr{E}_\phi$ components both point away from the axis, and thus could equally be considered $\mathscr{H}_\phi$ and $\mathscr{E}_r$ components. As such, the computed $\mathscr{H}_\phi$ and $\mathscr{E}_r$ fields at $r = \Delta r/2$ can be used to approximate these on-axis fields if necessary.

Finally, note that in addition to the axial correction described above, the 3D cylindrical coordinate problem has the same 2π discontinuity in ϕ which must be accounted for in the update equations, as in the 2D polar coordinate case earlier.

4.6.4   3D spherical coordinates

Discretization to find the FDTD update equations in spherical coordinates is somewhat more complicated than cylindrical coordinates, but follows the same principles. The six component Maxwell's equations in spherical coordinates are given below for simple media [e.g., 3]:

$$\epsilon \frac{\partial \mathscr{E}_r}{\partial t} + \sigma \mathscr{E}_r = \frac{1}{r \sin\theta} \left[\frac{\partial}{\partial \theta} (\mathscr{H}_\phi \sin\theta) - \frac{\partial \mathscr{H}_\theta}{\partial \phi} \right] \tag{4.43a}$$

$$\epsilon \frac{\partial \mathscr{E}_\theta}{\partial t} + \sigma \mathscr{E}_\theta = \frac{1}{r} \left[\frac{1}{\sin\theta} \frac{\partial \mathscr{H}_r}{\partial \phi} - \frac{\partial}{\partial r} (r \mathscr{H}_\phi) \right] \tag{4.43b}$$

$$\epsilon \frac{\partial \mathscr{E}_\phi}{\partial t} + \sigma \mathscr{E}_\phi = \frac{1}{r} \left[\frac{\partial}{\partial r} (r \mathscr{H}_\theta) - \frac{\partial \mathscr{H}_r}{\partial \theta} \right] \tag{4.43c}$$

$$\mu \frac{\partial \mathscr{H}_r}{\partial t} = -\frac{1}{r \sin\theta} \left[\frac{\partial}{\partial \theta} (\mathscr{E}_\phi \sin\theta) - \frac{\partial \mathscr{E}_\theta}{\partial \phi} \right] \tag{4.43d}$$

$$\mu \frac{\partial \mathscr{H}_\theta}{\partial t} = -\frac{1}{r} \left[\frac{1}{\sin\theta} \frac{\partial \mathscr{E}_r}{\partial \phi} - \frac{\partial}{\partial r} (r \mathscr{E}_\phi) \right] \tag{4.43e}$$

$$\mu \frac{\partial \mathscr{H}_\phi}{\partial t} = -\frac{1}{r} \left[\frac{\partial}{\partial r} (r \mathscr{E}_\theta) - \frac{\partial \mathscr{E}_r}{\partial \theta} \right]. \tag{4.43f}$$

There certainly exist circumstances where a 2D spherical coordinate system is useful; for example, the simulation shown in Figure 4.17 has azimuthal symmetry ($\partial/\partial\phi = 0$), but would be more accurately described in $r - \theta$ space rather than $r - z$ space. The method for creating a 2D simulation follows the method described above for the cylindrical coordinate case: set $\partial/\partial\phi = 0$ in the equations above, organize the resulting six equations into TE and TM modes, and then discretize as we have done throughout this chapter. In the interest of brevity, we will skip the 2D spherical coordinate case and jump directly into the 3D discretization process.

The derivation of the FDTD algorithm in spherical coordinates was first presented by R. Holland in 1983 [8]. Figure 4.18 shows the spherical coordinate system and the unit cell for FDTD. Noting that in spherical coordinates, the (i, j, k) subscripts refer to (r, θ, ϕ), respectively, the field vectors are all drawn pointing toward the increasing

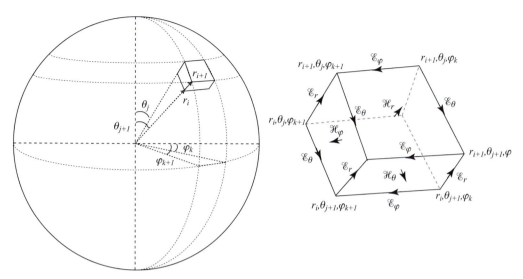

Figure 4.18 Spherical coordinate system used in FDTD. The left panel shows how the coordinate system is drawn; the right panel shows an expanded view of a unit cell. Adapted from [8].

coordinate direction. Then, the locations of field components follow the same logic from Table 4.1; i.e., the $\mathcal{E}_n$ components fall on the half-integer grid location for n and the integer grid location for the other two coordinates, and vice versa for the $\overline{\mathcal{H}}$ field components. For example, $\mathcal{E}_r$ is located at $(i + 1/2, j, k)$ and $\mathcal{H}_r$ is located at $(i, j + 1/2, k + 1/2)$. Note, however, that our cell size is not strictly $\Delta r \times \Delta\phi \times \Delta\theta$; those last two do not even have spatial units. Rather, the grid cell has dimensions Δr, $r\Delta\theta$, and $r \sin\theta \Delta\phi$. Thus, if we use constant Δr, $\Delta\theta$, and $\Delta\phi$, the grid cells grow with increasing r, and, to no surprise, we have another *nonuniform* grid.

As in the cylindrical coordinate case, we deal with the r and θ terms on the right side of Equations (4.43) by simply using the values of r and θ at the location that the equation is discretized. Similarly, the partial derivatives that include r or $\sin\theta$ must be treated in the same manner as Equation (4.31) earlier.

Applying the methods from this chapter to Equation (4.43a), and assuming $\sigma = 0$ for simplicity, we find its update equation:

$$\mathcal{E}_r\Big|_{i+1/2,j,k}^{n+1} = \mathcal{E}_r\Big|_{i+1/2,j,k}^{n}$$

$$+ \frac{\Delta t}{\epsilon \, r_{i+1/2} \sin\theta_j} \left[\frac{\mathcal{H}_\phi\Big|_{i+1/2,j+1/2,k}^{n+1/2} \sin\theta_{j+1/2} - \mathcal{H}_\phi\Big|_{i+1/2,j-1/2,k}^{n+1/2} \sin\theta_{j-1/2}}{\Delta\theta} \right.$$

$$\left. - \frac{\mathcal{H}_\theta\Big|_{i+1/2,j,k+1/2}^{n+1/2} - \mathcal{H}_\theta\Big|_{i+1/2,j,k-1/2}^{n+1/2}}{\Delta\phi} \right].$$

$$(4.44)$$

Notice that this equation requires the value of r at a half-step, coincident in the i index with the $\mathcal{E}_r$ field; and it also requires the values of θ at j and $j \pm 1/2$. Of course, these values are readily available and are simply defined by the geometry of the simulation space. For a regular grid, $r_i = i \Delta r$ and $\theta_j = j \Delta \theta$. The other update equations follow from similar discretization of Equations (4.43b–f).

Now, recall that in polar coordinates, we needed to take care of the 2π discontinuity in ϕ; the same issue exists in spherical coordinates, and must be accounted for in the same manner. In addition, recall that in the 2D polar and 3D cylindrical cases, special attention was given to field values at $r = 0$; this point is known as a *singularity*, since some component of the update equations has a zero in the denominator. In spherical coordinates, we have the same singularity at $r = 0$, which again requires special attention, which we shall deal with in a moment; in addition, however, there are problems at $\theta = 0$ and $\theta = \pi$, i.e., along the z-axis. Observe that in Equation (4.44), the term $\sin\theta$ appears in the denominator; since $\sin(0) = \sin(\pi) = 0$, we again have a singularity. As one might anticipate, this singularity also exists in the discretized equivalents of Equations (4.43b, d, and e), where $\sin\theta$ appears in the denominator.

Holland's [8] original formulation described a fix to the $\theta = 0$ ($j = 1$) and $\theta = \pi$ ($j = N_\theta$) problem. Note that the only field component that presents a problem on this axis is the $\mathcal{E}_r$ component. The solution is to apply the same Ampère's law contour integral around a small contour C enclosing the z-axis, either near the north pole or the south pole of the sphere in Figure 4.18, and evaluating the integral in ϕ as a summation,[4] we find the following corrected update equations at $\theta = 0$ and $\theta = \pi$:

$$\mathcal{E}_r\Big|_{i+1/2,1,k}^{n+1} = \mathcal{E}_r\Big|_{i+1/2,1,k}^{n} + \frac{\Delta t \, \sin\theta_{1/2}}{2\pi \, \epsilon r_{i+1/2}(1 - \cos\theta_{1/2})} \sum_{p=1}^{N_\phi} \mathcal{H}_\phi\Big|_{i+1/2,1,p}^{n+1/2} \Delta\phi \quad (4.45a)$$

$$\mathcal{E}_r\Big|_{i+1/2,N_\theta,k}^{n+1} = \mathcal{E}_r\Big|_{i+1/2,N_\theta,k}^{n} + \frac{\Delta t \, \sin\theta_{N_\theta-1/2}}{2\pi \epsilon \, r_{i+1/2}(1 + \cos\theta_{N_\theta-1/2})} \sum_{p=1}^{N_\phi} \mathcal{H}_\phi\Big|_{i+1/2,N_\theta,p}^{n+1/2} \Delta\phi$$

$$(4.45b)$$

where we note that $\theta_{1/2} = \Delta\theta/2$ and $\theta_{N_\theta-1/2} = \pi - \Delta\theta/2$.

Now, dealing with the origin ($r = 0$ or $i = 0$) is somewhat more complicated. If one imagines the unit cell in Figure 4.18 collapsed down to the origin, we end up with a small trapezoidal grid cell with an apex at the origin. Depending on where we put the origin, we may either have the triplet ($\mathcal{E}_r$, $\mathcal{H}_\theta$, $\mathcal{H}_\phi$) defined at the origin, or taking it a half-step further out in r, we have the complementary triplet ($\mathcal{H}_r$, $\mathcal{E}_\theta$, $\mathcal{E}_\phi$).

---

[4] Note in this case that the area of integration on the right-hand side of Ampère's law is the small "cap" on the top of a sphere, and not a simple circle. The area of a small cap on a sphere at radius r_i and angle θ_j is:

$$\int_0^{2\pi} \int_0^{\theta_j} r_i^2 \sin\theta \, d\theta d\phi = 2\pi r_i^2 (1 - \cos\theta_j).$$

A similar integral for the south pole cap of the sphere yields an analogous expression, except with a $(1 + \cos\theta_j)$ term. However, if we observe that $\cos(\pi - \theta_j) = -\cos\theta_j$, we find that the denominators in Equations (4.45) are the same.

In the rare cases where the origin is not part of the source radiator, the methods used to deal with this singularity involve solving a small region around the origin in Cartesian coordinates, and then mapping those fields back to spherical coordinates on the boundary of the small cube. The analysis of this method, however, is beyond the scope of our current discussion. We refer interested readers to the papers by Holland [8] and G. Liu et al. [9].

4.7 Summary

In this chapter, the interleaved leapfrog method from Chapter 3 was used to derive the FDTD algorithm in 1D, 2D, and 3D. In 1D, the FDTD algorithm for the TM mode is given by:

$$
\mathcal{H}_y\Big|_{i+1/2,j}^{n+1/2} = \mathcal{H}_y\Big|_{i+1/2,j}^{n-1/2} + \frac{\Delta t}{\mu_{i+1/2,j}\,\Delta x}\left[\mathcal{E}_z\Big|_{i+1,j}^{n} - \mathcal{E}_z\Big|_{i,j}^{n}\right]
$$

$$
\mathcal{E}_z\Big|_{i,j}^{n+1} = \mathcal{E}_z\Big|_{i,j}^{n} + \frac{\Delta t}{\epsilon_{i,j}\,\Delta x}\left[\mathcal{H}_y\Big|_{i+1/2,j}^{n+1/2} - \mathcal{H}_y\Big|_{i-1/2,j}^{n+1/2}\right],
$$

(1D TM mode)

with similar equations for the TE mode. In 2D, the FDTD algorithm for the TM mode is given by:

$$
\mathcal{H}_x\Big|_{i,j+1/2}^{n+1/2} = \mathcal{H}_x\Big|_{i,j+1/2}^{n-1/2} - \frac{\Delta t}{\mu_{i,j+1/2}\,\Delta y}\left[\mathcal{E}_z\Big|_{i,j+1}^{n} - \mathcal{E}_z\Big|_{i,j}^{n}\right]
$$

(2D TM mode)

$$
\mathcal{H}_y\Big|_{i+1/2,j}^{n+1/2} = \mathcal{H}_y\Big|_{i+1/2,j}^{n-1/2} + \frac{\Delta t}{\mu_{i+1/2,j}\,\Delta x}\left[\mathcal{E}_z\Big|_{i+1,j}^{n} - \mathcal{E}_z\Big|_{i,j}^{n}\right]
$$

$$
\mathcal{E}_z\Big|_{i,j}^{n+1} = \mathcal{E}_z\Big|_{i,j}^{n} + \frac{\Delta t}{\epsilon_{i,j}}\left[\frac{\mathcal{H}_y\Big|_{i+1/2,j}^{n+1/2} - \mathcal{H}_y\Big|_{i-1/2,j}^{n+1/2}}{\Delta x} - \frac{\mathcal{H}_x\Big|_{i,j+1/2}^{n+1/2} - \mathcal{H}_x\Big|_{i,j-1/2}^{n+1/2}}{\Delta y}\right],
$$

with similar equations for the TE mode. In 3D, there are no TE versus TM modes; the FDTD algorithm involves six update equations, given in Equations (4.18) and (4.19).

Update equations for lossy media were derived in Section 4.4; it was shown that the update equations for lossy media can be transcribed from the lossless equations, by simply replacing the coefficients C_1 and C_2 as given in Table 4.2.

Next, it was shown in Section 4.5 that the FDTD algorithm is divergence-free, i.e., it inherently satisfies Gauss's law that $\nabla \cdot \mathcal{D} = 0$. This is an important and appealing feature of the FDTD algorithm; zero divergence is not a feature that is inherent to any particular discretization scheme.

In Section 4.6 we introduced other orthogonal coordinate systems, namely cylindrical and spherical coordinates, and the discretization process that leads to FDTD update equations in these coordinate systems. We first introduced 2D polar and cylindrical coordinates, which are both commonly used where there is longitudinal or azimuthal symmetry in a problem. The discretization of Maxwell's equations in these coordinate systems again leads to TE and TM modes, but terms involving r or $1/r$ need to be handled carefully.

Similarly, the 3D cylindrical and spherical coordinate systems require special handling of r and $\sin\theta$ terms which appear in the update equations. In all cases, these terms lead to singularities at the origin or axis (where $r = 0$) and, in spherical coordinates, on the axis where $\theta = 0$ (i.e., the z-axis). These singularities are handled in the equations above by directly applying Ampère's law to a small contour around the z-axis.

4.8   Problems

4.1. **A lossy high-speed chip-to-chip interconnect.** Consider applying the inter-leaved leapfrog algorithm to solve for the voltage variation on a high-speed lossy interconnect, treating it as a lossy transmission line with $G = 0$, and $R = 150\ \Omega$-$(\text{cm})^{-1}$. Determine and plot the voltage distribution on the lossy line at $t = 0.5$ ns. Select appropriate values of Δx and Δt. Assume other line parameters and initial voltage distribution to be as in Problem 3.4 and use the same absorbing boundary condition specified therein. Comment on your results. Do you think this particular microstrip line is a good one to use for chip-to-chip transmission of 0.1 ns pulses over a 5-cm long interconnect on a circuit board? Repeat for $R = 50\ \Omega$-$(\text{cm})^{-1}$ and $R = 10\ \Omega$-$(\text{cm})^{-1}$; comment on the results.

Note: You are given $\mathcal{V}^0$, but no initial condition for $\mathcal{I}$. You will need to use another method to determine $\mathcal{I}^{1/2}$, the initial condition for the current, over the entire transmission line. This initial condition for $\mathcal{I}$ can have a considerable effect on the results.

4.2. **High-speed GaAs coplanar strip interconnect.** The per-unit-length parameters of a high-speed GaAs coplanar strip interconnect operating at 10 GHz are mea-sured to be $R \simeq 25\ \Omega$-$(\text{cm})^{-1}$, $C \simeq 0.902\ \text{pF}$-$(\text{cm})^{-1}$, $L \simeq 9.55\ \text{nH}$-$(\text{cm})^{-1}$, and $G = 0$. We shall study the propagation and attenuation on this line of a Gaussian pulse given by

$$\mathcal{V}^n_{i=1} = e^{-(n-3\tau)^2/\tau^2}$$

where $\tau\,\Delta t = 0.1$ ns.

(a) Use an appropriate algorithm and write a program to propagate the pulse on this line and measure the attenuation rate (in dB-m^{-1}). For time-harmonic signals, it is well known that the attenuation rate on a lossy transmission line is given by $\alpha = \mathfrak{Re}\{\sqrt{(R + j\omega L)(j\omega C)}\}$ (in np-m^{-1}), where ω is the frequency of operation. Compare the attenuation rate you measured with the time-harmonic attenuation rate (for the frequency range of your Gaussian pulse) and comment on your results.

(b) Assuming the coplanar strip interconnect to be 10-cm long, terminate it with a short circuit and run your simulation until after the Gaussian pulse is reflected and just before its front arrives back at the source point. Plot the variation with time of the propagating pulse as observed at fixed points on your grid, i.e., at $i = i_{max}/4$, $i = i_{max}/2$, and $i = 3i_{max}/4$, and compare its shape (by plotting them on top of one another) with that of the original Gaussian pulse. Note that the pulse will be observed twice at each of these locations, once before reflection and then again after reflection. Is the shape of the Gaussian pulse changing as it propagates? If so, why? Comment on your results.

(c) Repeat parts (a) and (b), but excite the simulation with the derivative of the Gaussian above. What effect does this have on results?

4.3. **1D wave incident on a dielectric slab.** Consider the propagation of a one-dimensional electromagnetic wave in a medium and its reflection from and transmission through a dielectric slab. Construct a 1D FDTD space with a lossless dielectric slab ($\epsilon_{r2} = 2$, $\mu_{r2} = 1$) at the center of your grid, free space ($\epsilon_{r1} = 1$, $\mu_{r1} = 1$) to the left of the slab, and another dielectric ($\epsilon_{r3} = 4$, $\mu_{r3} = 1$) to the right of the slab. You should allow for the length of the slab to be adjustable. Launch a Gaussian pulse from the left boundary of your FDTD space, which is given by

$$\left.\mathscr{E}_z\right|_{i=1}^n = e^{-(n-3\tau)^2/\tau^2}$$

where $\tau \Delta t = 0.8$ ns. Choose $\Delta x \geq \lambda_0/[20\sqrt{\epsilon_r}]$, where $\lambda_0 = 0.3$ m, $\Delta t = \Delta x/v_p$, and $i_{max}\Delta x = 3$ m. Once the tail end of the pulse leaves the source (i.e., when $n = 6\tau$), replace the source end with a simple boundary condition:

$$\left.\mathscr{E}_z\right|_1^{n+1} = \left.\mathscr{E}_z\right|_2^n + \left[\frac{v_p\Delta t - \Delta x}{v_p\Delta t + \Delta x}\right]\left[\left.\mathscr{E}_z\right|_2^{n+1} - \left.\mathscr{E}_z\right|_1^n\right].$$

Take the rightmost end of your FDTD space to be as far away as needed so that there are no reflections from that boundary, or implement a similar radiation boundary at $i = i_{max}$.

(a) Taking a slab width of $d = \lambda_0/[2\sqrt{\epsilon_{r2}}]$, write a program to propagate the wave on your FDTD space and observe the behavior of the pulse as it interacts with the slab. Can you measure the "reflection coefficient"? What is its value? What is the transmission coefficient? Plot the reflection and transmission coefficients as a function of frequency and compare to the analytical expression. Comment on the differences observed. Considering the finite precision of floating-point numbers on a computer, what is the effective "numerical bandwidth" of your pulse?

(b) Repeat (a) for a slab width of $d = \lambda_0/[4\sqrt{\epsilon_{r2}}]$. Comment on your results.

4.4. **2D FDTD solution of Maxwell's equations.** Write a program implementing the FDTD algorithm for the 2D TM mode. Assume square unit cells (i.e., $\Delta x = \Delta y$) and free space everywhere in the grid, and use a time step of $\Delta t = \Delta x/(v_p\sqrt{2})$. Assume the outer boundary of your grid to be surrounded by perfect electrically conducting sheets and terminate your mesh in electric field components set to be zero at all times. Excite a radially outgoing wave in the grid by implementing a hard source for a single electric field component at the center of the grid. For both of the sources specified below, display the electric and magnetic fields of the wave distribution across the grid at a number of time snapshots before and after the wave reaches the outermost grid boundary.

(a) First consider a Gaussian source waveform given as

$$
\mathcal{E}_z\Big|_{\text{imax}/2,\text{jmax}/2}^n = e^{-(n\Delta t - 1.8 \text{ ns})^2/(0.6 \text{ ns})^2}.
$$

Use a 50×50 grid with $\Delta x = \Delta y = 0.1\lambda_{\min}$, where $\lambda_{\min}$ is the minimum wavelength represented by the Gaussian source. How would you determine $\lambda_{\min}$?

(b) Now consider a sinusoidal source waveform which is instantly turned on at time $t = 0$, having a frequency of $f_0 = 30$ GHz. Let $\Delta x = \Delta y = \lambda_0/10$, and once again use a 50×50 grid. Why is the amplitude of the outgoing wave not constant?

(c) Repeat part (a) for the TE mode. To emulate PEC at the boundaries, note that the tangential $\overline{\mathcal{E}}$ field must be zero, and that this will be either $\mathcal{E}_x$ or $\mathcal{E}_y$ depending on the boundary in question.

Note that the two codes developed in this problem, for the TM and TE modes, will be useful in later problems in this and later chapters. The student will find it useful to write the code including parameters σ and σ_m, so that losses can be easily included later.

4.5. 3D rectangular resonator. Write a 3D FDTD algorithm using Equations (4.18) and (4.19) in a $100 \times 50 \times 20$ m box with perfectly conducting walls. Use $\Delta x = \Delta y = \Delta z = 1$ m.

(a) Excite this resonator with a sinusoidal $\mathcal{E}_z$ line source at the bottom center of the space with a frequency of 1.5 MHz. Make this source 10 grid cells long in z. Plot the magnitude of the $\mathcal{E}$ field in the plane perpendicular to the source excitation, and in a plane parallel to the excitation. What do you notice about the field evolution with time?

(b) Repeat part (a), but at frequencies of 3 MHz and 10 MHz. Compare to part (a).

(c) This time, excite the box with the derivative of a Gaussian source that has a pulsewidth τ equal to five time steps. Add the source to the update equations as a current source so that it does not interfere with reflected waves. Find the resonance frequencies for this resonator and compare to the theoretical resonances, which for the TE_{mnp} and TM_{mnp} modes are given by [3]:

$$
\omega_{mnp} = \frac{1}{\sqrt{\mu\epsilon}} \sqrt{\frac{m^2\pi^2}{a^2} + \frac{n^2\pi^2}{b^2} + \frac{p^2\pi^2}{d^2}}
$$

for a resonator with dimensions a, b, and d.

4.6. Antenna modeling. In this problem, we will use the 3D code from the previous problem to model two types of antennas. For each of these examples, measure the radiation pattern by monitoring points at a constant radius from the center of the antenna. Make the simulation space $100 \times 100 \times 100$ grid cells, and use 20 grid cells per wavelength; note that this implies the simulation boundaries are only 2.5 wavelengths from the antenna, so the results will be *near-field* radiation patterns.

(a) Model a half-wave dipole, similar to Figure 4.9, by making five grid cells perfect conductors above and below the source point. Drive the antenna with a sinusoidal source at the appropriate frequency.

(b) Model a monopole antenna above a ground plane by placing the source just above the bottom of the simulation space, and five grid cells of PEC above the source. Compare the radiation pattern to that found in (a).

In both cases above, watch out for reflections from the simulation boundaries: you will need to compute the radiation pattern before reflections corrupt the solution. Why does this problem not work with a small dipole, one grid cell in length? We will revisit and solve this problem in Chapter 7.

4.7. **2D polar coordinates.** Write a 2D TM mode simulation in polar coordinates that emulates the simulation space in Problem 4.4 above. Excite the simulation space with a sinusoidal $\mathscr{E}_z$ source at 30 GHz; compare the outgoing waves to those in Problem 4.4.

4.8. **2D cylindrical coordinates.** Create a 2D cylindrical simulation, choosing appropriate parameters Δr, Δz, and Δt to model the same wavelength as in Problem 4.6 above. Drive the simulation with the same two sources as in Problem 4.6, and compare the radiation patterns. How do the results differ from the previous problem? How does the simulation time compare?

4.9. **3D circular waveguide in cylindrical coordinates.** Create a simulation space for a circular waveguide in 3D cylindrical coordinates. Let the radius of the waveguide be 2.7 cm, and design it for a frequency of 10 GHz. Excite the waveguide at one end with a hard $\mathscr{E}$ source at a frequency of 10 GHz, where the source field is defined along a line through the cross-section of the waveguide. Plot a snapshot of the field patterns a few wavelengths down the waveguide; what modes exist in this waveguide? Repeat for frequencies of 8 GHz and 12 GHz and compare the results.

4.10. **Optical fiber.** Modify the code from the previous problem so that the waveguide consists of a dielectric with $n = 2$ with radius 2.0 cm, and a region of free space outside the dielectric. Terminate the simulation space outside the free space region with PEC, at a radius large enough so that evanescent fields outside the dielectric are small. Excite this fiber at one end as in the problem above, and plot snapshots of the field pattern a few wavelengths down the fiber.

4.11. **Cylindrical resonator.** Modify your 3D cylindrical coordinate code from the previous problems to model a cylindrical waveguide with radius 2 cm and thickness 2 cm (the z-dimension). Give the resonator a metallic outer shell, defined by a lossy material with $\sigma = 10^4$ S/m a few grid cells thick. Excite the resonator at 10 GHz with an $\mathscr{E}_z$ source at the center of the resonator, spanning the z-dimension, similar to Problem 4.5 above. The quality factor of a resonator is defined by the ratio of the energy stored in the resonant cavity to the energy dissipated (per cycle) in the lossy walls. Can you find a way to measure the quality factor Q of this resonator?

References

[1] K. Yee, "Numerical solution of initial boundary value problems involving Maxwell's equations in isotropic media," *IEEE Trans. on Ant. and Prop.*, vol. 14, no. 3, pp. 302–207, 1966.

[2] A. Taflove and S. Hagness, *Computational Electrodynamics: The Finite-Difference Time-Domain Method*, 3rd edn. Artech House, 2005.

[3] U. S. Inan and A. S. Inan, *Electromagnetic Waves*. Prentice-Hall, 2000.

[4] Y. Chen, R. Mittra, and P. Harms, "Finite-difference time-domain algorithm for solving Maxwell's equations in rotationally symmetric geometries," *IEEE Trans. Microwave Theory and Tech.*, vol. 44, pp. 832–839, 1996.

[5] H. Dib, T. Weller, M. Scardelletti, and M. Imparato, "Analysis of cylindrical transmission lines with the finite difference time domain method," *IEEE Trans. Microwave Theory and Tech.*, vol. 47, pp. 509–512, 1999.

[6] W. L. Stutzman and G. A. Thiele, *Antenna Theory and Design*. Wiley, 1998.

[7] H. Dib, T. Weller, and M. Scardelletti, "Analysis of 3-D cylindrical structures using the finite difference time domain method," *IEEE MTT-S International*, vol. 2, pp. 925–928, 1998.

[8] R. Holland, "THREDS: A finite-difference time-domain EMP code in 3D spherical coordinates," *IEEE Trans. Nuc. Sci.*, vol. NS-30, pp. 4592–4595, 1983.

[9] G. Liu, C. A. Grimes, and K. G. Ong, "A method for FDTD computation of field values at spherical coordinate singularity points applied to antennas," *IEEE Microwave and Opt. Tech. Lett.*, vol. 20, pp. 367–369, 1999.

5 Numerical stability of finite difference methods

In Chapter 3 we looked very briefly at the stability of numerical algorithms. By trial and error, we found the condition in Section 3.4.1 that, for a phase velocity of 1 ft ns^{-1}, we required $\Delta t \leq \Delta x$, or else the solution would rapidly go unstable as time progressed. This condition is known as the *Courant-Friedrichs-Lewy* or CFL stability condition.[1] In this chapter we will introduce a commonly used method for assessing the stability of finite difference methods. We will also investigate the source of instability and more formally derive the CFL condition.

The most commonly used procedure for assessing the stability of a finite difference scheme is the so-called *von Neumann* method, initially developed (like many other finite difference schemes) for fluid dynamics related applications. This stability analysis, based on spatial Fourier modes, was first proposed and used by J. von Neumann during World War II at Los Alamos National Laboratory [1, 2].

The von Neumann method is applied by first writing the initial spatial distribution of the physical property of interest (e.g., the voltage or the electric field) as a complex Fourier series. Then, we seek to obtain the exact solution of the finite difference equation (FDE) for a general spatial Fourier component of this complex Fourier series representation. If the exact solution of the FDE for the general Fourier component is bounded (either under all conditions or subject to certain conditions on Δx, Δt), then the FDE is said to be *stable*. If the solution for the general Fourier component is unbounded, then the FDE is said to be *unstable*. As we shall see, the stability criteria are generally also functions of ω and/or k, meaning the stability also depends on wavelength.

Determining the exact solution for a Fourier component is equivalent to finding the amplification factor q (or for a system of FDEs, the error amplification matrix $[\mathbf{q}]$) in a manner similar to that which was done for ODEs in Chapter 3. To illustrate, consider an arbitrary voltage distribution $\mathcal{V}(x, t)$ to be expressed in terms of a complex Fourier series in space, i.e.,

$$\mathcal{V}(x, t) = \sum_{m=-\infty}^{\infty} \mathcal{V}_m(x, t) = \sum_{m=-\infty}^{\infty} C_m(t) e^{j k_m x} \tag{5.1}$$

where $C_m(t)$ is dependent only on time, $j = \sqrt{-1}$ (not to be confused with the y-index), and $k_m = 2\pi / \lambda_m$ is the wavenumber corresponding to a wavelength λ_m. A general spatial

---

[1] The CFL condition is sometimes (perhaps unfairly) referred to simply as the *Courant* condition.

Fourier component is thus given by:

$$\mathcal{V}_m(x, t) = C_m(t) e^{jk_m x}. \tag{5.2}$$

For stability analysis, we will work with the discretized version of Equation (5.2) given by

$$\mathcal{V}_i^n = C(t^n) e^{jk i \Delta x} \tag{5.3}$$

where we have dropped the subscript m since we shall only work with a single component from here on, and k should not be confused with the z-index in space. Note that $C(t^n)$ is a constant at our discretized time step. Also note that the values of $\mathcal{V}_{i\pm1}^n$, those adjacent to $\mathcal{V}_i^n$ in the grid, are simply related to $\mathcal{V}_i^n$ in the Fourier domain by a phase shift:

$$\mathcal{V}_{i\pm1}^n = C(t^n) e^{jk(i\pm1)\Delta x} = \underbrace{C(t^n) e^{jk i \Delta x}}_{\mathcal{V}_i^n} e^{\pm jk \Delta x}$$

$$= \mathcal{V}_i^n e^{\pm jk \Delta x}. \tag{5.4}$$

The basic procedure for conducting a von Neumann stability analysis of a given finite difference equation (FDE) involves the following steps:

1. Substitute the discretized single Fourier component as given by Equation (5.3) and its shifted versions as given by Equation (5.4) into the FDE.
2. Express $e^{jk \Delta x}$ in terms of $\sin(k\Delta x)$ and $\cos(k\Delta x)$ and reduce the FDE to the form $\mathcal{V}_i^{n+1} = q \mathcal{V}_i^n$, so that the amplification factor q (or the amplification matrix $[\mathbf{q}]$) is determined as a function of $\sin(k\Delta x)$, $\cos(k\Delta x)$, and Δt.
3. Analyze q or $[\mathbf{q}]$ to determine the stability criteria for the FDE; that is, the conditions, if any, under which $|q| \leq 1$ or for which all the eigenvalues of the matrix $[\mathbf{q}]$ are bounded by unity.

We will now apply this procedure to some of the FDEs that we have worked with in previous lectures.

5.1   The convection equation

We first consider the 1D single convection Equation (3.63) that was considered in Chapter 3. Keep in mind that the stability criterion is a feature of the original partial differential equation, along with the particular choice of discretization; as such, the stability conditions depend critically on how we choose to discretize the PDE. This is, in fact, a key step in designing a numerical algorithm.

5.1.1   The forward-time centered space method

The first finite difference algorithm considered for this equation was the forward-time centered space method, the algorithm for which is repeated below for convenience:

$$\frac{\partial \mathcal{V}}{\partial t} + v_p \frac{\partial \mathcal{V}}{\partial x} = 0 \qquad \rightarrow \qquad \frac{\mathcal{V}_i^{n+1} - \mathcal{V}_i^n}{\Delta t} + v_p \frac{\mathcal{V}_{i+1}^n - \mathcal{V}_{i-1}^n}{2\Delta x} = 0 \qquad (5.5)$$

Forward-time centered space

$$\mathcal{V}_i^{n+1} = \mathcal{V}_i^n - \left(\frac{v_p \Delta t}{2\Delta x}\right) \left[\mathcal{V}_{i+1}^n - \mathcal{V}_{i-1}^n\right]. \qquad (5.6)$$

Substituting Equation (5.4) into Equation (5.6) we find:[2]

$$\mathcal{V}_i^{n+1} = \mathcal{V}_i^n - \left(\frac{v_p \Delta t}{2\Delta x}\right)\left[\mathcal{V}_i^n e^{+jk\Delta x} - \mathcal{V}_i^n e^{-jk\Delta x}\right]$$

$$\mathcal{V}_i^{n+1} = \underbrace{\left[1 - j\left(\frac{v_p \Delta t}{\Delta x}\right)\sin(k\Delta x)\right]}_{q} \mathcal{V}_i^n = q\,\mathcal{V}_i^n.$$

Thus, the amplification factor for this scheme is complex and is given by:

$$q = 1 - j\left(\frac{v_p \Delta t}{\Delta x}\right)\sin(k\Delta x). \qquad (5.7)$$

The magnitude of the amplification factor is then given by:

$$|q| = \sqrt{q\,q^*} = \sqrt{1 + \left[\left(\frac{v_p \Delta t}{\Delta x}\right)\sin(k\Delta x)\right]^2} > 1. \qquad (5.8)$$

Thus, for this method, the von Neumann stability condition (i.e., $|q| \leq 1$) cannot be satisfied for any nonzero value of k and for any time step Δt. The forward-time centered space differencing scheme is thus inherently and unconditionally *unstable* for the convection equation. It should be noted, however, that this same method may well be stable and well suited for the solution of other equations, such as the diffusion Equation (3.10), which is discretized as shown in Equation (3.61).

Figure 5.1 demonstrates the inherent instability of the forward-time centered space method. This 1D simulation uses $\Delta x = 0.01$ ft, $v_p = 1$ ft/ns, and $\Delta t = 0.01$ ns, which yields $v_p \Delta t / \Delta x = C = 1$. A Gaussian pulse with 0.2 ns half-width is input from the left edge. However, this simulation rapidly goes unstable, as is evident from the figure; an oscillation growing exponentially with time appears very rapidly. Figure 5.1 shows a snapshot of the waveform after 4 ns; the lower panel shows the waveform on a logarithmic scale. The wave has traveled exactly 4 feet in 4 ns because the *grid velocity* (or *lattice speed*) is exactly one spatial grid cell per time step.

---

[2] Notice that we don't actually need to substitute Equation (5.3) above, as it will simply cancel out in each of the terms.

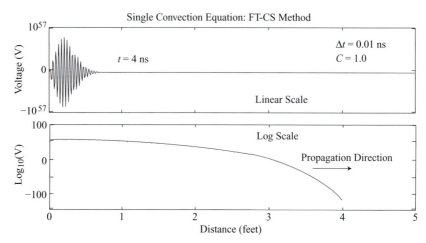

Figure 5.1 Stability of the forward-time centered space method, applied to the single convection equation. It is evident from this result that the method is unstable.

5.1.2   The Lax method

A minor modification to the forward-time centered space method does lead to a stable method for the convection equation even with only first-order differencing in time. This so-called Lax method [3] (not to be confused with the Lax-Wendroff methods in Chapter 3) is obtained by replacing $\mathcal{V}_i^n$ in Equation (5.6) with a spatial average:

$$\mathcal{V}_i^n = \frac{\mathcal{V}_{i+1}^n + \mathcal{V}_{i-1}^n}{2}.$$

so that the discretized convection equation becomes:

Lax method

$$\mathcal{V}_i^{n+1} = \underbrace{\frac{\mathcal{V}_{i+1}^n + \mathcal{V}_{i-1}^n}{2}}_{\cos(k\Delta x)} - \left(\frac{v_p \Delta t}{2\Delta x}\right) \underbrace{\left[\mathcal{V}_{i+1}^n - \mathcal{V}_{i-1}^n\right]}_{j2\sin(k\Delta x)}. \tag{5.9}$$

Substituting Equation (5.4) (i.e., $\mathcal{V}_{i\pm1}^n = \mathcal{V}_i^n e^{\pm jk\,\Delta x}$) into Equation (5.9) and manipulating, we find:

$$\mathcal{V}_i^{n+1} = \underbrace{\left[\cos(k\Delta x) - j\left(\frac{v_p \Delta t}{\Delta x}\right)\sin(k\Delta x)\right]}_{q} \mathcal{V}_i^n = q\,\mathcal{V}_i^n$$

so that

$$q = \cos(k\Delta x) - j\left(\frac{v_p \Delta t}{\Delta x}\right)\sin(k\Delta x). \tag{5.10}$$

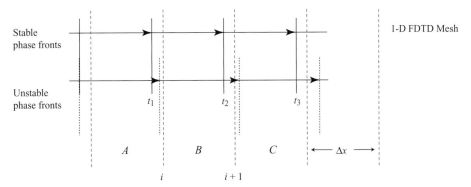

Figure 5.2 The CFL condition in one dimension. Stable phase fronts are shown with solid lines; unstable phase fronts, where $v_p \Delta t / \Delta x > 1$, are shown with dotted lines. Note that with these unstable phase fronts, cell B gets skipped between t_2 and t_3.

Notice the similarity with q from the forward-time centered space method in (5.7): the factor 1 has simply been replaced with a factor $\cos(k\Delta x)$. The square of the magnitude of the amplification factor is then given by

$$|q|^2 = q\, q^* = \cos^2(k\Delta x) + \left[\left(\frac{v_p \Delta t}{\Delta x} \right) \sin(k\Delta x) \right]^2$$

$$= 1 - \sin^2(k\Delta x) \left[1 - \left(\frac{v_p \Delta t}{\Delta x} \right)^2 \right] = 1 - K. \qquad (5.11)$$

For stability, we need $|q|^2 \leq 1$, so the second term on the right must be bounded by $0 \leq K \leq 2$. The $\sin^2(k\Delta x)$ term will always be positive and less than or equal to 1; we therefore need the term in the square brackets to be bounded by $[0, 2]$. It thus appears that the von Neumann stability condition (i.e., $|q|^2 \leq 1$) is satisfied for all wavenumbers k (i.e., for all wavelengths λ, since $k = 2\pi/\lambda$) as long as

$$\frac{|v_p \Delta t|}{\Delta x} \leq 1 \qquad \rightarrow \qquad \Delta t \leq \frac{\Delta x}{|v_p|}. \qquad (5.12)$$

This condition for stability on the time step Δt is another example of the CFL condition [4]. With careful consideration, we can see that the CFL condition makes perfect physical sense. Stability requires that we choose a time step smaller than the smallest characteristic physical time in the problem, which for the convective equation is simply the time with which we can move across one spatial mesh size at the speed v_p. Thus, the CFL condition simply requires that the *lattice speed* $\Delta x/\Delta t$ (which we must control) is less than the physical velocity v_p in the medium we are trying to simulate.

As another way of looking at it, in terms of Δx, one could say that the CFL condition requires Δx to be large enough, or Δt small enough, so that the propagating wave doesn't "skip" a grid cell in one time step. This is illustrated in Figure 5.2. The solid

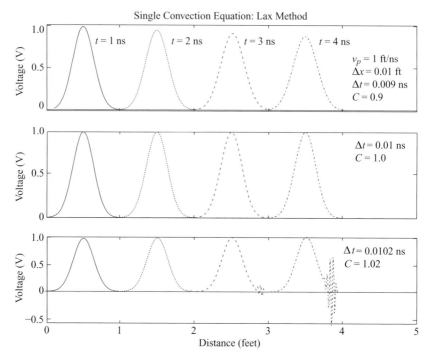

Figure 5.3 Stability of the Lax method, applied to the single convection equation. The wave is stable for $C \leq 1$, but goes unstable rapidly for values of C even slightly larger than 1.

vertical lines represent a stable propagating wave, with $\Delta t = \Delta x / v_p$. In this case the wave propagates exactly one Δx in one time step. The dotted vertical lines, however, represent a wave with $\Delta t > \Delta x / v_p$. Notice that between time steps t_1 and t_2, grid cell B was skipped entirely.

Figure 5.3 demonstrates the Lax method applied to the single convection equation. This simulation has the same parameters as Figure 5.1, but in the three panels $v_p \Delta t / \Delta x = C$ is varied from 0.9, to 1, to 1.02. Again, a Gaussian pulse with 0.2 ns half-width is input from the left edge. In each panel, snapshots of the waveform are shown at $t = 1, 2, 3$, and 4 ns.

Notice that for $C = 0.9$, the method is stable: the wave is not growing. Rather, the wave is *decaying*; this is an issue of accuracy, which will be discussed in great detail in Chapter 6. When $C = 1$, as in the second panel, the method is stable again, and the accuracy is much improved; the wave is not decaying with time and distance. When we increase C to 1.02, just barely above the CFL condition, the method begins to show signs of instability, as an oscillation appears at the front of the wave, starting at $t = 3$ ns. In fact, if we had chosen $C = 1.03$ rather than 1.02, the wave becomes unstable so fast that it reaches amplitudes of 30 V at $t = 4$ ns.

5.1.3 The leapfrog method

We will now examine the stability of the leapfrog algorithm applied to the convection equation. We derived the leapfrog finite difference algorithm in Equation (3.70) of Chapter 3, repeated here for convenience:

Leapfrog

$$\mathcal{V}_i^{n+1} = \mathcal{V}_i^{n-1} - \left(\frac{v_p \Delta t}{\Delta x}\right)\left[\mathcal{V}_{i+1}^n - \mathcal{V}_{i-1}^n\right]. \tag{5.13}$$

Substituting Equation (5.4) into Equation (5.13), and noting that $\mathcal{V}_i^n = q\,\mathcal{V}_i^{n-1}$, we find:

$$\mathcal{V}_i^{n+1} = \frac{\mathcal{V}_i^n}{q} - \left(\frac{v_p \Delta t}{\Delta x}\right)\left[\mathcal{V}_i^n e^{+jk\Delta x} - \mathcal{V}_i^n e^{-jk\Delta x}\right]$$

$$= \underbrace{\left[\frac{1}{q} - j2\left(\frac{v_p \Delta t}{\Delta x}\right)\sin(k\Delta x)\right]}_{q} \mathcal{V}_i^n = q\,\mathcal{V}_i^n.$$

Thus, the amplification factor q is given by:

$$q = \frac{1}{q} - j2\left(\frac{v_p \Delta t}{\Delta x}\right)\sin(k\Delta x)$$

$$\rightarrow \qquad q^2 + q\,j\,\underbrace{2\left(\frac{v_p \Delta t}{\Delta x}\right)\sin(k\Delta x)}_{2A} - 1 = 0$$

$$\rightarrow \qquad q^2 + q\,j\,2A - 1 = 0$$

$$\rightarrow \qquad q = -j\,A \pm \sqrt{1 - A^2} \tag{5.14}$$

where $A = (v_p \Delta t / \Delta x)\sin(k\Delta x)$. Note that q is complex, and the magnitude of q for both roots is identically equal to one, provided that the factor under the square root is positive; otherwise $|q| > 1$. Thus, we require $A \leq 1$, in which case we have:

$$|q|^2 = [\mathcal{R}e\{q\}]^2 + [\mathcal{I}m\{q\}]^2 = 1 - A^2 + A^2 = 1.$$

Therefore, to ensure the stability of this method for all wavenumbers k, we must have $1 - A^2 \geq 0$ or $|A| \leq 1$. The term $\sin(k\Delta x)$ is always less than or equal to 1, so we require:

$$\frac{v_p \Delta t}{\Delta x} \leq 1 \qquad \rightarrow \qquad \Delta t \leq \frac{\Delta x}{|v_p|}. \tag{5.15}$$

We thus once again arrive at the CFL condition. Note that if the factor under the square root in (5.14) is negative (i.e., if $|A| > 1$), then we have $|q| > 1$ and the method is unstable.

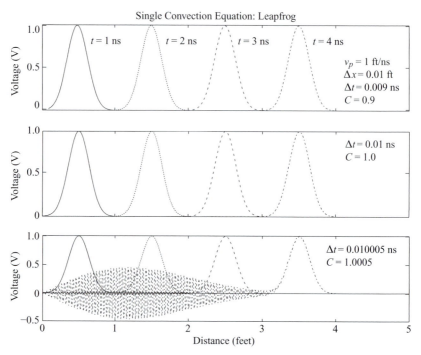

Figure 5.4 Stability of the leapfrog method, applied to the single convection equation. The wave is stable for $C \leq 1$, but goes unstable rapidly for values of C even slightly larger than 1.

Figure 5.4 demonstrates the stability of the leapfrog method applied to the single convection equation. This simulation has the same parameters as Figures 5.3, but in this case, $v_p \Delta t / \Delta x = C$ is varied from 0.9, to 1, to 1.0005 in the three panels.

Again, for $C = 0.9$, the method is stable, but unlike in the Lax method, the wave is not decaying. We will investigate the accuracy of these methods in Chapter 6; for now, the reader should note that the leapfrog method appears to be more accurate than the Lax method for $C \neq 1$. When $C = 1$, in the second panel, the method is stable and accurate again. Now, when we increase C to just 1.0005, very slightly above the CFL condition, the method shows signs of instability, as an oscillation appears at the back of the wave (the large oscillation from 0 to 3 feet is all part of the pulse at 4 ns). If we had chosen $C = 1.001$, the wave becomes unstable so rapidly that it reaches amplitudes of almost 100 V at $t = 4$ ns.

5.2   Two coupled first-order convection equations

The convection equation above provides a simple introduction to stability analysis, but for the FDTD algorithm, we are interested in coupled equations. The von Neumann stability analysis can be easily extended to systems of FDEs by using matrix notation. To illustrate the method, we consider the stability of some of the schemes discussed in

Section 3.5 of Chapter 3, for the solution of the telegrapher's equations (3.1), repeated below:

$$\frac{\partial \mathcal{V}}{\partial t} + \frac{1}{C}\frac{\partial \mathcal{I}}{\partial x} = 0 \tag{3.1a}$$

$$\frac{\partial \mathcal{I}}{\partial t} + \frac{1}{L}\frac{\partial \mathcal{V}}{\partial x} = 0. \tag{3.1b}$$

5.2.1 The forward-time centered space method

The forward-time centered space algorithm for the solution of Equation (3.1) was given in Equation (4.17), repeated below:

Forward-time centered space

$$\mathcal{V}_i^{n+1} = \mathcal{V}_i^n - \left(\frac{\Delta t}{2\,C\,\Delta x}\right)\left[\mathcal{I}_{i+1}^n - \mathcal{I}_{i-1}^n\right]$$

$$\mathcal{I}_i^{n+1} = \mathcal{I}_i^n - \left(\frac{\Delta t}{2\,L\,\Delta x}\right)\left[\mathcal{V}_{i+1}^n - \mathcal{V}_{i-1}^n\right]. \tag{5.16}$$

Any stability analysis of Equation (5.16) must consider both equations simultaneously, since they are obviously coupled. We do this by substituting general Fourier components for both $\mathcal{V}(x, t)$ and $\mathcal{I}(x, t)$, both expressed in a form similar to that given for $\mathcal{V}(x, t)$ in Equation (5.4); specifically, we substitute $\mathcal{V}_{i\pm1}^n = \mathcal{V}_i^n\,e^{\pm jk\,\Delta x}$ and $\mathcal{I}_{i\pm1}^n = \mathcal{I}_i^n\,e^{\pm jk\,\Delta x}$. Making the necessary substitutions in Equation (5.16) we find:

$$\mathcal{V}_i^{n+1} = \mathcal{V}_i^n - \left(\frac{\Delta t}{2\,C\,\Delta x}\right)\left[\mathcal{I}_i^n\,e^{+jk\,\Delta x} - \mathcal{I}_i^n\,e^{-jk\,\Delta x}\right]$$

$$= \mathcal{V}_i^n - j\,\mathcal{I}_i^n\left(\frac{\Delta t}{C\,\Delta x}\right)\left[\sin(k\Delta x)\right] \tag{5.17a}$$

$$\mathcal{I}_i^{n+1} = \mathcal{I}_i^n - \left(\frac{\Delta t}{2\,L\,\Delta x}\right)\left[\mathcal{V}_i^n\,e^{+jk\,\Delta x} - \mathcal{V}_i^n\,e^{-jk\,\Delta x}\right]$$

$$= \mathcal{I}_i^n - j\,\mathcal{V}_i^n\left(\frac{\Delta t}{L\,\Delta x}\right)\left[\sin(k\Delta x)\right]. \tag{5.17b}$$

Equations (5.17a) and (5.17b) can be written in matrix form as

$$\begin{bmatrix}\mathcal{V}_i^{n+1}\\\mathcal{I}_i^{n+1}\end{bmatrix} = \underbrace{\begin{bmatrix} 1 & -j\left(\dfrac{\Delta t}{C\,\Delta x}\right)\sin(k\Delta x)\\[2mm] -j\left(\dfrac{\Delta t}{L\,\Delta x}\right)\sin(k\Delta x) & 1 \end{bmatrix}}_{[\mathbf{q}]}\begin{bmatrix}\mathcal{V}_i^n\\\mathcal{I}_i^n\end{bmatrix}. \tag{5.18}$$

In order for the algorithm described by Equation (5.16) to be stable, the eigenvalues ζ of $[\mathbf{q}]$ must be bounded by unity. The eigenvalues ζ for any matrix $[\mathbf{q}]$ are found by setting the determinant of the matrix $\{[\mathbf{q}] - \zeta[\mathbf{I}]\}$ to zero, where $[\mathbf{I}]$ is the identity matrix. We can thus solve for the eigenvalues of $[\mathbf{q}]$ from

$$
\begin{vmatrix}
1 - \zeta & -j\left(\dfrac{\Delta t}{C\,\Delta x}\right)\sin(k\Delta x) \\
-j\left(\dfrac{\Delta t}{L\,\Delta x}\right)\sin(k\Delta x) & 1 - \zeta
\end{vmatrix} = 0 \qquad (5.19)
$$

which gives:

$$
(1 - \zeta)^2 + \frac{1}{LC}\left(\frac{\Delta t}{\Delta x}\right)^2 \sin^2(k\Delta x) = 0
$$

$$
\rightarrow \quad \zeta = 1 \pm j\sqrt{\frac{1}{LC}}\left(\frac{\Delta t}{\Delta x}\right)\sin(k\Delta x). \qquad (5.20)
$$

With $\sqrt{1/LC} = v_p$, note that the eigenvalues ζ have exactly the same form as in the case of the single convection equation, where q is given by Equation (5.7), except that here there are two eigenvalues, denoted by the $\pm$ sign. The magnitude of ζ is then given as:

$$
|\zeta| = \sqrt{[\mathcal{R}e\{\zeta\}]^2 + [\mathcal{I}m\{\zeta\}]^2} = \sqrt{1 + \frac{1}{LC}\left[\frac{\Delta t}{\Delta x}\sin(k\Delta x)\right]^2} > 1.
$$

Thus, we find that the eigenvalues ζ are always greater than one, and the finite difference scheme described by Equations (5.16) is unconditionally unstable (i.e., unstable for *any* values of Δt and Δx), just as it was for the single convection equation.

We will not provide an illustrative example here; however, the reader is encouraged to code this algorithm to demonstrate that it is unstable for any values of Δt and Δx.

5.2.2   The Lax method

Recall that for the single convection equation, the forward-time centered space method was unstable, but with a minor modification we arrived at the stable Lax method. We can conduct a similar procedure for the coupled convection equations. As was mentioned for the case of a single convection equation, the algorithm for the Lax method can be obtained from the forward-time centered space algorithm by adding a spatial approximation for the quantities to be evaluated at grid points (i, n). We thus have:

Lax method

$$
\mathcal{V}_i^{n+1} = \frac{\mathcal{V}_{i+1}^n + \mathcal{V}_{i-1}^n}{2} - \left(\frac{\Delta t}{2C\,\Delta x}\right)[\mathcal{I}_{i+1}^n - \mathcal{I}_{i-1}^n]
$$

$$
\mathcal{I}_i^{n+1} = \frac{\mathcal{I}_{i+1}^n + \mathcal{I}_{i-1}^n}{2} - \left(\frac{\Delta t}{2L\,\Delta x}\right)[\mathcal{V}_{i+1}^n - \mathcal{V}_{i-1}^n]. \qquad (5.21)
$$

Note once again that simple averaging is used to obtain Equations (5.21) from Equations (5.16), and appears to do the trick (i.e., bring about stability) as shown below. In this case, the stability analysis results in the amplification matrix:

$$[\mathbf{q}] = \begin{bmatrix} \cos(k\Delta x) & -j\left(\dfrac{\Delta t}{C\,\Delta x}\right)\sin(k\Delta x) \\[2ex] -j\left(\dfrac{\Delta t}{L\,\Delta x}\right)\sin(k\Delta x) & \cos(k\Delta x) \end{bmatrix} \qquad (5.22)$$

and the eigenvalues ζ are determined by:

$$\begin{vmatrix} \cos(k\Delta x) - \zeta & -j\left(\dfrac{\Delta t}{C\,\Delta x}\right)\sin(k\Delta x) \\[2ex] -j\left(\dfrac{\Delta t}{L\,\Delta x}\right)\sin(k\Delta x) & \cos(k\Delta x) - \zeta \end{vmatrix} = 0 \qquad (5.23)$$

which gives the equation for the eigenvalues:

$$(\cos(k\Delta x) - \zeta)^2 + \frac{1}{LC}\left(\frac{\Delta t}{\Delta x}\right)^2 \sin^2(k\Delta x) = 0$$

$$\rightarrow \quad \zeta = \cos(k\Delta x) \pm j\sqrt{\frac{1}{LC}}\left(\frac{\Delta t}{\Delta x}\right)\sin(k\Delta x) \qquad (5.24)$$

so that the magnitude of ζ is:

$$|\zeta| = \sqrt{\cos^2(k\Delta x) + \left[\sqrt{\frac{1}{LC}}\left(\frac{\Delta t}{\Delta x}\right)\sin(k\Delta x)\right]^2}$$

$$= \sqrt{1 - \sin^2(k\Delta x)\left[1 - \left(\frac{1}{LC}\right)\left(\frac{\Delta t}{\Delta x}\right)^2\right]}.$$

Since $\sin^2(k\Delta x) \geq 0$ for all k, it is clear that the magnitudes of the eigenvalues are less than unity as long as:

$$\underbrace{\left(\frac{1}{LC}\right)}_{v_p^2}\left(\frac{\Delta t}{\Delta x}\right)^2 \leq 1 \quad \rightarrow \quad \frac{v_p\Delta t}{\Delta x} \leq 1 \quad \rightarrow \quad \Delta t \leq \frac{\Delta x}{v_p}. \qquad (5.25)$$

In other words, we once again arrive at the CFL condition.

Figure 5.5 demonstrates the Lax method applied to the coupled convection equations. This simulation uses a capacitance $C = 96.8$ pF and inductance $L = 0.242$ μH, which yield $v_p = 2.06 \times 10^8$ m/s; $\Delta x = 0.01$ m, and Δt varying such that $v_p\Delta t/\Delta x = C$ is varied from 0.9, to 1, to 1.02. A Gaussian pulse with a half-width of $10\Delta t$ is input from the left edge. In each panel, the waveform is shown at four snapshots in time.

Notice that for $C = 0.9$, the method is stable, but the wave is *decaying*; again, this is an issue of accuracy, which will be discussed in great detail in Chapter 6. When

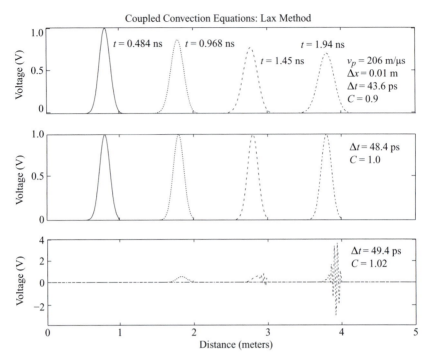

Figure 5.5 Stability of the Lax method, applied to the coupled convection equations. The wave is stable for $C \leq 1$, but goes unstable rapidly for values of C even slightly larger than 1. Note that the waveforms for the current $\mathcal{I}$ are identical, but scaled by a factor $1/Z_0$, where $Z_0 = \sqrt{L/C}$ is the impedance of the transmission line.

$C = 1$, as in the second panel, the method is stable again, and the accuracy is much improved; the wave is not decaying with time and distance. When we increase C to 1.02, just barely above the CFL condition, the method begins to show signs of instability, as an oscillation appears at the front of the wave, similar to the single convection equation.

5.2.3   The leapfrog method

We consider the stability of the leapfrog method for the solution of a system of two coupled hyperbolic PDEs. We rewrite the leapfrog algorithm from Chapter 3:

Leapfrog

$$\mathcal{V}_i^{n+1} = \mathcal{V}_i^{n-1} - \left(\frac{\Delta t}{C \, \Delta z} \right) \left[\mathcal{I}_{i+1}^n - \mathcal{I}_{i-1}^n \right]$$

$$\mathcal{I}_i^{n+1} = \mathcal{I}_i^{n-1} - \left(\frac{\Delta t}{L \, \Delta z} \right) \left[\mathcal{V}_{i+1}^n - \mathcal{V}_{i-1}^n \right].$$

$$(5.26)$$

We perform the usual stability analysis by substitution of the general Fourier components for both $\mathcal{V}(x, t)$ and $\mathcal{I}(x, t)$:

$$\mathcal{V}_i^{n+1} = 0\,\mathcal{V}_i^n - \left(\frac{\Delta t}{C\,\Delta x}\right) \left[\mathcal{I}_i^n\, e^{+jk\,\Delta x} - \mathcal{I}_i^n\, e^{-jk\,\Delta x}\right] + \mathcal{V}_i^{n-1}$$

$$= 0\,\mathcal{V}_i^n - j\,\mathcal{I}_i^n \left(\frac{2\Delta t}{C\,\Delta x}\right) \sin(k\Delta x) + \mathcal{V}_i^{n-1} \tag{5.27a}$$

$$\mathcal{I}_i^{n+1} = 0\,\mathcal{I}_i^n - \left(\frac{\Delta t}{L\,\Delta x}\right) \left[\mathcal{V}_i^n\, e^{+jk\,\Delta x} - \mathcal{V}_i^n\, e^{-jk\,\Delta x}\right] + \mathcal{I}_i^{n-1}$$

$$= 0\,\mathcal{I}_i^n - j\,\mathcal{V}_i^n \left(\frac{2\Delta t}{L\,\Delta x}\right) \sin(k\Delta x) + \mathcal{I}_i^{n-1}. \tag{5.27b}$$

Equations (5.27a) and (5.27b) can be written in matrix form as:

$$\begin{bmatrix} \mathcal{V}_i^{n+1} \\ \mathcal{I}_i^{n+1} \end{bmatrix} = -2j \underbrace{\begin{bmatrix} 0 & \left(\dfrac{\Delta t}{C\,\Delta x}\right) \sin(k\Delta x) \\ \left(\dfrac{\Delta t}{L\,\Delta x}\right) \sin(k\Delta x) & 0 \end{bmatrix}}_{[\mathbf{A}]} \begin{bmatrix} \mathcal{V}_i^n \\ \mathcal{I}_i^n \end{bmatrix} + \begin{bmatrix} \mathcal{V}_i^{n-1} \\ \mathcal{I}_i^{n-1} \end{bmatrix}.$$

$$\tag{5.28}$$

Although the matrix $[\mathbf{A}]$ is not quite the amplification matrix as we have defined it until now (because of the $-2j$ term and the vector at time $n-1$), it can be shown[3] that stability is ensured if the eigenvalues of $[\mathbf{A}]$ are bounded by unity. We thus have:

$$\begin{vmatrix} 0 - \zeta & \left(\dfrac{\Delta t}{C\,\Delta x}\right) \sin(k\Delta x) \\ \left(\dfrac{\Delta t}{L\,\Delta x}\right) \sin(k\Delta x) & 0 - \zeta \end{vmatrix} = 0 \tag{5.29}$$

which gives the equation for the eigenvalues:

$$(0 - \zeta)^2 - \frac{1}{LC} \left(\frac{\Delta t}{\Delta x}\right)^2 \sin^2(k\Delta x) = 0$$

$$\rightarrow \quad \zeta = \pm\sqrt{\frac{1}{LC}} \left(\frac{\Delta t}{\Delta x}\right) \sin(k\Delta x). \tag{5.30}$$

In order for $\zeta \leq 1$ for *all* Fourier components, regardless of k, we must have

$$\sqrt{\frac{1}{LC}} \left(\frac{\Delta t}{\Delta x}\right) \leq 1 \quad \rightarrow \quad \Delta t \leq \frac{\Delta x}{v_p} \tag{5.31}$$

which once again is the Courant-Friedrichs-Lewy (CFL) condition. Notice that the arguments here were somewhat different compared to the single convection equation

---

[3] See the problems at the end of this chapter.

case, but the result is the same. We will investigate the stability of the leapfrog algorithm in an example in Problem 5.2.

5.2.4   The interleaved leapfrog and FDTD method

Stability analysis of the interleaved leapfrog method, either for $\mathcal{V}$ and $\mathcal{I}$ (Equations 3.86) or for $\mathcal{E}$ and $\mathcal{H}$ (the FDTD algorithm, for example, Equations 4.8) is rather complicated using the von Neumann analysis method here. As such, we will not conduct the same analysis here, instead saving the derivation for the end of Chapter 6. To illustrate the problem, consider the coupled, interleaved leapfrog difference equations for the transmission line problem (where we have substituted x in place of z):

Interleaved Leapfrog

$$\mathcal{V}_i^{n+1} = \mathcal{V}_i^n - \left(\frac{\Delta t}{C \, \Delta x} \right) \left[\mathcal{I}_{i+1/2}^{n+1/2} - \mathcal{I}_{i-1/2}^{n+1/2} \right]$$

$$\mathcal{I}_{i+1/2}^{n+1/2} = \mathcal{I}_{i+1/2}^{n-1/2} - \left(\frac{\Delta t}{L \, \Delta x} \right) \left[\mathcal{V}_{i+1}^n - \mathcal{V}_i^n \right]. \tag{3.86}$$

We can follow the same analysis as above, by inserting $\mathcal{V}_{i+1}^n = \mathcal{V}_i^n \, e^{jk \, \Delta x}$ and $\mathcal{I}_{i-1/2}^{n+1/2} = \mathcal{I}_{i+1/2}^{n+1/2} \, e^{-jk \, \Delta x}$; after substitution and arranging into $\sin(\cdot)$ terms we find:

$$\mathcal{V}_i^{n+1} = \mathcal{V}_i^n - \left(\frac{\Delta t}{C \, \Delta x} \right) e^{-jk \, \Delta x/2} \, \mathcal{I}_{i+1/2}^{n+1/2} \, 2j \sin \left(\frac{k \Delta x}{2} \right) \tag{5.32a}$$

$$\mathcal{I}_{i+1/2}^{n+1/2} = \mathcal{I}_{i+1/2}^{n-1/2} - \left(\frac{\Delta t}{L \, \Delta x} \right) e^{jk \, \Delta x/2} \, \mathcal{V}_i^n \, 2j \sin \left(\frac{k \Delta x}{2} \right). \tag{5.32b}$$

At this point, we cannot write this system of equations in matrix notation as we did for the non-interleaved leapfrog algorithm, because of the $\mathcal{I}_{i+1/2}^{n-1/2}$ term in Equation (5.32b). On the other hand, we can substitute Equation (5.32b) into Equation (5.32a) and arrive at an equation for $\mathcal{V}_i^{n+1}$:

$$\mathcal{V}_i^{n+1} = \mathcal{V}_i^n - \left(\frac{\Delta t}{C \, \Delta x} \right) e^{-jk \, \Delta x/2} \, 2j \sin \left(\frac{k \Delta x}{2} \right)$$

$$\times \left[\mathcal{I}_{i+1/2}^{n-1/2} - \left(\frac{\Delta t}{L \, \Delta x} \right) e^{jk \, \Delta x/2} \, \mathcal{V}_i^n \, 2j \sin \left(\frac{k \Delta x}{2} \right) \right]$$

which has the form:

$$\mathcal{V}_i^{n+1} = q \, \mathcal{V}_i^n + f(\mathcal{I}_{i+1/2}^{n-1/2}).$$

Again, the $\mathcal{I}_{i+1/2}^{n-1/2}$ term causes trouble and prevents us from writing an equation yielding q or $[\mathbf{A}]$. As such, we will abandon the derivation of the stability criterion for the FDTD algorithm for now, and come back to it in Chapter 6. For now, the reader should be satisfied to discover that the stability criterion is yet again given by the CFL condition, i.e., in 1D, $v_p \Delta t / \Delta x \leq 1$.

5.3 Stability of higher dimensional FDTD algorithms

As illustrated in Figure 5.2, the CFL stability condition for a one-dimensional FDTD code ensures that the finite difference grid is causally connected and that the natural physical speed of the fundamental PDE(s) determine the rate at which information can move across the mesh. This physically based criterion is easily extended to higher dimensional FDTD algorithms. The stability criterion for a three-dimensional system is found to be:

$$\Delta t \leq \frac{1}{v_p \sqrt{\frac{1}{(\Delta x)^2} + \frac{1}{(\Delta y)^2} + \frac{1}{(\Delta z)^2}}}. \tag{5.33}$$

In 2D and 3D algorithms with square or cubic grid cells (i.e., $\Delta x = \Delta y = \Delta z$), the requirement for Δt becomes:

$$\Delta t \leq \underbrace{\frac{\Delta x}{\sqrt{2}\, v_p}}_{2D} \quad \text{or} \quad \Delta t \leq \underbrace{\frac{\Delta x}{\sqrt{3}\, v_p}}_{3D}. \tag{5.34}$$

We can generalize the CFL condition for a regular grid in D-dimensions:

$$\Delta t \leq \frac{\Delta x}{\sqrt{D}\, v_p} \tag{5.35}$$

where D can be one, two, or three dimensions. In other words, generally smaller time increments are required for higher dimensional algorithms. The reason for this requirement can be understood from Figure 5.6.[4] A uniform plane wave propagating at an angle θ to the x-axis so that its phase fronts span distances Δx and Δy respectively in the Cartesian directions is equivalent to 1D propagation along a mesh with cell size d, such that

$$\begin{aligned} \Delta x &= \frac{d}{\cos \theta} \\ \Delta y &= \frac{d}{\sin \theta} \end{aligned} \quad \rightarrow \quad \frac{1}{(\Delta x)^2} + \frac{1}{(\Delta y)^2} = \frac{\cos^2 \theta}{d^2} + \frac{\sin^2 \theta}{d^2} = \frac{1}{d^2}.$$

Thus, a smaller time must be used to ensure the stability on such an effectively smaller mesh.

---

[4] It is often misunderstood that the stability criterion in higher dimensions implies that Δx is replaced by the diagonal distance across the grid cell. This is incorrect, as it would imply a less stringent requirement, where Δx is replaced by $\sqrt{2}\, \Delta x$, in the square 2D case, rather than $\Delta x / \sqrt{2}$. Careful examination of Figure 5.6 is required to properly appreciate the stability criterion.

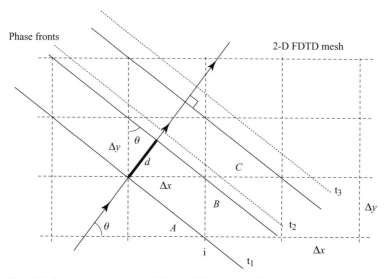

Figure 5.6 Interpretation of the CFL condition in two dimensions. For the dashed, unstable phase fronts, the wave jumps from cell A to cell C between t_1 and t_2, before being updated in cell B.

This "effective grid cell size" can be understood in the same context as Figure 5.2, where we interpreted the unstable case as having "skipped" a grid cell. Consider the dashed phase fronts in Figure 5.6, which have been given a Δt that is too large. In the direction of phase propagation, the phase front has jumped in one time step (from t_1 to t_2) from grid cell A to grid cell C (the first dashed phase front), skipping cell B.

While we have not strictly derived the stability criterion here, note that the analysis for the 2D case simply involves the examination of the 2D spatial-frequency spectrum or plane wave eigenmodes, which for the the 2D TE case can be written as:

$$\mathscr{E}_y\Big|_{i,j}^n = C e^{j[\omega_n n \Delta t + k_x i \Delta x + k_y j \Delta y]}. \tag{5.36}$$

In fact, the stability criteria for any method can be derived using the relation above (or the 1D or 3D versions where applicable). We will see this relation used in detail in the next chapter, when we derive the numerical dispersion for different methods.

Note that all the discussion of the CFL condition so far has assumed that we are operating in a simple medium with a constant v_p. As we will see in later chapters, the CFL condition can become more restrictive — and somewhat more difficult to define — in complex media, where the phase velocity may vary as a function of space, frequency, field direction, or even time.

5.4   Summary

In this chapter, we investigated the stability of a number of finite difference approximations to PDEs. For particular choices of parameters such as the time step Δt or spatial step Δx, certain finite difference methods can become unstable, meaning the solution grows exponentially with time in an unphysical way.

First, we introduced the von Neumann method for stability analysis. In this method, a general spatial Fourier component given by

$$\mathcal{V}_i^n = C(t^n)\, e^{jk\,i\Delta x}$$

is substituted into the FDE, along with its shifted versions (i.e., replacing i and n with $i \pm 1$ and $n \pm 1$) where applicable. The FDE is then written in the form $\mathcal{V}_i^{n+1} = q\, \mathcal{V}_i^n$, and the stability of the method is determined by the conditions under which the amplification factor $|q| \leq 1$.

We found that the forward-time centered space method applied to the convection equation was unconditionally unstable, but that the minor correction yielding the Lax method guaranteed stability for

$$\frac{v_p \Delta t}{\Delta x} \leq \frac{1}{\sqrt{D}} \qquad \text{(CFL condition)}$$

where D is the dimensionality of the problem. The leapfrog method applied to the convection equation yielded the same stability criterion.

For coupled differential equations such as the telegrapher's equations, the stability criteria are derived in a similar fashion, but in terms of an amplification matrix $[\mathbf{q}]$, where $\mathbf{V}_i^{n+1} = [\mathbf{q}]\mathbf{V}_i^n$, and the vector $\mathbf{V}_i^n = [\mathcal{V}_i^n ; \mathcal{I}_i^n]$. Despite this added complexity, the stability criteria for the coupled Lax and leapfrog methods are identical to those for the single convection equation.

5.5 Problems

5.1. Stability analysis for a lossy convection equation. Investigate the stability of the leapfrog algorithm for the lossy convection equation:

$$\frac{\partial \mathcal{V}}{\partial t} = -R\,\mathcal{V} - v_p \frac{\partial \mathcal{V}}{\partial x} \qquad \rightarrow$$

$$\mathcal{V}_i^{n+1} = \mathcal{V}_i^{n-1} - 2R\Delta t\, \mathcal{V}_i^n - \left(\frac{v_p \Delta t}{\Delta z}\right) \left[\mathcal{V}_{i+1}^n - \mathcal{V}_{i-1}^n\right]$$

where $v_p = (LC)^{-1/2}$. Carry out a stability analysis and assess the stability of the algorithm for the numerical values of the parameters (R, L, and C) given in Problem 4.2.

Hint: This problem can be solved analytically; but if you are having trouble with a complete mathematical analysis of the stability, just set up the stability equation and numerically plot out the imaginary part of k, the wavenumber, over a full range of parameters (e.g., frequency, Δt, R, and Δz) and investigate the conditions, if any, under which the algorithm is stable.

5.2. **Stability of the Runge-Kutta methods.** Derive the stability criteria for the second-order and fourth-order Runge Kutta methods, applied to ODEs. Consider the model equation $dy/dt = \lambda y$, and find the criteria for λ such that the methods are stable. Reproduce the diagrams in Figure 3.3.

5.3. **Stability of the coupled leapfrog algorithm.** Show that stability is ensured for Equation (5.28) if the eigenvalues of the matrix $\mathbf{A}$ are bounded by unity. Hint: You can write the matrix $\mathbf{A}$ as $\mathbf{A} = \mathbf{P}^{-1}\mathbf{DP}$, where $\mathbf{D}$ is a diagonal matrix whose eigenvalues are identical to those of $\mathbf{A}$. Then, find the condition on the eigenvalues of $\mathbf{D}$ such that the growth rate $q \leq 1$.

5.4. **Stability of the coupled leapfrog algorithm.** Investigate the stability of the leapfrog algorithm (Equations 5.26) in a simulation; you can reuse your code from Problem 3.5. Use $C = 96.8$ pF, $L = 0.242$ µH, and $\Delta x = 0.01$ m. Launch a Gaussian pulse (described by $\mathcal{V}(0, t) = V_0 e^{-(t-t_0)^2/\tau^2}$, where τ is the Gaussian pulse half-width, and should be at least 5–10 time steps. Choose t_0 such that the pulse slowly ramps up from zero; generally $t_0 \geq 3\tau$ is sufficient.

Vary the time step Δt, without changing the pulse shape, and investigate the stability of this algorithm. You will find it is most illustrative to look at the case where $v_p \Delta t/\Delta x$ is very close to one. How much larger than one can you make the CFL ratio before instability takes over? How does this depend on the length x of your simulation?

5.5. **Stability of the discretized wave equation.** Analyze the stability of the algorithm presented in Problem 3.4 using the von Neumann method. (This will require some steps similar to the leapfrog algorithm.) Does the stability criterion derived here agree with the results of Problem 3.4?

5.6. **Stability in the FDTD method.** Write a 1D FDTD simulation 400 grid cells long for a wavelength of 1 m, with $\Delta x = \lambda_0/20$. Launch a Gaussian pulse from the left end of the space with the same parameters as Problem 5.4 above. Now, increase Δt very slightly to excite an instability, similar to Figure 5.4. Can you measure the growth rate of this instability? Does this growth rate agree with q from Equation (5.14)?

5.7. **Stability in the 2D FDTD method.** Write a 2D TM mode FDTD simulation that is 100 grid cells in the y-dimension and 500 cells in the x-dimension. Use PEC at the far right boundary; at the top and bottom boundaries, set $\mathcal{E}_z|_{i,0} = \mathcal{E}_z|_{i,jmax}$ and $\mathcal{H}_x|_{i,jmax+1/2} = \mathcal{H}_x|_{i,1/2}$, i.e., replace the unknown fields with their counterparts on the opposite boundary. This is called a periodic boundary condition, which we will study in Chapter 12. Launch a plane wave from the leftmost boundary at a frequency suitable to the grid cell size.

Now, investigate the stability of this problem by varying Δt; how large can you make Δt before the simulation goes unstable? Can you surpass the CFL condition? Why or why not?

References

[1] J. von Neumann, "Proposal and analysis of a numerical method for the treatment of hydrody-namical shock problems," National Defense Research Committee, Tech. Rep. Report AM-551, 1944.

[2] J. von Neumann and R. D. Richtmeyer, "A method for numerical calculation of hydrodynamic shocks," *J. Appl. Phys.*, vol. 21, p. 232, 1950.

[3] P. D. Lax, "Weak solutions on non-linear hyperbolic equations and their numerical computa-tion," *Comm. on Pure and Appl. Math.*, vol. 7, p. 135, 1954.

[4] R. Courant, K. O. Friedrichs, and H. Lewy, "Uber die partiellen differenzengleichungen der mathematischen physik (translated as: On the partial differential equations of mathematical physics)," *Mathematische Annalen*, vol. 100, pp. 32–74, 1967.

6 Numerical dispersion and dissipation

The von Neumann stability analysis [1, 2] discussed in the previous chapter is widely applicable and enables the assessment of the stability of any finite difference scheme in a relatively simple manner. However, such an analysis reveals little about the detailed properties of the difference scheme, and especially the important properties of *dispersion* and *dissipation*. These two metrics together yield information about the *accuracy* of the finite difference algorithm.

In the continuous world, dispersion refers to the variation of the phase velocity v_p as a function of frequency or wavelength. Dispersion is present in all materials, although in many cases it can be neglected over a frequency band of interest. *Numerical dispersion* refers to dispersion that arises due to the discretization process, rather than the physical medium of interest. In addition, the discretization process can lead to *anisotropy*, where the phase velocity varies with propagation direction; this numerical anisotropy can also be separate from any real anisotropy of the medium.

In addition, as we saw in Figure 5.3, the finite difference scheme can lead to *dissipation*, which is nonphysical attenuation of the propagating wave. Recall that the example illustrated in Figure 5.3 was in a lossless physical medium, so any dissipation is due only to the discretization process.

In this chapter we will show how dispersion and dissipation arise in the finite difference algorithms that we have discussed so far, and show how to derive the *numerical dispersion relation* for any algorithm. In designing a finite difference algorithm, for an FDTD simulation or other types of finite difference simulations, it is most important to understand how much numerical dispersion is present is the algorithm of choice, and to assess how much dispersion the simulation can tolerate.

We begin by demonstrating how a dispersion relation is derived for a given algorithm. As in Chapter 5, the derivation of the dispersion relation, in either the discrete or continuous space, is rooted in Fourier analysis. In general, a partial differential equation couples points in space and time by acting as a transfer function, producing future values of a quantity distribution, e.g., $\mathcal{V}(x, t)$, from its initial values, e.g., $\mathcal{V}(x, 0)$. The properties of a PDE or a system of PDEs can be described by means of its effects on a single wave or Fourier mode in space and time:

$$\mathcal{V}(x, t) = C\, e^{j(\omega t + kx)} \tag{6.1}$$

where ω is the frequency of the wave, and k is the wavenumber corresponding to wavelength $\lambda = 2\pi/k$. Compare this to the Fourier modes that were used in Chapter 5, which had only spatial Fourier components. By inserting Equation (6.1) into a PDE and solving for ω, we obtain its *dispersion relation*,

$$\omega = f_1(k) \tag{6.2}$$

which relates the frequency ω, and thus corresponding time scale, to a particular wavelength (or wavenumber k) for the physical phenomenon described by the PDE. Note that, in general, the frequency ω may be real, when the PDE describes oscillatory or wave behavior, or it may be imaginary when the PDE describes the growth or decay of the Fourier mode, or a complex number in those cases where there is both oscillatory (wave-like) and dissipative behavior.

Numerical dispersion and dissipation occurs when the "transfer function" or the amplification factor of the corresponding FDE is not equal to that of the PDE, so that either phase (dispersion) or amplitude (dissipation) errors occur as a result of the finite difference approximation. Ideally, we assess the dispersive and dissipative properties of an FDE by obtaining the dispersion relation of the scheme, relating the frequency of a Fourier mode on the mesh to a particular wavelength λ (or wavenumber k):

$$\omega = f_2(k, \Delta x, \Delta t). \tag{6.3}$$

The applicability and accuracy of a finite difference scheme can be assessed and analyzed in detail by comparing Equations (6.2) and (6.3).

6.1 Dispersion of the Lax method

To illustrate the method, we consider once again the convection equation and the Lax method. First, the dispersion relation for the convection Equation (3.63) (the continuous PDE) is simply:

$$\omega = -v_p k. \tag{6.4}$$

In other words, the physical nature of the convection equation is such that ω can only be real, so that no damping (or growth) of any mode occurs. Furthermore, all wavenumbers have the same phase and group velocities, given by ω/k and $\partial\omega/\partial k$ respectively, so that there is no dispersion. The input property distribution is simply propagated in space, without any reduction in amplitude or distortion in phase. Note that the minus sign in Equation (6.4) is simply due the fact that we choose to use a Fourier mode with $+jkx$ in the exponent, which means that k has to be negative since the convection equation we have been using allows for waves traveling in the $+x$ direction.

To obtain the dispersion relation for the FDE, we rewrite the Lax method algorithm:

Lax method

$$V_i^{n+1} = \frac{V_{i+1}^n + V_{i-1}^n}{2} - \left(\frac{v_p \Delta t}{2\Delta x}\right)\left[V_{i+1}^n - V_{i-1}^n\right] \tag{6.5}$$

and consider a Fourier mode in time and space:

$$\mathcal{V}(x, t) = C\, e^{j(\omega t + kx)} \qquad \text{or} \qquad \mathcal{V}_i^n = C\, e^{j(wn\Delta t + ki\Delta x)} \tag{6.6}$$

where C is a constant. We now substitute Equation (6.6) into Equation (6.5) to find:

$$C e^{j[\omega(n+1)\Delta t + ki\Delta x]} = C e^{j(\omega n \Delta t + ki\Delta x)} \left[\frac{e^{jk\Delta x} + e^{-jk\Delta x}}{2} - \left(\frac{v_p \Delta t}{\Delta x}\right)\frac{e^{jk\Delta x} - e^{-jk\Delta x}}{2} \right].$$

Canceling $C e^{j\omega n \Delta t} e^{jki\Delta x}$ from each term, we find:

$$e^{j\omega \Delta t} = \cos(k\Delta x) - j\left(\frac{v_p \Delta t}{\Delta x}\right)\sin(k\Delta x). \tag{6.7}$$

This is the numerical dispersion relation for the Lax method, although it is not an explicit function of ω. Noting that, in general, the frequency ω can be complex, we can write it as $\omega = \omega_r + j\gamma$, and equate the amplitude and phase of both sides of Equation (6.7) to find explicit expressions for ω_r and γ. In other words,

$$e^{j(\omega_r + j\gamma)\Delta t} = \overbrace{e^{-\gamma \Delta t}}^{\text{Amplitude}} e^{j\omega_r \Delta t}$$

$$= \underbrace{\sqrt{\cos^2(k\Delta x) + \left(\frac{v_p \Delta t}{\Delta x}\right)^2 \sin^2(k\Delta x)}}_{\text{Amplitude}} \exp\left\{ j\,\tan^{-1}\left[-\left(\frac{v_p \Delta t}{\Delta x}\right)\frac{\sin(k\Delta x)}{\cos(k\Delta x)}\right] \right\}$$

$$\tag{6.8}$$

and thus the real and imaginary parts of ω are:

$$\tan(\omega_r \Delta t) = -\left(\frac{v_p \Delta t}{\Delta x}\right)\tan(k\Delta x)$$

$$\rightarrow \quad \omega_r = -\frac{1}{\Delta t}\tan^{-1}\left[\left(\frac{v_p \Delta t}{\Delta x}\right)\tan(k\Delta x)\right] \tag{6.9a}$$

$$e^{-2\gamma \Delta t} = \cos^2(k\Delta x) + \left(\frac{v_p \Delta t}{\Delta x}\right)^2 \sin^2(k\Delta x)$$

$$\rightarrow \quad \gamma = \frac{-1}{2\Delta t}\ln\left[\cos^2(k\Delta x) + \left(\frac{v_p \Delta t}{\Delta x}\right)^2 \sin^2(k\Delta x)\right]. \tag{6.9b}$$

Note that in the special case when $v_p \Delta t / \Delta x = 1$, we have $\gamma = 0$, and $\omega_r = -v_p k$, which is identical to the dispersion relation (6.4) for the PDE; for this reason, the choice of Δt such that $v_p \Delta t / \Delta x = 1$ is often known as the "magic" time step. In general, however, γ is nonzero for the Lax scheme and numerical dissipation occurs on the finite difference mesh. Also, since ω_r is not simply proportional to k, the phase and group velocities are both functions of the wavenumber, so that numerical dispersion occurs.

Plots of ω_r and γ for selected values of $v_p \Delta t / (\Delta x)$ are shown in Figure 6.1 as a function of $k\Delta x$. Apart from the Δt and Δx scaling factors, these are examples of what are known as ω–k diagrams, which are ubiquitous in the continuous world of

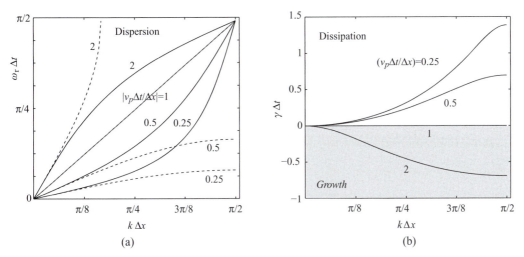

Figure 6.1 Dispersion and dissipation for the Lax and leapfrog methods, applied to the convection equation. The Lax method is shown in solid lines while the leapfrog method is shown in dashed lines. Note that the leapfrog method is non-dissipative.

electromagnetics for analyzing the dispersion relation in a medium. In our discretized world, they are extremely useful in comparing the numerical dispersion to the physical dispersion.

From Figure 6.1 it is clear that the effects of dissipation increase rapidly with increasing wavenumber, resulting in a large deviation of the amplitude of the solution from that of the PDE. At smaller wavenumbers, with the corresponding wavelengths being much larger than the spatial mesh size Δx, γ is small. Dispersion effects must be assessed by looking at the difference between the plots in Figure 6.1(a) and the ideal dispersion curve (i.e., $\omega_r = v_p k$). It appears that dispersion is present at all wavenumbers, and becomes smaller only when $v_p \Delta t / \Delta x$ is brought closer to unity.

A very important note must be made here on stability: the astute reader will observe from Equations (6.9) that the stability criterion has been inherently derived through this dispersion analysis. In Figure 6.1, when $\gamma < 0$, we have *growth* in the simulation, which is not physical, and results in instability. From Equation (6.9b), we see that $\gamma < 0$ if the argument of $\ln(\cdot)$ is greater than one; through arguments similar to Chapter 5, we see that this is the case for $v_p \Delta t / \Delta x > 1$, which is the CFL condition. In fact, the analyses presented in Chapter 5 can be replaced by the current method, which derives both stability and dispersion in one. This is true for each of the dispersion analyses we will conduct in this chapter.

Numerical dispersion and dissipation, especially for short wavelengths compared to a given spatial mesh size, result from replacing the PDEs with the corresponding FDE approximations. These numerical phenomena are particularly severe in finite difference schemes using first-order integrations in time, such as the Lax method. Note that the Lax method is stable subject to the CFL condition, but the time differentiation is still approximated with first-order differencing, using only $\mathcal{V}^n$ and $\mathcal{V}^{n+1}$. Methods which use higher accuracy in time, such as the leapfrog method, have fewer of these numerical problems, as we shall see below.

6.2   Dispersion of the leapfrog method

We now consider the dispersion and dissipation properties of the leapfrog scheme, repeated below:

$$\text{Leapfrog}$$

$$V_i^{n+1} = V_i^{n-1} - \left(\frac{v_p \Delta t}{\Delta z}\right)\left[V_{i+1}^n - V_{i-1}^n\right]. \tag{6.10}$$

Substituting the Fourier mode described by Equation (6.6) into Equation (6.10) above, and canceling common terms on both sides, we find:

$$e^{j\omega \Delta t} = e^{-j\omega \Delta t} - \left(\frac{v_p \Delta t}{\Delta x}\right)\left[e^{jk\Delta x} - e^{-jk\Delta x}\right]. \tag{6.11a}$$

Multiplying through by $e^{j\omega \Delta t}$:

$$e^{j\omega \Delta t}\left\{ e^{j\omega \Delta t} = e^{-j\omega \Delta t} - \left(\frac{v_p \Delta t}{\Delta x}\right)\left[e^{jk\Delta x} - e^{-jk\Delta x}\right] \right\} \tag{6.11b}$$

$$[e^{j\omega \Delta t}]^2 + j 2\left(\frac{v_p \Delta t}{\Delta x}\right)\sin(k\Delta x)[e^{j\omega \Delta t}] - 1 = 0. \tag{6.11c}$$

Equation (6.11c) is a quadratic equation in $[e^{j\omega \Delta t}]$, with the solution

$$[e^{j\omega \Delta t}] = \underbrace{\pm\sqrt{1 - \left(\frac{v_p \Delta t}{\Delta x}\right)^2 \sin^2(k\Delta x)}}_{\mathscr{R}e\{e^{j\omega \Delta t}\}} - \underbrace{j\left(\frac{v_p \Delta t}{\Delta x}\right)\sin(k\Delta x)}_{\mathscr{I}m\{e^{j\omega \Delta t}\}} \tag{6.12}$$

where we have implied the stability criterion $v_p \Delta t / \Delta x \leq 1$, otherwise the term under the square root is not necessarily real! We note that the magnitude of $[e^{j\omega \Delta t}]$ is unity for all values of k, since

$$\left|e^{j\omega \Delta t}\right| = \sqrt{\left[1 - \left(\frac{v_p \Delta t}{\Delta x}\right)^2 \sin^2(k\Delta x)\right] + \left[\left(\frac{v_p \Delta t}{\Delta x}\right)\sin(k\Delta x)\right]^2} = 1, \tag{6.13}$$

revealing an important feature of the leapfrog scheme, i.e., that it is non-dissipative. It is one of the few finite difference schemes that propagates an initial wave property without decreasing its amplitude. In other words, the frequency ω for this scheme is purely real, i.e., $\gamma = 0$, as long as the stability criterion is met. Note that if the stability criterion is *not* met, the term $e^{j\omega \Delta t}$ will be non-unity and complex, implying growth; it is straightforward to show that such scenarios always lead to growth rather than dissipation.

Going back to Equation (6.11a), we can rewrite it as:

$$e^{j\omega\Delta t} - e^{-j\omega\Delta t} = -\left(\frac{v_p\Delta t}{\Delta x}\right)\left[e^{jk\Delta x} - e^{-jk\Delta x}\right] \tag{6.14a}$$

$$\sin(\omega\Delta t) = -\left(\frac{v_p\Delta t}{\Delta x}\right)\sin(k\Delta x). \tag{6.14b}$$

Equation (6.14b) is the numerical dispersion relation for the leapfrog scheme in terms of real frequency ω. To assess the dispersive properties of this scheme, we need to compare it to the ideal dispersion relation for the convection PDE, namely Equation (6.4). This can be done by plotting ω versus k as was done in Figure 6.1 for the Lax method.

We can solve for ω from Equation (6.14b) as

$$\omega = \frac{-1}{\Delta t}\sin^{-1}\left[\left(\frac{v_p\Delta t}{\Delta x}\right)\sin(k\Delta x)\right] \tag{6.15}$$

and expand Equation (6.15) into a power series[1] for small $(k\Delta x)$ as:

$$\omega = -v_p k\left[1 - \frac{1-(v_p\Delta t/\Delta x)^2}{6}(k\Delta x)^2\right.$$
$$\left. + \frac{1-10(v_p\Delta t/\Delta x)^2 + 9(v_p\Delta t/\Delta x)^4}{120}(k\Delta x)^4 + \cdots\right]. \tag{6.16}$$

We can then write an approximate expression for the numerical phase velocity $\bar{v}_p \equiv \omega/k$ from Equation (6.16):

$$\bar{v}_p = \frac{-1}{k\Delta t}\sin^{-1}\left[\left(\frac{v_p\Delta t}{\Delta x}\right)\sin(k\Delta x)\right] \simeq -v_p\left[1 - \frac{1-(v_p\Delta t/\Delta x)^2}{6}(k\Delta x)^2\right]. \tag{6.17}$$

Thus we note that for $(k\Delta x) \simeq 0$, the first term is indeed in agreement with Equation (6.4), as must be true for any consistent difference scheme. It is also evident from Equation (6.17) that the errors increase in this scheme as $(k\Delta x)^2$. In general, the phase velocity is a function of wavenumber k, which is precisely the effect of dispersion.

The numerical dispersion effects can sometimes be better understood in terms of the numerical group velocity $\bar{v}_g \equiv d\omega/dk$. Differentiating (6.14b) we can write

$$\Delta t\cos(\omega\Delta t)d\omega = -\Delta x\left(\frac{v_p\Delta t}{\Delta x}\right)\cos(k\Delta x)dk, \tag{6.18}$$

---

[1] Note that:

$$\sin^{-1}x = x + \frac{x^3}{2\cdot 3} + \frac{1\cdot 3\,x^5}{2\cdot 4\cdot 5} + \frac{1\cdot 3\cdot 5\,x^7}{2\cdot 4\cdot 6\cdot 7} + \cdots$$
$$\sin x = x - \frac{x^3}{3!} + \frac{x^5}{5!} - \frac{x^7}{7!} + \cdots.$$

which gives

$$\bar{v}_g = -v_p \frac{\cos(k\Delta x)}{\cos(\omega \Delta t)}. \tag{6.19}$$

This result indicates that the effects of the discretization in space and time affect the group velocity in opposite ways. For small Δx and Δt, spatial discretization tends to decrease $|v_g|$, while discretization in time tends to increase it. Noting that we must have $\Delta t \leq \Delta x / v_p$ for stability, the first effect dominates, and the numerical group velocity is smaller than the group velocity for the PDE.

Using Equation (6.15), we can eliminate $\omega \Delta t$ in Equation (6.19) to write:

$$\bar{v}_g = \frac{-v_p \cos(k\Delta x)}{\sqrt{1 - (v_p \Delta t / \Delta x)^2 \sin^2(k\Delta x)}} \simeq -v_p \left[1 - \frac{1 - (v_p \Delta t / \Delta x)^2}{2} (k\Delta x)^2\right]. \tag{6.20}$$

Comparing Equations (6.17) and (6.20) with Equation (6.4), we find that both $\bar{v}_p$ and $\bar{v}_g$ are less than v_p in magnitude, but that $\bar{v}_g$ is different by a significantly larger amount than $\bar{v}_p$ — by a factor of three, in fact.

The dispersive properties of a finite difference scheme can also be analyzed by considering the evolution in time of a given initial property distribution. As an example, assume an initial voltage distribution given by:

$$\mathcal{V}(x, 0) = A \sin(\pi x). \tag{6.21}$$

Note that this corresponds to a wavenumber $k = \pi$ rad-(ft)$^{-1}$, with x given in feet. As a specific numerical example, consider a simple case with $v_p \simeq 1$ ft-(ns)$^{-1}$. The analytical solution of the convection equation tells us that the voltage at any future time is given by:

$$\mathcal{V}(x, t) = A \sin[\pi (x - v_p t)].$$

As an example, the voltage at $t = 10$ ns is given by:

$$\mathcal{V}(x, t = 10 \text{ ns}) = A \sin[\pi (x - 10)]. \tag{6.22}$$

To assess the dispersive property of the finite difference scheme, we may compare the numerical solution at time $t = 10$ ns to Equation (6.22). For this purpose, we need only to consider the amplification factor for the scheme, which for the leapfrog method is given in Equation (5.14) as:

$$q = \pm \sqrt{1 - \left[\left(\frac{v_p \Delta t}{\Delta x}\right) \sin(k\Delta x)\right]^2} + j \left(\frac{v_p \Delta t}{\Delta x}\right) \sin(k\Delta x) = e^{j\psi}$$

where

$$\psi = \tan^{-1} \left[\frac{\left(\dfrac{v_p \Delta t}{\Delta x}\right) \sin(k\Delta x)}{\pm \sqrt{1 - \left[\left(\dfrac{v_p \Delta t}{\Delta x}\right) \sin(k\Delta x)\right]^2}}\right]$$

and the magnitude of q is always unity for this scheme (i.e., non-dissipative).

Consider a propagating wave with a wavelength of 2 ft, solved with a discretization scheme using $\Delta t = 0.025$ ns and $\Delta x = 0.05$ ft (i.e., 40 grid cells per wavelength). In this case we have $k\Delta x = 0.05\pi$, and $[v_p \Delta t/(\Delta x)] = 0.5$, so that we have:

$$\psi = \tan^{-1}\left[\frac{(0.5)\sin(0.05\pi)}{\sqrt{1 - [0.5\,\sin(0.05\pi)]^2}}\right] \simeq 0.024923\pi.$$

The solution at $t = 10$ ns can be found by repeating the finite difference scheme $[10/(\Delta t)] = 400$ times. Thus we have

$$\bar{\mathcal{V}}(x, 10\text{ ns}) \simeq q^{400}\,\mathcal{V}(x, 0) = \left[e^{j0.024923\pi}\right]^{400} A\sin(\pi x) = A\sin[\pi(x - 9.9688)],$$

(6.23)

which is only slightly different from the ideal solution in Equation (6.22). This is because this particular wavelength, i.e., $\lambda = 2\pi/k = 2$ ft, is quite nicely resolved with a spatial mesh size of 0.05 ft. If, instead, we had chosen $\Delta x = 0.25$ ft and kept the ratio $v_p \Delta t/\Delta x = 0.5$ (so that $\Delta t = 0.125$ ns), we find $\psi = 0.115027\pi$, and at $t = 10$ ns the wave has the form:

$$\bar{\mathcal{V}}(x, 10\text{ ns}) \simeq q^{80}\,\mathcal{V}(x, 0) = \left[e^{j0.115027\pi}\right]^{80} A\sin(\pi x) = A\sin[\pi(x - 9.2021)]$$

where we note that for $\Delta t = 0.125$ ns, the wave has propagated 10 ft in 80 time steps rather than 400 as before. So, with the more coarse resolution of $\Delta x = 0.25$ ft, corresponding to eight grid cells per wavelength, the error that has accumulated after 10 ns is almost 0.8 ft or 0.4λ.

Going back to Equation (6.12) for the solution of $[e^{j\omega\Delta t}]$, we can explicitly solve for ω to find:

$$\omega_{\pm} = -\frac{1}{\Delta t}\tan^{-1}\left[\frac{\left(\dfrac{v_p\Delta t}{\Delta x}\right)\sin(k\Delta x)}{\pm\sqrt{1 - \left[\left(\dfrac{v_p\Delta t}{\Delta x}\right)\sin(k\Delta x)\right]^2}}\right]$$

(6.24)

which indicates that, for a given value of k, the leapfrog solution supports two different wave speeds, $\bar{v}_p^{\pm} = \omega_{\pm}/k$ that are the negative of one another. Since the frequency supported by the actual corresponding PDE is negative (for positive k) as given in Equation (6.4), the "correct" root in Equation (6.24) is ω_+, which is also called the *principal* root. The other root ω_- is called the *parasitic* root, since it is an artifact of the differencing scheme and propagates in the direction opposite to the principal mode. This effect is due to the effective up-coupling of the odd and even meshes in the leapfrog scheme, which can even be seen by looking back at Chapter 3 on the use of the leapfrog method for integration of a first-order ODE. In general, care must be taken in starting the leapfrog scheme to minimize the excitation of the undesirable mode.

We note in passing that some of the other schemes that were briefly introduced in Chapter 3, especially the two-step Lax-Wendroff method, do not introduce any extraneous numerical modes and may thus be desirable in cases where more accuracy is desired. These schemes do have a small amount of dissipation, only to fourth order in $(k\Delta x)$, which actually is sometimes useful in smoothing discontinuities on the mesh.

6.3   Dispersion relation for the FDTD algorithm

The numerical dispersion effects in the interleaved leapfrog (or FDTD) scheme are expected to be less severe than the single-equation leapfrog scheme since the second-order differencing is done across one step size rather than two steps (i.e., across Δt and Δx rather than $2\Delta t$ and $2\Delta x$). We now derive the numerical dispersion relation for the interleaved leapfrog algorithm, repeated below for convenience:

Interleaved leapfrog

$$\mathcal{V}_i^{n+1} = \mathcal{V}_i^n - \left(\frac{\Delta t}{C\,\Delta z}\right)\left[\mathcal{I}_{i+1/2}^{n+1/2} - \mathcal{I}_{i-1/2}^{n+1/2}\right]$$

$$\mathcal{I}_{i+1/2}^{n+1/2} = \mathcal{I}_{i+1/2}^{n-1/2} - \left(\frac{\Delta t}{L\,\Delta z}\right)\left[\mathcal{V}_{i+1}^n - \mathcal{V}_i^n\right].$$

(6.25)

The dispersion analysis for a system of FDEs is performed by substituting the general Fourier mode in space and time for both of the dependent variables:

$$\mathcal{V}(x,t) = C_1\,e^{j(\omega t + kx)} \qquad \mathcal{V}_i^n = C_1\,e^{j(wn\Delta t + ki\Delta x)}$$

$$\mathcal{I}(x,t) = C_2\,e^{j(\omega t + kx)} \qquad \mathcal{I}_i^n = C_2\,e^{j(wn\Delta t + ki\Delta x)}.$$

(6.26)

Substituting Equation (6.26) into Equation (6.25) and manipulating yields:

$$\begin{bmatrix} 2j\dfrac{\sin(\omega\Delta t/2)}{\Delta t} & 2j\dfrac{\sin(k\Delta x/2)}{C\,\Delta x} \\ 2j\dfrac{\sin(k\Delta x/2)}{L\,\Delta x} & 2j\dfrac{\sin(\omega\Delta t/2)}{\Delta t} \end{bmatrix}\begin{bmatrix} C_1 \\ C_2 \end{bmatrix} = \begin{bmatrix} 0 \\ 0 \end{bmatrix}.$$

(6.27)

The numerical dispersion relation is then given by the determinant of the multiplying matrix:

$$\begin{vmatrix} 2j\dfrac{\sin(\omega\Delta t/2)}{\Delta t} & 2j\dfrac{\sin(k\Delta x/2)}{C\,\Delta x} \\ 2j\dfrac{\sin(k\Delta x/2)}{L\,\Delta x} & 2j\dfrac{\sin(\omega\Delta t/2)}{\Delta t} \end{vmatrix} = 0$$

$$\rightarrow \qquad \sin^2\left(\frac{\omega\Delta t}{2}\right) = \frac{1}{LC}\left(\frac{\Delta t}{\Delta x}\right)^2\sin^2\left(\frac{k\Delta x}{2}\right)$$

$$\rightarrow \qquad \sin\left(\frac{\omega\Delta t}{2}\right) = \pm\left(\frac{v_p\,\Delta t}{\Delta x}\right)\sin\left(\frac{k\Delta x}{2}\right)$$

(6.28)

which is to be compared with Equation (6.14b). Note that the system of equations (6.25) that are an approximation for (i.e., the transmission line equations) allow propagation of waves in both directions.

As was done before for the leapfrog case (see 6.15), we can explicitly solve for ω from Equation (6.28), or we could explicitly solve for k:

$$k = \frac{2}{\Delta x} \sin^{-1} \left[\left(\frac{\Delta x}{v_p \, \Delta t} \right) \sin \left(\frac{\omega \Delta t}{2} \right) \right]$$

or

$$\omega = \frac{2}{\Delta t} \sin^{-1} \left[\left(\frac{v_p \, \Delta t}{\Delta x} \right) \sin \left(\frac{k \Delta x}{2} \right) \right] \tag{6.29}$$

so that the numerical phase velocity is given by

$$\bar{v}_p \equiv \frac{\omega}{k} = \frac{\omega \Delta x}{2 \sin^{-1} \left[\left(\frac{\Delta x}{v_p \, \Delta t} \right) \sin \left(\frac{\omega \Delta t}{2} \right) \right]}$$

$$= \frac{2}{k \Delta t} \sin^{-1} \left[\left(\frac{v_p \, \Delta t}{\Delta x} \right) \sin \left(\frac{k \Delta x}{2} \right) \right]. \tag{6.30}$$

It is comforting to note that, for $\Delta t \to 0$ and $\Delta x \to 0$, Equation (6.30) reduces to

$$\bar{v}_p \simeq \pm \frac{\omega}{k} = \pm v_p. \tag{6.31}$$

Furthermore, for $v_p \Delta t = \Delta x$, Equation (6.28) reduces to the dispersion relation of the associated PDE, namely $\omega = \pm v_p \, k$. To see this, note that:

$$\sin \left(\frac{\omega \Delta t}{2} \right) = \pm \left(\frac{v_p \, \Delta t}{v_p \, \Delta t} \right) \sin \left(\frac{k \, v_p \, \Delta t}{2} \right) \tag{6.32}$$

$$\to \quad \omega \Delta t = \pm k \, v_p \, \Delta t$$

$$\to \quad \omega = \pm v_p k.$$

Thus, we see once again that the FDTD algorithm exhibits no numerical dispersion when the algorithm is operated at the CFL limit, or at the "magic" time step.

6.3.1 Group velocity

The expressions above derived the numerical phase velocity $\bar{v}_p$ for the finite difference scheme, which is a measure of the numerical dispersion. Often, we are also interested in the effect of numerical dispersion on the group velocity v_g, defined as $v_g = \partial \omega / \partial k$. An expression for the numerical group velocity $\bar{v}_g$ can be derived by differentiating both

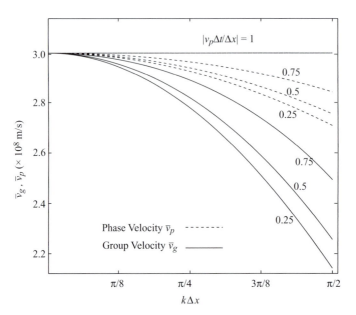

Figure 6.2 Numerical phase and group velocity for the FDTD algorithm. Both $\bar{v}_g$ and $\bar{v}_p$ tend toward the real value of 3×10^8 m/s when $v_p \Delta t / \Delta x$ approaches 1, or when k approaches zero.

sides of Equation (6.28) to find:

$$\frac{\Delta t}{2} \cos\left(\frac{\omega \Delta t}{2}\right) d\omega = \pm \left(\frac{\Delta x}{2}\right) \left(\frac{v_p \Delta t}{\Delta x}\right) \cos\left(\frac{k \Delta x}{2}\right) dk$$

$$\cos\left(\frac{\omega \Delta t}{2}\right) d\omega = \pm v_p \cos\left(\frac{k \Delta x}{2}\right) dk$$

$$\bar{v}_g \equiv \frac{d\omega}{dk} = \pm v_p \frac{\cos\left(\dfrac{k \Delta x}{2}\right)}{\cos\left(\dfrac{\omega \Delta t}{2}\right)}. \tag{6.33}$$

Using (6.28), we can rewrite the group velocity expression in Equation (6.33) as:

$$\bar{v}_g = \pm v_p \frac{\cos\left(\dfrac{k \Delta x}{2}\right)}{\sqrt{1 - \left(\dfrac{v_p \Delta t}{\Delta x}\right)^2 \sin^2\left(\dfrac{k \Delta x}{2}\right)}}. \tag{6.34}$$

Once again, note that at the CFL limit where $v_p \Delta t = \Delta x$, this expression reduces to $\bar{v}_g = \pm v_p$; that is, the numerical group velocity is equal to the real phase velocity. In the simple media we are considering here, the phase velocity is equal to the group velocity, so this is equivalent to writing $\bar{v}_g = v_g$, and hence there is no numerical dispersion.

When $v_p \Delta t < \Delta x$, the group velocity deviates somewhat from the physical scenario where $v_g = v_p$. Figure 6.2 shows the numerical group and phase velocities

$\bar{v}_p$ and $\bar{v}_g$ as a function of $k\Delta x$. It is apparent that both $\bar{v}_g$ and $\bar{v}_p$ tend toward the real value of 3×10^8 m/s when $v_p\Delta t/\Delta x$ approaches 1, or when k approaches zero. For $v_p\Delta t/\Delta x < 1$, the numerical phase and group velocities both decrease with increasing k (or increasing Δx), and $\bar{v}_g$ falls faster than $\bar{v}_p$.

6.3.2 Dispersion of the wave equation

The numerical dispersion relation in Equation (6.28) for the FDTD algorithm can also be derived directly from the second-order wave equation, which for the 1D case, say for the TE mode, is given by

$$\frac{\partial^2 \mathscr{E}_y}{\partial x^2} - \frac{1}{v_p^2}\frac{\partial^2 \mathscr{E}_y}{\partial t^2} = 0. \tag{6.35}$$

The FDTD discretization of Maxwell's curl equations, carried out on a staggered mesh with leapfrog integration in time, is numerically equivalent to discretization of the second derivatives with second-order, three-point, centered time, and centered space discretization of Equation (6.35) as follows:

$$\frac{\mathscr{E}_y\Big|_{i+1}^n - 2\mathscr{E}_y\Big|_i^n + \mathscr{E}_y\Big|_{i-1}^n}{(\Delta x)^2} - \frac{1}{v_p^2}\frac{\mathscr{E}_y\Big|_i^{n+1} - 2\mathscr{E}_y\Big|_i^n + \mathscr{E}_y\Big|_i^{n-1}}{(\Delta t)^2} = 0. \tag{6.36}$$

The dispersion relation in Equation (6.28) is obtained by substitution into Equation (6.36) with a Fourier mode given by:

$$\mathscr{E}_y(x, t) = C\, e^{j(\omega t + kx)} \qquad \mathscr{E}_y\Big|_i^n = C\, e^{j(\omega n\Delta t + ki\Delta x)}. \tag{6.37}$$

Inserting this expression into Equation (6.36) and canceling like terms, we find:

$$e^{j\omega\Delta t} - 2 + e^{-j\omega\Delta t} = \left(\frac{v_p\Delta t}{\Delta x}\right)^2 \left(e^{jk\Delta x} - 2 + e^{-jk\Delta x}\right)$$

$$2\left(\cos(\omega\Delta t) - 1\right) = \left(\frac{v_p\Delta t}{\Delta x}\right)^2 2\left(\cos(k\Delta x) - 1\right).$$

Using the trigonometric identity $1 - \cos(2A) = \sin^2(A)$ we find:

$$\sin\left(\frac{\omega\Delta t}{2}\right)^2 = \left(\frac{v_p\Delta t}{\Delta x}\right)^2 \sin\left(\frac{k\Delta x}{2}\right)^2 \tag{6.38}$$

which is exactly the dispersion relation that was found in Equation (6.28) for the interleaved leapfrog method, which is identical to the FDTD method.

6.3.3 Dispersion relation in 2D and 3D

The numerical dispersion relation is derived for either 2D or 3D problems in a very similar manner to that described above. Here we will focus on the 2D problem to illustrate the method. Just as above, the dispersion relation can be derived either from the coupled first-order interleaved leapfrog equations (for example, Equations 4.17 for the TM mode), or from the wave equation. Let us first consider the wave equation, which is a much simpler derivation. The wave equation in 2D is valid for each of the electric field components and is given for the 2D case as:

$$\nabla^2 \overline{\mathscr{E}} - \frac{1}{v_p^2} \frac{\partial^2 \overline{\mathscr{E}}}{\partial t^2} = 0 \qquad \rightarrow \qquad \frac{\partial^2 \mathscr{E}_y}{\partial x^2} + \frac{\partial^2 \mathscr{E}_y}{\partial y^2} - \frac{1}{v_p^2} \frac{\partial^2 \mathscr{E}_y}{\partial t^2} = 0 \qquad (6.39)$$

the discretized version of which is

$$\frac{\mathscr{E}_y\Big|_{i+1,j}^n - 2\mathscr{E}_y\Big|_{i,j}^n + \mathscr{E}_y\Big|_{i-1,j}^n}{(\Delta x)^2} + \frac{\mathscr{E}_y\Big|_{i,j+1}^n - 2\mathscr{E}_y\Big|_{i,j}^n + \mathscr{E}_y\Big|_{i,j-1}^n}{(\Delta y)^2}$$

$$- \frac{1}{v_p^2} \frac{\mathscr{E}_y\Big|_{i,j}^{n+1} - 2\mathscr{E}_y\Big|_{i,j}^n + \mathscr{E}_y\Big|_{i,j}^{n-1}}{(\Delta t)^2} = 0. \qquad (6.40)$$

In two dimensions, we are interested in wave-like modes propagating in both x and y directions given by

$$j = \sqrt{-1}$$
$$\downarrow$$

$$\mathscr{E}_y(x, y, t) = C\, e^{j(\omega t + k_x x + k_y y)} \qquad \mathscr{E}_y\Big|_{i,j}^n = C\, e^{j(\omega n \Delta t + k_x i \Delta x + k_y j \Delta y)} \qquad (6.41)$$
$$\uparrow$$
$$j = \text{mesh index}$$

where k_x and k_y are the wavenumbers respectively in the x and y directions. Substituting Equation (6.41) into Equation (6.40) we find (after some manipulation):

$$\sin^2\left(\frac{\omega \Delta t}{2}\right) = \left(\frac{v_p \Delta t}{\Delta x}\right)^2 \sin^2\left(\frac{k_x \Delta x}{2}\right) + \left(\frac{v_p \Delta t}{\Delta y}\right)^2 \sin^2\left(\frac{k_y \Delta y}{2}\right) \qquad (6.42)$$

which is a straightforward step in understanding from the 1D case. The reader should be easily convinced that the 3D dispersion relation is given by:

$$\sin^2\left(\frac{\omega \Delta t}{2}\right) = \left(\frac{v_p \Delta t}{\Delta x}\right)^2 \sin^2\left(\frac{k_x \Delta x}{2}\right) \qquad (6.43)$$

$$+ \left(\frac{v_p \Delta t}{\Delta y}\right)^2 \sin^2\left(\frac{k_y \Delta y}{2}\right) + \left(\frac{v_p \Delta t}{\Delta z}\right)^2 \sin^2\left(\frac{k_z \Delta z}{2}\right).$$

It can be shown (by expanding the $\sin(\cdot)$ terms into a power series and dropping all except the first terms) that as Δx, Δy, $\Delta z \to 0$, and $\Delta t \to 0$, Equations (6.42) and (6.43) reduce to $\omega = \pm v_p \sqrt{k_x^2 + k_y^2}$ in 2D and $\omega = \pm v_p \sqrt{k_x^2 + k_y^2 + k_z^2}$ in 3D, which are the dispersion relations for the continuous case (PDE) in two and three dimensions. Thus, numerical dispersion can be reduced to any desired amount by using a fine enough FDTD mesh.

Note that the dispersion relations in 2D and 3D could also have been derived along similar lines to Equations (6.25)–(6.28), i.e., directly from the FDTD update equations. We leave this derivation up to the reader in the problems at the end of this chapter.

Once again, we see that the stability criterion is inherently contained in the dispersion analysis. Consider the case where $\Delta x = \Delta y = \Delta z$; if $v_p \Delta t / \Delta x > 1/\sqrt{3}$, it will lead to a scenario where the right-hand side of Equation (6.43) is greater than one (the same is true for Equation (6.42) if $v_p \Delta t / \Delta x > 1/\sqrt{2}$). The left-hand side can only be greater than one for complex ω, which implies either growth or dissipation; again, one can show that this scenario always leads to growth.

Equation (6.42) also shows that the numerical phase velocity $\bar{v}_p$ for the 2D case is a function of angle of propagation through the FDTD mesh. To see this, assume wave propagation at an angle θ with respect to the positive x-axis, in which case $k_x = k \cos \theta$ and $k_y = k \sin \theta$, where $k = \sqrt{k_x^2 + k_y^2}$ is the wavenumber. The numerical phase velocity is then given by

$$\sin\left(\frac{\omega \Delta t}{2}\right) = \pm v_p \Delta t \sqrt{\frac{1}{(\Delta x)^2} \sin^2\left(\frac{k_x \Delta x}{2}\right) + \frac{1}{(\Delta y)^2} \sin^2\left(\frac{k_y \Delta y}{2}\right)}$$

$$\bar{v}_p = \frac{\omega}{k} = \frac{\pm 2}{k \Delta t} \sin^{-1}\left\{ v_p \Delta t \sqrt{\frac{1}{(\Delta x)^2} \sin^2\left(\frac{k \cos \theta \Delta x}{2}\right) + \frac{1}{(\Delta y)^2} \sin^2\left(\frac{k \sin \theta \Delta y}{2}\right)} \right\}.$$

$$(6.44)$$

The dependence of $\bar{v}_p$ on propagation angle θ is known as *grid anisotropy* and is a source of additional numerical dispersion effects. To illustrate this grid anisotropy, consider the case where $\Delta x = \Delta y$. The equation for the numerical phase velocity then becomes:

$$\bar{v}_p = \pm \frac{2}{k \Delta t} \sin^{-1}\left\{ \frac{v_p \Delta t}{\Delta x} \sqrt{\sin^2\left(\frac{k \cos \theta \Delta x}{2}\right) + \sin^2\left(\frac{k \sin \theta \Delta x}{2}\right)} \right\}. \quad (6.45)$$

Note that in 2D, the CFL condition is such that $v_p \Delta t / \Delta x \leq 1/\sqrt{2}$. This function is plotted in Figure 6.3 as a function of wavelength, for variations in grid cells per wavelength (equivalent to $k \Delta x$; 6.3a) and CFL condition $v_p \Delta t / \Delta x$ (6.3b). The most obvious trends are as expected: first, with increased *resolution*, i.e., more grid cells per wavelength, the anisotropy is reduced. Second, the anisotropy is reduced closer to the CFL limit of $v_p \Delta t / \Delta x = 1/\sqrt{2}$. However, two other features are of note: first, the anisotropy is generally very small. Even for only five cells per wavelength, the anisotropy has a maximum of only 6%, and at a more realistic resolution of 10 cells per wavelength, it is only 1.5%. Second, however, the anisotropy is never fully eliminated, even at the CFL limit, since we are limited to a finite number of grid cells per wavelength.

6.3.4 Numerical dispersion of lossy Maxwell's equations

In most cases of real interest, of course, dispersion and/or dissipation may be present in the physical medium, and this dispersion/dissipation is separate from the numerical

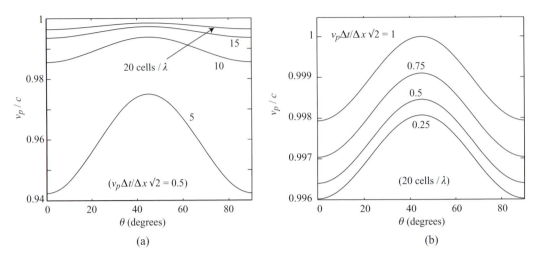

Figure 6.3 Grid anisotropy in the 2D FDTD method. Panel (a) shows the numerical phase velocity $\bar{v}_p$ as a function of propagation direction θ for varied sampling density, in grid cells per wavelength. Observe that higher sampling resolution reduces the anisotropy. Panel (b) shows the variation with $v_p \Delta t / \Delta x$ for 20 cells per wavelength; again, notice that closer to the CFL limit, the anisotropy is the smallest.

dispersion due to the differencing scheme. We will introduce dispersive materials in Chapter 10; in the meantime, consider the 1D Maxwell's equations in lossy media, modified from Equations (4.9) for the TM mode:

$$
\mathcal{H}_y \Big|_{i+1/2,j}^{n+1/2} = \left[\frac{2\mu - \sigma_m \Delta t}{2\mu + \sigma_m \Delta t} \right] \mathcal{H}_y \Big|_{i+1/2,j}^{n-1/2} + \left[\frac{2\Delta t}{2\mu + \sigma_m \Delta t} \right] \frac{1}{\Delta x} \left[\mathcal{E}_z \Big|_{i+1,j}^{n} - \mathcal{E}_z \Big|_{i,j}^{n} \right]
$$

(6.46a)

$$
\mathcal{E}_z \Big|_{i,j}^{n+1} = \left[\frac{2\epsilon - \sigma \Delta t}{2\epsilon + \sigma \Delta t} \right] \mathcal{E}_z \Big|_{i,j}^{n} + \left[\frac{2\Delta t}{2\epsilon + \sigma \Delta t} \right] \frac{1}{\Delta x} \left[\mathcal{H}_y \Big|_{i+1/2,j}^{n+1/2} - \mathcal{H}_y \Big|_{i-1/2,j}^{n+1/2} \right].
$$

(6.46b)

Substituting expressions similar to Equation (6.37) for both $\mathcal{H}_y$ and $\mathcal{E}_z$ we find:

$$
\mathcal{H}_y \Big|_i^{n} e^{j\omega\Delta t/2} = C_1 \mathcal{H}_y \Big|_i^{n} e^{-j\omega\Delta t/2} + \frac{C_2}{\Delta x} \mathcal{E}_z \Big|_i^{n} 2j \sin\left(\frac{k_x \Delta x}{2} \right)
$$

(6.47a)

$$
\mathcal{E}_z \Big|_i^{n} e^{j\omega\Delta t/2} = C_1' \mathcal{E}_z \Big|_i^{n} e^{-j\omega\Delta t/2} + \frac{C_2'}{\Delta x} \mathcal{H}_y \Big|_i^{n} 2j \sin\left(\frac{k_x \Delta x}{2} \right)
$$

(6.47b)

where we have used C_1, C_2, C_1', and C_2' to denote the multiplying constants just as in Table 4.2. Simplifying, we find the dispersion relation:

$$
\left(e^{j\omega\Delta t/2} - C_1 e^{-j\omega\Delta t/2} \right) \left(e^{j\omega\Delta t/2} - C_1' e^{-j\omega\Delta t/2} \right) = \frac{C_2 C_2'}{(\Delta x)^2} \left[2j \sin\left(\frac{k_x \Delta x}{2} \right) \right]^2.
$$

(6.48)

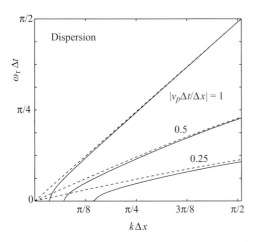
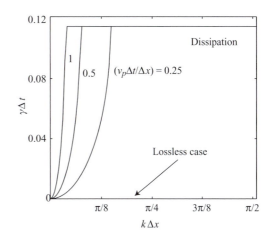

Figure 6.4 Dispersion and dissipation for the FDTD method, for lossless and lossy media. The FDTD method for the lossless case is shown in dashed lines while the solid lines denote a lossy algorithm with $\sigma = 0.002$ S/m. Note that the FDTD algorithm for the lossless case is non-dissipative.

This equation unfortunately does not simplify down to simple sinusoids as in the lossless version above. However, we can investigate the dispersion in this method visually, by solving for ω (this is rather involved) and plotting the real and imaginary parts, as shown in Figure 6.4. Dashed lines show the dispersion for the case $\sigma = \sigma_m = 0$, i.e., the lossless case; solid lines show the dispersion and dissipation for the lossy case with $\sigma = 0.002$ S/m and $\sigma_m = 0$. Note that there is no dissipation for the lossless case.

When $\sigma = 0$, the dispersion looks very similar to the leapfrog case shown in Figure 6.1. For the lossy case, the dispersion has non-zero values of k for zero w, but tends toward the lossless case as k increases. Note that if σ is very large, say 0.01, the dispersion profile will not approach the lossless curve within realistic values of k; for that reason, with large σ very small time steps Δt are required.

Note that the dissipation $\gamma \Delta t$ starts at zero for $k = 0$ and then tends toward a constant value for increasing k. This is as expected, since we input a constant σ; we would expect the dissipation to be a constant function of k, and independent of Δt. The ramp up to this constant value, however, is evidence of numerical dissipation due to the grid. Note that the dissipation ramps up to its constant value faster for larger values of $v_p \Delta t / \Delta x$, i.e., smaller Δx, up to the limit of the CFL condition.

6.4 Numerical stability of the FDTD algorithm revisited

Recall that at the end of Chapter 5, we postponed the derivation of the stability criterion for the FDTD algorithm. Here we will provide a very simple analysis that shows that the CFL condition is, in fact, the stability criterion, despite the fact that we did not use the von Neumann method to derive it.

Consider the 3D equivalent of Equation (6.42), written in a slightly different form:

$$\sin\left(\frac{\omega \Delta t}{2}\right)$$

$$= \pm v_p \Delta t \sqrt{\left[\frac{1}{\Delta x} \sin\left(\frac{k_x \Delta x}{2}\right)\right]^2 + \left[\frac{1}{\Delta y} \sin\left(\frac{k_y \Delta y}{2}\right)\right]^2 + \left[\frac{1}{\Delta z} \sin\left(\frac{k_z \Delta z}{2}\right)\right]^2}.$$

$$(6.49)$$

Solving for ω, we find:

$$\omega = \pm \frac{2}{\Delta t} \sin^{-1}\left(v_p \Delta t \sqrt{A_x^2 + A_y^2 + A_z^2}\right)$$

$$= \pm \frac{2}{\Delta t} \sin^{-1}(\zeta) \qquad (6.50)$$

where we have used the shorthand notation:

$$A_m = \frac{1}{\Delta m} \sin\left(\frac{k_m \Delta m}{2}\right) \quad , \quad m = x, y, z.$$

Consider the functional form of the field components in Equation (6.1). Clearly, if ω becomes complex, then the functional form of any field component takes on the form $\mathcal{V}(x, t) = C\, e^{\alpha t}\, e^{j(\beta t + kx)}$, where β and α are the real and imaginary parts of ω, respectively. If $\alpha > 0$, we have exponential growth, which implies instability. Hence, for the general case we require that ω in Equation (6.50) remain purely real at all times.

The term under the square root sign in Equation (6.50) is always positive, but if the larger term ζ becomes greater than 1, then ω will be complex. Thus, we require ζ to be bounded by $0 \le \zeta \le 1$. This factor ζ will be maximum when the $\sin(\cdot)$ terms are unity, giving $A_m \to 1/\Delta m$, so this requirement is equivalent to:

$$v_p \Delta t \sqrt{\frac{1}{(\Delta x)^2} + \frac{1}{(\Delta y)^2} + \frac{1}{(\Delta z)^2}} \le 1 \qquad (6.51)$$

which in terms of Δt is:

$$\Delta t \le \frac{1}{v_p \sqrt{\dfrac{1}{(\Delta x)^2} + \dfrac{1}{(\Delta y)^2} + \dfrac{1}{(\Delta z)^2}}}$$

which is exactly the stability criterion we found in Equation (5.33).

6.5   Summary

In this chapter, we introduced the concept of numerical dispersion and its relationship to the discretization of a PDE. While a continuous PDE can be described by its dispersion equation $\omega = f_1(k)$, the discrete FDE has its own numerical dispersion equation $\omega = f_2(k)$. Ideally, these two would be the same; but in reality, f_2 is a function of the discretization parameters Δt and Δx, among other parameters.

Like stability analysis, the derivation of the numerical dispersion relation starts from the application of the Fourier component, but this time in both space and time:

$$\mathcal{V}(x, t) = C\, e^{j(\omega t + kx)} \qquad \text{or} \qquad \mathcal{V}_i^n = C\, e^{j(wn\Delta t + ki\Delta x)}.$$

By substituting this component into the FDE, one can then solve for ω and a function of k or vice versa. Alternatively, one can solve for the phasor $e^{j\omega t}$, written in terms of real and imaginary components:

$$e^{j\omega t} = e^{j(\omega_r + j\gamma)\Delta t} = e^{-\gamma \Delta t}\, e^{j\omega_r \Delta t},$$

from which one can easily extract the dispersion ω_r and the dissipation γ. Note that if $\gamma > 0$, then the FDE exhibits growth, and is unstable.

In this chapter we derived the numerical dispersion relations for the Lax method and the leapfrog method applied to the single convection equation, which are plotted in Figure 6.1. In either case, the dispersion and dissipation are minimized for small values of $k\Delta x$; in essence, this means the frequencies in the simulation must be well sampled ($\Delta x \leq\sim \lambda/10$) to minimize numerical dispersion (note that the leapfrog method exhibits no dissipation for any values of $k\Delta x$).

We next derived the numerical dispersion relation for the FDTD algorithm, i.e., the interleaved leapfrog method. In 2D, the resulting dispersion relation is:

$$\left[\sin\left(\frac{\omega \Delta t}{2}\right)\right]^2 = \left(\frac{v_p \Delta t}{\Delta x}\right)^2 \sin^2\left(\frac{k_x \Delta x}{2}\right) + \left(\frac{v_p \Delta t}{\Delta y}\right)^2 \sin^2\left(\frac{k_y \Delta y}{2}\right).$$

The dispersion relation in 1D and 3D follow a similar form. We note that as $\Delta x, \Delta t \to 0$, the numerical dispersion relation approaches the real dispersion relation where $\omega = v_p |k|$, where $|k| = \sqrt{k_x^2 + k_y^2}$ in 2D.

Similarly, we found expressions for the numerical phase and group velocities as functions of k, Δt, and Δx, and showed that these also tend toward the continuous values as Δt and Δx tend toward zero. However, it is important to note that in general both $\bar{v}_p$ and $\bar{v}_g$ are functions of k.

Finally, at the end of this chapter we returned to the stability criterion for the FDTD algorithm, which could not be analytically derived by the methods of Chapter 5. In this chapter, using the numerical dispersion relation above, the stability criterion could be established by solving for ω, and finding the conditions under which the argument of $\sin(\cdot)$ is less than one, making ω strictly real. In this way we re-derived the CFL condition for the FDTD algorithm.

6.6   Problems

6.1. **Numerical dispersion.** In this problem we investigate numerical dispersion properties of the leapfrog approximation as applied to the convection equation by considering the propagation of selected initial voltage distributions. In all cases, assume $v_p = 1$ ft-(ns)$^{-1}$. Simulate the propagation of the following initial voltage distributions on an infinitely long 1D mesh:

(a) Using $\Delta x = 0.01$ ft, $\Delta t = 0.004$ ns,

$$\mathcal{V}(x, 0) = e^{-(8x)^2} \sin(250 x)).$$

Run your simulation up to $t = 2$ ns and determine the speed of propagation of the peak of the Gaussian envelope and interpret your result in terms of phase and group velocities.

(b) Using $\Delta x = 0.01$ ft, $\Delta t = 0.005$ ns,

$$\mathcal{V}(x, 0) = \frac{1}{2} e^{-50(x-1/2)^2} (1 + \sin(100 x)).$$

Run your simulation up to $t = 4$ ns and comment on your observations, interpreting the results in terms of phase and group velocities.

(c) Using $\Delta x = 0.00625$ ft, $\Delta t = 0.0025$ ns,

$$\mathcal{V}(x, 0) = e^{-3200(x-1/2)^2}.$$

Run your simulation up to $t = 2$ ns and comment on your observations. Quantitatively interpret your results in terms of phase and group velocities.

6.2. **Numerical dispersion in the frequency domain.** Use your code from Problem 5.6 to propagate a Gaussian pulse along the 1D grid. Monitor the pulse as it passes through a point some distance x from the left end; use this pulse to plot the propagation speed as a function of frequency. Compare to the analytical and numerical dispersion relations. Repeat for $v_p \Delta t / \Delta x = 1, 0.9, 0.7$, and 0.5.

6.3. **Grid anisotropy.** Use your 2D code from Problem 4.4 to investigate grid anisotropy in the FDTD method. Run the simulation only long enough to get a good estimate of the propagation speed, without allowing reflections from the outer boundary to corrupt the solution.

(a) Launch a sinusoidal wave from the center of the 2D space at a frequency of 30 GHz, and find a way to measure the propagation velocity as a function of angle in the grid. For the most accurate measurement, interpolate to find the time when a crest of the wave passes through points at a constant radius.

(b) Repeat part (a) for frequencies of 2, 5, 10, and 20 GHz. Compare the angular dispersion characteristics (propagation velocity as a function of angle).

(c) Repeat using the Gaussian source described in Problem 4.4(a), and measure the propagation velocity at angles of 0, 15, 30, and 45 degrees, each as a function of frequency. How well do the results of this method match up with the discrete frequencies in part (b)?

6.4. **Dispersion of the fourth-order method.** Use the methods of this chapter to derive the numerical dispersion relation for the algorithm in Equation (3.91) (along with its counterpart current update equation).

6.5. **Dispersion of the fourth-order method.** Repeat Problem 3.7, using fourth-order spatial differencing and second-order leapfrog time integration. Launch a Gaussian pulse from the left edge and monitor it as it passes some distance x from the source, as in Problem 6.2 above. Measure the propagation speed as a function of frequency, and compare to the analytical result from the previous problem.

References

[1] J. von Neumann, "Proposal and analysis of a numerical method for the treatment of hydro-dynamical shock problems," National Defense Research Committee, Tech. Rep. Report AM-551, 1944.

[2] J. von Neumann and R. D. Richtmeyer, "A method for numerical calculation of hydrodynamic shocks," *J. Appl. Phys.*, vol. 21, p. 232, 1950.

7 Introduction of sources

In Chapter 4 we presented the Yee algorithm for the discretization of Maxwell's equations on a regular, orthogonal grid. In Chapter 6 we discussed the numerical *accuracy* of the discretization, which depends on the choices of Δx, Δy, Δz, and Δt, as well as the medium of interest (ϵ, μ, and σ) and the frequencies to be simulated. However, the accuracy of the method is further restricted by how we choose to *drive* the simulation. The choice of a *source* is thus integral to accurate simulation in the FDTD method.

Sources can be either internal or external to the simulation. *Internal* sources are those that are driven at one or more grid cells within the simulation space. An example of an internal source might be a hard antenna, whose parameters we wish to investigate in a particular medium. *External* sources are those which are assumed to be outside of the simulation space. For example, we may wish to investigate the scattering pattern of an object when excited by a plane wave, but the source of that plane wave may be far from the scattering object. The problem, of course, is that in a numerical simulation, all sources must be applied to real grid cells, which means they are by definition "internal." However, as we will see beginning in Section 7.2, methods have been derived with great success to make sources appear external. First, however, we will discuss some possible methods of applying internal sources.

7.1 Internal sources

7.1.1 Hard sources

A common and simple method for introduction of internal sources is to use the so-called *hard source*, which is set up by simply assigning a desired temporal variation to specific electric or magnetic field components at a single or a few grid points. This method is quick and simple, and most useful for testing the discretized equations before applying more accurate sources. A few examples of hard sources for 1D TE modes are the continuous sinusoid:

$$\mathscr{E}_y \Big|_{i_s}^{n} = E_0 \, \sin(2\pi f_0 n \Delta t)$$

the Gaussian pulse:

$$\mathscr{E}_y \Big|_{i_s}^{n} = E_0 \, e^{-(n\Delta t - 3n_\tau \Delta t)^2/(n_\tau \Delta t)^2}$$

and the modulated Gaussian pulse:

$$\mathcal{E}_y\Big|_{i_s}^{n} = E_0\, e^{-(n\Delta t - 3n_\tau \Delta t)^2/(n_\tau \Delta t)^2}\, \sin\left[2\pi f_0(n - 3n_\tau)\Delta t\right]$$

where n is the time step and $n_\tau \Delta t$ is the characteristic half-width (in time) of the Gaussian pulse. Note that choosing the total duration of the Gaussian to be at least $6n_\tau$ time steps ensures a smooth transition both from and back to zero.

The major issue with hard sources is that they reflect. By hard coding the electric field at a given grid point (or, sometimes, at a number of grid points), we are effectively treating that point in the grid as a perfect conductor, and any scattered fields in the simulation space will reflect off this hard source. In some cases, this might be desirable: such reflection occurs, for example, from real antennas. However, in other cases we may be interested in accurately measuring the scattered field without having it interrupted by the source.

For simple hard sources, there are at least two ways to circumvent this problem:

1. In the case of a pulsed source (Gaussian or modulated Gaussian), one can turn the source "off" after the pulse is completely within the space, i.e., after $n > 6n_\tau$. This can usually be successfully implemented with a simple "if" statement. Of course, this will not work with a source that has a long duration or a sinusoidal variation.
2. Alternatively, rather than applying the hard source to a point in space, one can *add* the source to the applicable update equation. For example, for an $\mathcal{E}_y$ source at time $n + 1$ and grid point i_s, one can modify Equation (4.8) as follows:

$$\mathcal{E}_y\Big|_{i_s}^{n+1} = \mathcal{E}_y\Big|_{i_s}^{n} - \frac{\Delta t}{\epsilon\, \Delta x}\left[\mathcal{H}_z\Big|_{i_s+1/2}^{n+1/2} - \mathcal{H}_z\Big|_{i_s-1/2}^{n+1/2}\right] + \mathcal{E}_{y,\text{source}}\Big|_{i_s}^{n+1} \tag{7.1}$$

where $\mathcal{E}_{y,\text{source}}\Big|_{i_s}^{n+1}$ is the source field at time $n + 1$ applied to the grid point i_s. In the case of a pulsed source, for example, when the source is completely within the simulation space and $\mathcal{E}_{y,\text{source}}\Big|_{i_s}^{n+1} = 0$, the source no longer affects the update equation.

7.1.2 Current and voltage sources

Typically, when a hard source is required, it is a physical element such as an antenna that is driving the fields. These sources are more often described by currents and voltages than they are by electric and magnetic fields, of course. An internal source can be specified as a driven current, for example as a current density $\mathcal{J}_{\text{source}}(\mathbf{r}, t)$ which is specified in terms of its temporal form at one or more grid points. This source current works to generate the electromagnetic fields via the curl equation:

$$\left[\frac{\partial \overline{\mathcal{E}}}{\partial t}\right]^{n+1/2} = \frac{1}{\epsilon}[\nabla \times \overline{\mathcal{H}}]^{n+1/2} - \frac{1}{\epsilon}[\overline{\mathcal{J}}_{\text{source}}]^{n+1/2} \tag{7.2}$$

where the source current is located at the same spatial grid location as the electric field but at the same time point as the magnetic field. Note that this is nearly equivalent to the electric field applied in Equation (7.1), if we assume that $\epsilon \mathcal{E}_{y,\text{source}} = \Delta t \, \overline{\mathcal{J}}_{\text{source}}$. However, in the case of the current source, we can apply a known physical driving current, rather than a somewhat arbitrary electric field.

Another simple source is a driven voltage applied across a gap, such as at the terminals of a wire antenna. Assuming a voltage source applied over a one-cell wide gap in the y direction, we can write:

$$\mathcal{E}_y\Big|_{i_s, j_s}^n = \frac{V(t)}{\Delta y} = \frac{V(n\Delta t)}{\Delta y}$$

Once again, any given temporal variation can be imposed, by simply specifying the functional form of $V(t)$. Note that the antenna wire itself may extend over a number of mesh points $(i_s, j_s \pm m)$ on both sides of (i_s, j_s), and its presence is simply incorporated into the FDTD calculation by forcing all $\mathcal{E}_y$ values to be zero at these mesh points, as was done in the example in Figure 4.17.

7.1.3   The thin-wire approximation

A problem arises with the use of current and voltage hard sources described above. Consider the case where we are driving an antenna, and the frequency of interest is in the ~100 MHz range. The free-space wavelength at 100 MHz is 3 meters; for a grid cell of $\lambda/10$, which is a commonly used resolution, this means our grid cell is 30 cm. However, 100 MHz antennas are not usually 30 cm wide! Things are even worse at lower frequencies: grid cells for simulations in the VLF range at ~20 kHz are typically on the order of 1 km! This poor resolution of the antenna itself usually results in significantly less accurate calculations of the antenna near field, while the far field is less affected.

One method for accurately simulating a real antenna is to subgrid the simulation space, so that the grid cells are considerably smaller near the antenna, and grow up to $\lambda/10$ in the far field. We will introduce the concept of non-regular grids in Chapter 15. This method can provide much improved accuracy, but suffers from the grid design complexity and some amount of extra computational requirements, although this may be small.

A much simpler method to simulate a "thin" (relative to the grid cell) antenna is to use the thin-wire approximation. This method relies on a simple technique for modeling sub-cell structures which will be discussed in greater detail in Chapter 12. For now, we refer to Figure 7.1 to understand the problem. Figure 7.1 shows an antenna in the x–z plane, zoomed in to look at one grid cell. Note that if this is truly a "thin wire," this must be a 3D simulation; in 2D Cartesian coordinates, our antenna is actually a thin "plane" antenna.[1]

We start by drawing a contour C as shown in Figure 7.1. The fields can be accurately modeled near the antenna by noting that (a) $\mathcal{H}$ varies as $1/x$ in the enclosed cell, (b) $\mathcal{E}_x$ likewise varies as $1/x$ in the cell, and (c) z-directed fields are average values over the

---

[1] In cylindrical coordinates, of course, this thin antenna can be accurately modeled in 2D, as discussed in Chapter 4.

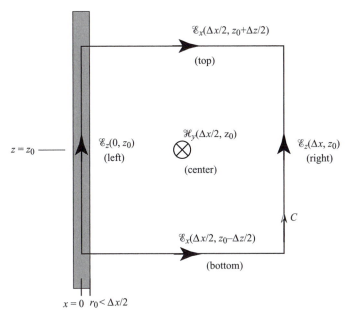

Figure 7.1 The thin-wire approximation.

Δx interval. These imply, in turn, that the fields at relevant locations in the cell shown are given analytically by:

$$\mathcal{H}_y(x, z) = \mathcal{H}_y\Big|_{\Delta x/2, z_0} \frac{\Delta x/2}{x}$$

$$\mathcal{E}_x(x, z_0 \pm \Delta z/2) = \mathcal{E}_x\Big|_{\Delta x/2, z_0 \pm \Delta/2} \frac{\Delta x/2}{x}$$

$$\mathcal{E}_z(x \leq r_0, z) = 0$$

$$\mathcal{E}_z(\Delta x, z) = \mathcal{E}_z\Big|_{\Delta x, z_0}.$$

Next, consider applying Faraday's law in integral form, Equation (2.1), to C and the surface S enclosed by C, rather than the typical update equations:

$$\oint_C \overline{\mathcal{E}} \cdot d\mathbf{l} = -\int_S \frac{\partial \overline{\mathcal{B}}}{\partial t} \cdot d\mathbf{s}$$

$$\mathcal{E}_z\Big|_{\text{right}}^n \Delta z - \underbrace{\mathcal{E}_z\Big|_{\text{left}}^n}_{=0} - \int_{r_0}^{\Delta x} \mathcal{E}_x\Big|_{\text{top}}^n \frac{\Delta x/2}{x} dx + \int_{r_0}^{\Delta x} \mathcal{E}_x\Big|_{\text{bottom}}^n \frac{\Delta x/2}{x} dx$$

$$= \frac{\mu}{\Delta t} \int_{z_0-\Delta z/2}^{z_0+\Delta z/2} \int_{r_0}^{\Delta x} \left(\mathcal{H}_y\Big|_{\text{center}}^{n+1/2} - \mathcal{H}_y\Big|_{\text{center}}^{n-1/2} \right) \frac{\Delta x/2}{x} dx \, dz$$

$$\mathcal{E}_z\Big|_{\text{right}}^n \Delta z + \frac{\Delta x}{2} \ln \frac{\Delta x}{r_0} \left(\mathcal{E}_x\Big|_{\text{top}}^n - \mathcal{E}_x\Big|_{\text{bottom}}^n \right) = \frac{\mu \Delta z}{\Delta t} \frac{\Delta x}{2} \ln \frac{\Delta x}{r_0} \left(\mathcal{H}_y\Big|_{\text{center}}^{n+1/2} - \mathcal{H}_y\Big|_{\text{center}}^{n-1/2} \right)$$

where the subscripts (left, right, top, bottom, and center) refer to the locations in the grid cell in Figure 7.1. Finally, simplifying and adding in the proper subscripts, we find the modified update equation for $\mathcal{H}_y$ one half-cell from the axis of the antenna:

$$\mathcal{H}_y\Big|_{\Delta x/2, z_0}^{n+1/2} = \mathcal{H}_y\Big|_{\Delta x/2, z_0}^{n-1/2} + \frac{\Delta t}{\mu \Delta z \frac{\Delta x}{2} \ln \frac{\Delta x}{r_0}} \tag{7.3}$$

$$\times \left[\left(\mathcal{E}_z\Big|_{\Delta x/2, z_0 - \Delta z/2}^{n} - \mathcal{E}_z\Big|_{\Delta x/2, z_0 + \Delta z/2}^{n} \right) \cdot \frac{\Delta x}{2} \ln \frac{\Delta x}{r_0} + \mathcal{E}_z\Big|_{\Delta x, z_0}^{n} \Delta z \right].$$

A similar procedure must be applied to the other $\mathcal{H}$ components surrounding the wire in three dimensions, but the update equations are very similar to Equation (7.3) above. Note that the thin-wire approximation does not actually apply a source, but rather sets the electric field to zero on the wire; as such, a voltage needs to be applied at the antenna gap, as described in Section 7.1.2 above.

7.2   External sources: total and scattered fields

The methods described above can be very useful for applying internal sources to the simulation space, but are useless for applying, for example, an incident plane wave that may have been generated a great distance away. In this latter scenario, we wish to make it appear to our simulation as if the plane wave source is *external* to the simulation space. The solution to this problem relies on the total-field/scattered-field formulation, which is the subject of the remainder of this chapter.

Our ability to separate the fields into *incident* and *scattered* components derives from the linearity of Maxwell's equations;[2] that is, if the fields $\overline{\mathcal{E}}_1$ and $\overline{\mathcal{E}}_2$ separately satisfy Maxwell's equations, then the sum of these two $\overline{\mathcal{E}}_{sum} = \overline{\mathcal{E}}_1 + \overline{\mathcal{E}}_2$ also satisfies Maxwell's equations. Thus, we additionally use the fact that the incident and scattered components separately satisfy the two curl equations, (2.1) and (2.3), which are modeled in an FDTD algorithm.

In general terms we can think of a free-space region within which an *incident* field is specified analytically, being defined as the field which is present in this region in the absence of any objects (dielectric or conductor) or "scatterers." The incident field thus satisfies Maxwell's equations for free space, everywhere in the regions being modeled. The *scattered* wave arises because of the presence of the scatterer, and in response to the need that electromagnetic boundary conditions must be satisfied as the incident field interacts with the object. The scattered fields must satisfy Maxwell's equations with the medium parameters of the scatterer, namely ϵ, μ, σ, and σ_m, as well as satisfying

---

[2] Note that there are materials which are nonlinear, such as some crystals, whose responses involve higher powers of the electric field such as $\overline{\mathcal{E}}^2$ or $\overline{\mathcal{E}}^3$, typically due to the permittivity and/or permeability depending on the field intensities. We will not address such materials in this book, but the reader should be aware of the assumption of linearity in the total-field / scattered-field formulation.

Maxwell's equations for free space immediately outside the scatterer. The *total* field is simply the superposition of the incident and scattered fields. Within the scatterer, the total fields must satisfy the equations:

$$\nabla \times \overline{\mathscr{E}}_{\text{tot}} = -\mu \frac{\partial \overline{\mathscr{H}}_{\text{tot}}}{\partial t} - \sigma_m \overline{\mathscr{H}}_{\text{tot}} \tag{7.4a}$$

$$\nabla \times \overline{\mathscr{H}}_{\text{tot}} = \epsilon \frac{\partial \overline{\mathscr{E}}_{\text{tot}}}{\partial t} + \sigma \overline{\mathscr{E}}_{\text{tot}} \tag{7.4b}$$

where $\epsilon \neq \epsilon_0$ and $\mu \neq \mu_0$. Meanwhile, the incident fields satisfy

$$\nabla \times \overline{\mathscr{E}}_{\text{inc}} = -\mu_0 \frac{\partial \overline{\mathscr{H}}_{\text{inc}}}{\partial t} \tag{7.5a}$$

$$\nabla \times \overline{\mathscr{H}}_{\text{inc}} = \epsilon_0 \frac{\partial \overline{\mathscr{E}}_{\text{inc}}}{\partial t} \tag{7.5b}$$

since the incident fields are not subject to the material parameters ϵ, μ, σ, and σ_m. We can rewrite Equation (7.4), with $\overline{\mathscr{H}}_{\text{tot}} = \overline{\mathscr{H}}_{\text{inc}} + \overline{\mathscr{H}}_{\text{scat}}$, as:

$$\nabla \times \left[\overline{\mathscr{E}}_{\text{inc}} + \overline{\mathscr{E}}_{\text{scat}} \right] = -\mu \frac{\partial \left[\overline{\mathscr{H}}_{\text{inc}} + \overline{\mathscr{H}}_{\text{scat}} \right]}{\partial t} - \sigma_m \left[\overline{\mathscr{H}}_{\text{inc}} + \overline{\mathscr{H}}_{\text{scat}} \right] \tag{7.6a}$$

$$\nabla \times \left[\overline{\mathscr{H}}_{\text{inc}} + \overline{\mathscr{H}}_{\text{scat}} \right] = \epsilon \frac{\partial \left[\overline{\mathscr{E}}_{\text{inc}} + \overline{\mathscr{E}}_{\text{scat}} \right]}{\partial t} + \sigma \left[\overline{\mathscr{E}}_{\text{inc}} + \overline{\mathscr{E}}_{\text{scat}} \right]. \tag{7.6b}$$

Subtracting Equations (7.5) from (7.6) gives the equations for the scattered fields within the scatterer:

$$\nabla \times \overline{\mathscr{E}}_{\text{scat}} = -\mu \frac{\partial \overline{\mathscr{H}}_{\text{scat}}}{\partial t} - \sigma_m \overline{\mathscr{H}}_{\text{scat}} - \left[(\mu - \mu_0) \frac{\partial \overline{\mathscr{H}}_{\text{inc}}}{\partial t} + \sigma_m \overline{\mathscr{H}}_{\text{inc}} \right] \tag{7.7a}$$

$$\nabla \times \overline{\mathscr{H}}_{\text{scat}} = \epsilon \frac{\partial \overline{\mathscr{E}}_{\text{scat}}}{\partial t} + \sigma \overline{\mathscr{E}}_{\text{scat}} + \left[(\epsilon - \epsilon_0) \frac{\partial \overline{\mathscr{E}}_{\text{inc}}}{\partial t} + \sigma \overline{\mathscr{E}}_{\text{inc}} \right]. \tag{7.7b}$$

In regions outside the scatterer, both the total fields and scattered fields must satisfy Maxwell's equations:

$$\nabla \times \overline{\mathscr{E}}_{\text{tot}} = -\mu_0 \frac{\partial \overline{\mathscr{H}}_{\text{tot}}}{\partial t} \tag{7.8a}$$

$$\nabla \times \overline{\mathscr{H}}_{\text{tot}} = \epsilon_0 \frac{\partial \overline{\mathscr{E}}_{\text{tot}}}{\partial t} \tag{7.8b}$$

and

$$\nabla \times \overline{\mathscr{E}}_{\text{scat}} = -\mu_0 \frac{\partial \overline{\mathscr{H}}_{\text{scat}}}{\partial t} \tag{7.9a}$$

$$\nabla \times \overline{\mathscr{H}}_{\text{scat}} = \epsilon_0 \frac{\partial \overline{\mathscr{E}}_{\text{scat}}}{\partial t}. \tag{7.9b}$$

Note that Equations (7.9) could have been obtained from Equations (7.7) by simply substituting zero for σ and σ_m, μ_0 for μ, and ϵ_0 for ϵ. Thus, it is possible to construct an FDTD code only in terms of the scattered fields based on (7.7), which can be rewritten with the time derivatives on the left-hand side more suitable for discretization:

$$\frac{\partial \overline{\mathcal{H}}_{scat}}{\partial t} = -\frac{1}{\mu} \nabla \times \overline{\mathcal{E}}_{scat} - \frac{\sigma_m}{\mu} \overline{\mathcal{H}}_{scat} - \frac{\sigma_m}{\mu} \overline{\mathcal{H}}_{inc} - \frac{\mu - \mu_0}{\mu} \frac{\partial \overline{\mathcal{H}}_{inc}}{\partial t} \tag{7.10a}$$

$$\frac{\partial \overline{\mathcal{E}}_{scat}}{\partial t} = \frac{1}{\epsilon} \nabla \times \overline{\mathcal{H}}_{scat} - \frac{\sigma}{\epsilon} \overline{\mathcal{E}}_{scat} - \frac{\sigma}{\epsilon} \overline{\mathcal{E}}_{inc} - \frac{\epsilon - \epsilon_0}{\epsilon} \frac{\partial \overline{\mathcal{E}}_{inc}}{\partial t} \tag{7.10b}$$

which is valid in both free-space and scatterer regions, depending on what we substitute for σ, σ_m, μ, and ϵ.

7.2.1   Total-field and pure scattered-field formulations

Using an FDTD formulation based on Equations (7.10), with analytically specified incident fields $\overline{\mathcal{H}}_{inc}$ and $\overline{\mathcal{E}}_{inc}$, amounts to the so-called pure scattered-field formulation [1]. This is an alternative to the total-field formulations that we have discussed so far, such as the FDTD expressions based essentially on Equations (7.4).

Note that the standard interleaved leapfrog discretization of Equations (7.10) results in:

$$\overline{\mathcal{H}}_{scat}\Big|^{n+1/2} = \left[\frac{\mu}{\mu + \sigma_m \Delta t}\right] \left\{ \overline{\mathcal{H}}_{scat}\Big|^{n-1/2} - \frac{\sigma_m}{\mu}\left[\nabla \times \overline{\mathcal{E}}_{scat}\right]^n \right.$$

$$\left. - \sigma_m \overline{\mathcal{H}}_{inc}\Big|^n - \left[\frac{(\mu - \mu_0)\Delta t}{\mu}\right]\left[\frac{\partial \overline{\mathcal{H}}_{inc}}{\partial t}\right]^n \right\}$$

$$\overline{\mathcal{E}}_{scat}\Big|^{n+1} = \left[\frac{\epsilon}{\epsilon + \sigma \Delta t}\right] \left\{ \overline{\mathcal{E}}_{scat}\Big|^n + \frac{\sigma}{\epsilon}\left[\nabla \times \overline{\mathcal{H}}_{scat}\right]^{n+1/2} \right.$$

$$\left. - \sigma \overline{\mathcal{E}}_{inc}\Big|^{n+1/2} - \left[\frac{(\epsilon - \epsilon_0)\Delta t}{\epsilon}\right]\left[\frac{\partial \overline{\mathcal{E}}_{inc}}{\partial t}\right]^{n+1/2} \right\}$$

so that while the FDTD algorithm can simply proceed at free-space points by using the first two terms in the braces in each of the above equations, the incident fields must be calculated at each time step at all of the grid points within the scatterer regions. This requirement doubles the necessary computational resources.

Many electromagnetic problems involve the interaction of dielectric, magnetic, or metallic objects with an incident (i.e., externally produced) plane wave. In general, FDTD problems involving incident plane waves can be formulated in terms of either the total or scattered fields. Total-field formulations have the advantage of allowing more straightforward handling of perfect-electric-conductor boundaries and penetrable materials. The perfect-electric-conductor (PEC) boundary in a total-field formulation is simply handled by setting the tangential electric field $\mathcal{E}_{tan} = 0$, either by maintaining the initial conditions throughout a calculation or by resetting the appropriate electric field components to zero at each time step. In a pure-scattered-field formulation, however,

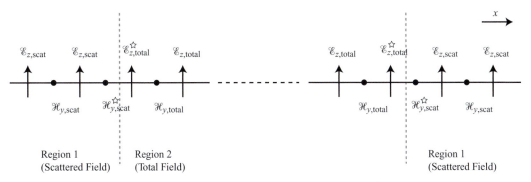

Figure 7.2 Total-field/scattered-field formulation in 1D. The grey dashed lines show the boundary between the total-field region and the scattered-field regions; the fields marked with a star are those that need to have their updates modified.

PEC boundaries must be driven with the negative of the instantaneous incident field, in order to cancel out the field to zero. Also, it is necessary to evaluate the incident fields at any and all points where $\epsilon \neq \epsilon_0$ or $\mu \neq \mu_0$, as is clear from (7.10). Furthermore, the use of Absorbing Boundary Conditions (ABCs, to be described in Chapter 8) at the edges of the FDTD grid, designed to remove waves from the problem space, is in direct conflict with the excitation of the incident plane wave in a total-field formulation. These drawbacks together make the pure-scattered field formulation derived above a cumbersome approach.

The alternative solution is to use the so-called total-field / scattered-field (TF/SF) formulation, described next, which is based on using equations (7.4) within the so-called total-field region (Region 1) and Equations (7.9) in the so-called scattering-field region (Region 2); however, appropriate connecting conditions on the adjoining surfaces are necessary. For the TF/SF formulation, the incident fields only need to be known at the boundary between the total-field region (Region 2) and the scattered-field region (Region 1). This incident field distribution can be introduced by means of an auxiliary 1D grid as described in Figure 7.2. The significant advantage of this formulation is that the incident plane wave does not interact with the absorbing boundary conditions, thus minimizing the computational load on these ABC or PML algorithms, which are often tasked to provide for 30 to 80 dB reduction of numerical reflections.

7.3   Total-field/scattered-field formulation

A typical setup of an 2D FDTD space implementing a TF/SF formulation is shown in Figure 7.3a. For the case shown, the incident plane wave is launched at the bottom edge of Region 2, noting that the incident fields can simply be kept in 1D arrays ($i = i_0, \ldots, i_1$, $j = j_0$). If the FDTD space is not aligned so that the incident wave is along one of the principal axes, the incident fields can be analytically computed at each cell position or determined by means of look-up tables and/or interpolation schemes; detailed methods for doing so are described well in [2, Ch. 5].

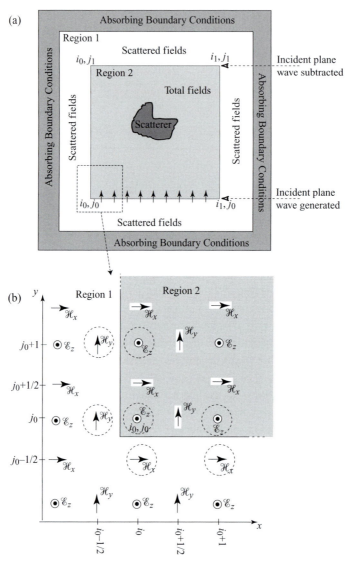

Figure 7.3 Total-field/scattered-field formulation in 2D. (a) FDTD problem space with various regions identified and showing a particular y-directed incident plane wave. (b) The lower left corner of the FDTD problem space showing the location of the field components for a 2D TM problem. The field components in dashed circles are those that require modified update equations.

Now consider the lower left corner of the FDTD problem space as shown in Figure 7.3b, for the case of an arbitrary (i.e., not necessarily y-directed) incident plane wave. For this example, we shall consider a 2D TM wave, the grid locations of the fields for which are shown. As we step the FDTD algorithm in time, it is clear that the updating of the field components proceeds in an accurate manner as long as the next value of the field component is determined in terms of *only* total or *only* scattered values. For the case shown, it is clear that the only field components for which we would have a problem are those which are circled. For example, consider the updating of the $\mathscr{E}_z$ component at the lower left edge, i.e., at the $(i = i_0, j = j_0)$ location:

$$
\mathscr{E}_z^{\text{tot}}\Big|_{i_0,j_0}^{n+1} = \mathscr{E}_z^{\text{tot}}\Big|_{i_0,j_0}^{n} + \frac{\Delta t}{\epsilon} \left[\frac{\overbrace{\mathscr{H}_y\Big|_{i+1/2,j_0}^{n+1/2}}^{\text{total}} - \overbrace{\mathscr{H}_y\Big|_{i_0-1/2,j_0}^{n+1/2}}^{\text{scattered}}}{\Delta x} - \frac{\overbrace{\mathscr{H}_x\Big|_{i_0,j_0+1/2}^{n+1/2}}^{\text{total}} - \overbrace{\mathscr{H}_x\Big|_{i_0,j_0-1/2}^{n+1/2}}^{\text{scattered}}}{\Delta y} \right].
$$

$$(7.11)$$

Note that $\mathscr{E}_z$ itself is a total field, since it is in Region 2.

The above expression is *incorrect*, since the spatial derivatives are evaluated using a mix of total- and scattered-field components. What we require instead is an equation equivalent to Equation (7.4), i.e., only in terms of the total fields, since we are evaluating Maxwell's equations at (i_0, j_0) within Region 2. However, the fix is quite simple, using the linearity of Maxwell's equations, and noting that:

$$
\mathscr{H}_y^{\text{tot}}\Big|_{i-1/2,j}^{n+1/2} = \mathscr{H}_y^{\text{inc}}\Big|_{i-1/2,j}^{n+1/2} + \mathscr{H}_y^{\text{scat}}\Big|_{i-1/2,j}^{n+1/2}
$$

$$(7.12)$$

Thus, we can determine the correct $\mathscr{E}_z$ component at (i_0, j_0) by modifying Equation (7.11) as:

$$
\mathscr{E}_z^{\text{tot}}\Big|_{i_0,j_0}^{n+1} = \mathscr{E}_z^{\text{tot}}\Big|_{i_0,j_0}^{n} + \frac{\Delta t}{\epsilon} \left[\frac{\mathscr{H}_y^{\text{tot}}\Big|_{i+1/2,j_0}^{n+1/2} - \mathscr{H}_y^{\text{scat}}\Big|_{i_0-1/2,j_0}^{n+1/2}}{\Delta x} - \frac{\mathscr{H}_x^{\text{tot}}\Big|_{i_0,j_0+1/2}^{n+1/2} - \mathscr{H}_x^{\text{scat}}\Big|_{i_0,j_0-1/2}^{n+1/2}}{\Delta y} \right]
$$
$$
- \frac{\Delta t}{\epsilon \Delta x} \mathscr{H}_y^{\text{inc}}\Big|_{i_0-1/2,j_0}^{n+1/2} - \frac{\Delta t}{\epsilon \Delta y} \mathscr{H}_x^{\text{inc}}\Big|_{i_0,j_0-1/2}^{n+1/2}.
$$

$$(7.13)$$

Note that the correction needs to be added *only* at the grid points around the boundary of Region 2 and can be added after first applying the FDTD algorithm (i.e., time marching) at *all* grid points; i.e., one would calculate the field $\mathscr{E}_z^{\text{tot}}\big|_{i_0,j_0}^{n+1}$ everywhere in the space by first using the FDTD algorithm as before, and then applying the correction at the boundary of Region 2, which at (i_0, j_0) is:

$$
\mathscr{E}_z^{\text{tot}}\Big|_{i_0,j_0}^{n+1} = \underbrace{\left[\mathscr{E}_z^{\text{tot}}\Big|_{i_0,j_0}^{n+1} \right]}_{\text{From, e.g., eq. (7.11)}} - \frac{\Delta t}{\epsilon \Delta x} \mathscr{H}_y^{\text{inc}}\Big|_{i_0-1/2,j_0}^{n+1/2} - \frac{\Delta t}{\epsilon \Delta y} \mathscr{H}_x^{\text{inc}}\Big|_{i_0,j_0-1/2}^{n+1/2}.
$$

Note that the incident field data necessary for this correction only need to be kept at grid points displaced by one half-mesh-size from Region 2, i.e., at $(i_0 - 1/2, j)$ and $(i, j_0 - 1/2)$.

The two correction terms (one for $\mathcal{H}_x$ and another for $\mathcal{H}_y$) are only necessary for the corner point. At other points on the boundary, we only need a single ($\mathcal{H}_x$ or $\mathcal{H}_y$) correction term. For example, for any other point $(i_0, j_0 < j < j_1)$ on the left edge of Region 2 we have:

$$
\mathcal{E}_z^{\mathrm{tot}}\Big|_{i_0,j}^{n+1} = \mathcal{E}_z^{\mathrm{tot}}\Big|_{i_0,j}^{n} + \frac{\Delta t}{\epsilon}\left[\frac{\mathcal{H}_y^{\mathrm{tot}}\Big|_{i+1/2,j}^{n+1/2} - \mathcal{H}_y^{\mathrm{scat}}\Big|_{i_0-1/2,j}^{n+1/2}}{\Delta x} - \frac{\mathcal{H}_x^{\mathrm{tot}}\Big|_{i_0,j+1/2}^{n+1/2} - \mathcal{H}_x^{\mathrm{tot}}\Big|_{i_0,j-1/2}^{n+1/2}}{\Delta y}\right]
$$

$$
- \frac{\Delta t}{\epsilon\,\Delta x}\,\mathcal{H}_y^{\mathrm{inc}}\Big|_{i_0-1/2,j}^{n+1/2}. \tag{7.14}
$$

The FDTD algorithm in the scattered-field region (Region 1) will operate on scattered-field components which are available, and will only need special corrections for those field components whose derivatives require the use of field components from Region 2. In Figure 7.3b, these Region 1 field components requiring special handling are circled, being the $\mathcal{H}_y\big|_{i_0-1/2,j}$ on the left edge and the $\mathcal{H}_y\big|_{i,j_0-1/2}$ on the bottom edge. The corrected update equations for these components can be written with analogy to Equations (7.12) through (7.14):

$$
\mathcal{H}_x^{\mathrm{scat}}\Big|_{i_0,j-1/2}^{n+1/2} = \mathcal{H}_x^{\mathrm{scat}}\Big|_{i_0,j-1/2}^{n-1/2} - \frac{\Delta t}{\mu\,\Delta y}\left[\mathcal{E}_z^{\mathrm{tot}}\Big|_{i_0,j}^{n} - \mathcal{E}_z^{\mathrm{scat}}\Big|_{i_0,j-1}^{n}\right] + \frac{\Delta t}{\mu\,\Delta y}\mathcal{E}_z^{\mathrm{inc}}\Big|_{i_0,j}^{n} \tag{7.15a}
$$

$$
\mathcal{H}_y^{\mathrm{scat}}\Big|_{i_0-1/2,j}^{n+1/2} = \mathcal{H}_y^{\mathrm{scat}}\Big|_{i_0-1/2,j}^{n+1/2} + \frac{\Delta t}{\mu\,\Delta x}\left[\mathcal{E}_z^{\mathrm{tot}}\Big|_{i_0,j}^{n} - \mathcal{E}_z^{\mathrm{scat}}\Big|_{i_0-1,j}^{n}\right] - \frac{\Delta t}{\mu\,\Delta x}\mathcal{E}_z^{\mathrm{inc}}\Big|_{i_0,j}^{n}. \tag{7.15b}
$$

Note that there is no corner issue for the $\overline{\mathcal{H}}$ field components. Similar correction terms for the $\overline{\mathcal{E}}$ and $\overline{\mathcal{H}}$ fields at the other three corners and the other three sides simply follow.

7.3.1   Example 2D TF/SF formulations

Figure 7.4 shows an example total-field scattered-field simulation in 2D, to help illustrate the concept behind the method. The top right panel shows the incident field, in which a sinusoidal point source is excited in the top corner of the space, and absorbing boundaries (described in Chapter 8) are used to absorb the fields at the edges of the space.[3] This panel shows simply the "incident" field; the top left panel shows the total-field scattered-field simulation space. The central box is the "total" field, and the outer region is the scattered field. Note that no scattered fields are present, since we have not yet included

---

[3] Note that the fields near the top-right and bottom-left corners have been poorly modeled; this is due to the poor performance of the Mur boundary at grazing angles, as will be discussed in Chapter 8.

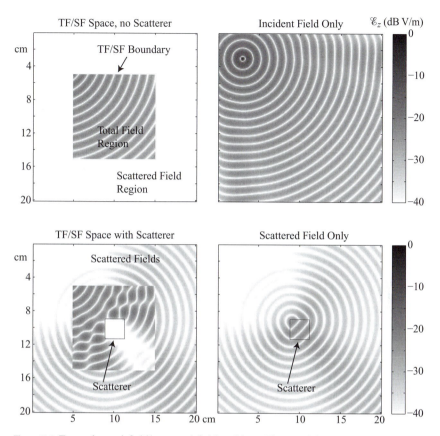

Figure 7.4 Example total-field/scattered-field problem. The top right panel show the incident field, excited by a sinusoidal point source, and absorbed at the boundaries by a Mur boundary condition (Chapter 8). The top left panel shows the TF/SF space with no scatterer, and the bottom left panel shows the TF/SF panel with a scattering object. The bottom right panel shows the complete scattered field, after subtracting the incident field from the total-field region.

a scatterer. The incident field is simply added at the top and left TF/SF boundaries, and then subtracted again at the bottom and right boundaries.

Now, in the lower left panel, a small box of perfectly conducting material has been imposed at the center of the simulation space. Now, the total-field region contains both the incident and scattered fields, resulting in an interference pattern that would be difficult to interpret. In the scattered-field region, however, the incident field is removed, leaving only the fields directly scattered by the object. It is evident that the scattered field forms a spherical outgoing wave, but with variations in the propagation angle. It would be quite simple, then, to determine the scattering pattern in 2D for this object. In the lower right panel, we have subtracted the incident field from the total-field region only, resulting in the complete scattered-field pattern, which can then be used for any desired analyses.

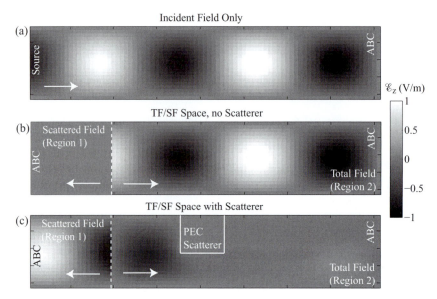

Figure 7.5 Total-field/scattered-field in a waveguide. (a) The incident field is calculated at all points in the space, with a sinusoidally varying source at the leftmost edge. An absorbing boundary condition (ABC) (Chapter 8) is applied at the right edge of all three spaces. (b) The TF/SF space; incident fields in (a) are applied as the correction terms in (b) at the dashed line. Any "scattered" fields thus appear to propagate to the left in the scattered-field region; however, as there are no scatterers, there are no scattered fields, and instead we see the incident field propagate unhindered to the right. In (c) a scatterer has been introduced, which is a PEC box filling half the width of the waveguide. In this case, fields reflect from the scatterer and appear propagating to the left in Region 1. Note that, unlike Figures 7.2 and 7.3, there is no "Region 1" on the right.

Figure 7.5 shows a second example of a 2D total-field scattered-field simulation. In this simulation we have modeled a rectangular air-filled waveguide 2 cm wide (in the y-direction) and 8 cm long. We wish to introduce a TE mode sinusoidal source at a frequency of 11.25 GHz. Note that for a 2 cm waveguide, the TE mode[4] cutoff frequencies are given by $f_{c_m} = m/2a\sqrt{\mu\epsilon} = 7.5m$ GHz [3, p. 269], and so our setup restricts propagation to only the TE_1 mode (as there is no TE_0 mode). Since the tangential $\mathcal{E}_z$ must equal zero at the waveguide walls, we excite a source with the TE_1 variation in y, i.e., $\mathcal{E}_z|_{0,y} = \mathcal{E}_{z,\text{max}} \sin(\pi y/a)$.

Special care must be undertaken in the introduction of plane sources in waveguide systems due to the multiplicity of propagating modes, presence of a cutoff frequency near which the group velocity approaches zero, and inherent dispersion. One possible way to excite a desired waveguide mode is to externally maintain the transverse field distribution of that particular mode across the waveguide aperture at the input plane of the waveguide (e.g., at $z = 0$). Such hard-source examples for the TE_{10} and TM_{11} modes

---

[4] Here we refer to the waveguide sense of TE, i.e., our space is in the x–y plane and we have an $\mathcal{E}_z$ component only; this is our FDTD TM mode.

are given below (for 3D waveguides):

$$\text{TE}_{10} \qquad \mathcal{H}_z = A \frac{1}{\eta} \cos\left(\frac{\pi}{a} x\right) \sin \omega t \qquad \text{for } 0 \le x \le a \text{ and } 0 \le y \le b$$

$$\text{TM}_{11} \qquad \mathcal{E}_z = A \sin\left(\frac{\pi}{a} x\right) \sin\left(\frac{\pi}{b} y\right) \sin \omega t \qquad \text{for } 0 \le x \le a \text{ and } 0 \le y \le b.$$

For the example in Figure 7.5, we used the expression similar to the TE_{10} expression above, but the variation is in y.

Figure 7.5a shows the incident field, calculated using the simple 2D FDTD algorithm, without worrying about the total-field / scattered-field formulation. Note, however, that we have applied a simple absorbing boundary condition (ABC) (discussed in Chapter 8) to the rightmost wall. The field appears from the left wall, propagates to the right, and is absorbed by the ABC at the right wall.

In Figure 7.5b, the TF/SF algorithm has been applied at the dashed line by using the incident fields at the corresponding locations in (a) as the corrections to Equations (7.14) and (7.15). What we see are the incident fields appearing in the "total field" region (Region 2), unimpeded, since we have not introduced scattering objects. In Figure 7.5c, a 1 cm by 1 cm PEC block has been introduced by setting the $\mathcal{E}_z$ components to zero everywhere inside the block. Now, we see scattered fields appearing in Region 1, propagating to the left; note that an ABC at the left edge of the TF/SF space acts to absorb these scattered fields. Notice that very little of the incident field gets past the scatterer, and almost all of the field gets reflected. This is because the scatterer acts to "shrink" the waveguide to half its width over a 1 cm region, wherein the cutoff frequency for the TE_1 mode is now 15 GHz; as such, our incident wave is strongly attenuated in this region.

7.4 Total-field/scattered-field in three dimensions

The total-field/scattered-field formulation was described in the previous section for the 2D case, requiring two correction terms for the electric field components (for the TM case) located at the corners of the FDTD grid. It is important to note that in a 3D grid, only a single correction term is required for most of the components, except for the electric field components lying along the edges of the total-field/scattered-field boundary, which require two correction terms.

The vicinity of one of the corners ($i = i_0$, $j = j_0$, $k = k_0$) of the 3D FDTD grid is shown in Figure 7.6. Here, the field components that reside inside or at the edges of the Region 2 FDTD grid (the total field region) are indicated with solid-filled arrows, while those that reside outside this region and inside Region 1 (the scattered-field region) are indicated with empty arrowheads. The shaded gray regions show where the box at the bottom of the figure has been cut out of the top portion. Note that since there are no field components at the corner of the boundary at (i_0, j_0, k_0), there is no need for a three-component "corner" correction. The components of the $\overline{\mathcal{E}}$ and $\overline{\mathcal{H}}$ fields that will require modifications to their update equations are shown in dashed circles.

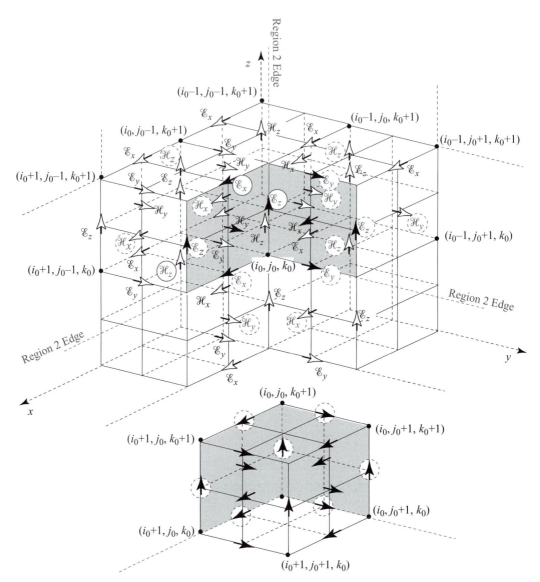

Figure 7.6 Three-dimensional total-field/scattered-field formulation. Only the vicinity of one of the corners of Region 1 ($i = i_0$, $j = j_0$, $k = k_0$) is shown. The filled arrows indicate total-field components in Region 1, while the empty arrows indicate scattered-field components in Region 2. The field components that need to be "corrected" are marked with a dashed circle.

Electric field components located on the faces (but not on the edges) of Region 2 require only a single correction term. Note that these fields are total fields, but their update equations involve a single scattered-field component from Region 1. For example, consider $\mathscr{E}_x^{\text{tot}}$ at location ($i_0 + 1/2, j_0, k_0 + 1$), which is in a solid circle in Figure 7.6. This component is surrounded by three total-field components, $\mathscr{H}_y^{\text{tot}}$ at ($i_0 + 1/2, j_0, k_0 + 1/2$) (below it), $\mathscr{H}_z^{\text{tot}}$ at ($i_0 + 1/2, j_0 + 1/2, k_0 + 1$) (to the right and forward, in the cutout cube), and $\mathcal{H}_y^{\text{tot}}$ at $(i_0 + 1/2, j_0, k_0 + 3/2)$ (above the large cube, not shown). In addition, our $\mathcal{E}_x$ component is surrounded by one scattered-field component, $\mathcal{H}_z^{\text{tot}}$ $(i_0 + 1/2, j_0 - 1/2, k_0 + 1)$, behind and to the left. This means the "normal" update equation will need to be corrected by adding the incident field at $\mathcal{H}_z^{\text{inc}}$ $(i_0 + 1/2, j_0 - 1/2, k_0 + 1)$. Thus, its update equation is:

$$
\mathcal{E}_x \Big|_{i_0+1/2, j_0, k_0+1}^{n+1} = \underbrace{\left[\mathcal{E}_x \Big|_{i_0+1/2, j_0, k_0+1}^{n+1}\right]}_{\text{Normal FDTD update, Eq. (4.18a)}} - \frac{\Delta t}{\epsilon \, \Delta y} \, \mathcal{H}_z^{\text{inc}} \Big|_{i_0+1/2, j_0-1/2, k_0+1}^{n+1/2}. \quad (7.16)
$$

This modification to the update equation applies to all of the $\mathcal{E}_x$ components along the face at j_0; similar update equation modifications need to be made to the $\mathcal{E}_y$ components along the i_0 face, and the $\mathcal{E}_z$ components along both the i_0 and j_0 faces, with the exception of the terms along the edge at (i_0, j_0, k), which need to be dealt with separately, as we will show presently.

From Figure 7.6, the magnetic field components at the faces of Region 2 are total-field components surrounded by total-field electric field components, so that their update equations do not need correction terms. However, scattered-field $\overline{\mathcal{H}}$ components located a half-mesh size outside Region 2 require correction terms. For example, consider $\mathcal{H}_z^{\text{scat}}$ at $(i_0 + 1/2, j_0 - 1/2, k_0)$, in a solid circle at the left of Figure 7.6. This component is surrounded by three scattered-field components, $\mathcal{E}_y^{\text{scat}}$ at $(i_0 + 1, j_0 - 1/2, k_0)$, at $\mathcal{E}_y^{\text{scat}}$ $(i_0, j_0 - 1/2, k_0)$, and $\mathcal{E}_x^{\text{scat}}$ $(i_0 + 1/2, j_0 - 1, k_0)$. It is also adjacent to one total-field component, $\mathcal{E}_x^{\text{tot}}$ at $(i_0 + 1/2, j_0, k_0)$, and as such we will need to subtract the incident field at that location from the update equation (since it is an update equation in terms of scattered fields). The update equation for this component is then:

$$
\mathcal{H}_z \Big|_{i_0+1/2, j_0-1/2, k_0}^{n+1} = \underbrace{\left[\mathcal{H}_z \Big|_{i_0+1/2, j_0-1/2, k_0}^{n+1}\right]}_{\text{Normal FDTD update, Eq. (4.19c)}} - \frac{\Delta t}{\epsilon \, \Delta y} \, \mathcal{E}_x^{\text{inc}} \Big|_{i_0+1/2, j_0, k_0}^{n+1/2}. \quad (7.17)
$$

This update equation applies to all $\mathcal{H}_z$ components along the $(j_0 - 1/2)$ face. Similar update modifications can be made in the same fashion for $\mathcal{H}_z$ components along the $(i_0 - 1/2)$ face, as well as $\mathcal{H}_x$ and $\mathcal{H}_y$ components along the $(k_0 - 1/2)$ face. Note, however, that the $\mathcal{H}_z$ components located along the line diagonally displaced from the z-axis edge of Region 2 (i.e., $\mathcal{H}_z^{\text{scat}}$ at $(i_0 - 1/2, j_0 - 1/2, k \geq k_0)$) do not need to be corrected, since these are surrounded by only scattered-field electric fields.

Along the edges of the boundary, things are slightly more complicated, but equally straightforward to deal with, and follow the same logic as the corner points in the 2D case. Consider, for example, the $\mathcal{E}_z^{\text{tot}}$ $(i_0, j_0, k_0 + 1/2)$ component in the center of Figure 7.6, in the solid circle, which is a total-field component. This field component is surrounded by two total-field components to its left and right (and slightly in front), namely $\mathcal{H}_x^{\text{tot}}$ at $(i_0, j_0 + 1/2, k_0 + 1/2)$ and $\mathcal{H}_y^{\text{tot}}$ at $(i_0 + 1/2, j_0, k_0 + 1/2)$. It is also surrounded by two scattered-field components to the left and right and slightly behind, namely $\mathcal{H}_x^{\text{scat}}$ at $(i_0, j_0 - 1/2, k_0 + 1/2)$ and $\mathcal{H}_y^{\text{scat}}$ at $(i_0 - 1/2, j_0, k_0 + 1/2)$. Thus, the incident field at these two locations will need to be added to the update equation. Thus, the modified updating equation should be:

$$
\mathscr{E}_z\Big|_{i_0,j_0,k_0+1/2}^{n+1} = \left[\mathscr{E}_z\Big|_{i_0,j_0,k_0+1/2}^{n+1}\right] \Bigg\}\; \text{Normal FDTD update, Eq. (4.18c)}
$$

$$
-\frac{\Delta t}{\epsilon\,\Delta x}\,\mathscr{H}_y^{\text{inc}}\Big|_{i_0-1/2,j_0,k_0+1/2}^{n+1/2} + \frac{\Delta t}{\epsilon\,\Delta y}\,\mathscr{H}_x^{\text{inc}}\Big|_{i_0,j_0-1/2,k_0+1/2}^{n+1/2}. \tag{7.18}
$$

Note that the updating equation is similar for all the other $\mathscr{E}_z$ components along the edge of Region 1, namely $\mathscr{E}_z^{\text{tot}}\,(i_0, j_0, k > k_0)$. Expressions similar to (7.18) can be written in a straightforward manner for the other electric field components at the edges of Region 2, namely $\mathscr{E}_x^{\text{tot}}\,(i > i_0, j_0, k_0)$ and $\mathscr{E}_y^{\text{tot}}\,(i_0, j > j_0, k_0)$.

7.4.1 Calculating the incident field

While our example in Figure 7.4 used a hard point source as an excitation, the total-field scattered-field formulation is more valuable, and more commonly used, with a plane wave excitation, evaluated directly on the TF/SF boundary. A plane wave incident normally from one side of the simulation can be introduced rather trivially; however, when a plane wave is sought propagating at some angle, one must take care to keep track of the incident fields very carefully. The computation of field components on the boundary is straightforward but somewhat cumbersome. We refer the readers to the paper by Umashankar and Taflove [4], and the summary in Taflove and Hagness [2].

7.5 FDTD calculation of time-harmonic response

The FDTD method is inherently a time domain solution of Maxwell's equations, which does not explicitly reveal numerical results that can be interpreted or quantified in comparison with the time-harmonic (sinusoidal steady-state) response of the configurations being modeled. Such comparisons are often the only means by which absolute assessment of the accuracy of FDTD results can be benchmarked.

One way to determine the time-harmonic response would be to use a sinusoidal source function at a particular frequency ω, run the FDTD algorithm until we observe steady state, and determine the resulting amplitude and phase of whatever quantity is of interest at every point of interest within the FDTD grid, for that excitation frequency ω. Although this procedure would in principle work, one would then have to repeat the procedure for every frequency of interest.

However, we know from the linearity of Maxwell's equations and theory of linear systems that the response of the system at every frequency is revealed by its response to an impulse excitation. This means that we can use a Gaussian pulse excitation, by choosing its width to be narrow enough to contain all of the frequencies of interest, so that it resembles an impulse.[5] We could then run the FDTD algorithm until the excitation pulse dies out, and take the Fourier transform of the field components of interest, to determine the amplitude and phase of the time-harmonic response at any frequency of interest.

At first thought, it might appear that such a procedure is a daunting task, since the entire time history of the field quantity of interest would have to be stored from the beginning to the end of the FDTD algorithm run. However, it turns out that the desired time-harmonic response can be obtained by maintaining only two buffers for every frequency of interest, if we carry out the Fourier transformation integral as the FDTD algorithm progresses.

Assume that our goal is to determine the magnitude and phase of a time-harmonic electric field component $\mathcal{E}$ at a given frequency f_0 at some given grid point i. We have

$$\mathcal{E}_i(f_0) = \int_0^{t_{\text{end}}} \mathcal{E}_i(t)\, e^{-j2\pi f_0 t}\, dt \tag{7.19}$$

where the upper limit t_{end} is the time at which the FDTD run is stopped. Converting (7.19) into finite difference form:

$$\mathcal{E}_i(f_0) = \sum_{n=0}^{T} \mathcal{E}_i(n\Delta t)\, e^{-j2\pi f_0 n \Delta t} \tag{7.20}$$

where $T = t_{\text{end}}/\Delta t$. We can express the Fourier component $\mathcal{E}_i(f_0)$ in terms of its real and imaginary components:

$$\mathcal{E}_i(f_0) = \underbrace{\sum_{n=0}^{T} \mathcal{E}_i(n\Delta t)\, \cos(2\pi f_0 n \Delta t)}_{\mathcal{R}e\{\mathcal{E}_i(f_0)\}} - j \underbrace{\sum_{n=0}^{T} \mathcal{E}_i(n\Delta t)\, \sin(2\pi f_0 n \Delta t)}_{\mathcal{I}m\{\mathcal{E}_i(f_0)\}}. \tag{7.21}$$

So that the quantity $\mathcal{E}_i(f_0)$ can be calculated by simply maintaining the two buffers, for its real and imaginary parts, which are incrementally determined by adding the new contributions at each time step.

Note that the amplitude and phase of $\mathcal{E}_i(f_0)$ can simply be determined from its real and imaginary parts. If we are interested in the response at more than one frequency, we need to maintain a pair of buffers at each frequency of interest.

7.6   Summary

In this chapter we have introduced methods for applying sources in the FDTD method. The simplest, yet least realistic method for applying sources is the hard source, where the field(s) at one or more grid locations are replaced with the desired input field. This method can be effective for testing an FDTD simulation, but is not particularly useful for real simulations.

---

[5] For better accuracy it is prudent to use the derivative of a Gaussian pulse, rather than a Gaussian directly, to avoid grid capacitance effects.

Next we introduced two other methods for applying internal sources. First, a current or voltage source can be added to Maxwell's equations as shown:

$$\left[\frac{\partial \overline{\mathscr{E}}}{\partial t}\right]^{n+1/2} = \frac{1}{\epsilon}[\nabla \times \overline{\mathscr{H}}]^{n+1/2} - \frac{1}{\epsilon}[\overline{\mathscr{I}}_{\text{source}}]^{n+1/2}.$$

The second method, the thin-wire approximation, enables the simulation of a thin antenna in 3D.

The remainder of this chapter focused on the total-field scattered-field (TF/SF) method for applying *external* sources to the FDTD simulation. This allows for the accurate simulation of plane wave sources that may be far from the scattering object. The TF/SF method relies on the linearity of Maxwell's equations, so that fields can be broken into incident and scattered components:

$$\overline{\mathscr{E}}_{\text{tot}} = \overline{\mathscr{E}}_{\text{inc}} + \overline{\mathscr{E}}_{\text{scat}}$$

where the incident, scattered, and total fields each satisfy Maxwell's equations. In the TF/SF method, the simulation space is broken into *total-field* regions and *scattered-field* regions, and update equations are written for each region in terms of the field components there; i.e.,

$$\nabla \times \overline{\mathscr{E}}_{\text{tot}} = -\mu\frac{\partial \overline{\mathscr{H}}_{\text{tot}}}{\partial t} \qquad \text{in the total-field region}$$

$$\nabla \times \overline{\mathscr{E}}_{\text{scat}} = -\mu\frac{\partial \overline{\mathscr{H}}_{\text{scat}}}{\partial t} \qquad \text{in the scattered-field region}$$

and similarly for the $\overline{\mathscr{E}}$ update equations. Then, at the boundary between the regions, a correction must be applied so that the equations are consistent. This correction involves adding or subtracting components of the *incident* field, which is determined analytically or from a separate FDTD simulation without scattering objects. This has the desired effect of applying the incident field at the boundary, but leaving it transparent to scattered fields returning across the boundary.

7.7   Problems

7.1. **Parallel-plate waveguide with scatterer.** Model a parallel-plate waveguide of height $x = a = 20$ mm, and of infinite extent in the z and y directions (i.e., make it long enough in the z-direction so that reflections from the end of the space do not return the scatterer). A thin perfectly conducting strip of length 10 mm in the z direction is parallel to the plates and is halfway between the two plates (at $x = 10$ mm), as shown in Figure 7.7.

(a)   Using a total-field/scattered-field formulation, launch an incident TE$_1$ mode[6] wave from the left-hand side of the guide at a frequency of

---

[6] Note that the TE$_1$ mode referred to here is the classical context, having one transverse electric field component and two magnetic field components, one transverse and one axial (along the propagation direction, i.e., z).

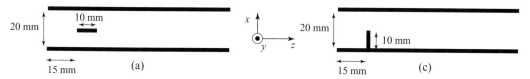

Figure 7.7 Waveguide used for Problem 7.1a (left) and Problem 7.1c (right).

$f = (f_{c_{TE_1}} + f_{c_{TE_2}})/2$, where $f_{c_{TE_1}}$ and $f_{c_{TE_2}}$ are respectively the cutoff frequencies for the TE$_1$ and TE$_2$ modes. What can you say about the reflected and transmitted waves?

(b) Repeat (a) but for a wave frequency $f = 1.2 f_{c_{TE_2}}$.

(c) Repeat (a) and (b) for the vertical strip case as shown in Figure 7.7.

7.2. **Total-field/scattered-field in 2D.** Consider implementing a 2D TM code similar to the example in Figure 7.4, except the scattering object will be a cylinder of radius 10 grid cells, rather than a square box. Make the space large enough so that fields do not reflect back into the space from the outer boundary.

(a) First, implement this scattering problem without the TF/SF method. In a separate simulation, compute the "incident" field by removing the scatterer. Calculate the scattered field by subtracting the incident field from the "total" field. Measure the scattering pattern by monitoring a number of points around the object at some distance.

(b) Now, implement this problem using the TF/SF method. You can use the same two simulations from part (a), and simply use the "incident" field at the boundary of the TF/SF interface. Measure the scattering pattern as you did in part (a). How do they compare? Comment on the relative merits of these two methods.

7.3. **A current source.** Repeat Problem 4.4, the 2D TM mode simulation, but excite a current source $\mathcal{J}_z$ at the center of the space rather than a hard electric field source. Use Equation (7.2) to discretize the current source; note that it is located at the same grid location in space as $\overline{\mathcal{E}}$, but the same time step as $\overline{\mathcal{H}}$. Compare the propagating fields to those of Problem 4.4. What is the relationship between the source current in Ampères and the fields in V/m?

7.4. **Thin-wire antenna in 3D.** Repeat Problem 4.6, exciting a half-wave dipole and monopole in a 3D Cartesian grid; however, this time, model the antenna using the thin-wire approximation for the conductive parts of the antenna. Compare the radiation patterns in the near field; how do they compare to the method in Chapter 4?

7.5. **3D rectangular waveguide.** Write a simulation for a 3D rectangular waveguide with cross-section dimensions 10.922 × 5.461 cm and length ~40 cm.

In other words, such a TE mode would correspond to a 2D TM$_y$ mode in the context of FDTD modeling, with the y-axis being the axis over which there are no variations of any quantity.

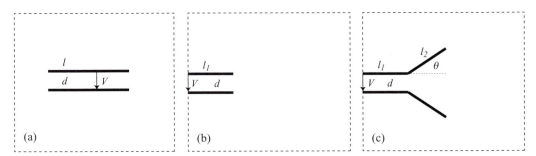

Figure 7.8 (a) Leaky capacitor in Problem 7.7; (b) waveguide source in Problem 7.8; (c) horn antenna in Problem 7.9. The dashed square represents the simulation boundary.

(a) Excite the waveguide at one end with a TE_{10} mode (in the waveguide sense) at frequency 2 GHz and monitor the wave as it passes through a point 20 cm down the waveguide. Use 20 grid cells per wavelength.

(b) Repeat part (a) for the TE_{11} and TE_{21} modes. How do you go about exciting these modes?

(c) Repeat part (a) for the TM_{11} mode.

(d) Repeat part (a), but model the waveguide walls with a thin layer of conductive material, varying σ from 1 to 10^4 S/m. How does the amplitude at the observation point depend on the conductivity? For each conductivity value, plot the amplitude as a function of distance; what is the loss in nepers/m?

7.6. 3D cylindrical waveguide. Write a 3D simulation in cylindrical coordinates to model a cylindrical waveguide. Give the waveguide a radius of 0.27 cm, and a length of 10 cm, and excite the waveguide with a frequency of 40 GHz. (How will you go about exciting this waveguide?) Monitor the fields at 5 cm down the waveguide; can you determine what modes are present? Is this what you would expect from waveguide theory?

7.7. Leakage fields in a capacitor. Consider the simulation of a capacitor in 3D Cartesian coordinates, as shown in Figure 7.8a. Create two parallel square conducting plates in the center of the space with length $l = 10$ cm and a free-space gap between them of $d = 1$ cm. Excite the capacitor by applying a sinusoidal voltage between the plates at frequency 1 MHz, pointing from one plate to the other, evenly distributed over the area of the plates.

Observe the fields that propagate away from the capacitor. What is the "radiation pattern" of this capacitor? How does it depend on frequency and the dimensions of the capacitor?

7.8. A waveguide radiator. Create a 2D TE mode simulation to observe the radiation leaving a waveguide aperture, as illustrated in Figure 7.8b. The waveguide should be 10 cm wide and 20 cm long. Excite the waveguide at the left end with a Gaussian pulse; observe what happens as the fields exit the aperture.

7.9. Horn antenna. Modify your 2D TE simulation code to excite a horn antenna, as illustrated in Figure 7.8c. Excite the antenna with an $\mathscr{E}_y$ source at the left end of the antenna, with a given voltage across the antenna plates (note that this 2D horn is really the opening of a 2D waveguide).

(a) Measure the antenna radiation pattern some distance from the antenna at a range of frequencies. (You can do this either by exciting a Gaussian pulse, or by running the simulation multiple times with a sinusoidal source at different frequencies.) Vary the length l_2 and the angle θ of the horn antenna and compare the results.

(b) Measure the characteristic driving-point impedance of this horn / waveguide combination, and compare to the impedance of the waveguide alone in the previous problem. Comment on the effect of the horn in this context.

You will need to take special care when handling the boundary conditions on the angled section of the horn, since the "normal" and "tangential" fields are no longer along grid edges; this will result in a stair-stepped horn aperture.

7.10. **A thin slot.** Create a 2D TM mode simulation that is 100×200 grid cells, designed for a frequency of 300 MHz. Place a vertical "wall" of PEC material across the center of the simulation space, but with a hole in the center of a few grid cells. Excite the simulation with an $\mathcal{E}_z$ source to the left of the wall, and measure the field intensity that leaks through the hole. Vary the size of the hole from 20 down to one grid cell; how does the field intensity to the right of the wall depend on the hole size? Can you think of a way to do this simulation with a hole that is less than one grid cell wide? We will revisit this problem in Chapter 12.

References

[1] R. Holland, "Threde: a free-field EMP coupling and scattering code," *IEEE Trans. Nuclear Science*, vol. 24, pp. 2416–2421, 1977.
[2] A. Taflove and S. Hagness, *Computational Electrodynamics: The Finite-Difference Time-Domain Method*, 3rd edn. Artech House, 2005.
[3] U. S. Inan and A. S. Inan, *Electromagnetic Waves*. Prentice-Hall, 2000.
[4] K. R. Umashankar and A. Taflove, "A novel method to analyze electromagnetic scattering of complex objects," *IEEE Trans. Electromag. Compat.*, vol. 24, pp. 397–405, 1982.

8 Absorbing boundary conditions

The necessarily finite nature of any FDTD spatial grid is a most important limitation in the FDTD method. Consider the scenario in Figure 8.1a, which sketches the scenario that we modeled in Figure 7.5c. In the real world, this waveguide would have a wave introduced at the left edge, which in our TF/SF space appears at the TF/SF boundary between Region 1 (left) and Region 2 (right). Waves will scatter from the scattering object (which is not necessarily PEC), both forward down the waveguide and back toward the source (labeled R0 in the figure). Since in our "ideal" scenario we have no other scattering objects, these scattered fields will continue to propagate forever, and so they should be unaffected by the ends of our simulation space.

Now consider the scenario in Figure 8.1b, where our numerical space, in which we have modeled a 2D parallel-plate waveguide, is truncated at the ends by PEC boundaries (by setting the electric field $\mathscr{E}_z = 0$ at the limiting grid points). The source wave enters the space at the TF/SF boundary as before, and scatters from the scattering object (R0 in the figure). However, at some time t_1, corresponding to the length of Region 2 divided by v_p, the source wave will reflect from the far right PEC boundary (R1). This reflected wave will scatter from the object as well (R2) at time $t_2 > t_1$; the reflected wave R2 will again reflect from the far right PEC, and the process will continue until a standing-wave pattern is established. In addition, the first reflection R1 will add to the "real" reflection from the scatterer R0, and propagate toward the leftmost PEC boundary, whereupon the combined wave will reflect again (R3). This wave will reflect from the scatterer again and continue to reflect back and forth between the scatterer and leftmost PEC.

Obviously, this scenario is far from the physical system in Figure 8.1a that we wish to model. Hence, the goal of absorbing boundaries is to emulate the scenario in Figure 8.1a as accurately as possible.

The problem at the boundaries of the FDTD grid arises because the field components at the outer edge of a finite FDTD space are not completely surrounded by the field components required for the update equation. Accordingly, there is not enough information to correctly update these components during the implementation of the FDTD algorithm. If we want the boundary to act as a perfect electric conductor (PEC), we can simply set the boundary points to $\mathscr{E}_{tan} = 0$, as was done for the top and bottom in Figure 7.5.

An absorbing boundary condition (ABC), on the other hand, is a means to approximately estimate the missing field components just outside the FDTD grid, in order to

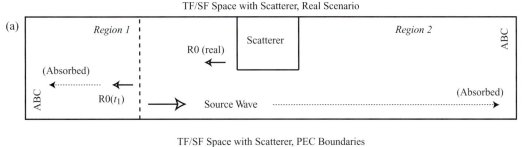

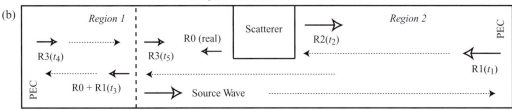

Figure 8.1 The need for absorbing boundaries in a 2D waveguide. (a) The ideal scenario has an incident wave scattering from the object, and then being absorbed at either end of the waveguide. (b) In the space with PEC walls, where no ABC has been used, numerical reflections render the fields inside the waveguide meaningless.

emulate an "infinite" space. Such an approximation typically involves assuming that a plane wave is incident on the boundary and estimating the fields just outside the boundary by using the fields just inside the boundary. However, this cannot be done without error, since in most cases the wave arriving at the boundary is not exactly a plane wave and is not normally incident. The ABCs are thus in general approximations, and reflect some of the waves back into the FDTD space. *First-order ABCs* are those which estimate the value of the fields outside the boundary by looking back one step in time and one grid cell in space, whereas higher order ABCs may look back over more steps in time and more grid cells in space. In general, different ABCs are better suited for different applications and the choice of the particular ABC is also made by considering its numerical efficiency and stability properties.

   Among the different types of absorbing boundary conditions, we shall limit our attention in this course to two particular ones, namely (i) those based on one-way wave equations, and (ii) those based on surrounding the FDTD domain with a layer of absorbing material, essentially creating a numerical *anechoic chamber*.[1] The latter of these two, known as *Perfectly Matched Layers*, will be considered in the next chapter.

---

[1] Microwave dark rooms, or anechoic chambers, are constructed with walls completely covered with absorbing materials, often in corrugated form (for tapered impedance matching), designed to minimize reflections and thus make the rooms suitable for testing antennas and other radiating devices. One absorbing material is a lossy mixture of high-permeability ferrite material and a high-permittivity barium titanate material known as the ferrite-titanate medium, for which both $\mu_c = \mu' - j\mu''$ and $\epsilon_c = \epsilon' - j\epsilon''$ are complex but the magnitude of the ratio $|\mu_c/\epsilon_c| = \mu_0/\epsilon_0$.

8.1 ABCs based on the one-way wave equation

The first category of ABCs relies on the fact that the solution of Maxwell's two coupled curl equations via an FDTD algorithm is equivalent to the solution of the second-order wave equation for any one of the field components. Although the wave equation naturally supports waves propagating in both forward and backward directions, it can be factored into two *one-way wave equations*, each of which supports waves in only one direction. This property provides the basis for an algorithmic method by which the fields can be "propagated out" of the FDTD domain, minimizing reflections back into the numerical space. A first-order scheme of this type is first discussed below, using the 1D wave equation, for simple one-way wave equations. This scheme is known as the *first-order Mur boundary condition*, and is quite effective in the removal of the plane wave fields normally incident on an FDTD boundary, being particularly suitable for 1D problems.

8.1.1 First-order Mur boundary

The first-order Mur boundary condition is one of the simplest boundary conditions available, but it can be very effective for 1D simulations, as it relies on normal incidence of the wave on the boundary. This Mur ABC is based on the one-way wave equations, or the convection equations, that we have used many times already in the book, having first derived them as factors of the wave equation in Chapter 3. These are repeated below:

$$\frac{\partial \mathcal{E}_z}{\partial t} + v_p \frac{\partial \mathcal{E}_z}{\partial x} = 0 \quad \rightarrow \quad \mathcal{E}_z(x, t) = f(x - v_p t) \tag{8.1a}$$

$$\frac{\partial \mathcal{E}_z}{\partial t} - v_p \frac{\partial \mathcal{E}_z}{\partial x} = 0 \quad \rightarrow \quad \mathcal{E}_z(x, t) = g(x + v_p t). \tag{8.1b}$$

The solutions of Equations (8.1a) and (8.1b) are waves propagating respectively in the positive $(+x)$ and negative $(-x)$ directions. These equations can respectively be used to simulate open-region boundaries at the right and left sides of a 1D FDTD grid, naturally propagating (or radiating) the incident waves away; hence the term *radiation boundary conditions*.

The ABC algorithms for these one-way wave equations can be derived respectively from (8.1a) and (8.1b), by simple two-point centered difference discretization. To derive the boundary condition for the wave propagating to the left boundary, we rewrite (8.1b) in terms of $\mathcal{E}_z$ and discretize it at $x = \Delta x/2$ (i.e., at mesh point $i = 3/2$, just inside the boundary)[2] and at time $t = (n + 1/2)\Delta t$ (i.e., at time index $n + 1/2$):

$$\frac{\partial \mathcal{E}_z}{\partial t} = v_p \frac{\partial \mathcal{E}_z}{\partial x} \quad \rightarrow \quad \frac{\mathcal{E}_z \Big|_{3/2}^{n+1} - \mathcal{E}_z \Big|_{3/2}^{n}}{\Delta t} = v_p \frac{\mathcal{E}_z \Big|_{2}^{n+1/2} - \mathcal{E}_z \Big|_{1}^{n+1/2}}{\Delta x}. \tag{8.2}$$

---

[2] Note that we have chosen to write our simulations in the Matlab style, where the boundary point $x = 0$ corresponds to $i = 1$, since Matlab uses 1 as the first index into an array. Other authors, probably C/C++ users, will label the first index as $i = 0$, which has the advantage that $x = i\Delta x$, whereas in our definition, $x = (i - 1)\Delta x$.

However, note that in our 1D TM mode FDTD algorithm (Equations 4.9), the $\mathcal{E}_z$ component is defined at integer time and spatial points, and so is not available at the half grid point $i = 3/2$ nor at the half time increment $n + 1/2$. As such, we implement a simple spatial average for the former and a simple time average for the latter. We can thus rewrite (8.2) as:

$$\frac{\left(\mathcal{E}_z\Big|_1^{n+1} + \mathcal{E}_z\Big|_2^{n+1}\right) - \left(\mathcal{E}_z\Big|_1^{n} + \mathcal{E}_z\Big|_2^{n}\right)}{2\Delta t} = v_p \frac{\left(\mathcal{E}_z\Big|_2^{n+1} + \mathcal{E}_z\Big|_2^{n}\right) - \left(\mathcal{E}_z\Big|_1^{n+1} + \mathcal{E}_z\Big|_1^{n}\right)}{2\Delta x}.$$

(8.3)

Solving for the boundary field at time $n + 1$, i.e., $\mathcal{E}_z\Big|_1^{n+1}$, we find:

$$\mathcal{E}_z\Big|_1^{n+1} = \mathcal{E}_z\Big|_2^{n} + \left[\frac{v_p \Delta t - \Delta x}{v_p \Delta t + \Delta x}\right]\left[\mathcal{E}_z\Big|_2^{n+1} - \mathcal{E}_z\Big|_1^{n}\right].$$

(8.4)

Note from this equation that at the CFL limit of $v_p \Delta t / \Delta x = 1$, the boundary field is equal to the field immediately to its right at the previous time step, i.e.,

$$\mathcal{E}_z\Big|_1^{n+1} = \mathcal{E}_z\Big|_2^{n} \quad \text{when} \quad \frac{v_p \Delta t}{\Delta x} = 1.$$

(8.5)

Otherwise, when $v_p \Delta t / \Delta x < 1$, there is a correction to the boundary field involving the previous value at the boundary and the current value to the right of the boundary. As such, this implies that the boundary update must be made *after* the rest of the simulation space has been updated.

At the rightmost edge of our 1D simulation space, with $i = i_{\text{max}}$, an equivalent version of the first-order Mur boundary condition is:

$$\mathcal{E}_z\Big|_{i_{\text{max}}}^{n+1} = \mathcal{E}_z\Big|_{i_{\text{max}}-1}^{n} + \left[\frac{v_p \Delta t - \Delta x}{v_p \Delta t + \Delta x}\right]\left[\mathcal{E}_z\Big|_{i_{\text{max}}-1}^{n+1} - \mathcal{E}_z\Big|_{i_{\text{max}}}^{n}\right].$$

(8.6)

8.1.2 Higher dimensional wave equations: second-order Mur

The first-order Mur boundary derived above works very well for 1D problems, where the waves at both ends of the space are, by definition, propagating normal to the boundary. In higher dimensional problems, wave fields are generally incident on the FDTD grid boundaries at arbitrary angles. In those cases, we must work with the 2D or 3D wave equation, as the problem requires. Here we consider the 2D case, written (for example) for the $\mathcal{E}_z$ component of a TM wave:

$$\frac{\partial^2 \mathcal{E}_z}{\partial x^2} + \frac{\partial^2 \mathcal{E}_z}{\partial y^2} - \frac{1}{v_p^2}\frac{\partial^2 \mathcal{E}_z}{\partial t^2} = 0$$

$$\left[\frac{\partial^2}{\partial x^2} + \frac{\partial^2}{\partial y^2} - \frac{1}{v_p^2}\frac{\partial^2}{\partial t^2}\right]\mathcal{E}_z = 0.$$

(8.7)

Note that the choice of wave equation depends on the simulation space, and which field component(s) is (are) defined at the boundary. In other words, an ABC is only

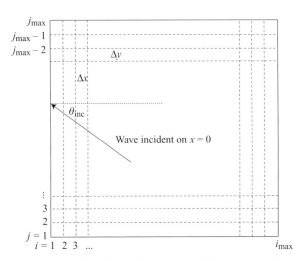

Figure 8.2 Absorbing boundary conditions in 2D. A wave incident on the $x=0$ ($i = 1$) boundary at angle θ_{inc} will partially reflect; effective ABCs in 2D will partially reduce these reflections.

required for those field components for which the normal update equations will not work, due to points falling outside the boundary. Consider the 2D TE case, as illustrated in Figure 4.5. Along the $y = 0$ boundary (i.e., $j = 1$) we find only the $\mathcal{E}_x$ field component, so a boundary condition based on a wave equation for $\mathcal{E}_x$ will be required; but on the $x = 0$ boundary, we find an $\mathcal{E}_y$ field component, so a boundary condition based on a wave equation for $\mathcal{E}_y$ will be required. Alternatively, we could define the boundary at the half-point where $\mathcal{H}_z$ falls, requiring a boundary condition only for $\mathcal{H}_z$ at all boundaries.

Here we will consider the 2D TM case where $\mathcal{E}_z$ falls on all boundaries. In our 2D space, we will need to consider waves incident on four walls of the numerical space, as well as corners. First, consider waves incident on the left and right walls as illustrated in Figure 8.2 (i.e., at $i = 1$ and $i = i_{\text{max}}$). We can factor Equation (8.7) specifically for these waves as follows:

$$\left[\frac{\partial^2}{\partial x^2} - \left(\frac{1}{v_p^2} \frac{\partial^2}{\partial t^2} - \frac{\partial^2}{\partial y^2} \right) \right] \mathcal{E}_z = 0$$

$$\underbrace{\left(\frac{\partial}{\partial x} + \sqrt{\frac{1}{v_p^2} \frac{\partial^2}{\partial t^2} - \frac{\partial^2}{\partial y^2}} \right)}_{\mathcal{L}^+} \underbrace{\left(\frac{\partial}{\partial x} - \sqrt{\frac{1}{v_p^2} \frac{\partial^2}{\partial t^2} - \frac{\partial^2}{\partial y^2}} \right)}_{\mathcal{L}^-} \mathcal{E}_z = 0. \qquad (8.8)$$

Note that $\mathcal{L}^+$ and $\mathcal{L}^-$ are two operators successively applied to the field component $\mathcal{E}_z$, similar to the expansion of the 1D wave Equation (3.18b).

Application of each of the operators $\mathcal{L}^+$ and $\mathcal{L}^-$ separately to $\mathcal{E}_z$ results in a one-way wave equation, albeit in 2D. For example, on the rightmost edge of a 2D FDTD grid, we can use:

$$\mathcal{L}^+ \mathcal{E}_z = \left(\frac{\partial}{\partial x} + \sqrt{\frac{1}{v_p^2} \frac{\partial^2}{\partial t^2} - \frac{\partial^2}{\partial y^2}} \right) \mathcal{E}_z = 0. \tag{8.9}$$

In contemplating the actual implementation of Equation (8.9), we must note that $\mathcal{L}^+$ and $\mathcal{L}^-$ are pseudo-differential operators that mix a square root operation with partial derivative operations. To interpret such an operator, consider the Fourier transformation of the quantity $\mathcal{E}_z(x, y, t)$ with respect to y and t. We have:

$$\mathcal{E}_z(x, y, t) = \mathcal{F}^{-1} \left\{ \tilde{\mathcal{E}}_z(x, k_y, \omega) \right\} \tag{8.10}$$

where $\tilde{\mathcal{E}}_z(x, k_y, \omega)$ is the 2D Fourier transform (in y and t) of $\mathcal{E}_z(x, y, t)$. The derivatives with respect to t and with respect to y can then be interpreted as:

$$\frac{\partial \mathcal{E}_z}{\partial t} = \mathcal{F}^{-1} \{ j\omega \tilde{\mathcal{E}}_z \} \quad \text{and} \quad \frac{\partial \mathcal{E}_z}{\partial y} = \mathcal{F}^{-1} \{ jk_y \tilde{\mathcal{E}}_z \}, \tag{8.11}$$

allowing us to define the square root operator in (8.9) as:

$$\left(\sqrt{\frac{1}{v_p^2} \frac{\partial^2}{\partial t^2} - \frac{\partial^2}{\partial y^2}} \right) \mathcal{E}_z \equiv \mathcal{F}^{-1} \left\{ \left[\left(\frac{j\omega}{v_p} \right)^2 - (jk_y)^2 \right]^{1/2} \tilde{\mathcal{E}}_z(x, k_y, \omega) \right\}. \tag{8.12}$$

Unfortunately, Equation (8.12) is not usable in a numerical algorithm, since the inverse Fourier transform is a global operation, requiring the knowledge of the values of the function $\tilde{\mathcal{E}}_z(x, k_y, \omega)$ for all k_y and ω. Thus, we seek to develop practical one-way wave equations which approximate the square root operation in Equation (8.9); for this approximation we follow the work of [1, 2]. For compactness of notation, we can write the operator $\mathcal{L}^+$ as

$$\mathcal{L}^+ = \frac{\partial}{\partial x} + \sqrt{\frac{1}{v_p^2} \frac{\partial^2}{\partial t^2} - \frac{\partial^2}{\partial y^2}} = \frac{\partial}{\partial x} + \frac{1}{v_p} \frac{\partial}{\partial t} \sqrt{1 - S^2} \tag{8.13a}$$

where

$$S \equiv \frac{v_p (\partial/\partial y)}{(\partial/\partial t)}. \tag{8.13b}$$

The parameter S is a unitless quantity that compares the spatial derivative (in y) with the time derivative. Note that the quantity S is actually a measure of the degree to which the incidence of the wave on the rightmost wall is normal. A very small value of S means that the partial derivative with respect to y of the incident wave (which, for a plane wave having the functional form $e^{j(\omega t + k_y y + k_x x)}$, is proportional to $|k_y|$) is much smaller than the time derivative (proportional to ω) as scaled by v_p. Since $k_x^2 + k_y^2 = \omega^2 / v_p^2$ is the eigenvalue equation for the wave Equation (8.1a), this in turn means that $k_x \gg k_y$, i.e., the wave is incident nearly normally on the rightmost wall defined by $x = i_{max} \Delta x$. Higher values of S correspond to incidence at larger incidence angles $\theta_{inc} = \tan^{-1}(k_y/k_x)$, as measured from the normal to the wall.

To implement the $\mathcal{L}^+$ operator in an FDTD algorithm, we seek an approximation for S that can be discretized in the methods we have derived in Chapter 3. As is typical for an expression of the form $(1 + x)^{1/2}$, we expand the square root term in a Taylor series around $S = 0$ (corresponding to normal incidence):

$$\sqrt{1 - S^2} = 1 - \frac{S^2}{2} - \frac{S^4}{8} - \frac{S^6}{16} - \cdots .$$ (8.14)

The simplest approximation, known as the first-order Mur method, is obtained by keeping only the unity term in Equation (8.14), in which case we have:

$$\mathcal{L}^+ \simeq \frac{\partial}{\partial x} + \frac{1}{v_p}\frac{\partial}{\partial t}$$

giving the one-way wave equation:

$$\frac{\partial \mathcal{E}_z}{\partial x} + \frac{1}{v_p}\frac{\partial \mathcal{E}_z}{\partial t} \simeq 0$$

exactly as derived in the previous section. The implementation at the rightmost wall is exactly as given by Equation (8.6), repeated below for the 2D case:

$$\mathcal{E}_z\Big|_{i_{max},j}^{n+1} = \mathcal{E}_z\Big|_{i_{max}-1,j}^{n} + \left[\frac{v_p\Delta t - \Delta x}{v_p\Delta t + \Delta x}\right]\left[\mathcal{E}_z\Big|_{i_{max}-1,j}^{n+1} - \mathcal{E}_z\Big|_{i_{max},j}^{n}\right]$$ (8.15)

The implementation of the first-order Mur boundary condition on the other walls follows from Equation (8.15) in a straightforward manner. Note that in this equation we have kept the index j to denote that we are operating in a 2D FDTD space; however, this first-order Mur implementation depends only on values along the x-axis (i.e., it does not depend on values at $j - 1$ or $j + 1$). This means the first-order Mur boundary condition above is only appropriate and accurate for waves normally incident on the walls, with $\theta_{inc} \simeq 0$; we will investigate the reflections from this boundary as a function of θ_{inc} later in this chapter. In order to have a boundary condition which can absorb waves arriving at a wider range of angles, we need to keep additional terms in Equation (8.14). Keeping the $S^2/2$ term we have:

$$\mathcal{L}^+ \simeq \frac{\partial}{\partial x} + \frac{1}{v_p}\frac{\partial}{\partial t}\left(1 - \frac{S^2}{2}\right)$$

$$= \frac{\partial}{\partial x} + \frac{1}{v_p}\frac{\partial}{\partial t} - \frac{1}{2}\frac{1}{v_p}\frac{\partial}{\partial t}\left[\frac{v_p(\partial/\partial y)}{(\partial/\partial t)}\right]^2$$

$$= \frac{\partial}{\partial x} + \frac{1}{v_p}\frac{\partial}{\partial t} - \frac{v_p}{2}\frac{\partial^2/\partial y^2}{\partial/\partial t}.$$ (8.16)

Substituting Equation (8.16) into Equation (8.9) and multiplying by $\partial/\partial t$, we have:

$$\frac{\partial^2 \mathscr{E}_z}{\partial x \partial t} + \frac{1}{v_p} \frac{\partial^2 \mathscr{E}_z}{\partial t^2} - \frac{v_p}{2} \frac{\partial^2 \mathscr{E}_z}{\partial y^2} \simeq 0. \tag{8.17}$$

The corresponding equation for the $x = 0$ boundary (using the $\mathcal{L}^-$ operator) is:

$$\frac{\partial^2 \mathscr{E}_z}{\partial x \partial t} - \frac{1}{v_p} \frac{\partial^2 \mathscr{E}_z}{\partial t^2} + \frac{v_p}{2} \frac{\partial^2 \mathscr{E}_z}{\partial y^2} \simeq 0. \tag{8.18}$$

These two equations constitute approximate one-way wave equations in 2D. Note that the Taylor series expansion of $(1 - S^2)^{1/2}$ is most accurate when S is small, so we can expect this 2nd-order boundary to be most effective for small angles. We can now discretize Equation (8.18) using the second-order differencing techniques that we studied in Chapter 3. The first term is handled by discretizing first in time using a two-point centered difference, and then discretizing the resulting spatial derivatives in a similar manner:

$$\frac{\partial^2 \mathscr{E}_z}{\partial x \partial t}\bigg|^n_{3/2,j} = \frac{1}{2\Delta t} \left(\frac{\partial \mathscr{E}_z}{\partial x}\bigg|^{n+1}_{3/2,j} - \frac{\partial \mathscr{E}_z}{\partial x}\bigg|^{n-1}_{3/2,j} \right)$$

$$= \frac{1}{2\Delta t} \left(\frac{\mathscr{E}_z\big|^{n+1}_{2,j} - \mathscr{E}_z\big|^{n+1}_{1,j}}{\Delta x} - \frac{\mathscr{E}_z\big|^{n-1}_{2,j} - \mathscr{E}_z\big|^{n-1}_{1,j}}{\Delta x} \right). \tag{8.19}$$

The third term in Equation (8.18), the second derivative in y, can be simply discretized using three-point centered differences as in Equation (3.54), after averaging in space across the $i = 1$ and $i = 2$ points:

$$\left[\frac{\partial^2 \mathscr{E}_z}{\partial y^2} \right]^n_{3/2,j} \simeq \frac{1}{2} \left(\frac{\partial^2 \mathscr{E}_z}{\partial y^2}\bigg|^n_{2,j} + \frac{\partial^2 \mathscr{E}_z}{\partial y^2}\bigg|^n_{1,j} \right)$$

$$\simeq \frac{1}{2} \left(\frac{\mathscr{E}_z\big|^n_{2,j+1} - 2\mathscr{E}_z\big|^n_{2,j} + \mathscr{E}_z\big|^n_{2,j-1}}{(\Delta y)^2} - \frac{\mathscr{E}_z\big|^n_{1,j+1} - 2\mathscr{E}_z\big|^n_{1,j} + \mathscr{E}_z\big|^n_{1,j-1}}{(\Delta y)^2} \right) \tag{8.20}$$

and similarly for the second term in Equation (8.18), the time derivative:

$$\left[\frac{\partial^2 \mathscr{E}_z}{\partial t^2} \right]^n_{3/2,j} \simeq \frac{1}{2} \left(\frac{\partial^2 \mathscr{E}_z}{\partial t^2}\bigg|^n_{1,j} + \frac{\partial^2 \mathscr{E}_z}{\partial t^2}\bigg|^n_{2,j} \right)$$

$$\simeq \frac{1}{2} \left(\frac{\mathscr{E}_z\big|^{n+1}_{2,j} - 2\mathscr{E}_z\big|^n_{2,j} + \mathscr{E}_z\big|^{n-1}_{2,j}}{(\Delta t)^2} - \frac{\mathscr{E}_z\big|^{n+1}_{1,j} - 2\mathscr{E}_z\big|^n_{1,j} + \mathscr{E}_z\big|^{n-1}_{1,j}}{(\Delta t)^2} \right). \tag{8.21}$$

Finally, substituting Equations (8.19), (8.20), and (8.21) into Equation (8.18) and solving for $\mathscr{E}_z\big|_{1,j}^{n+1}$, we find the update equation for the left boundary of our 2D space:

Second-order Mur boundary condition in 2D: left edge

$$
\begin{aligned}
\mathscr{E}_z\big|_{1,j}^{n+1} &= -\mathscr{E}_z\big|_{2,j}^{n-1} - \frac{\Delta x - v_p\Delta t}{\Delta x + v_p\Delta t}\left[\mathscr{E}_z\big|_{2,j}^{n+1} + \mathscr{E}_z\big|_{1,j}^{n-1}\right] + \frac{2\Delta x}{\Delta x + v_p\Delta t}\left[\mathscr{E}_z\big|_{1,j}^{n} + \mathscr{E}_z\big|_{2,j}^{n}\right] \\
&+ \frac{\Delta x(v_p\Delta t)^2}{2(\Delta y)^2(\Delta x + v_p\Delta t)}\left[\mathscr{E}_z\big|_{1,j+1}^{n} - 2\mathscr{E}_z\big|_{1,j}^{n} + \mathscr{E}_z\big|_{1,j-1}^{n} + \mathscr{E}_z\big|_{2,j+1}^{n}\right. \\
&\left. - 2\mathscr{E}_z\big|_{2,j}^{n} + \mathscr{E}_z\big|_{2,j-1}^{n}\right]
\end{aligned}
$$

$$(8.22)$$

Similar update equations can be derived for the other three boundaries of the 2D space, or can be simply transposed from Equation (8.22) by modifying subscripts. For example, the update on the bottom boundary (at $j = 1$) can be written as:

Second-order Mur boundary condition in 2D: bottom edge

$$
\begin{aligned}
\mathscr{E}_z\big|_{i,1}^{n+1} &= -\mathscr{E}_z\big|_{i,2}^{n-1} - \frac{\Delta x - v_p\Delta t}{\Delta x + v_p\Delta t}\left[\mathscr{E}_z\big|_{i,2}^{n+1} + \mathscr{E}_z\big|_{i,1}^{n-1}\right] + \frac{2\Delta x}{\Delta x + v_p\Delta t}\left[\mathscr{E}_z\big|_{i,1}^{n} + \mathscr{E}_z\big|_{i,2}^{n}\right] \\
&+ \frac{\Delta x(v_p\Delta t)^2}{2(\Delta y)^2(\Delta x + v_p\Delta t)}\left[\mathscr{E}_z\big|_{i+1,1}^{n} - 2\mathscr{E}_z\big|_{i,1}^{n} + \mathscr{E}_z\big|_{i-1,1}^{n} + \mathscr{E}_z\big|_{i+1,2}^{n}\right. \\
&\left. - 2\mathscr{E}_z\big|_{i,2}^{n} + \mathscr{E}_z\big|_{i-1,2}^{n}\right]
\end{aligned}
$$

$$(8.23)$$

An obvious disadvantage of the second-order Mur condition is the fact that, in addition to using the previous field values at time step n, prior values at time step $(n-1)$ are required, and thus must be stored in memory during the simulation.

The grid points used in the Mur first- and second-order boundaries in 2D are shown in Figure 8.3. In the first-order Mur, apart from the boundary point itself at the previous time step, only the point adjacent to the boundary is used; as such it is no surprise that this boundary condition is only effective at normal incidence. In the second-order Mur boundary, all five surrounding points are used to update the boundary point.

8.1.3   Higher-order Mur boundaries

The first-order and second-order Mur boundaries derived above leave reflections as shown in Figure 8.4. It is possible, of course, to reduce these reflections to the degree required by a particular problem by including higher-order terms of the Taylor series expansion of $\sqrt{1 - S^2}$ in Equation (8.14). However, as we include S^4 or higher terms of

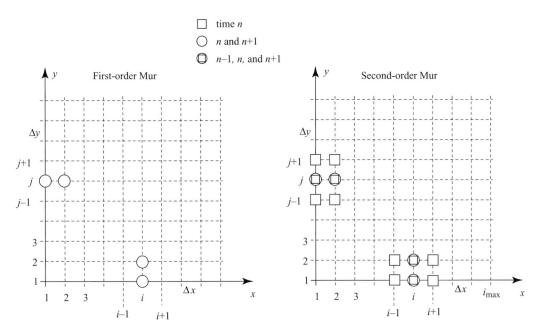

Figure 8.3 Mesh diagram for the Mur boundary conditions. Similar to Figure 3.10, the points used to update the Mur first- and second-order boundaries are shown.

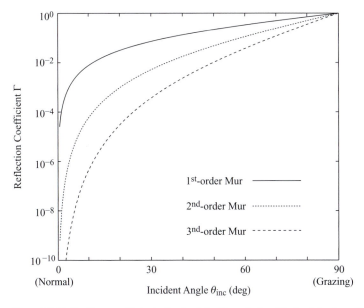

Figure 8.4 Reflection coefficients for the Mur boundaries. Each of the boundaries absorbs perfectly at normal incidence and reflects perfectly at grazing incidence, and the performance improves with increasing order.

the expansion, the implementation in discrete space becomes exceedingly cumbersome, in particular requiring the storage of fields at more time points (i.e., $n - 2$, $n - 3$, etc.).

Instead, a method derived by Trefethen and Halpern [3] provides increased accuracy to the Mur boundary without resorting to the Taylor series expansion. Rather, their method relies on a different approximation to $\sqrt{1 - S^2}$, given by:

$$\sqrt{1 - S^2} \simeq \frac{p_0 + p_2 S^2}{q_0 + q_2 S^2} \tag{8.24}$$

where the coefficients p_0, p_2, q_0, and q_2 can be modified to suit the needs of the problem. One can see that the first-order Mur boundary results from $p_0 = q_0 = 1$, $p_2 = q_2 = 0$, and the second-order Mur boundary results from $p_0 = q_0 = 1$, $p_2 = -1/2$, and $q_2 = 0$. The approximation in Equation (8.24) yields the following general one-way wave equation at the boundary:

$$q_0 v_p \frac{\partial^3 \mathcal{E}_z}{\partial x \partial t^2} + q_2 v_p^3 \frac{\partial^3 \mathcal{E}_z}{\partial x \partial y^2} - p_0 \frac{\partial^3 \mathcal{E}_z}{\partial t^3} - p_2 v_p^2 \frac{\partial^3 \mathcal{E}_z}{\partial t \partial y^2} = 0 \tag{8.25}$$

which can then be discretized with relative ease, using the same methodology as for the second-order Mur boundary. The reader should also note that by using the correct constants for p_0, p_2, q_0, and q_2, one can easily re-derive the one-way wave equations used for the Mur first- and second-order boundaries.

Second-order radiation-operator boundaries are those for which $q_0 = 1$ and $q_2 = 0$, such as the Mur boundary we studied earlier. Third-order boundaries are those for which $q_2 \neq 0$. Detailed discussion of a variety of second- and third-order boundaries is given by [4]. As a simple example, the third-order equivalent of the Mur boundary results from $p_0 = 1$, $p_2 = -3/4$ (rather than $-1/2$ in the second-order Mur), and $q_0 = 1$, $q_2 = -1/4$, resulting in:

$$\sqrt{1 - S^2} \simeq \frac{4 - 3S^2}{4 - S^2}.$$

The resulting reflection coefficient as a function of incident angle is shown in Figure 8.4 along with the first- and second-order reflections; in the next section we will derive the theoretical performance of these boundary conditions by deriving their reflection coefficients.

8.1.4   Performance of the Mur boundaries

To assess the performance of the ABCs given in Equations (8.15) and (8.22), we can examine their performance in absorbing a given propagating Fourier mode, similar to our assessment of dispersion in Chapter 6. Consider a given discrete Fourier mode which satisfies the finite-differenced form of the second-order wave equation:

$$\mathcal{E}_z^{\text{inc}}\Big|_{i,j}^n = e^{j(\omega n \Delta t)} e^{j(k_x(i-1)\Delta x)} e^{-j(k_y(j-1)\Delta y)}, \tag{8.26}$$

where the "inc" superscript refers to its incidence on the boundary; later we will have "reflected" and "total" waves. The reader should once again note the difference between the index j and the imaginary number $j = \sqrt{-1}$. Now consider the implementation of an ABC at the FDTD space boundary at $x = 0$ (i.e., $i = 1$, using our Matlab notation). The uniform plane wave described by Equation (8.26) propagates in the $-x$ and $+y$ directions, and is incident on the $x = 0$ boundary at an angle θ_{inc} with respect to the $+x$ axis, given by

$$\tan \theta_{inc} = \frac{k_y}{k_x} \tag{8.27}$$

assuming that k_x and k_y are positive. At the boundary, the imperfect nature of the absorbing boundary condition will lead to a reflected wave traveling in the $+x$ and $+y$ directions, matching the y variation of the incident wave but reversing the direction of its propagation along the x-axis:

$$\mathscr{E}_z^{ref}\Big|_{i,j}^n = \Gamma e^{j(\omega n \Delta t)} e^{-j(k \cos \theta_{inc}(i-1)\Delta x)} e^{-j(k \sin \theta_{inc}(j-1)\Delta y)} \tag{8.28}$$

where the reflection coefficient Γ is what we are aiming to evaluate, and we have written the equation in terms of the angle θ_{inc}. Thus, the total wave in the vicinity of the boundary is given by:

$$\mathscr{E}_z^{tot}\Big|_{i,j}^n = e^{j(\omega n \Delta t)} \left[e^{j(k \cos \theta_{inc}(i-1)\Delta x)} + \Gamma e^{-j(k \cos \theta_{inc}(i-1)\Delta x)} \right] e^{-j(k \sin \theta_{inc}(j-1)\Delta y)}. \tag{8.29}$$

The performance of the ABC can be assessed via the magnitude of the reflection coefficient, i.e., $|\Gamma|$. An expression for Γ can be found by substituting (8.29) into the ABC algorithms, Equations (8.15) and (8.22), and then evaluating the resulting equation at the $x = 0$ ($i = 1$) boundary. However, we can also derive a general reflection coefficient by substituting Equation (8.29) into the general form of the one-way wave Equation (8.25) and simplifying, resulting in the following general third-order reflection coefficient, as derived by [5]:

$$\Gamma = \frac{q_0 \cos \theta_{inc} + q_2 \cos \theta_{inc} \sin^2 \theta_{inc} - p_0 - p_2 \sin^2 \theta_{inc}}{q_0 \cos \theta_{inc} + q_2 \cos \theta_{inc} \sin^2 \theta_{inc} + p_0 + p_2 \sin^2 \theta_{inc}}. \tag{8.30}$$

In this relation, note that the approximation $\omega/k = v_p$ has been made for simplicity, which eliminates the frequency dependence of the reflection coefficient. The reader should be aware, however, that in reality the reflection coefficient is dependent on frequency. For the third-order Mur boundary in Equation (8.28), the reflection coefficient is:

Third-order Mur boundary reflection coefficient

$$\boxed{\Gamma = \frac{4 \cos \theta_{inc} - \cos \theta_{inc} \sin^2 \theta_{inc} - 4 + 3 \sin^2 \theta_{inc}}{4 \cos \theta_{inc} - \cos \theta_{inc} \sin^2 \theta_{inc} + 4 - 3 \sin^2 \theta_{inc}}.} \tag{8.31}$$

The second-order Mur boundary reflection can be derived by simply setting $q_2 = 0$ in Equation (8.30), resulting in:

Second-order Mur boundary reflection coefficient

$$\Gamma = \left| \frac{2\cos\theta_{\text{inc}} - 2 + \sin^2\theta_{\text{inc}}}{2\cos\theta_{\text{inc}} + 2 - \sin^2\theta_{\text{inc}}} \right|. \qquad (8.32)$$

The first-order Mur boundary can be similarly derived by simplification to find:

First-order Mur boundary reflection coefficient

$$\Gamma = \left| \frac{\cos\theta_{\text{inc}} - 1}{\cos\theta_{\text{inc}} + 1} \right|. \qquad (8.33)$$

These three relations are plotted together in Figure 8.4. Notice that, as expected, the boundaries perform better with increasing order. In all cases, the boundaries absorb essentially perfectly at normal incidence ($\theta_{\text{inc}} = 0$), and reflect perfectly at grazing incidence ($\theta_{\text{inc}} = 90$). For this reason, it is generally prudent to design an FDTD simulation space such that grazing angles do not occur; typically, the placement of scatterers should be designed so that waves are not incident on the boundaries at angles greater than 45 degrees. However, this is not always possible, and so other boundaries may need to be used, as will be described in this and the next chapter.

Note that other combinations of the constants q_0, p_0, q_1, and p_1 can be chosen in Equation (8.24); the result is that the numerical boundary will absorb perfectly at different angles. As such, different combinations can be chosen for the particular application in question. For example, the choice in a second-order boundary of $p_0 = q_0 = p_2 = 1$ and $q_2 = 0$ results in perfect absorption both at normal incidence ($\theta_{\text{inc}} = 0$) and at grazing incidence ($\theta_{\text{inc}} = 90$), with nonzero reflections at all other angles. For a detailed list of second- and third-order boundaries, we refer the reader to [5].

8.1.5   Mur boundaries in 3D

So far, we have considered only the 2D Mur boundary conditions; however, the extension to 3D is straightforward. Here we will provide a brief introduction to the method. Consider a 3D space, with a wave incident on the $x = 0$ boundary, on which the $\mathscr{E}_z$ component is situated, referring to Figure 4.1. Note that the $\mathscr{E}_y$ and $\mathscr{H}_x$ components are situated on the same plane; we will need another boundary condition for the $\mathscr{E}_y$ component, but not for the $\mathscr{H}_x$ component, since the fields required for its update equation are available at the same $x = 0$ plane. At the other five boundaries, other boundary conditions will be required depending on which fields are missing components for their update equations.

For the $\mathscr{E}_z$ component at $x = 0$, we start from the wave equation in 3D:

$$\frac{\partial^2 \mathscr{E}_z}{\partial x^2} + \frac{\partial^2 \mathscr{E}_z}{\partial y^2} + \frac{\partial^2 \mathscr{E}_z}{\partial z^2} - \frac{1}{v_p^2}\frac{\partial^2 \mathscr{E}_z}{\partial t^2} = 0$$

$$\left[\frac{\partial^2}{\partial x^2} + \frac{\partial^2}{\partial y^2} + \frac{\partial^2}{\partial z^2} - \frac{1}{v_p^2}\frac{\partial^2}{\partial t^2}\right]\mathscr{E}_z = 0. \tag{8.34}$$

Similar to the 2D case, we can factor the operator into $\mathscr{L}^+$ and $\mathscr{L}^-$ as in Equation (8.8); the $\mathscr{L}^-$ operator (to be applied at $x = 0$) is now given by:

$$\mathscr{L}^- = \frac{\partial}{\partial x} - \sqrt{\frac{1}{v_p^2}\frac{\partial^2}{\partial t^2} - \frac{\partial^2}{\partial y^2} - \frac{\partial^2}{\partial z^2}} \tag{8.35}$$

$$= \frac{\partial}{\partial x} - \frac{1}{v_p}\frac{\partial}{\partial t}\sqrt{1 - S^2}$$

where S is now given by:

$$S \equiv \sqrt{\frac{v_p^2(\partial^2/\partial y^2)}{(\partial^2/\partial t^2)} + \frac{v_p^2(\partial^2/\partial z^2)}{(\partial^2/\partial t^2)}}.$$

We can now find the approximate one-way wave equations for the first- and second-order Mur boundary conditions. For the first-order Mur boundary, recall that $\sqrt{1 - S^2} \simeq 1$, and so the one-way wave equation is the same as for the 2D case (Equation 8.15, with a k index included). For the second-order Mur boundary, $\sqrt{1 - S^2} \simeq 1 - S^2/2$, giving the following one-way wave equation:

$$\mathscr{L}^-\mathscr{E}_z = \left[\frac{\partial}{\partial x} - \frac{1}{v_p}\frac{\partial}{\partial t} + \frac{v_p(\partial^2/\partial y^2)}{2(\partial/\partial t)} + \frac{v_p(\partial^2/\partial z^2)}{2(\partial/\partial t)}\right]\mathscr{E}_z \simeq 0$$

which when multiplied through by $\partial/\partial t$ becomes:

$$\frac{\partial^2 \mathscr{E}_z}{\partial x \partial t} - \frac{1}{v_p}\frac{\partial^2 \mathscr{E}_z}{\partial t^2} + \frac{v_p}{2}\frac{\partial^2 \mathscr{E}_z}{\partial y^2} + \frac{v_p}{2}\frac{\partial^2 \mathscr{E}_z}{\partial z^2} \simeq 0. \tag{8.36}$$

Note that this equation is identical to Equation (8.18), with the addition of the $\partial^2/\partial z^2$ term, as expected. This equation can thus be discretized exactly as Equation (8.18), resulting in an update equation for the $\mathscr{E}_z$ component on the boundary similar to Equation (8.22).

The higher-order boundary conditions listed in [5] can also be used in 3D; the difference equations are derived in a similar manner using Equation (8.24) as the approximation for $\sqrt{1 - S^2}$, and result in one-way wave equations analogous to Equation (8.25), with $\partial/\partial z$ terms included.

8.2   Other radiation operators as ABCs

The Mur boundary condition described above is just one of a number of examples of using *radiation operators* as absorbing boundaries; $\mathscr{L}^-$ and $\mathscr{L}^+$ are the radiation operators themselves, which are applied in this case to the electric field $\overline{\mathscr{E}}$ at the boundary.

In general, as above for the Mur boundaries, the idea in using radiation operators is to construct a linear partial differential operator from a weighted sum of three types of derivatives of the field components: (i) spatial derivatives in the direction outward from the FDTD space ($\partial/\partial x$ in the Mur derivation above), (ii) spatial derivatives in the direction transverse to the outward direction ($\partial/\partial y$), and (iii) temporal derivatives ($\partial/\partial t$). Apart from the Mur ABCs, a number of other radiation operator ABCs have been developed in the short history of the FDTD method, a few of which will be described in this section.

8.2.1   Bayliss-Turkel operators

The so-called Bayliss-Turkel operators [6] are based on the time domain expressions for the radiation field (i.e., the far-field) solutions of the wave equation. However, rather than using the exact equations, they are presented in series expansion form, for both spherical and cylindrical coordinates. Here we briefly describe the two-dimensional (polar) case.

The 2D cylindrical wave equation, for example for the wave component $\mathscr{E}_z$ for a 2D TM mode,[3] is given as:

$$\nabla^2 \mathscr{E}_z - \frac{1}{v_p^2} \frac{\partial^2 \mathscr{E}_z}{\partial t^2} = 0$$

$$\frac{\partial^2 \mathscr{E}_z}{\partial r^2} + \frac{1}{r} \frac{\partial \mathscr{E}_z}{\partial r} + \frac{1}{r^2} \frac{\partial^2 \phi \mathscr{E}_z}{\partial \phi^2} - \frac{1}{v_p^2} \frac{\partial^2 \mathscr{E}_z}{\partial t^2} = 0 \qquad (8.37)$$

since we assume no variations in z in this 2D system. Note that $v_p = (\mu\epsilon)^{-1/2}$.

Those solutions of Equation (8.37) which propagate in directions outward from the origin (i.e., in the $\hat{r}$ direction for a cylindrical system) can be expanded in a convergent series as:

$$\mathscr{E}_z(r, \phi, t) = \sum_{m=1}^{\infty} \frac{\mathscr{E}_{z,m}(v_p t - r, \phi)}{r^{m-1/2}} = \frac{\mathscr{E}_{z,1}(v_p t - r, \phi)}{r^{1/2}} + \frac{\mathscr{E}_{z,2}(v_p t - r, \phi)}{r^{3/2}} + \cdots$$

$$(8.38)$$

Here $\mathscr{E}_{z,m}$ are functions of r and ϕ, but are all propagating in the radial direction, i.e., have time dependences of the form $(v_p t - r)$. It is clear from the $r^{-(m-1/2)}$ dependence that the leading terms in Equation (8.38) will dominate at large distances from the origin;

---

[3] Note that in cylindrical coordinates, unlike in Cartesian coordinates, we must define which coordinate is independent. This is most typically either the z coordinate as in the current discussion (in which case the 2D coordinates are known as polar coordinates) or the ϕ coordinate, in which case the problem is azimuthally symmetric. The "TM" designation in the FDTD method, then, is defined by which field is in the plane of the problem; in the case above it is the $\overline{\mathscr{H}}$ field in the $\phi - r$ plane.

we can thus envision creating a boundary that is far enough from the origin so that the higher-order terms in Equation (8.38) become small compared to our desired error. We now form the differential operator as shown below:

$$\mathcal{L}^+ = \frac{1}{v_p}\frac{\partial}{\partial t} + \frac{\partial}{\partial r}. \tag{8.39}$$

Note the similarity between this operator and the operator $\mathcal{L}^+$ used in the Mur boundary condition earlier, which is simply the Cartesian coordinate version of Equation (8.39). This operator is in fact known as the *Sommerfeld radiation condition* [7].

Note that in our 2D polar coordinate system, the only boundary we need to worry about is the $r = r_{max}$ boundary; as such, we do not need an $\mathcal{L}^-$ operator at $r = 0$, nor do we require any boundary condition for the ϕ direction, since it obviously has no boundary.

We next apply this operator to (8.37), considering just the first two terms of the series:

$$\mathcal{L}^+[\mathcal{E}_z(r,\phi,t)] = \left[\frac{1}{v_p}\frac{\mathcal{E}'_{z,1}\,v_p}{r^{1/2}} + \frac{\mathcal{E}'_{z,1}\,(-1)}{r^{1/2}} + \frac{\mathcal{E}_{z,1}\,(-1/2)}{r^{3/2}}\right]$$

$$+ \left[\frac{1}{v_p}\frac{\mathcal{E}'_{z,2}\,v_p}{r^{3/2}} + \frac{\mathcal{E}'_{z,2}\,(-1)}{r^{3/2}} + \frac{\mathcal{E}_{z,2}\,(-3/2)}{r^{5/2}}\right] + \cdots$$

$$= \underbrace{\frac{\mathcal{E}_{z,1}\,(-1/2)}{r^{3/2}} + \frac{\mathcal{E}_{z,2}\,(-3/2)}{r^{5/2}} + \cdots}_{O(r^{-3/2})}$$

where the prime denotes differentiation of the function with respect to its argument. Note that terms proportional to $r^{-1/2}$ have canceled out. Using the exact form of $\mathcal{L}^+\mathcal{E}_z$ from Equation (8.39) together with the above expression, we find the approximate one-way wave equation:

$$\mathcal{L}^+[\mathcal{E}_z(r,\phi,t)] = \frac{1}{v_p}\frac{\partial\mathcal{E}_z}{\partial t} + \frac{\partial\mathcal{E}_z}{\partial r} = \underbrace{\frac{\mathcal{E}_{z,1}\,(-1/2)}{r^{3/2}} + \frac{\mathcal{E}_{z,2}\,(-3/2)}{r^{5/2}} + \cdots}_{O(r^{-3/2})} \tag{8.40}$$

which can be rewritten in the following form:

$$\frac{\partial\mathcal{E}}{\partial r} = -\frac{1}{v_p}\frac{\partial\mathcal{E}}{\partial t} + O(r^{-3/2}). \tag{8.41}$$

Thus, in principle we can use Equation (8.41) at the outer boundary of the FDTD space to estimate the spatial derivative $\partial\mathcal{E}/\partial r$ from known values of the interior fields, assuming that the $O(r^{-3/2})$ term can be neglected. Doing so would be equivalent to using the one-way wave equation in Cartesian coordinates with a single term Taylor series approximation of $\sqrt{1-S^2}$ as was done for the Mur boundary condition. However, this method would in general only work if the fields are propagating *exactly* in the radial direction, meaning they are normally incident on the FDTD boundary. Alternatively, if the fields are not normal, one would have to use extensive computational and memory resources to make the outer boundary very far from the origin (which is presumably the location of the object of study, e.g., the scatterer), so that neglecting the higher order terms can be justified.

This limitation of normal incidence or very large computational space can be overcome with a slight modification to the operator $\mathcal{L}^+$ as given in Equation (8.39). Making a modification to the radiation operator is somewhat analogous to the "modification" that was made in Equation (8.9) to include the $\partial/\partial y$ component in the Mur boundary condition. The modification in cylindrical coordinates was introduced by Bayliss and Turkel [6, 8]. They noticed that (8.32) above has as its highest-order remainder term $-\mathcal{E}_{z,1}/2r^{3/2}$; thus, to cancel this term and leave only the $r^{-5/2}$ remainder term, one could add a term $1/2r$ to the $\mathcal{L}^+$ operator (hereafter denoted as $\mathcal{L}_1$):

$$\mathcal{B}_1 = \mathcal{L}_1 + \frac{1}{2r} = \frac{1}{v_p}\frac{\partial}{\partial t} + \frac{\partial}{\partial r} + \frac{1}{2r}. \tag{8.42}$$

Applying Equation (8.42) to Equation (8.38) and canceling terms gives:

$$\mathcal{B}_1\left[\mathcal{E}_z(r,\phi,t)\right] = \frac{1}{v_p}\frac{\partial\mathcal{E}_z}{\partial t} + \frac{\partial\mathcal{E}_z}{\partial r} + \frac{\mathcal{E}}{2r} = \underbrace{-\frac{\mathcal{E}_{z,2}}{r^{5/2}} - \frac{2\mathcal{E}_{z,3}}{r^{7/2}}}_{O(r^{-5/2})} + \cdots. \tag{8.43}$$

The one-way wave equation that then results from applying the operator in Equation (8.42) is:

$$\frac{\partial\mathcal{E}_z}{\partial r} = -\frac{1}{v_p}\frac{\partial\mathcal{E}_z}{\partial t} - \frac{\mathcal{E}_z}{2r} - O(r^{-5/2}). \tag{8.44}$$

Thus, by simply adding the $1/2r$ term to the operator in Equation (8.39), we are able to increase the order of accuracy of the one-way equation to $O(r^{-5/2})$. In other words, the remainder term in Equation (8.44) decreases to zero as $r^{-5/2}$ instead of the $r^{-3/2}$ dependence of the remainder in Equation (8.33).

The operator $\mathcal{B}_1$ in Equation (8.42) is known as the first-order Bayliss-Turkel radiation operator. Even higher accuracy can be achieved by the use of the second-order Bayliss-Turkel radiation operator, which is given by:

$$\mathcal{B}_2 = \left[\mathcal{L}_1 + \frac{5}{2r}\right]\overbrace{\left[\mathcal{L}_1 + \frac{1}{2r}\right]}^{\mathcal{B}_1} \tag{8.45}$$

which, when applied to $\mathcal{E}_z$ and simplified, leads to:

$$\mathcal{B}_2\mathcal{E}_z = \left(\frac{1}{v_p}\frac{\partial}{\partial t} + \frac{\partial}{\partial r} + \frac{5}{2r}\right)\left(\frac{1}{v_p}\frac{\partial}{\partial t} + \frac{\partial}{\partial r} + \frac{1}{2r}\right)\mathcal{E}_z = \underbrace{\frac{2\mathcal{E}_{z,3}}{r^{9/2}} + \frac{6\mathcal{E}_{z,4}}{r^{11/2}}}_{O(r^{-9/2})}. \tag{8.46}$$

In this case, as $r \to \infty$, the error term goes to zero r^{-3} times faster than that in the zero-order case in Equation (8.41). However, note that the extraction of the spatial derivative $\partial\mathcal{E}/\partial r$ from Equation (8.45) requires more complexity in the implementation of the boundary condition.

The importance of the Bayliss-Turkel type of radiation operators is that they are constructed without any knowledge of the angular dependence of the partial wavefunctions.

In practice, the second-order operator $\mathcal{B}_2$ has been most commonly used, representing a good balance between accuracy and complexity.

Spherical coordinates

The Bayliss-Turkel operators can also be constructed in cylindrical coordinates for an r–z coordinate system, or in 3D cylindrical or spherical coordinates, using analogous methods. For details, we refer the reader to [8] and [4] and provide only a brief introduction to the spherical coordinate case here. The field component, for example $\mathcal{E}_\phi$, is expanded in the following form:

$$\mathcal{E}_\phi(r, \theta, \phi, t) = \sum_{m=1}^{\infty} \frac{\mathcal{E}_{\phi,m}(v_p t - r, \theta, \phi)}{r^m}$$

$$= \frac{\mathcal{E}_{\phi,1}(v_p t - r, \theta, \phi)}{r} + \frac{\mathcal{E}_{\phi,2}(v_p t - r, \theta, \phi)}{r^2} + \cdots . \qquad (8.47)$$

The Bayliss-Turkel operators are then constructed, similar to those for the cylindrical case in Equations (8.42) and (8.45) except for constant factors:

$$\mathcal{B}_1 = \mathcal{L}_1 + \frac{1}{r} \quad \text{and} \quad \mathcal{B}_2 = \left[\mathcal{L}_1 + \frac{3}{r}\right]\left[\mathcal{L}_1 + \frac{1}{r}\right]. \qquad (8.48)$$

Because of the r^{-1} dependence in the expansion in (8.47) rather than $r^{-1/2}$ in the polar coordinate case, there is an improvement of $r^{1/2}$ in the accuracy of each operator. Specifically, the $\mathcal{L}_1$ operator used alone gives an error of order $O(r^{-2})$; the first-order operator $\mathcal{B}_1$ gives an error of order $O(r^{-3})$, and the second-order operator $\mathcal{B}_2$ gives an error of order $O(r^{-5})$.

8.2.2 Higdon operators

The radiation boundary conditions discussed so far are effective at absorbing waves in specific scenarios. In particular, the Mur boundaries are effective at near normal incidence, and the Bayliss-Turkel operators are effective in cylindrical or spherical coordinates. We also introduced particular cases, related to the Mur boundary, where absorption is most effective at particular angles of incidence, through choices of the constants in Equation (8.24). However, those angles cannot be chosen. In this section, we introduce a method by which absorption can be optimized at a specific choice of incident angles.

This radiation operator technique, introduced by R. L. Higdon [9], relies on a series of linear partial differential equations for cancellation of outgoing numerical waves, and is particularly suited for absorbing plane waves propagating at specific angles in a Cartesian grid. Consider 2D plane waves propagating at speed v_p toward the $x = 0$ boundary in a 2D Cartesian FDTD space. Assuming that component waves constituting an overall plane wave structure are propagating at angles $\theta_1, \theta_2, \ldots, \theta_M$ from the x-axis, the analytical expression for the superposition of such a wave structure is:

$$\mathscr{E}(x, y, t) = \sum_{m=1}^{M} \left[f_m(v_p t + x \cos \theta_m + y \sin \theta_m) + g_m(v_p t + x \cos \theta_m - y \sin \theta_m) \right]$$

(8.49)

where $-\pi/2 \le \theta_m \le \pi/2$, and $f(\cdot)$ and $g(\cdot)$ are arbitrary functions. The numerical operator proposed by Higdon for use in such a case is:

$$\left[\prod_{m=1}^{M} \left(\cos \theta_m \frac{\partial}{\partial t} - v_p \frac{\partial}{\partial x} \right) \right] \mathscr{E} = 0$$

(8.50)

for the numerical space boundary at $x = 0$.

The particular advantage of the absorbing boundary condition (8.50) is the fact that any combination of plane waves propagating toward the $x = 0$ wall at the chosen discrete angles θ_m are *completely* absorbed, with no reflection. For a sinusoidal numerical plane wave propagating at an incidence angle of $\theta \ne \theta_m$, the theoretical reflection coefficient is given by:

$$\Gamma = -\prod_{m=1}^{M} \frac{\cos \theta_m - \cos \theta}{\cos \theta_m + \cos \theta}.$$

(8.51)

For any given problem at hand, the exact absorption angles θ_m can be chosen to optimize the overall performance of the ABC, and Equation (8.51) can be used as a measure of its effectiveness at other angles.

It is useful to examine the first and second Higdon operators, obtained respectively for $M = 1$ and $M = 2$. For $M = 1$, we have:

$$\cos \theta_1 \frac{\partial \mathscr{E}}{\partial t} - v_p \frac{\partial \mathscr{E}}{\partial x} = 0.$$

(8.52)

Note that for $\theta_1 = 0$, Equation (8.52) reduces to the one-way wave equation in Section 8.1.2 resulting from a single term Taylor expansion of $\sqrt{1 - S^2}$; i.e., at $\theta_1 = 0$, the first-order Higdon operator is identical to the first-order Mur operator. More generally, Equation (8.52) is a perfect boundary condition to use in problems where the geometry dictates that the wave arriving at the $x = 0$ wall will do so at or near a given incidence angle θ_1.

What about a scenario when waves are incident on the boundary wall from not one but two discrete angles? For $M = 2$, we have from Equation (8.50):

$$\left(\cos \theta_1 \frac{\partial}{\partial t} - v_p \frac{\partial}{\partial x} \right) \left(\cos \theta_2 \frac{\partial}{\partial t} - v_p \frac{\partial}{\partial x} \right) \mathscr{E} = 0$$

$$\cos \theta_1 \cos \theta_2 \frac{\partial^2 \mathscr{E}}{\partial t^2} - v_p (\cos \theta_1 + \cos \theta_2) \frac{\partial^2 \mathscr{E}}{\partial x \partial t} + v_p^2 \frac{\partial^2 \mathscr{E}}{\partial x^2} = 0.$$

(8.53)

Note that Equation (8.53) can be discretized in the same manner as Equation (8.18) for the second-order Mur boundary, to arrive at an equation similar to Equation (8.22). For better comparison with Equation (8.18), we can express the last term $(\partial^2 \mathscr{E}/\partial x^2)$ in terms of the corresponding second derivative with respect to y, namely $\partial^2 \mathscr{E}/\partial y^2$, using the 2D wave equation:

$$\frac{\partial^2 \mathcal{E}}{\partial x^2} = \frac{1}{v_p^2} \frac{\partial^2 \mathcal{E}}{\partial t^2} - \frac{\partial^2 \mathcal{E}}{\partial y^2}.$$

Resulting in the following expression:

$$\frac{\partial^2 \mathcal{E}}{\partial x \partial t} - \underbrace{\frac{1}{v_p} \left[\frac{1 + \cos\theta_1 \cos\theta_2}{\cos\theta_1 + \cos\theta_2} \right]}_{p_0} \frac{\partial^2 \mathcal{E}}{\partial t^2} + v_p \underbrace{\left[\frac{1}{\cos\theta_1 + \cos\theta_2} \right]}_{-p_2} \frac{\partial^2 \mathcal{E}}{\partial y^2} = 0 \qquad (8.54)$$

we now see that Equation (8.18) is simply another form of Equation (8.54) for the special case of

$$p_0 = \frac{1 + \cos\theta_1 \cos\theta_2}{\cos\theta_1 + \cos\theta_2} = 1 \qquad \text{and} \qquad p_2 = -\frac{1}{\cos\theta_1 + \cos\theta_2} = -\frac{1}{2}. \qquad (8.55)$$

Thus, the second-order Higdon operator can be made identical to the second-order Mur boundary condition, by simply selecting the numerical values of the angles θ_1 and θ_2 so that Equation (8.55) holds. In particular, the only angle that solves Equations (8.55) is $\theta_1 = \theta_2 = 0$, the case of normal incidence; this agrees with the theoretical reflection coefficient for the second-order Mur boundary plotted in Figure 8.4. More generally, one can show quite easily that the equations for p_0 and p_2 above agree with the angles of exact absorption presented in tabular form in [5].

To demonstrate the use of the Higdon operators, Figure 8.5 shows the reflection coefficients for a couple of possible pairs of angles. These reflection coefficients are calculated by first choosing the pairs of angles, and then using Equation (8.55) to determine p_0 and p_2 to use in Equation (8.30).

Consider a scenario in which a square simulation space is excited at its center with a hard source. The Mur boundary will give zero reflection at normal incidence, and the worst reflection will occur at 45 degree incidence, where the wave hits the corners of the space. In fact, these reflections will propagate directly back toward the source. A Higdon operator can be easily devised for $\theta_1 = 0$ and $\theta_2 = 45°$, resulting in $p_0 = 1$ and $p_2 = -0.586$. The resulting reflection coefficient is plotted in Figure 8.5, and demonstrates the effect as expected. Of course, the trade-off is that, compared to the second-order Mur boundary, the reflections are somewhat higher between 0 and 30 degrees incidence; but everywhere in the 0- to 45-degree range of interest, $\Gamma < 10^{-2}$. For comparison, Figure 8.5 also shows the reflection coefficient for a scenario in which the angles of interest are 30 and 60 degrees.

8.3 Summary

In this chapter we have introduced methods for terminating the FDTD simulation space, where the normal FDTD update equations cannot be implemented due to missing field components. These radiation boundary conditions or absorbing boundary conditions

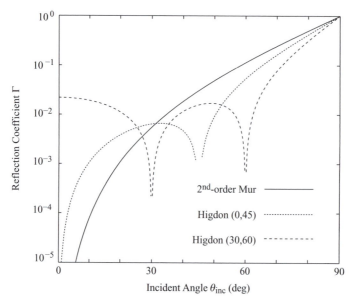

Figure 8.5 Reflection coefficients for the Higdon operators. The second-order Mur boundary is shown for comparison. With the Hidgon operators we can choose two angles at which zero reflection will occur; these examples show $\theta_1, \theta_2 = 0$ and 45 degrees or 30 and 60 degrees.

allow us to keep the simulation space small – thus saving computation time – without having to worry about unphysical reflections from the simulation space boundary.

We first introduced the relatively simple Mur boundaries, which are derived from the one-way wave equations. These one-way wave equations are found by factoring the wave equation into two operators, each describing waves propagating in opposite directions in a given dimension. For example, in 2D, the wave equation factors into

$$
\underbrace{\left(\frac{\partial}{\partial x} + \sqrt{\frac{1}{v_p^2} \frac{\partial^2}{\partial t^2} - \frac{\partial^2}{\partial y^2}} \right)}_{\mathscr{L}^+} \underbrace{\left(\frac{\partial}{\partial x} - \sqrt{\frac{1}{v_p^2} \frac{\partial^2}{\partial t^2} - \frac{\partial^2}{\partial y^2}} \right)}_{\mathscr{L}^-} \mathscr{E}_z = 0
$$

where the two operators describe waves propagating in the $-x$ and $+x$ directions, respectively. At the $x = 0$ boundary, we approximate the square root term using a Taylor series expansion:

$$
\mathscr{L}^+ = \frac{\partial}{\partial x} + \frac{1}{v_p} \frac{\partial}{\partial t} \sqrt{1 - S^2} = \frac{1}{v_p} \frac{\partial}{\partial t} \left(1 - \frac{S^2}{2} - \frac{S^4}{8} - \frac{S^6}{16} - \cdots \right)
$$

where $S \equiv v_p (\partial / \partial y)/(\partial / \partial t)$. Depending on where the Taylor series expansion is truncated, different levels of accuracy (i.e., correspondingly lower reflection coefficients) are achieved. The first-order ($\sqrt{1 - S^2} \simeq 1$) and second-order ($\sqrt{1 - S^2} \simeq 1 - S^2/2$) Mur boundaries were derived in Section 8.1. Alternatively, by instead making the simplification

$$\sqrt{1 - S^2} \simeq \frac{p_0 + p_2 S^2}{q_0 + q_2 S^2},$$

we can tailor the reflection coefficient as a function of angle of incidence, through particular choices of the constants, to mimimize reflections from the boundary at particular angles.

In Section 8.2 we introduced variations of absorbing boundary conditions, including the Bayliss-Turkel operators and Higdon operators. The Bayliss-Turkel operators are an improvement to the equivalent of the first-order Mur boundary in cylindrical or spherical coordinates, involving the one-way wave equation:

$$\underbrace{\left(\frac{\partial}{\partial r} + \frac{1}{v_p} \frac{\partial}{\partial t} \right)}_{\mathcal{L}_1} \mathcal{E} = O(r^{-3/2}) \qquad \text{(cylindrical coordinates)}$$

where the error term (i.e., the reflection coefficient) decreases with increasing distance r as $r^{-3/2}$. The Bayliss-Turkel operator improves this error term by modifying the $\mathcal{L}_1$ operator. The first-order Bayliss-Turkel operator gives:

$$\underbrace{\left(\frac{\partial}{\partial r} + \frac{1}{v_p} \frac{\partial}{\partial t} + \frac{1}{2r} \right)}_{\mathcal{B}_1} \mathcal{E} = O(r^{-5/2}) \qquad \text{(Bayliss-Turkel first order)}$$

where now the error goes as $r^{-5/2}$. The second-order Bayliss-Turkel operator, while much more complex, increases the accuracy to $O(r^{-9/2})$, thus requiring a smaller simulation space.

The Higdon operator, introduced in Section 8.2.2, allows us to minimize the reflection coefficient in Cartesian coordinates at particular angles, similar to the expansion of $\sqrt{1 - S^2}$ in Equation (8.24), but in this case the angle can be directly specified and tuned. The Higdon operator is given by:

$$\left[\prod_{m=1}^{M} \left(\cos \theta_m \frac{\partial}{\partial t} - v_p \frac{\partial}{\partial x} \right) \right] \mathcal{E} = 0 \qquad \text{(Higdon operator)}$$

and the reflection coefficient is analytically specified by Equation (8.51). In this way, an operator can be constructed for an arbitrary number m of incident angles at which the reflection is to be minimized.

8.4   Problems

8.1. **Mur boundaries in corner regions.** The second-order Mur boundary in Equations (8.22) and (8.23) will not work in the corners of a 2D simulation, or in the corners and edges of a 3D simulation. Derive a Mur boundary condition for the 2D corner regions, both (a) first-order, and (b) second-order.

8.2. Second-order Mur in 2D. In 2D, program the second-order Mur absorbing boundary condition for the case you considered in Problem 4.4, replacing the perfectly conducting boundaries with the second-order Mur ABC. Set up the problem space as described in part (a) of Problem 4.4.

(a) Assess the effects of the Mur ABC on the FDTD solution, by comparing this calculation with that of an FDTD space of 200×200 for a source placed in its center. Both calculations should be run until the wave in the 200×200 FDTD space hits the edges. The 200×200 run thus represents a perfect ABC for the 50×50 run. Calculate the global error of the problem over this range of time steps.

(b) Calculate the reflection coefficient for various angles of incidence of the outgoing wave on the boundary. The reflection coefficient is the difference in the two calculations from part (a). Do this for incident angles of: 0, 15, 30, and 45 degrees. Comment on the results. Does this agree with the $\Gamma(\omega, \theta)$ formula for the second-order Mur ABC as given in Equation (8.32)? Should it? Why or why not?

8.3. Bayliss-Turkel operator in polar coordinates. Modify Problem 4.7 to implement Bayliss-Turkel operators at the outer boundary. Implement the (a) first-order and (b) second-order Bayliss-Turkel operator, and measure the reflection coefficient; how do the two methods compare?

8.4. Bayliss-Turkel operator in cylindrical coordinates. Derive the Bayliss-Turkel operator in an r–z coordinate system. Implement this boundary condition for Problem 4.8 and measure the reflections from the outer boundary.

8.5. Higher-order Mur boundary. Implement a third-order Mur boundary for the 2D simulation in Problem 8.2. Use the coefficients given in Section 8.1.3. Compare the reflections to those in Problem 8.2.

8.6. Higdon operator reflection coefficient. Plot the reflection coefficient for the Higdon operator, Equation (8.51), for $M = 2$ and for $M = 4$. In the $M = 2$ case use angles of 0 and 45 degrees, and in the $M = 4$ case use angles of 0, 30, 45, and 60 degrees. In the nonzero reflection regions, how do the two operators compare? Derive an expression similar to Equation (8.54) for the $M = 4$ Higdon operator.

8.7. Second-order Higdon operator in 2D. Repeat Problem 8.2 using the Higdon operator, using zero-reflection angles of (a) 0 and 45 degrees, and (b) 30 and 60 degrees. How do the reflection coefficients compare to the theoretical values plotted in Figure 8.5? How do the absolute reflections compare to the Mur case in Problem 8.2?

8.8. 2D polar simulation. Repeat Problem 7.2, the cylindrical scatterer, only this time in 2D polar coordinates. In this case, the circular scattering object is much more accurately modeled in cylindrical coordinates, and the total-field/scattered-field boundary is also circular. Use your code from Problem 8.3, including the second-order Bayliss-Turkel boundary condition. Compare the scattering pattern to that found in the Cartesian problem.

8.9. **A screw in a waveguide.** Modify your code from Problem 7.5 to include the total-field/scattered-field formulation. Include scattered-field regions of 10 grid cells at either end of the waveguide, and Mur boundary conditions at either end. Now, simulate a small "screw" in the top of the waveguide by adding a few cells of PEC material halfway down the waveguide, one grid cell wide in x and y. Measure the field intensity that reflects back toward the source; how does this vary with the length of the screw?

8.10. **Thin slot revisited.** Repeat Problem 7.10, but include a second-order Mur boundary on each of the walls. Compare the fields leaking through the hole to the simulation without boundary conditions; how do they compare?

8.11. **2D Total-field/scattered-field.** Modify your code from Problem 7.2, the cylindrical scatterer, to include a second-order Mur boundary condition on each of the simulation boundaries.

8.12. **The Luneberg lens.** The simulation shown on the cover of this book is known as a Luneberg lens. The lens is defined by a permittivity $\epsilon = \epsilon_0[2 - (r/R)^2]$, where R is the radius of the lens; the permittivity smoothly transitions to that of free space at its edge.

(a) Replicate the simulation on the cover for a 2D TM mode, at a frequency of 2 GHz with a lens of radius 1 m. Use a second-order Mur boundary on all edges. How well does the lens produce a plane wave as you vary the input frequency?

(b) Wrap the Luneberg lens with a total-field/scattered-field boundary, and run the simulation in reverse: excite the space with a plane at the left edge of the TF/SF boundary. Observe how well the Luneberg lens brings the incident plane wave to a focus. Compare its performance as you vary the input frequency.

References

[1] B. Enquist and A. Majda, "Absorbing boundary conditions for the numerical simulation of waves," *Math. Comp.*, vol. 31, pp. 629–651, 1977.

[2] G. Mur, "Absorbing boundary conditions for the finite-difference approximation of the time-domain electromagnetic field equations," *IEEE Trans. Elec. Compat.*, vol. EMC-23, pp. 377–382, 1981.

[3] L. N. Trefethen and L. Halpern, "Well-posedness of one-way wave equations and absorbing boundary conditions," *Math. Comp.*, vol. 47, pp. 421–435, 1986.

[4] A. Taflove and S. Hagness, *Computational Electrodynamics: The Finite-Difference Time-Domain Method*, 3rd edn. Artech House, 2005.

[5] T. G. Moore, J. G. Blaschak, A. Taflove, and G. A. Kriegsmann, "Theory and application of radiation boundary operators," *IEEE Trans. Ant. and Prop.*, vol. 36, pp. 1797–1812, 1988.

[6] A. Bayliss and E. Turkel, "Radiation boundary conditions for wave-like equations," *Comm. Pure Appl. Math.*, vol. 23, pp. 707–725, 1980.

[7] A. Sommerfeld, *Partial Differential Equations in Physics*. New York: Academic Press, 1949.

[8] A. Bayliss, M. Gunzburger, and E. Turkel, "Boundary conditions for the numerical solution of elliptic equation in exterior regions," *SIAM J. Appl. Math*, vol. 42, pp. 430–451, 1982.

[9] R. L. Higdon, "Numerical absorbing boundary conditions for the wave equation," *Math. Comp.*, vol. 49, pp. 65–90, 1987.

9　The perfectly matched layer

The methods described in the previous chapter for analytically treating outgoing waves and preventing reflections can be very effective under certain circumstances. However, a number of drawbacks are evident. The reflection coefficient that results from these analytical ABCs is a function of incident angle, and can be very high for grazing angles, as shown in Figure 8.4. Furthermore, as can be seen from the Higdon operators, for higher accuracy the one-way wave equation that must be applied at the boundary becomes increasingly complex, and requires the storage of fields at previous time steps (i.e., $n - 1$). One of the great advantages of the FDTD method is that it does not require the storage of any fields more than one time step back, so one would prefer a boundary method that also utilizes this advantageous property.

Those methods in the previous chapter should most accurately be referred to as "radiation" boundary conditions. This type of boundary emulates a one-way wave equation at the boundary, as a method for circumventing the need for field values outside the boundary, which would be required in the normal update equation. Strictly speaking, however, they are not "absorbing" boundary conditions.

The family of methods that will be discussed in this chapter are absorbing boundaries. These methods involve modifying the medium of the simulation in a thin layer around the boundary, as shown in Figure 9.1, so that this layer becomes an artifically "absorbing" or *lossy* medium. The boundary layer is designed so that it absorbs enough of the outgoing wave so that reflections from the actual boundary are acceptably low. In addition, however, the boundary layer must be designed to prevent reflections from the interface between the actual medium and this boundary medium (i.e., $\Gamma_1 = 0$, as shown in Figure 9.1). This means the two media must be impedance-matched to very high accuracy. For this reason, the family of methods is known as perfectly matched layers or PMLs.

The first effective PML was introduced by J. P. Bérenger in 1994 [1]. Bérenger's PML method for absorbing waves incident on the boundaries of an FDTD grid is based on reflection and refraction of uniform plane waves at the interface between a lossless dielectric and a general lossy medium. In the next section we briefly review the physical concepts of waves incident on a lossy dielectric material, before introducing the PML formulation in Section 9.2.

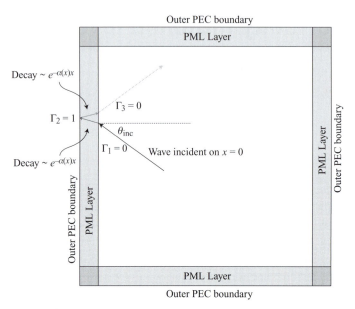

Figure 9.1 Application of a perfectly matched layer. An incident wave finds $\Gamma_1 = 0$ upon its first interaction with the PML, but the PML is lossy, reducing the amplitude considerably before the wave reaches the outer PEC boundary. It then reflects (Γ_2) and decays further before returning to the simulation space (with $\Gamma_3 = 0$), preferably with an amplitude $< 10^{-6}$ relative to the incident field.

9.1   Oblique incidence on a lossy medium

The analysis of a plane wave incident obliquely on a lossy medium is well treated in a number of introductory textbooks on electromagnetics. For a detailed treatment we refer the reader to [2, Sec. 3.8], and we provide only a brief overview here.

Note that there are some notation differences between the analytical treatment and that given in the rest of this chapter, which will be geared toward FDTD modeling. Specifically, in the analytical analysis, the plane of incidence is typically chosen to be the x–z plane and the interface between the two media is typically placed at $z = 0$. In this scenario, the reflection coefficient is defined typically in terms of the electric field, i.e.,

$$\Gamma \equiv \frac{E_{r0}}{E_{i0}}$$

where E_{i0} and E_{r0} are respectively the magnitudes of the incident and reflected wave electric fields at the $z = 0$ interface. In FDTD analysis, on the other hand, the reflection coefficient is typically defined in terms of the singular wave component, i.e., $\mathcal{H}_z$ for TE waves or $\mathcal{E}_z$ for TM waves.

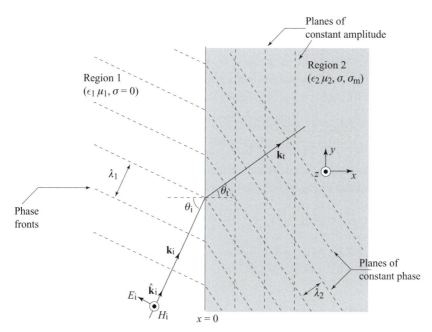

Figure 9.2 Oblique incidence on a lossy medium: Constant amplitude and phase fronts. In Region 2, the planes of constant amplitude are parallel to the interface, while the planes of constant phase are perpendicular to $\mathbf{k}_t$, which is at an angle θ_t from the normal to the interface.

In the same vein, the terms "perpendicular" versus "parallel" polarization in the traditional context usually refer to the orientation of the wave electric field with respect to the plane of incidence. In this case, the perpendicular case is defined as having an electric field $\overline{\mathscr{E}}$ perpendicular to the x–z plane of the page, i.e., $\overline{\mathscr{E}} = \mathscr{E}_y$, and the $\overline{\mathscr{H}}$ field spans the other two dimensions. The opposite is true for parallel polarization. On the other hand, in FDTD analysis of a 2D system, it is more convenient to think in terms of the singular wave component ($\mathscr{H}_z$ for TE waves or $\mathscr{E}_z$ for TM waves), and choose the plane of incidence to be the x–y plane. The reader can then consider the TM case in the FDTD interpretation to be equivalent to the perpendicular polarization case in the analytical treatment, except that we are working with an $\mathscr{E}_z$ field in the x–y plane.

We now proceed with the analysis of the problem of a uniform plane wave incident from a lossless dielectric medium (Region 1) to a lossy half-space (Region 2), as shown in Figure 9.2. In our lossy Region 2, we will allow for both electric and magnetic losses, as represented respectively by σ and σ_m; the reader should keep in mind that neither of these need to be physically meaningful conductivities. In this analysis we will show that by careful choice of the medium parameters σ, σ_m, ϵ, and μ, we can achieve perfect matching between Region 1 and Region 2, but only for waves incident normally at the interface between the two regions.

9.1.1　　Uniform plane wave incident on general lossy media

Consider the geometry shown in Figure 9.2, with a lossy medium occupying the half-space $x > 0$. The incident wave magnetic field phasor for a 2D TE_z wave (the subscript indicating that the system is infinite in the z-direction) is pointed in the z-direction, while the electric field phasor has x and y components. We consider wave solution of the type

$$(\cdots)\,e^{-j\mathbf{k}\cdot\mathbf{r}}$$

where $|\mathbf{k}| = k = \omega\sqrt{\mu\epsilon}$ is the propagation constant. In a general lossy medium, the propagation constant $\mathbf{k}$ can be complex, as we shall see shortly.

Returning to Figure 9.2, the incident fields in Region 1 are given by:

$$\mathbf{H}_i(x, y) = \hat{\mathbf{z}}\, H_0 e^{-jk_1(x\cos\theta_i + y\sin\theta_i)} \tag{9.1a}$$

$$\mathbf{E}_i(x, y) = \left[-\hat{\mathbf{x}}\, \underbrace{\frac{k_1\sin\theta_i}{\omega\epsilon_1}}_{k_{1y}/(\omega\epsilon_1)} + \hat{\mathbf{y}}\, \underbrace{\frac{k_1\cos\theta_i}{\omega\epsilon_1}}_{k_{1x}/(\omega\epsilon_1)} \right] H_0 e^{-jk_1(x\cos\theta_i + y\sin\theta_i)}. \tag{9.1b}$$

Note that $k_1 = \omega\sqrt{\mu_1\epsilon_1}$ is the propagation constant in Region 1 and $\eta_1 = k_1/(\omega\epsilon_1) = \sqrt{\mu_1/\epsilon_1}$ is the intrinsic impedance of Region 1. The quantity $\mathbf{H}_i$ (for example) is the phasor for the time-harmonic real magnetic field $\mathcal{H}_i$, being related to it as:

$$\mathcal{H}_i(x, y, t) = \mathcal{R}e\left\{\mathbf{H}_i(x, y)\,e^{j\omega t}\right\}. \tag{9.2}$$

Assuming the presence of a reflected wave $\mathbf{H}_r$ related to the incident one, via a magnetic field reflection coefficient defined as:

$$\Gamma \equiv \frac{|\mathbf{H}_r(0, y)|}{|\mathbf{H}_i(0, y)|}. \tag{9.3}$$

The reflected field $\mathbf{H}_r$ is then given by

$$\mathbf{H}_r = \hat{\mathbf{z}}\,\Gamma\, H_0 e^{-jk_1(-x\cos\theta_i + y\sin\theta_i)}$$

where the sign has changed in x, as the reflected wave propagates in the $-x$ direction. The total fields in Region 1 can be written as:

$$\mathbf{H}_1(x, y) = \hat{\mathbf{z}}\left[1 + \Gamma e^{j2k_{1x}x}\right] H_0\, e^{-j(k_{1x}x + k_{1y}y)} \tag{9.4a}$$

$$\mathbf{E}_1(x, y) = \left[-\hat{\mathbf{x}}\frac{k_{1y}}{\omega\epsilon_1}\left(1 + \Gamma e^{j2k_{1x}x}\right) + \hat{\mathbf{y}}\frac{k_{1x}}{\omega\epsilon_1}\left(1 - \Gamma e^{j2k_{1x}x}\right)\right] H_0\, e^{-j(k_{1x}x + k_{1y}y)} \tag{9.4b}$$

where $k_{1x} = k_1\cos\theta_i$ and $k_{1y} = k_1\sin\theta_i$.

Now consider the fields in Region 2. Even though we know[1] that the nature of the electromagnetic field in this region is rather complex, we could simply write algebraic expressions for the fields that are analogous to Equation (9.1). For this purpose, we note

---

[1] See, for example, Inan and Inan [2, Sec. 3.8].

that time-harmonic Maxwell's equations for general lossy media are:

$$\nabla \times \mathbf{E} = -j\omega\mu\,\mathbf{H} - \sigma_m\,\mathbf{H} = -j\omega\,\underbrace{\mu\left(1 + \frac{\sigma_m}{j\omega\mu}\right)}_{\mu_{\text{eff}}}\mathbf{H} \tag{9.5a}$$

$$\nabla \times \mathbf{H} = j\omega\epsilon\,\mathbf{E} + \sigma\,\mathbf{E} = j\omega\,\underbrace{\epsilon\left(1 + \frac{\sigma}{j\omega\epsilon}\right)}_{\epsilon_{\text{eff}}}\mathbf{E} \tag{9.5b}$$

so that all we need to do to generate the field expressions for Region 2 from Equation (9.1) is to use the effective complex permittivity and permeability (of Region 2) instead of ϵ_1 and μ_1, respectively. Note that σ_m is a fictitious magnetic conductivity that is not necessarily physical.[2] The transmitted fields in Region 2 can thus be written as:

$$\mathbf{H}_2(x, y) = \hat{\mathbf{z}}\,\mathcal{T}\,H_0\,e^{-j(k_{2x}x + k_{2y}y)} \tag{9.6a}$$

$$\mathbf{E}_2(x, y) = \left[-\hat{\mathbf{x}}\,\frac{k_{2y}}{\omega\epsilon_2\left(1 + \dfrac{\sigma}{j\omega\epsilon_2}\right)} + \hat{\mathbf{y}}\,\frac{k_{2x}}{\omega\epsilon_2\left(1 + \dfrac{\sigma}{j\omega\epsilon_2}\right)}\right]H_0\,\mathcal{T}\,e^{-j(k_{2x}x + k_{2y}y)} \tag{9.6b}$$

where $k_{2x} = k_2 \cos\theta_t$ and $k_{2y} = k_2 \sin\theta_t$: note that $\theta_i \neq \theta_t$. $\mathcal{T}$ is the magnetic field transmission coefficient defined as:

$$\mathcal{T} \equiv \frac{|\mathbf{H}_t(0, y)|}{|\mathbf{H}_i(0, y)|}. \tag{9.7}$$

Note that $\mathbf{H}_t = \mathbf{H}_2$ since there is no "reflected" wave in Region 2. We also note that the wave equation is valid in Region 2 so that we must have:

$$k_{2x}^2 + k_{2y}^2 = k_2^2 = \omega^2\epsilon_{\text{eff}}\mu_{\text{eff}} = \omega^2\epsilon_2\mu_2\left(1 + \frac{\sigma}{j\omega\epsilon_2}\right)\left(1 + \frac{\sigma_m}{j\omega\mu_2}\right). \tag{9.8}$$

At the interface $x = 0$, we can equate $\mathbf{H}_1 = \mathbf{H}_2$ since both are tangential components. Using Equations (9.4a) and (9.6a), we have

$$[1 + \Gamma]\,H_0\,e^{-jk_{1y}y} = \mathcal{T}\,H_0\,e^{-jk_{2y}y}.$$

For this equation to be valid for all y, we must have

$$k_{2y} = k_{1y} = k_1 \sin\theta_i \tag{9.9}$$

which implies that k_{2y} is purely real. Thus, since k_{1x} is real, from Equation (9.8), k_{2x} must in general be complex, and is given from Equation (9.8) by:

$$k_{2x} = \sqrt{\omega^2\epsilon_2\mu_2\left(1 + \frac{\sigma}{j\omega\epsilon_2}\right)\left(1 + \frac{\sigma_m}{j\omega\mu_2}\right) - k_{2y}^2}. \tag{9.10}$$

---

[2] While the magnetic conductivity is most often unphysical, recall from Chapter 1 that such magnetic losses may arise from magnetic relaxation effects in material media for which the permeability may be complex, i.e., $\mu_c = \mu' - j\mu''$.

The reflection and transmission coefficients are given[3] by:

$$\Gamma = \frac{\eta_1 \cos \theta_i - \eta_2 \cos \theta_t}{\eta_1 \cos \theta_i + \eta_2 \cos \theta_t} = \frac{\dfrac{k_{1x}}{\omega \epsilon_1} - \dfrac{k_{2x}}{\omega \epsilon_2 \left[1 + \sigma/(j\omega \epsilon_2) \right]}}{\dfrac{k_{1x}}{\omega \epsilon_1} + \dfrac{k_{2x}}{\omega \epsilon_2 \left[1 + \sigma/(j\omega \epsilon_2) \right]}} \tag{9.11}$$

and

$$\mathcal{T} = 1 + \Gamma = \frac{\dfrac{2k_{1x}}{\omega \epsilon_1}}{\dfrac{k_{1x}}{\omega \epsilon_1} + \dfrac{k_{2x}}{\omega \epsilon_2 \left[1 + \sigma/(j\omega \epsilon_2) \right]}}. \tag{9.12}$$

Note that $\cos \theta_t$ is a complex quantity, but we do not have to determine explicitly the complex angle θ_t in our analysis here.

For the purposes of designing an absorbing boundary, we would like to see how this reflection coefficient performs as a function of angle and how it can be designed to minimize reflections. For an arbitrary incidence angle $\theta_i \neq 0$, the reflection coefficient is in general a complex nonzero number. However, for normal incidence with $\theta_i = 0$, we have $\theta_t = 0$ also, and $k_{1x} = k_1$ and $k_{2x} = k_2$ since neither has y components. In this case the reflection coefficient becomes:

$$\Gamma = \frac{\eta_1 - \eta_2}{\eta_1 + \eta_2} \tag{9.13}$$

where

$$\eta_1 = \frac{k_1}{\omega \epsilon_1} = \sqrt{\frac{\mu_1}{\epsilon_1}} \quad \text{and} \quad \eta_2 = \sqrt{\frac{\mu_{\text{eff}}}{\epsilon_{\text{eff}}}} = \sqrt{\frac{\mu_2 \left[1 + \sigma_{\text{m}}/(j\omega \mu_2) \right]}{\epsilon_2 \left[1 + \sigma/(j\omega \epsilon_2) \right]}}.$$

It is clear that we can have zero reflection ($\Gamma = 0$) by choosing values of the various parameters so that $\eta_1 = \eta_2$. For example, let $\epsilon_1 = \epsilon_2$ and $\mu_1 = \mu_2$ (i.e., we "match" Region 2 to Region 1) and choose σ_{m} such that:

$$\frac{\sigma_{\text{m}}}{\mu_1} = \frac{\sigma}{\epsilon_1} \quad \text{or} \quad \sigma_{\text{m}} = \frac{\sigma \mu_1}{\epsilon_1} = \sigma \eta_1^2, \tag{9.14}$$

in which case we have $\eta_1 = \eta_2$, resulting in $\Gamma = 0$. Also, we note from Equation (9.10) that we now have

$$k_{2x} = \omega \sqrt{\mu_1 \epsilon_1} \left[1 + \frac{\sigma}{j\omega \epsilon_1} \right] = \underbrace{\omega \sqrt{\mu_1 \epsilon_1}}_{k_1} - j\sigma \eta_1. \tag{9.15}$$

---

[3] Note that Equation (9.11) is analogous to Equation (3.72) from Inan and Inan [2], but is slightly different since Γ here is a magnetic field reflection coefficient.

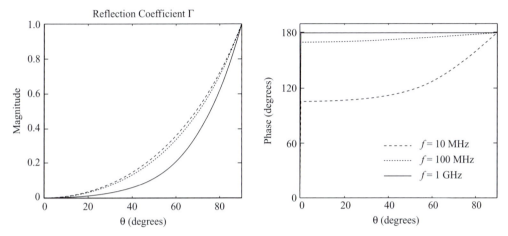

Figure 9.3 Oblique incidence on a lossy medium. The magnitude and phase of the reflection coefficient in Equation (9.11) are shown, for $\sigma = 0.01$ S/m and σ_{m} satisfying Equation (9.14).

Noting that $k_{2y} = 0$ (since $\theta_i = 0$, i.e., we have normal incidence), the electric and magnetic field phasors in Region 2 are given by:

$$\mathbf{H}_2(x, y) = \hat{\mathbf{z}} \, H_0 \, e^{-jk_1 x} e^{-\sigma \eta_1 x} \tag{9.16a}$$

$$\mathbf{E}_2(x, y) = \hat{\mathbf{z}} \, \eta_1 \, H_0 \, e^{-jk_1 x} e^{-\sigma \eta_1 x}. \tag{9.16b}$$

Thus we see a rather simple wave, propagating in the x-direction without *any* dispersion; i.e., the phase velocity $v_p = k_1/\omega = (\mu_1 \epsilon_1)^{-1}$ is independent of frequency. Furthermore, the wave is attenuating exponentially in the x-direction. Since $\Gamma = 0$, and thus $\mathcal{T} = 1$, we note that Region 1 is *perfectly matched* to Region 2, albeit only for normal incidence. Unfortunately, at angles $\theta_i \neq 0$, the reflection coefficient is nonzero. The reflection coefficient magnitude and phase are plotted in Figure 9.3 for three different frequencies, for $\sigma = 0.01$ S/m and σ_{m} satisfying Equation (9.14). Similar to the Mur boundary, this simple absorbing boundary yields unacceptably high reflections for any angle off the normal. As such, this method is unsuitable for an absorbing boundary in a 2D or 3D simulation.

9.2 The Bérenger PML medium

As mentioned at the beginning of this chapter, the problem of nonzero reflection at arbitrary incident angles was addressed by J. P. Bérenger in 1994. The Bérenger or split-field perfectly matched layer method is based on field-splitting of Maxwell's equations and selectively choosing different values for the "conductivities" (σ and σ_{m}) in different directions. In this section we will introduce the field-splitting method.

Within Region 2, where $\sigma \neq 0$ and $\sigma_m \neq 0$, we express Maxwell's curl equations as follows for the TM and TE modes (in the FDTD sense):

TM   TE

$$\epsilon_2 \frac{\partial \mathcal{E}_x}{\partial t} + \sigma_y \mathcal{E}_x = \frac{\partial \mathcal{H}_z}{\partial y} \qquad\qquad \mu_2 \frac{\partial \mathcal{H}_x}{\partial t} + \sigma_{m,y} \mathcal{H}_x = -\frac{\partial \mathcal{E}_z}{\partial y} \qquad (9.17a)$$

$$\epsilon_2 \frac{\partial \mathcal{E}_y}{\partial t} + \sigma_x \mathcal{E}_y = -\frac{\partial \mathcal{H}_z}{\partial x} \qquad\qquad \mu_2 \frac{\partial \mathcal{H}_y}{\partial t} + \sigma_{m,x} \mathcal{H}_y = \frac{\partial \mathcal{E}_z}{\partial x} \qquad (9.17b)$$

$$\mu_2 \frac{\partial \mathcal{H}_{zx}}{\partial t} + \sigma_{m,x} \mathcal{H}_{zx} = -\frac{\partial \mathcal{E}_y}{\partial x} \qquad\qquad \epsilon_2 \frac{\partial \mathcal{E}_{zx}}{\partial t} + \sigma_x \mathcal{E}_{zx} = \frac{\partial \mathcal{H}_y}{\partial x} \qquad (9.17c)$$

$$\mu_2 \frac{\partial \mathcal{H}_{zy}}{\partial t} + \sigma_{m,y} \mathcal{H}_{zy} = \frac{\partial \mathcal{E}_x}{\partial y} \qquad\qquad \epsilon_2 \frac{\partial \mathcal{E}_{zy}}{\partial t} + \sigma_y \mathcal{E}_{zy} = -\frac{\partial \mathcal{H}_x}{\partial y} \qquad (9.17d)$$

$$\mathcal{H}_z = \mathcal{H}_{zx} + \mathcal{H}_{zy} \qquad\qquad \mathcal{E}_z = \mathcal{E}_{zx} + \mathcal{E}_{zy}. \qquad (9.17e)$$

Note that Equations (9.17) reduce to the regular Maxwell's equations for $\sigma_x = \sigma_y = \sigma$ and $\sigma_{m,x} = \sigma_{m,y} = \sigma_m$.

However, separating the fields and conductivities into two parts provides for the possibility of selectively attenuating waves in one direction. For example, if we have $\sigma_x = \sigma_{m,x} = 0$ while $\sigma_y \neq 0$ and $\sigma_{m,y} \neq 0$, then waves with field components $\mathcal{E}_x$ and $\mathcal{H}_{zy}$ propagating along the the y-direction are absorbed while those waves with field components $\mathcal{E}_y$ and $\mathcal{H}_{zx}$ propagating along the the x-direction are not. At this point, we can begin to see the perfectly matched layer property. For example, if the nonzero conductivity pairs σ_y and $\sigma_{m,y}$ at the Region 1/Region 2 interface perpendicular to the y-direction are used, we can attenuate the components of waves propagating only in the y-direction.

Note that the fields $\mathcal{H}_{zx}$, $\mathcal{H}_{zy}$, $\mathcal{E}_{zx}$, and $\mathcal{E}_{zy}$ are not physical fields, but are rather convenient representations of the physical fields $\mathcal{H}_z$ and $\mathcal{E}_z$ for the purpose of building our absorbing boundary. However, once these split fields are added in Equation (9.17e), they become physically meaningful.

We now consider the TM mode, and proceed to write the left column of Equations (9.17) in time-harmonic form:

$$j\omega\epsilon_2 \underbrace{\left(1 + \frac{\sigma_y}{j\omega\epsilon_2}\right)}_{s_y} E_x = \frac{\partial}{\partial y}\left(H_{zx} + H_{zy}\right) \qquad (9.18a)$$

$$j\omega\epsilon_2 \underbrace{\left(1 + \frac{\sigma_x}{j\omega\epsilon_2}\right)}_{s_x} E_y = -\frac{\partial}{\partial x}\left(H_{zx} + H_{zy}\right) \qquad (9.18b)$$

$$j\omega\mu_2 \underbrace{\left(1 + \frac{\sigma_{m,x}}{j\omega\mu_2}\right)}_{s_{m,x}} H_{zx} = -\frac{\partial E_y}{\partial x} \qquad (9.18c)$$

$$j\omega\mu_2 \underbrace{\left(1 + \frac{\sigma_{m,y}}{j\omega\mu_2}\right)}_{s_{m,y}} H_{zy} = \frac{\partial E_x}{\partial y}. \qquad (9.18d)$$

We now proceed to derive the wave equation, so that we can determine the form of plane wave solutions in the Bérenger medium (Region 2). For this purpose, we differentiate Equation (9.18a) with respect to y and Equation (9.18b) with respect to x, and then substitute for $\partial E_y / \partial x$ and $\partial E_x / \partial y$ from Equations (9.18c) and (9.18d) to find:

$$-\omega^2 \mu_2 \epsilon_2 H_{zx} = \frac{1}{s_{m,x}} \frac{\partial}{\partial x} \frac{1}{s_x} \frac{\partial}{\partial x} \left(H_{zx} + H_{zy} \right) \tag{9.19a}$$

$$-\omega^2 \mu_2 \epsilon_2 H_{zy} = \frac{1}{s_{m,y}} \frac{\partial}{\partial y} \frac{1}{s_y} \frac{\partial}{\partial y} \left(H_{zx} + H_{zy} \right). \tag{9.19b}$$

Assuming for now that the four conductivities do not vary with x, and adding Equations (9.19) together, we find the wave equation:

$$\frac{1}{s_x s_{m,x}} \frac{\partial^2 H_z}{\partial x} + \frac{1}{s_y s_{m,y}} \frac{\partial^2 H_z}{\partial y} + \omega^2 \mu_2 \epsilon_2 H_z = 0 \tag{9.20}$$

which has the solution

$$H_z = \mathcal{T} H_0 \, e^{-j \sqrt{s_x s_{m,x}} \, k_{2x} x - j \sqrt{s_y s_{m,y}} \, k_{2y} y} \tag{9.21}$$

subject to the condition

$$k_{2x}^2 + k_{2y}^2 = \omega^2 \mu_2 \epsilon_2 \qquad \rightarrow \qquad k_{2x} = \sqrt{\omega^2 \epsilon_2 \mu_2 - k_{2y}^2}. \tag{9.22}$$

The corresponding electric field components can be found from Equations (9.18a) and (9.18b):

$$E_x = -\mathcal{T} H_0 \frac{k_{2y}}{\omega \epsilon_2} \sqrt{\frac{s_{m,y}}{s_y}} \, e^{-j \sqrt{s_x s_{m,x}} \, k_{2x} x - j \sqrt{s_y s_{m,y}} \, k_{2y} y} \tag{9.23a}$$

$$E_y = \mathcal{T} H_0 \frac{k_{2x}}{\omega \epsilon_2} \sqrt{\frac{s_{m,x}}{s_x}} \, e^{-j \sqrt{s_x s_{m,x}} \, k_{2x} x - j \sqrt{s_y s_{m,y}} \, k_{2y} y}. \tag{9.23b}$$

We now need to enforce the required continuity of the tangential electric and magnetic fields across the interface between Region 1 and Region 2. Assuming the interface to be at $x = 0$ as shown in Figure 9.4, and using Equation (9.4a) together with Equation (9.21) above, we can write:

$$[1 + \Gamma] \, H_0 \, e^{-jk_{1y} y} = \mathcal{T} H_0 \, e^{-j \sqrt{s_y s_{m,y}} \, k_{2y} y} \tag{9.24}$$

which, in order to be valid at all values of y, requires $s_y = s_{m,y} = 1$, or $\sigma_y = \sigma_{m,y} = 0$, and the phase matching condition $k_{2y} = k_{1y} = k_1 \sin \theta_i$.

Equating the tangential components across the interface also reveals the reflection and transmission coefficients:

$$\Gamma = \frac{\dfrac{k_{1x}}{\omega \epsilon_1} - \dfrac{k_{2x}}{\omega \epsilon_2} \sqrt{\dfrac{s_{m,x}}{s_x}}}{\dfrac{k_{1x}}{\omega \epsilon_1} + \dfrac{k_{2x}}{\omega \epsilon_2} \sqrt{\dfrac{s_{m,x}}{s_x}}} \tag{9.25}$$

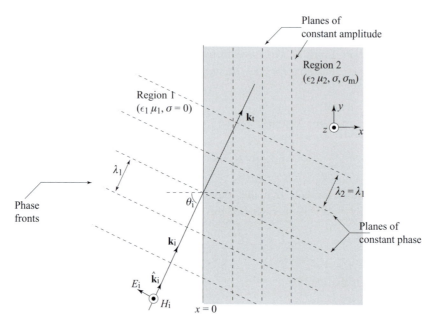

Figure 9.4 Oblique incidence on the Bérenger PML medium: Constant amplitude and phase fronts. (Note that Region 2 is now anisotropic, since it has different loss rates in the x and y directions.)

and

$$\mathcal{T} = 1 + \Gamma = \frac{\dfrac{2k_{1x}}{\omega\epsilon_1}}{\dfrac{k_{1x}}{\omega\epsilon_1} + \dfrac{k_{2x}}{\omega\epsilon_2}\sqrt{\dfrac{s_{m,x}}{s_x}}}. \tag{9.26}$$

The zero-reflection matching condition can now be established by analogy with Equation (9.14). We assume $\epsilon_1 = \epsilon_2$, $\mu_1 = \mu_2$, and $s_x = s_{m,x}$, which gives us $\sigma_x/\epsilon_1 = \sigma_{m,x}/\mu_1$ (similar to 9.14). Since $k_{2y} = k_{1y}$, we must then have $k_{2x} = k_{1x}$ from Equation (9.22). This implies that the wave in Region 2 propagates at the same angle as in Region 1, as shown in Figure 9.4. Substituting these conditions in Equation (9.25) gives $\Gamma = 0$, *regardless* of the incident angle θ_i.

With the assumed medium parameter choices, the field component expressions (9.21) and (9.23) can be written as:

$$H_z = H_0\, e^{-js_x k_{1x}\, x + jk_{1y}\, y} = H_0 e^{-jk_{1x}\, x - jk_{1y}\, y}\, e^{-\sigma_x x \eta_1 \cos\theta_i} \tag{9.27a}$$

$$E_x = -H_0\eta_1 \sin\theta_i\, e^{-jk_{1x}\, x - jk_{1y}\, y}\, e^{-\sigma_x x \eta_1 \cos\theta_i} \tag{9.27b}$$

$$E_y = H_0\eta_1 \cos\theta_i\, e^{-jk_{1x}\, x - jk_{1y}\, y}\, e^{-\sigma_x x \eta_1 \cos\theta_i} \tag{9.27c}$$

since $s_x = 1 - j[\sigma_x/(\omega\epsilon_2)]$. The perfectly matched property of the fields is apparent upon close examination. The wave in Region 2 propagates in the same direction as that in Region 1 (as defined by $e^{-jk_{1x}x-jk_{1y}y}$), while simultaneously decaying exponentially in the x-direction. In other words, the depiction of the wave in the second medium as shown in Figure 9.2 is no longer correct, due to the choice of different conductivity values in the x and y directions. The planes of constant phase and amplitude in Region 2 are now as shown in Figure 9.4. Another interesting property is the fact that the attenuation rate $\sigma_x \eta_1 \cos \theta_i$ is *independent* of frequency. Thus, we truly have *perfect* matching between Region 1 and Region 2 across the interface at $x = 0$.

9.2.1 Bérenger split-field PML in 3D

The derivation of the split-field method above was for the 2D TM mode (in the FDTD sense); similar equations for the 2D TE mode are given in Equations (9.17) in the right-hand column. In the TE case, the $\mathscr{E}_z$ component is split into $\mathscr{E}_{zx}$ and $\mathscr{E}_{zy}$, along with both conductivities. In 3D, both fields need to be split, and conductivity components are required for each of the x, y, and z directions. The resulting equations, analogous to Equations (9.17), are given by:

$$\epsilon_2 \frac{\partial \mathscr{E}_{xy}}{\partial t} + \sigma_y \mathscr{E}_{xy} = \frac{\partial}{\partial y} \left(\mathscr{H}_{zx} + \mathscr{H}_{zy} \right) \qquad \mu_2 \frac{\partial \mathscr{H}_{xy}}{\partial t} + \sigma_{m,y} \mathscr{H}_{xy} = -\frac{\partial}{\partial y} \left(\mathscr{E}_{zx} + \mathscr{E}_{zy} \right)$$
$$(9.28a)$$

$$\epsilon_2 \frac{\partial \mathscr{E}_{xz}}{\partial t} + \sigma_z \mathscr{E}_{xz} = -\frac{\partial}{\partial z} \left(\mathscr{H}_{yx} + \mathscr{H}_{yz} \right) \qquad \mu_2 \frac{\partial \mathscr{H}_{xz}}{\partial t} + \sigma_{m,z} \mathscr{H}_{xz} = \frac{\partial}{\partial z} \left(\mathscr{E}_{yx} + \mathscr{E}_{yz} \right)$$
$$(9.28b)$$

$$\epsilon_2 \frac{\partial \mathscr{E}_{yz}}{\partial t} + \sigma_z \mathscr{E}_{yz} = \frac{\partial}{\partial z} \left(\mathscr{H}_{xy} + \mathscr{H}_{xz} \right) \qquad \mu_2 \frac{\partial \mathscr{H}_{yz}}{\partial t} + \sigma_{m,z} \mathscr{H}_{yz} = -\frac{\partial}{\partial z} \left(\mathscr{E}_{xy} + \mathscr{E}_{xz} \right)$$
$$(9.28c)$$

$$\epsilon_2 \frac{\partial \mathscr{E}_{yx}}{\partial t} + \sigma_x \mathscr{E}_{yx} = -\frac{\partial}{\partial x} \left(\mathscr{H}_{zx} + \mathscr{H}_{zy} \right) \qquad \mu_2 \frac{\partial \mathscr{H}_{yx}}{\partial t} + \sigma_{m,x} \mathscr{H}_{yx} = \frac{\partial}{\partial x} \left(\mathscr{E}_{zx} + \mathscr{E}_{zy} \right)$$
$$(9.28d)$$

$$\epsilon_2 \frac{\partial \mathscr{E}_{zx}}{\partial t} + \sigma_x \mathscr{E}_{zx} = \frac{\partial}{\partial x} \left(\mathscr{H}_{yx} + \mathscr{H}_{yz} \right) \qquad \mu_2 \frac{\partial \mathscr{H}_{zx}}{\partial t} + \sigma_{m,x} \mathscr{H}_{zx} = -\frac{\partial}{\partial x} \left(\mathscr{E}_{yx} + \mathscr{E}_{yz} \right)$$
$$(9.28e)$$

$$\epsilon_2 \frac{\partial \mathscr{E}_{zy}}{\partial t} + \sigma_y \mathscr{E}_{zy} = -\frac{\partial}{\partial y} \left(\mathscr{H}_{xy} + \mathscr{H}_{xz} \right) \qquad \mu_2 \frac{\partial \mathscr{H}_{zy}}{\partial t} + \sigma_{m,y} \mathscr{H}_{zy} = \frac{\partial}{\partial y} \left(\mathscr{E}_{xy} + \mathscr{E}_{xz} \right).$$
$$(9.28f)$$

These equations can be discretized as usual, and similar conditions on σ_x, $\sigma_{m,x}$ and the other conductivities provide perfect matching at the appropriate boundaries. For example, at the $z = 0$ boundary, we would set $\sigma_x = \sigma_{m,x} = \sigma_y = \sigma_{m,y} = 0$, $\sigma_z/\epsilon_1 = \sigma_{m,z}/\mu_1$, with $\epsilon_2 = \epsilon_1$ and $\mu_2 = \mu_1$. How the particular values of σ and σ_m are chosen will be discussed next.

9.2.2   Grading the PML

The Bérenger split-field PML, as well as other types of PML that we shall discuss later in this chapter, require two basic principles in order to function as desired: first, they must provide zero reflection at the interface, and second, they must absorb the incident wave as it propagates into the PML. To first-order, the actual reflection that will be achieved by any PML can be calculated. Consider the scenario in Figure 9.1. The incident wave will enter the PML with no reflection, will attenuate at rate α_{PML} (in np/m) over the thickness of the PML, denoted by Δ_{PML}. Upon reaching the end of the PML, the wave will reflect perfectly from the PEC boundary, and continue to attenuate with the same rate α_{PML} over the thickness of the PML Δ_{PML} again, at which point it will re-enter the simulation space. So, in essence, the wave will have attenuated by a factor $e^{-2\alpha_{PML}\Delta_{PML}}$. In practice of course it is not quite so simple, as the distance propagated will depend on incident angle, and the rates of attenuation may also depend on angle and frequency. In the case where the conductivity σ_x varies with x, a reflection factor can be given as

$$R(\theta) = e^{-2\eta \cos\theta \int_0^d \sigma_x(x)dx}.$$ (9.29)

A larger issue arises simply from the discretization of the space, which prevents zero reflection from occurring. In essence, waves incident on the PML boundary run into a "brick wall" of lossy material, albeit with the same impedance. Furthermore, since $\mathcal{E}$ and $\mathcal{H}$ are interleaved by half a spatial step, they will encounter the PML boundary differently. To reduce the resulting reflection error, Bérenger proposed *grading* the PML conductivity smoothly from zero to some maximum value at the outer boundary; hence the $\sigma_x(x)$ in Equation (9.29) above.

This grading has been successfully implemented with either a polynomial variation or a geometric variation. For polynomial grading, the conductivity profile is given by

$$\sigma_x(x) = \left(\frac{x}{d}\right)^m \sigma_{x,\max}$$ (9.30a)

$$\kappa_x(x) = 1 + (\kappa_{x,\max} - 1) \cdot \left(\frac{x}{d}\right)^m$$ (9.30b)

where κ_x is used in some PMLs for futher optimization of the s parameters,[4] and d is the PML thickness, so that $\sigma_x(d) = \sigma_{x,\max}$. Using these parameters, the reflection factor given in Equation (9.29) can be evaluated as:

$$R(\theta) = e^{-2\eta\sigma_{x,\max}d \cos\theta/(m+1)}$$ (9.31)

The PML is typically taken to be about 10 cells thick, in which case the thickness d is $10\Delta x$; then m and $\sigma_{x,\max}$ are left to be defined. For most FDTD simulations, $3 \leq m \leq 4$ has been found through experiment to be optimal, and the maximum conductivity is

---

[4] The parameter κ is used sometimes in the PML as

$$s_x = \kappa_x + \frac{\sigma_x}{j\omega\epsilon}$$

to allow for the possibility of a non-unity real part of s_x. In the presentation above, $\kappa_x = 1$.

given by

$$\sigma_{x,\max} = -\frac{(m+1)\ln[R(0)]}{2\eta d} \tag{9.32}$$

where $R(0)$ is the desired reflection at normal incidence; this is typically taken to be on the order of 10^{-6}.

For geometric grading, the conductivity profile can be defined as

$$\sigma_x(x) = \left(g^{1/\Delta}\right)^x \sigma_{x,0} \tag{9.33a}$$

$$\kappa_x(x) = \left[(\kappa_{x,\max})^{1/d} g^{1/\Delta}\right]^x. \tag{9.33b}$$

Here, $\sigma_{x,0}$ is the PML conductivity at the start of the PML, g is the geometric scaling factor, and $\Delta = \Delta x$ is the grid spacing. Similar analysis can be applied to find the reflection factor:

$$R(\theta) = e^{-2\eta\sigma_{x,0}\Delta(g^{d/\Delta}-1)\cos\theta/\ln g}. \tag{9.34}$$

Typically, $2 \leq g \leq 3$ has been found to be optimal for most FDTD simulations; from this, the initial conductivity can be determined as before:

$$\sigma_{x,0} = -\frac{\ln[R(0)]\ln g}{2\eta\Delta\left(g^{d/\Delta}-1\right)}. \tag{9.35}$$

The polynomial and geometric gradings described here are simply examples of how the PML might be graded; there is nothing special about these particular analytical descriptions of the grading. They are convenient choices that are easy to implement in code; however, similar gradings could be easily designed, which may or may not have analytical descriptions, but which will yield comparable results.

9.2.3   Example split-field simulation

Figure 9.5 gives the results of an example simulation utilizing Bérenger's split-field PML. In this case, we are simulating a 2D TE mode (i.e., the $\mathscr{E}_z$ component is split) in a 6 m × 3 m space, with a PML defined on the upper and lower boundaries. This way, we can compare the absorbed waves in the y-direction with the unperturbed waves in the x-direction.

We excite a sinusoidal wave with an amplitude of 1 V/m and frequency $f_0 = 600$ MHz, or $\lambda_0 = 0.5$ m, at the center of the space. The PML is 10 cells thick with $\sigma_{\max}$ designed for reflections of $R(0) = 10^{-6}$; using a polynomial grading with order $m = 4$, and $\Delta x = \Delta y = \lambda_0/10 = 5$ cm, from Equation (9.32) we find $\sigma_{\max} = 0.183$ S/m. The magnetic conductivity $\sigma_{m,\max}$ is then scaled by ϵ_0/μ_0.

The simulation is run long enough for any reflected waves from the outer boundary (in y) to return to the center of the simulation space. Then, we calculate the average error (over time) at a point at the edge of the PML, starting from the time the wave first reaches the PML. Using this method we find an average error of $\sim$0.3%, or a reflection coefficient of $\sim 3 \times 10^{-3}$.

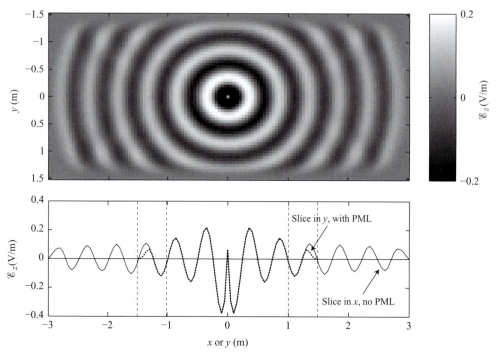

Figure 9.5 Example simulation with Bérenger split-field PML. Top: 2D distribution at $t = 10$ ns of a sinusoidal wave of 600 MHz ($\lambda = 0.5$ m). Bottom: 1D distribution along the center in x (solid) and y (dashed); if both were extended to infinity, they should overlap perfectly.

Observe that the wave incident on the upper and lower boundary is absorbed quite well, but not to our specification of 10^{-6}. The theoretical reflection coefficient from Equation (9.32) is based on an analytical treatment of the PML as a continuous lossy medium with thickness d; however, the discretization process limits the performance of the PML. As one might expect, the PML performance increases with the number of cells. In this simple simulation, if we increase the PML thickness to 20 cells, the reflection is reduced to $\sim 10^{-4}$. However, that comes at the expense of a larger simulation space and thus increased computation time. Typically, a 10-cell PML is considered adequate for most purposes.

9.3   Perfectly matched uniaxial medium

It was briefly mentioned above that the important property of the Bérenger PML medium was that it hosted different conductivities in the x and y directions, i.e., it was inherently an anisotropic medium. However, the Bérenger PML medium cannot be considered a physical medium, because different conductivities are acting on non-physical split

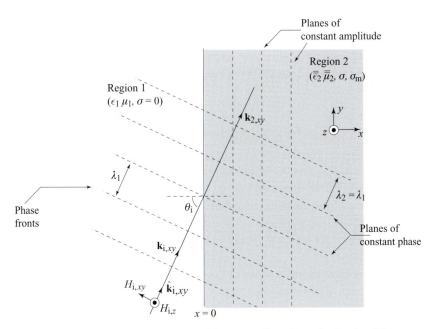

Figure 9.6 Oblique incidence on the uniaxial PML medium. Note that Region 2 is now anisotropic, since it has different loss rates in the x and y directions, hence the permittivity and permeability are denoted by $\bar{\bar{\epsilon}}_2$ and $\bar{\bar{\mu}}_2$.

fields $\mathcal{H}_{zx}$ and $\mathcal{H}_{zy}$ (in the TM mode case). In this section, we discuss an alternative formulation of the perfectly matched layer (PML) absorbing boundary condition in terms of a physical anisotropic medium[5] having tensor permittivity and permeability.

Consider a time-harmonic plane wave incident on the $x = 0$ wall, with its arbitrarily polarized magnetic field given by:

$$\mathbf{H}_i = \mathbf{H}_{i,xy} + \hat{\mathbf{z}}\,H_{i,z} = \mathbf{H}_0\,e^{-j\left(k_{1x}x + k_{1y}y + k_{1z}z\right)} \tag{9.36}$$

where we note that $\mathbf{H}_0 = \hat{\mathbf{x}}H_{0x} + \hat{\mathbf{y}}H_{0y} + \hat{\mathbf{z}}H_{0z}$ is a constant vector and $\mathbf{H}_{i,xy}$ represents the component of $\mathbf{H}_i$ lying in the x–y plane, as shown in Figure 9.6. Note that the propagation direction of this incident wave is given by the $\hat{\mathbf{k}}_i$ vector given by:

$$\hat{\mathbf{k}}_i = \frac{\hat{\mathbf{x}}k_{1x} + \hat{\mathbf{y}}k_{1y} + \hat{\mathbf{z}}k_{1z}}{\sqrt{k_{1x}^2 + k_{1y}^2 + k_{1z}^2}} = \hat{\mathbf{k}}_{i,xy} + \hat{\mathbf{z}}\,\frac{k_{1z}}{k_1} \tag{9.37}$$

where $\hat{\mathbf{k}}_{i,xy}$ is the component of the unit vector $\hat{\mathbf{k}}_i$ lying in the x–y plane, as shown in Figure 9.6. Note that the total magnetic field in Region 1 is in general given by

---

[5] An anisotropic medium is one in which the medium parameters (ϵ, μ, etc.) depend on the direction of the field vectors; we will discuss FDTD modeling in general anisotropic media in detail in Chapter 11.

Equation (9.36) plus a reflected wave, with an expression similar to Equation (9.4a), depending on the particular polarization of the wave.

The half-space region $x > 0$, referred to as Region 2, is characterized by the tensor permittivities and permeabilities given by:

$$\bar{\bar{\epsilon}}_2 = \epsilon_2 \begin{bmatrix} a_1 & 0 & 0 \\ 0 & a_2 & 0 \\ 0 & 0 & a_2 \end{bmatrix} \quad \text{and} \quad \bar{\bar{\mu}}_2 = \mu_2 \begin{bmatrix} b_1 & 0 & 0 \\ 0 & b_2 & 0 \\ 0 & 0 & b_2 \end{bmatrix} \tag{9.38}$$

where ϵ_2 and μ_2 are the scalar permittivity and permeability, which may or may not equal that of free space; and a_1, a_2, b_1, and b_2 are parameters used to describe the anisotropy of the medium. Note that when we consider arrival of plane waves on the $x = 0$ wall, it is appropriate to consider Region 2 to be rotationally symmetric about the x-axis, hence the fact that $\bar{\bar{\epsilon}}_2$ and $\bar{\bar{\mu}}_2$ have components along the y and z axes that are equal.

Plane electromagnetic waves in Region 2 satisfy Maxwell's curl equations, which can be written in tensor form as:

$$\hat{k}_2 k_2 \times \mathbf{E}_2 = \omega \bar{\bar{\mu}}_2 \mathbf{H}_2 \tag{9.39a}$$

$$\hat{k}_2 k_2 \times \mathbf{H}_2 = -\omega \bar{\bar{\epsilon}}_2 \mathbf{E}_2 \tag{9.39b}$$

where $\hat{k}_2 k_2 = \hat{x} k_{2x} + \hat{y} k_{2y} + \hat{z} k_{2z}$ is the wave vector in Region 2. A wave equation can be derived from Equations (9.39) as:

$$\hat{k}_2 k_2 \times \left[\left(\bar{\bar{\epsilon}}_2 \right)^{-1} \hat{k}_2 k_2 \right] \times \mathbf{H}_2 + \omega^2 \bar{\bar{\mu}}_2 \mathbf{H}_2 = 0 \tag{9.40}$$

which can be written in matrix form as:

$$\begin{bmatrix} b_1 \omega^2 \mu_2 \epsilon_2 - k_{2y}^2 a_2^{-1} & k_{2x} k_{2y} a_2^{-1} & 0 \\ k_{2x} k_{2y} a_2^{-1} & b_2 \omega^2 \mu_2 \epsilon_2 - k_{2x}^2 a_2^{-1} & 0 \\ 0 & 0 & b_2 \omega^2 \mu_2 \epsilon_2 - k_{2x}^2 a_2^{-1} - k_{2y}^2 a_1^{-1} \end{bmatrix} \begin{bmatrix} H_x \\ H_y \\ H_z \end{bmatrix} = 0 \tag{9.41}$$

Note that the dispersion relation (i.e., an equation equivalent to Equation (9.8) giving the relationship between the wave numbers k_{2x}, k_{2y}, and k_{2z} that must hold so that the wave equation is satisfied for nontrivial solutions) for this medium is found by setting the determinant of the matrix to zero. It is interesting to note from Equation (9.41) that because of the form of the matrix, TM solutions (with H_x, $H_y = 0$) are uncoupled from TE solutions (with $H_z = 0$).

Considering now the specific case of TE waves (with nonzero H_z), inspection of the bottom row of the matrix in (9.41) gives the dispersion relation:

$$b_2 \omega^2 \mu_2 \epsilon_2 - k_{2x}^2 a_2^{-1} - k_{2y}^2 a_1^{-1} = 0$$

$$\rightarrow \quad \omega^2 \mu_2 \epsilon_2 - k_{2x}^2 a_2^{-1} b_2^{-1} - k_{2y}^2 a_1^{-1} b_2^{-1} = 0. \tag{9.42}$$

The total fields in Region 1 are given by Equation (9.4), repeated below for convenience:

$$\mathbf{H}_1(x, y) = \hat{\mathbf{z}} \left[1 + \Gamma e^{j2k_{1x}x}\right] H_0 e^{-j(k_{1x}x + k_{1y}y)} \tag{9.4a}$$

$$\mathbf{E}_1(x, y) = \left[-\hat{\mathbf{x}} \frac{k_{1y}}{\omega\epsilon_1} \left(1 + \Gamma e^{j2k_{1x}x}\right) + \hat{\mathbf{y}} \frac{k_{1x}}{\omega\epsilon_1} \left(1 - \Gamma e^{j2k_{1x}x}\right)\right] H_0 e^{-j(k_{1x}x + k_{1y}y)} \tag{9.4b}$$

where Γ is the reflection coefficient. The wave transmitted into Region 2 can be written in a manner analogous to Equation (9.6) as:

$$\mathbf{H}_2(x, y) = \hat{\mathbf{z}} \, \mathcal{T} \, H_0 \, e^{-j(k_{2x}x + k_{2y}y)} \tag{9.43a}$$

$$\mathbf{E}_2(x, y) = \left[-\hat{\mathbf{x}} \frac{k_{2y}}{\omega\epsilon_2 a_1} + \hat{\mathbf{y}} \frac{k_{2x}}{\omega\epsilon_2 a_2}\right] H_0 \, \mathcal{T} \, e^{-j(k_{2x}x + k_{2y}y)} \tag{9.43b}$$

where $\mathcal{T}$ is the transmission coefficient.

To derive the transmission and reflection coefficients $\mathcal{T}$ and Γ, we equate the tangential field components across the interface (i.e., E_y and H_z). First, using Equations (9.4a) and (9.43a) at $x = 0$ for the H_z component:

$$(1 + \Gamma) H_0 e^{-jk_{1y}y} = \mathcal{T} H_0 e^{-jk_{2y}y} \tag{9.44}$$

which, in order to hold for all y, implies $k_{1y} = k_{2y}$ and thus $1 + \Gamma = \mathcal{T}$. Next, using Equations (9.4b) and (9.43b) at $x = 0$ for the $\hat{\mathbf{y}}$ component:

$$\frac{k_{1x}}{\omega\epsilon_1} (1 - \Gamma) H_0 e^{-jk_{1y}y} = \frac{k_{2x}}{\omega\epsilon_2 a_2} \mathcal{T} H_0 e^{-jk_{2y}y}. \tag{9.45}$$

Using $\mathcal{T} = 1 + \Gamma$ and $k_{1y} = k_{2y}$ in Equation (9.45), we find the reflection coefficient:

$$\frac{k_{1x}}{\omega\epsilon_1} (1 - \Gamma) = \frac{k_{2x}}{\omega\epsilon_2 a_2} (1 + \Gamma)$$

$$\rightarrow \quad \Gamma = \frac{k_{1x}\epsilon_2 a_2 - k_{2x}\epsilon_1}{k_{1x}\epsilon_2 a_2 + k_{2x}\epsilon_1} \tag{9.46}$$

and from $\mathcal{T} = 1 + \Gamma$ we have the transmission coefficient:

$$\rightarrow \quad \mathcal{T} = \frac{2k_{1x}\epsilon_2 a_2}{k_{1x}\epsilon_2 a_2 + k_{2x}\epsilon_1}. \tag{9.47}$$

Substituting the result that $k_{1y} = k_{2y}$ into the dispersion Equation (9.42) and solving for k_{2x}, we find:

$$k_{2x} = \sqrt{\omega^2 \epsilon_2 \mu_2 a_2 b_2 - k_{1y}^2 a_1^{-1} a_2}. \tag{9.48}$$

In order to have zero reflection at the $x = 0$ interface, we note from Equation (9.46) that we must have $\epsilon_1 = \epsilon_2$ and $k_{2x} = a_2 k_{1x}$. To realize this goal, we let

$$\epsilon_1 = \epsilon_2 \quad ; \quad \mu_1 = \mu_2 \quad ; \quad b_2 = a_2 \quad ; \quad a_2 = a_1^{-1} \tag{9.49}$$

in which case we have $\omega\sqrt{\mu_2\epsilon_2} = \omega\sqrt{\mu_1\epsilon_1} = k_1$, and Equation (9.48) reduces to:

$$k_{2x} = \sqrt{k_1^2 b^2 - (k_{1y})^2 b^2} = b\sqrt{k_1^2 - (k_{1y})^2} = b\,k_{1x} \tag{9.50}$$

which is the desired result, giving $\Gamma = 0$ from Equation (9.46).

Note that this no-reflection property is completely independent of the incidence angle θ_i, or the ratio k_{1x}/k_{1y}, or even of frequency. Furthermore, this result is valid for any values of the tensor entries a_1, a_2, b_1, and b_2, regardless of whether they are complex or real, as long as the conditions in Equation (9.49) are satisfied. For this TE case we do not have any restrictions on the parameter b_1. However, similar analysis (see Problem 9.1) of the TM case reveals that we must have $b_2 = b_1^{-1}$ in order to have $\Gamma = 0$.

In order to have zero reflection at the $x = 0$ wall for both TE and TM modes, the permittivity and permeability tensors must thus be in the form:

$$\bar{\bar{\epsilon}}_2 = \epsilon_1 \begin{bmatrix} s_x^{-1} & 0 & 0 \\ 0 & s_x & 0 \\ 0 & 0 & s_x \end{bmatrix} \quad \text{and} \quad \bar{\bar{\mu}}_2 = \mu_1 \begin{bmatrix} s_x^{-1} & 0 & 0 \\ 0 & s_x & 0 \\ 0 & 0 & s_x \end{bmatrix}. \tag{9.51}$$

Once again we have $\Gamma = 0$ for *any* value of s_x, whether complex or real, and at this point we have made no assumptions about the form of s_x. In order for Region 2 to work effectively as a boundary condition for FDTD modeling, in addition to providing zero reflection it must also rapidly attenuate the waves that are transmitted into it. To this end, we choose s_x, similar to the split-field method, to be given by:

$$s_x = 1 + \frac{\sigma_x}{j\omega\epsilon_1} \quad \rightarrow \quad k_{2x} = \left(1 - j\frac{\sigma_x}{\omega\epsilon_1}\right)k_{1x} = k_{1x} - j\,\sigma_x\eta_1\cos\theta_i. \tag{9.52}$$

Note that the real part of k_{2x} is identical to k_{1x}. With k_{2y} also being equal to k_{1y} from earlier, the transmitted wave enters Region 2 without any bending (refraction) and propagates with the same phase velocity and has the same wavelength as the incident wave, as shown in Figure 9.6. The full expression for the fields in Region 2 is:

$$\mathbf{H}_2 = \hat{\mathbf{z}}\,H_0\,e^{-js_x k_{1x}x + jk_{1y}y} = \hat{\mathbf{z}}\,H_0 e^{-jk_{1x}x + jk_{1y}y}\,e^{-\sigma_x x\eta_1\cos\theta_i} \tag{9.53a}$$

$$\mathbf{E}_2 = (-\hat{\mathbf{x}}\,\eta_1\sin\theta_i + \hat{\mathbf{y}}\,\eta_1\cos\theta_i)\,H_0\,e^{-jk_{1x}x + jk_{1y}y}\,e^{-\sigma_x x\eta_1\cos\theta_i}. \tag{9.53b}$$

Once again the attenuation factor is independent of frequency, so that all frequency components of a plane wave packet are transmitted across the interface without reflection and attenuate together at the same rate. At the same time, it should be noted that the attenuation factor for the waves in the second medium is dependent on the incidence angle θ_i.

Note that the x component of the electric field in Equation (9.53b) is different from that found using Bérenger's split-field technique, as given by Equation (9.27). Otherwise, the two methods give quite similar results, as expected.

Results for incidence on other walls of a 2D FDTD space can be obtained via a very similar analysis. For example, on the $y = 0$ and $y = y_{max}$ boundaries we define the $\bar{\bar{\epsilon}}_2$

and $\bar{\bar{\mu}}_2$ tensors such that:

$$\bar{\bar{\epsilon}}_2 = \epsilon_1 \begin{bmatrix} s_y & 0 & 0 \\ 0 & s_y^{-1} & 0 \\ 0 & 0 & s_y \end{bmatrix} \quad \text{and} \quad \bar{\bar{\mu}}_2 = \mu_1 \begin{bmatrix} s_y & 0 & 0 \\ 0 & s_y^{-1} & 0 \\ 0 & 0 & s_y \end{bmatrix}$$

where s_y is defined similar to s_x in Equation (9.52), but with σ_y replacing σ_x.

In a general 3D FDTD simulation space, we can define general tensors $\bar{\bar{\epsilon}}_2$ and $\bar{\bar{\mu}}_2$ that apply at each of the boundaries (and corners) with different choices of conductivities. The general tensors are given by the product of the three s matrices:

$$\bar{\bar{\epsilon}}_2 = \epsilon_1 \begin{bmatrix} s_x^{-1} & 0 & 0 \\ 0 & s_x & 0 \\ 0 & 0 & s_x \end{bmatrix} \begin{bmatrix} s_y & 0 & 0 \\ 0 & s_y^{-1} & 0 \\ 0 & 0 & s_y \end{bmatrix} \begin{bmatrix} s_z & 0 & 0 \\ 0 & s_z & 0 \\ 0 & 0 & s_z^{-1} \end{bmatrix}$$

$$= \epsilon_1 \begin{bmatrix} s_x^{-1} s_y s_z & 0 & 0 \\ 0 & s_x s_y^{-1} s_z & 0 \\ 0 & 0 & s_x s_y s_z^{-1} \end{bmatrix} \tag{9.54}$$

and similarly for $\bar{\bar{\mu}}_2$. Then, to implement this UPML, one simply chooses the nonzero conductivities at appropriate boundaries. For example, at the $x = 0$ and $x = x_{\text{max}}$ boundaries, we set $\sigma_y = \sigma_z = 0$, and thus $s_y = s_z = 1$, and the matrix in Equation (9.54) reduces to the one in Equation (9.51). At an "edge" of the simulation space, where $x = 0$ and $y = 0$, for example, we set $\sigma_z = 0$ and thus $s_z = 1$, but use nonzero σ_x and σ_y.

Finally, note that the grading of the UPML follows the same procedure as the split-field PML, described in Section 9.2.2 above.

9.3.1 Bérenger's PML as an anisotropic medium

It should be noted here that the setup of the uniaxial PML medium, in terms of the permittivity and permeability tensors in equation (9.51), is simply an alternate means of achieving the same fundamental result that was obtained by the field-splitting technique described earlier. In fact, it may have been obvious to the alert reader that the medium described by the split-field technique is indeed also a uniaxial medium with permittivity and permeability tensors given by:

$$\bar{\bar{\epsilon}}_2 = \epsilon_1 \begin{bmatrix} 1 + \dfrac{\sigma_x}{j\omega\epsilon_1} & 0 & 0 \\ 0 & 1 + \dfrac{\sigma_y}{j\omega\epsilon_1} & 0 \\ 0 & 0 & 1 + \dfrac{\sigma_z}{j\omega\epsilon_1} \end{bmatrix} ;$$

$$\bar{\bar{\mu}}_2 = \mu_1 \begin{bmatrix} 1 + \dfrac{\sigma_{m,x}}{j\omega\mu_1} & 0 & 0 \\ 0 & 1 + \dfrac{\sigma_{m,y}}{j\omega\mu_1} & 0 \\ 0 & 0 & 1 + \dfrac{\sigma_{m,z}}{j\omega\mu_1} \end{bmatrix}$$

with the additional constraint that

$$\sigma_{m,x} = \sigma_x \, \eta_1^2 \quad ; \quad \sigma_{m,y} = \sigma_y \, \eta_1^2 \quad ; \quad \sigma_{m,z} = \sigma_z \, \eta_1^2 \tag{9.55}$$

Thus we see that the Bérenger PML is a simple example of an anisotropic medium, although it is not as flexible as the uniaxial PML method.

9.4   FDTD implementation of the UPML

In this section, we describe the discretization of the partial differential equations governing the behavior of the waves within the UPML region. This is not straightforward, since the UPML development has been presented in the frequency domain, whereas an efficient FDTD implementation should be in the form of time-domain difference equations. If we are not careful, we may end up having to take inverse Fourier transforms of terms involving ω, thus ending up with convolutions in the time domain, which are computationally intensive to implement.

The UPML uses anisotropic tensors for the permittivity and permeability. In the previous section, we derived the anisotropic tensors for incidence on the $x = 0$ boundary. In the general 3D setup, these anisotropic tensors are given by

$$\bar{\bar{\epsilon}}_2 = \epsilon_1 \bar{\bar{s}}$$
$$\bar{\bar{\mu}}_2 = \mu_1 \bar{\bar{s}} \tag{9.56}$$

where $\bar{\bar{s}}$ is the 3D diagonal tensor given by Equation (9.54) above, wherein the terms $s_i = 1 + \sigma_i/(j\omega\epsilon_i)$. Note that because of the constraint that $\sigma_i/\epsilon_i = \sigma_{m,i}/\mu_i$, the anisotropicity matrices for $\bar{\bar{\epsilon}}_2$ and $\bar{\bar{\mu}}_2$ are the same. This is equivalent to the constraints on a_1, a_2, b_1, and b_2 in the description of the UPML.

The general 3D time-harmonic Maxwell's equations in the UPML medium are given by

$$\nabla \times \mathbf{H} = j\omega\epsilon\bar{\bar{s}}\,\mathbf{E} \tag{9.57a}$$
$$\nabla \times \mathbf{E} = -j\omega\mu\bar{\bar{s}}\,\mathbf{H}. \tag{9.57b}$$

We begin our demonstratation of the FDTD implementation of the PML equations by considering the 2D TM mode (E_z, H_x, H_y) across the $x = 0$ interface, with Region 1 being free space ($\epsilon_1 = \epsilon_0$, $\mu_1 = \mu_0$) (later we will consider the general 3D implementation). At this $x = 0$ interface, we choose the following values of s_i:

$$s_x = \left[1 + \frac{\sigma_x(x)}{j\omega\epsilon_1}\right] \quad ; \quad s_y = 1 \quad ; \quad s_z = 1 \tag{9.58}$$

such that the $\bar{\bar{s}}$ tensor is given by:

$$\bar{\bar{s}} = \begin{bmatrix} s_x^{-1}s_y s_z & 0 & 0 \\ 0 & s_x s_y^{-1} s_z & 0 \\ 0 & 0 & s_x s_y s_z^{-1} \end{bmatrix} = \begin{bmatrix} s_{xx} & 0 & 0 \\ 0 & s_{yy} & 0 \\ 0 & 0 & s_{zz} \end{bmatrix} = \begin{bmatrix} s_x^{-1} & 0 & 0 \\ 0 & s_x & 0 \\ 0 & 0 & s_x \end{bmatrix} \tag{9.59}$$

just as we derived earlier for the $x = 0$ boundary. Using Equations (9.57), Maxwell's equations in the PML medium can be written as:

$$\underbrace{j\omega\epsilon_0 \left[1 + \frac{\sigma_x(x)}{j\omega\epsilon_0}\right] E_z}_{j\omega\epsilon_0 s_{zz} E_z} = \underbrace{\left(\frac{\partial H_y}{\partial x} - \frac{\partial H_x}{\partial y}\right)}_{\nabla \times \mathbf{H}} \tag{9.60a}$$

$$\underbrace{j\omega\mu_0 \left[1 + \frac{\sigma_x(x)}{j\omega\epsilon_0}\right]^{-1} H_x}_{j\omega\mu_0 s_{xx} H_x} = -\frac{\partial E_z}{\partial y} \tag{9.60b}$$

$$\underbrace{j\omega\mu_0 \left[1 + \frac{\sigma_x(x)}{j\omega\epsilon_0}\right] H_y}_{j\omega\mu_0 s_{yy} H_y} = \frac{\partial E_z}{\partial x} \tag{9.60c}$$

where we have explicitly noted that σ_x may be a function of x (anticipating that it will be graded in the PML). We now proceed with the discretization, noting that multiplication of the phasor quantities with the factor $j\omega$ in the frequency domain is equivalent to taking the time derivative in the time domain. In other words, we can expand Equation (9.60a) and take its inverse Fourier transform to obtain:

$$j\omega\epsilon_0 \left[1 + \frac{\sigma_x(x)}{j\omega\epsilon_0}\right] E_z = \left(\frac{\partial H_y}{\partial x} - \frac{\partial H_x}{\partial y}\right)$$

$$j\omega\epsilon_0 E_z + \sigma_x(x) E_z = \left(\frac{\partial H_y}{\partial x} - \frac{\partial H_x}{\partial y}\right)$$

Inverse transform $\quad\rightarrow\quad$ $\epsilon_0 \dfrac{\partial \mathcal{E}_z}{\partial t} + \sigma_x(x) \mathcal{E}_z = \left(\dfrac{\partial \mathcal{H}_y}{\partial x} - \dfrac{\partial \mathcal{H}_x}{\partial y}\right)$

which can then be discretized to find:

$$\epsilon_0 \left[\frac{\mathcal{E}_z\big|_{i,j}^{n+1} - \mathcal{E}_z\big|_{i,j}^{n}}{\Delta t}\right] + \sigma_x\big|_i \overbrace{\left[\frac{\mathcal{E}_z\big|_{i,j}^{n+1} + \mathcal{E}_z\big|_{i,j}^{n}}{2}\right]}^{\text{Avg. to get value at } n+\frac{1}{2}}$$

$$= \left[\frac{\mathcal{H}_y\big|_{i+1/2,j}^{n+1/2} - \mathcal{H}_y\big|_{i-1/2,j}^{n+1/2}}{\Delta x} - \frac{\mathcal{H}_x\big|_{i,j+1/2}^{n+1/2} - \mathcal{H}_x\big|_{i,j-1/2}^{n+1/2}}{\Delta y}\right]. \tag{9.61}$$

Note that this equation has the same form as we have seen many times earlier, i.e., the FDE for a lossy medium with conductivity σ_x (which can vary with x, hence the index i). The update equation can be written from Equation (9.61) as:

$$
\mathcal{E}_z\Big|_{i,j}^{n+1} = \overbrace{\left[\frac{2\epsilon_0 - \Delta t\,\sigma_x\big|_i}{2\epsilon_0 + \Delta t\,\sigma_x\big|_i}\right]}^{G_1(i)}\mathcal{E}_z\Big|_{i,j}^{n}
$$

$$
+ \underbrace{\left[\frac{2\Delta t}{2\epsilon_0 + \Delta t\,\sigma_x\big|_i}\right]}_{G_2(i)}\left[\frac{\mathcal{H}_y\big|_{i+1/2,j}^{n+1/2} - \mathcal{H}_y\big|_{i-1/2,j}^{n+1/2}}{\Delta x} - \frac{\mathcal{H}_x\big|_{i,j+1/2}^{n+1/2} - \mathcal{H}_x\big|_{i,j-1/2}^{n+1/2}}{\Delta y}\right]
$$

$$\tag{9.62}$$

which is then the final discretized update equation for Equation (9.60a). The discretization of Equation (9.60c) follows in an almost identical manner; we find:

$$
\mathcal{H}_y\Big|_{i+1/2,j}^{n+1/2} = \overbrace{\left[\frac{2\epsilon_0 - \Delta t\,\sigma_x\big|_{i+1/2}}{2\epsilon_0 + \Delta t\,\sigma_x\big|_{i+1/2}}\right]}^{F_1(i+1/2)}\mathcal{H}_y\Big|_{i+1/2,j}^{n-1/2}
$$

$$
+ \frac{1}{\mu_0}\overbrace{\left[\frac{2\epsilon_0\Delta t}{2\epsilon_0 + \Delta t\,\sigma_x\big|_{i+1/2}}\right]}^{F_2(i+1/2)}\left[\frac{\mathcal{E}_z\big|_{i+1,j}^{n} - \mathcal{E}_z\big|_{i,j}^{n}}{\Delta x}\right]. \tag{9.63}
$$

Discretization of Equation (9.60b) requires considerably more attention due to the fact that the s_x term appears in the denominator. We proceed by rewriting (9.60b) as:

$$
j\omega\mu_0 H_x = -\left[\frac{\partial E_z}{\partial y} + \frac{\sigma_x(x)}{\epsilon_0}\frac{1}{j\omega}\frac{\partial E_z}{\partial y}\right]. \tag{9.64}
$$

Unfortunately, we now have an unavoidable scenario involving an integration, since multiplication with $(j\omega)^{-1}$ in the frequency domain corresponds to integration in the time domain. The time domain equation corresponding to (9.64) is then:

$$
\mu_0\frac{\partial \mathcal{H}_x}{\partial t} = -\frac{\partial \mathcal{E}_z}{\partial y} - \frac{\sigma_x(x)}{\epsilon_0}\int_0^t\frac{\partial \mathcal{E}_z}{\partial y}\,dt. \tag{9.65}
$$

This equation can be discretized according to the methods we have seen thus far, except for the integral term. However, we can discretize the integral as a summation, resulting in:

$$
\mu_0\left[\frac{\mathcal{H}_x\big|_{i,j+1/2}^{n+1/2} - \mathcal{H}_x\big|_{i,j+1/2}^{n-1/2}}{\Delta t}\right] = -\left[\frac{\mathcal{E}_z\big|_{i,j+1}^{n} - \mathcal{E}_z\big|_{i,j}^{n}}{\Delta y} + \frac{\sigma_x\big|_i}{\epsilon_0}\Delta t\sum_0^n\left(\frac{\mathcal{E}_z\big|_{i,j+1}^{n} - \mathcal{E}_z\big|_{i,j}^{n}}{\Delta y}\right)\right].
$$

$$\tag{9.66}$$

We can write the corresponding update equation for $\mathcal{H}_x$ as:

$$\mathcal{H}_x\Big|_{i,j+1/2}^{n+1/2} = \mathcal{H}_x\Big|_{i,j+1/2}^{n-1/2} + \frac{\Delta t}{\mu_0 \Delta y}\left[\mathcal{E}_z\Big|_{i,j+1}^n - \mathcal{E}_z\Big|_{i,j}^n\right] + \left(\frac{\sigma_x\big|_i\,\Delta t^2}{\mu_0\epsilon_0}\right)I_{i,j+1/2}^n \quad (9.67)$$

where $I_{i,j+1/2}^n$ is an integral buffer, being accumulated as follows:

$$I_{i,j+1/2}^n = I_{i,j+1/2}^{n-1} + \left[\mathcal{E}_z\Big|_{i,j+1}^n - \mathcal{E}_z\Big|_{i,j}^n\right]. \quad (9.68)$$

Note that the terms $G_1(\cdot)$, $G_2(\cdot)$, $F_1(\cdot)$, and $F_2(\cdot)$ are simply calculated either from a known (stored) array of values of $\sigma_x\big|_i$, or by means of a polynomial or geometric grading function, as described earlier. We can see that this method becomes slightly more computationally expensive because we must store and update this integral buffer I at each point in the PML. However, the PML is only some 10 cells thick at each boundary, so this extra computational requirement is not as heavy as it might seem.

Things become more complicated in the corner regions of this 2D space, where we must have attenuation in both the x and y directions. This can be realized by choosing the s parameters as:

$$s_x = \left[1 + \frac{\sigma_x(x)}{j\omega\epsilon_0}\right] \quad ; \quad s_y = \left[1 + \frac{\sigma_y(y)}{j\omega\epsilon_0}\right] \quad ; \quad s_z = 1. \quad (9.69)$$

We can then write Maxwell's equations (i.e., the equivalents to Equations 9.60) as:

$$\underbrace{j\omega\epsilon_0\left[1 + \frac{\sigma_x(x)}{j\omega\epsilon_0}\right]\left[1 + \frac{\sigma_y(y)}{j\omega\epsilon_0}\right]E_z}_{j\omega\epsilon_0 s_{zz}E_z} = \underbrace{\left(\frac{\partial H_y}{\partial x} - \frac{\partial H_x}{\partial y}\right)}_{\nabla\times\mathbf{H}} \quad (9.70a)$$

$$\underbrace{j\omega\mu_0\left[1 + \frac{\sigma_x(x)}{j\omega\epsilon_0}\right]^{-1}\left[1 + \frac{\sigma_y(y)}{j\omega\epsilon_0}\right]H_x}_{j\omega\mu_0 s_{xx}^{-1}H_x} = -\frac{\partial E_z}{\partial y} \quad (9.70b)$$

$$\underbrace{j\omega\mu_0\left[1 + \frac{\sigma_x(x)}{j\omega\epsilon_0}\right]\left[1 + \frac{\sigma_y(y)}{j\omega\epsilon_0}\right]^{-1}H_y}_{j\omega\mu_0 s_{yy}H_y} = \frac{\partial E_z}{\partial x}. \quad (9.70c)$$

The discretization of these equations follows along similar lines. The second and third equations turn out to be a combination of Equations (9.60a) and (9.60b) in that they contain both a conductivity term and a running integral. To see this, we can write Equation (9.70b) as follows:

$$j\omega\mu_0 H_x + \underbrace{\sigma_y(y)\frac{\mu_0}{\epsilon_0}}_{\sigma_{m,y}(y)}H_x = -\frac{\partial E_z}{\partial y} - \frac{\sigma_x(x)}{j\omega\epsilon_0}\frac{\partial E_z}{\partial y}. \quad (9.71)$$

One can see that this equation looks just like Equation (9.64), and so requires a similar running integral for the term on the right; but in addition, the loss term on the left requires an averaging as in Equation (9.61). A similar process is used for the discretization of Equation (9.70c).

Discretization of Equation (9.70a) is slightly more complicated. Multiplying out the left-hand side, we find:

$$\left[j\omega\epsilon_0 + \sigma_x(x) + \sigma_y(y) + \frac{\sigma_x(x)\sigma_y(y)}{j\omega\epsilon_0} \right] E_z = \left(\frac{\partial H_y}{\partial x} - \frac{\partial H_x}{\partial y} \right). \tag{9.72}$$

Taking the inverse transform we have:

$$\epsilon_0 \frac{\partial \mathcal{E}_z}{\partial t} + \left(\sigma_x(x) + \sigma_y(y) \right) \mathcal{E}_z + \frac{\sigma_x(x)\sigma_y(y)}{\epsilon_0} \int_0^t \mathcal{E}_z\, dt = \left(\frac{\partial \mathcal{H}_y}{\partial x} - \frac{\partial \mathcal{H}_x}{\partial y} \right). \tag{9.73}$$

Discretization will thus involve a running integral term, which is a part of the updated field $\mathcal{E}_z$. We leave the derivation of the final result as a problem at the end of this chapter.

9.5   Alternative implementation via auxiliary fields

One can see from the derivations above that the implementation of the UPML is a bit cumbersome, involving running integrals that must be stored for each grid cell in the PML. A much simpler, alternative implementation of the uniaxial PML medium exists, as described by [3]. This implementation is based on the introduction of intermediate fields **D** and **B**, which are related[6] respectively to **E** and **H**, although this relationship is different from the constitutive relationships in Chapter 2. The FDTD updating then proceeds as a four-step process rather than a two-step process. First, **D** is determined from the curl of **H**, and then **E** is found from **D** through the relationship that we will define momentarily. Then, **B** is found from the curl equation for **E**, and finally **H** is found from **B**. Here we present the 3D formulation, from which the 2D cases are easily derived.

Starting from the 3D Maxwell's equations in the UPML medium (Equations 9.57), and the full $\bar{\bar{s}}$ tensor in Equation (9.59), we can write Ampère's law, Equation (9.57a), in matrix form as:

$$\begin{bmatrix} \dfrac{\partial H_z}{\partial y} - \dfrac{\partial H_y}{\partial z} \\[2mm] \dfrac{\partial H_x}{\partial z} - \dfrac{\partial H_z}{\partial x} \\[2mm] \dfrac{\partial H_y}{\partial x} - \dfrac{\partial H_x}{\partial y} \end{bmatrix} = j\omega\epsilon \begin{bmatrix} s_x^{-1}s_y s_z & 0 & 0 \\ 0 & s_x s_y^{-1} s_z & 0 \\ 0 & 0 & s_x s_y s_z^{-1} \end{bmatrix} \begin{bmatrix} E_x \\ E_y \\ E_z \end{bmatrix}. \tag{9.74}$$

We now choose the s_i parameters in a general media, where here we also include the possibility of non-unity κ:

$$s_x = \kappa_x + \frac{\sigma_x}{j\omega\epsilon_0} \quad ; \quad s_y = \kappa_y + \frac{\sigma_y}{j\omega\epsilon_0} \quad ; \quad s_x = \kappa_z + \frac{\sigma_z}{j\omega\epsilon_0}. \tag{9.75}$$

Now, to avoid convolutions or integral buffers in the implementation of the above equations, we define a relationship between **E** and **D** in the PML medium which conveniently

---

[6] The **D** and **B** fields that will be introduced are not physical fields, as can be seen by careful examination of Equations (9.76); the s_i terms multiplying them render these fields unphysical.

decouples the frequency-dependent terms. We define **D** in the PML as follows:

$$D_x = \epsilon \frac{s_z}{s_x} E_x \quad ; \quad D_y = \epsilon \frac{s_x}{s_y} E_y \quad ; \quad D_z = \epsilon \frac{s_y}{s_z} E_z. \tag{9.76}$$

It is important to reiterate at this point that **D** above (and the associated **B**) is not strictly the **D** field defined through the constitutive relationships in Chapter 2; however, it is convenient that when $\sigma = 0$ and $\kappa = 1$, Equations (9.76) reduce to the true constitutive relation between **E** and **D**. Now, the matrix Equation (9.74) then becomes:

$$\begin{bmatrix} \dfrac{\partial H_z}{\partial y} - \dfrac{\partial H_y}{\partial z} \\[2mm] \dfrac{\partial H_x}{\partial z} - \dfrac{\partial H_z}{\partial x} \\[2mm] \dfrac{\partial H_y}{\partial x} - \dfrac{\partial H_x}{\partial y} \end{bmatrix} = j\omega \begin{bmatrix} s_y & 0 & 0 \\ 0 & s_z & 0 \\ 0 & 0 & s_x \end{bmatrix} \begin{bmatrix} D_x \\ D_y \\ D_z \end{bmatrix}. \tag{9.77}$$

This leaves us with three equations that are simple to convert back to time domain, due to direct conversions from $j\omega \rightarrow \partial/\partial t$. They are also easy to implement in FDTD, and follow the same lines as the simple lossy medium described in Chapter 4. For example, plugging in the expression for s_y from (9.75), the first equation in (9.77) converted to time domain is

$$\frac{\partial \mathcal{H}_z}{\partial y} - \frac{\partial \mathcal{H}_y}{\partial z} = \kappa_y \frac{\partial \mathcal{D}_x}{\partial t} + \frac{\sigma_y}{\epsilon_0} \mathcal{D}_x \tag{9.78}$$

where κ_y is assumed not to be time dependent. This equation is then very simply discretized in the same manner as the lossy Maxwell's equations to find an update equation for $\mathcal{D}_x$:

$$\mathcal{D}_x \Big|_{i+1/2,j,k}^{n+1} = \left(\frac{2\epsilon_0 \kappa_y - \sigma_y \Delta t}{2\epsilon_0 \kappa_y + \sigma_y \Delta t} \right) \mathcal{D}_x \Big|_{i+1/2,j,k}^{n} \tag{9.79}$$

$$+ \left(\frac{2\epsilon_0 \Delta t}{2\epsilon_0 \kappa_y + \sigma_y \Delta t} \right) \left(\frac{\partial \mathcal{H}_z}{\partial y} \Big|_{i+1/2,j,k}^{n+1/2} - \frac{\partial \mathcal{H}_y}{\partial z} \Big|_{i+1/2,j,k}^{n+1/2} \right)$$

where the spatial derivatives of $\mathcal{H}_z$ and $\mathcal{H}_y$ with respect to y and z, respectively, are discretized as usual. Similar equations follow for the other two in the system of Equations (9.77).

Now, we must turn our attention to the relationships given between the frequency domain **D** and **E** in Equations (9.76). Given that we have found $\mathcal{D}$ from $\mathcal{H}$, we need to use these relationships to get back $\mathcal{E}$, which in turn will be used to update $\mathcal{H}$ (through $\mathcal{B}$, as we'll see later).

Fortunately, the relationship between **D** and **E** given in Equation (9.76) lends itself well to the time domain; we can write, for example for the x components,

$$D_x = \epsilon \frac{s_z}{s_x} E_x \qquad \rightarrow \qquad s_x D_x = \epsilon s_z E_x$$

$$\left(\kappa_x + \frac{\sigma_x}{j\omega\epsilon_0} \right) D_x = \epsilon \left(\kappa_z + \frac{\sigma_z}{j\omega\epsilon_0} \right) E_x. \tag{9.80}$$

Multiplying through by $j\omega$ and converting to time domain, we have

$$\kappa_x \frac{\partial \mathcal{D}_x}{\partial t} + \frac{\sigma_x}{\epsilon_0} \mathcal{D}_x = \epsilon \left[\kappa_z \frac{\partial \mathcal{E}_x}{\partial t} + \frac{\sigma_z}{\epsilon_0} \mathcal{E}_x \right]. \tag{9.81}$$

This equation is now easily discretized in the FDTD formulation, yielding the update equation:

$$\mathcal{E}_x \Big|_{i+1/2,j,k}^{n+1} = \left(\frac{2\epsilon_0 \kappa_z - \sigma_z \Delta t}{2\epsilon_0 \kappa_z + \sigma_z \Delta t} \right) \mathcal{E}_x \Big|_{i+1/2,j,k}^{n} \tag{9.82}$$

$$+ \frac{1}{\epsilon} \left(\frac{2\epsilon_0 \kappa_x + \sigma_x \Delta t}{2\epsilon_0 \kappa_z + \sigma_z \Delta t} \right) \mathcal{D}_x \Big|_{i+1/2,j,k}^{n+1} - \frac{1}{\epsilon} \left(\frac{2\epsilon_0 \kappa_x - \sigma_x \Delta t}{2\epsilon_0 \kappa_z + \sigma_z \Delta t} \right) \mathcal{D}_x \Big|_{i+1/2,j,k}^{n}$$

where spatial averages have been used where appropriate. Similar equations can be formulated for the relationships between $\mathcal{D}_y \to \mathcal{E}_y$ and $\mathcal{D}_z \to \mathcal{E}_z$. This has now transformed the $\overline{\mathcal{E}}$ update equations to a two-step process, requiring first solving for the $\mathcal{D}$ field from the $\mathcal{H}$ field, and then calculating $\mathcal{E}$ from $\mathcal{D}$. However, it should be noted that, computationally, this is no worse than the integral buffer method in the previous section, which involves calculating an extra "field" that is the integral buffer in Equation (9.68).

As one might expect, a similar two-step procedure is required to evaluate $\mathcal{H}$ from $\mathcal{E}$. This should be obvious on comparison between the two Maxwell's Equations (9.57). Fortunately, due to the symmetry of these equations and the use of the same $\overline{\overline{s}}$ tensor, the formulation is the same. Here we give the update equations for the $\mathcal{B}_x$ and $\mathcal{H}_x$ components as an example. $\mathcal{B}_x$ is updated from $\mathcal{E}_z$ and $\mathcal{E}_y$ as follows:

$$\mathcal{B}_x \Big|_{i,j+1/2,k+1/2}^{n+1/2} = \left(\frac{2\epsilon_0 \kappa_y - \sigma_y \Delta t}{2\epsilon_0 \kappa_y + \sigma_y \Delta t} \right) \mathcal{B}_x \Big|_{i,j+1/2,k+1/2}^{n-1/2} + \left(\frac{2\epsilon_0 \Delta t}{2\epsilon_0 \kappa_y + \sigma_y \Delta t} \right) \tag{9.83}$$

$$\times \left(\frac{\mathcal{E}_z \Big|_{i,j+1,k+1/2}^{n} - \mathcal{E}_z \Big|_{i,j,k+1/2}^{n}}{\Delta y} - \frac{\mathcal{E}_y \Big|_{i,j+1/2,k+1}^{n} - \mathcal{E}_y \Big|_{i,j+1/2,k}^{n}}{\Delta z} \right).$$

Then, $\mathcal{H}_x$ is found from $\mathcal{B}_x$ similar to Equations (9.80)–(9.82):

$$s_x B_x = \mu s_z H_x$$

$$\left(\kappa_x + \frac{\sigma_x}{j\omega\epsilon_0} \right) B_x = \mu \left(\kappa_z + \frac{\sigma_z}{j\omega\epsilon_0} \right) H_x$$

Inverse Transform $\quad\to\quad$ $\kappa_x \dfrac{\partial B_x}{\partial t} + \dfrac{\sigma_x}{\epsilon_0} B_x = \mu \left[\kappa_z \dfrac{\partial \mathcal{H}_x}{\partial t} + \dfrac{\sigma_z}{\epsilon_0} \mathcal{H}_x \right].$ $\quad$ (9.84)

Finally, this equation can be easily discretized in the FDTD formulation, arriving at the update equation for $\mathcal{H}_x$:

$$\mathcal{H}_x \Big|_{i,j+1/2,k+1/2}^{n+1/2} = \left(\frac{2\epsilon_0 \kappa_z - \sigma_z \Delta t}{2\epsilon_0 \kappa_z + \sigma_z \Delta t} \right) \mathcal{H}_x \Big|_{i,j+1/2,k+1/2}^{n-1/2} + \frac{1}{\mu} \left(\frac{2\epsilon_0 \kappa_x + \sigma_x \Delta t}{2\epsilon_0 \kappa_z + \sigma_z \Delta t} \right) \tag{9.85}$$

$$\times \mathcal{B}_x \Big|_{i,j+1/2,k+1/2}^{n+1/2} - \frac{1}{\mu} \left(\frac{2\epsilon_0 \kappa_x - \sigma_x \Delta t}{2\epsilon_0 \kappa_z + \sigma_z \Delta t} \right) \mathcal{B}_x \Big|_{i,j+1/2,k+1/2}^{n-1/2}.$$

Equations (9.79), (9.82), (9.83), and (9.85), together with the equations for the other field components, thus form a four-step method for updating the fields in the PML boundary regions.

9.6   Convolutional perfectly matched layer (CPML)

While the Bérenger split-field PML and the uniaxial PML described above are both very robust and efficient at terminating most FDTD simulation spaces, both suffer from the inability to absorb evanescent waves. This means the PML must be placed sufficiently far from any scattering objects so that all evanescent waves have decayed sufficiently before reaching the PML. This restriction can lead to simulation spaces that are larger than desired – yet, after all, the purpose of the PML is to reduce the size of the simulation space. The trouble lies with the PML setup being "weakly causal": in the constitutive relations given by Equations (9.76), for example, the time domain $\overline{\mathcal{D}}(t)$ depends on future values of $\overline{\mathcal{E}}(t)$.

In order to absorb evanescent waves (and thus allow a reduction of the simulation space), a strictly causal form of the PML has been derived by Kuzuoglu and Mittra [4]. This method involves shifting the frequency-dependent pole of the s terms off the real axis and into the negative imaginary plane,[7] by making the simple substitution $j\omega\epsilon \rightarrow \alpha + j\omega\epsilon$, where α is a positive real number. This is known as the complex-frequency-shifted (CFS) PML.

The most effective implementation of the CFS-PML formulation was derived by Roden and Gedney in 2000 [5], using a recursive convolution technique. Their technique is commonly referred to as the convolutional PML or CPML. In a general lossy medium, the x-component of $\mathcal{E}$ of Ampère's law in frequency domain form is

$$j\omega\epsilon E_x + \sigma E_x = \frac{1}{s_y}\frac{\partial H_z}{\partial y} - \frac{1}{s_z}\frac{\partial H_y}{\partial z}. \tag{9.86}$$

Note that σ here is the real conductivity accounting for loss in the simulation space, whereas σ_i (within the s_i) are the PML conductivities. Here the s_i are the same terms we have used in the Bérenger split-field and UPML formulations above,

$$s_i = 1 + \frac{\sigma_i}{j\omega\epsilon_0}. \tag{9.87}$$

As presented in Equation (9.86), the s_i terms are typically known as "stretched coordinates" metrics [6], in that they apply to the ∂x, ∂y, and ∂z spatial parameters; for instance, in the first term on the right in Equation (9.86), ∂y has been stretched to $\partial y \rightarrow s_y \partial y$.

---

[7] The UPML method can also be made causal, and thus can absorb evanescent waves, by shifting its pole off the imaginary axis. The derivation is somewhat more involved, but such implementations have been used with success.

In the CFS-PML (and thus CPML) formulation, a more general stretched coordinate is introduced:

$$s_i = \kappa_i + \frac{\sigma_i}{\alpha_i + j\omega\epsilon_0} \tag{9.88}$$

where κ_i (always real and ≥ 1) has been introduced to allow for a non-unity real part of s_i, and α_i (positive real, along with σ_i) is introduced to move the pole off the real axis and into the imaginary half-space. Now, Equation (9.89) can be directly transformed into the time domain:

$$\epsilon\frac{\partial \mathcal{E}_x}{\partial t} + \sigma\mathcal{E}_x = \bar{s}_y * \frac{\partial \mathcal{H}_z}{\partial y} - \bar{s}_z * \frac{\partial \mathcal{H}_y}{\partial z} \tag{9.89}$$

where, for simplicity, we have defined $\bar{s}_i(t)$ as the inverse Laplace transform of the *inverse* of the stretching parameter, $s_i(\omega)^{-1}$. It can be shown [5] that $\bar{s}_i(t)$ has the impulse response

$$\bar{s}_i(t) = \frac{\delta(t)}{\kappa_i} - u(t)\frac{\sigma_i}{\epsilon_0\kappa_i^2}e^{-\left(\frac{\sigma_i}{\epsilon_0\kappa_i} + \frac{\alpha_i}{\epsilon_0}\right)t}$$

$$= \frac{\delta(t)}{\kappa_i} + \zeta_i(t) \tag{9.90}$$

where $\delta(t)$ and $u(t)$ are the impulse and step functions, respectively. Convolution with a delta function simply returns the original function, so the time domain Ampère's law (Equation 9.89) now becomes

$$\epsilon_r\epsilon_0\frac{\partial \mathcal{E}_x}{\partial t} + \sigma\mathcal{E}_x = \frac{1}{\kappa_y}\frac{\partial \mathcal{H}_z}{\partial y} - \frac{1}{\kappa_z}\frac{\partial \mathcal{H}_y}{\partial z} + \zeta_y(t) * \frac{\partial \mathcal{H}_z}{\partial y} - \zeta_z(t) * \frac{\partial \mathcal{H}_y}{\partial z}. \tag{9.91}$$

Here we have explicitly separated the convolution terms from the remaining terms. Now, the *discrete* impulse response of $\zeta_i(t)$ is

$$Z_{0_i}(m) = \int_{m\Delta t}^{(m+1)\Delta t} \zeta_i(\tau)d\tau$$

$$= -\frac{\sigma_i}{\epsilon_0\kappa_i^2}\int_{m\Delta t}^{(m+1)\Delta t} e^{-\left(\frac{\sigma_i}{\epsilon_0\kappa_i} + \frac{\alpha_i}{\epsilon_0}\right)\tau}d\tau$$

$$= a_i e^{\left(\frac{\sigma_i}{\kappa_i} + \alpha_i\right)\left(\frac{m\Delta t}{\epsilon_0}\right)} \tag{9.92}$$

where the term a_i is used as shorthand, defined as

$$a_i = \frac{\sigma_i}{\sigma_i\kappa_i + \kappa_i^2\alpha_i}\left(e^{-(\sigma_i/\kappa_i + \alpha_i)(\Delta t/\epsilon_0)} - 1\right).$$

Notice that a_i is a scalar value for any given grid location i, as is the term $Z_{0_i}(m)$ given by Equation (9.92). Now, we can discretize Equation (9.91) using the standard staggered Yee cell, resulting in

$$
\epsilon_r \epsilon_0 \frac{\mathcal{E}_x \Big|_{i+1/2,j,k}^{n+1} - \mathcal{E}_x \Big|_{i+1/2,j,k}^{n}}{\Delta t} + \sigma \frac{\mathcal{E}_x \Big|_{i+1/2,j,k}^{n+1} + \mathcal{E}_x \Big|_{i+1/2,j,k}^{n}}{2} \tag{9.93}
$$

$$
= \frac{\mathcal{H}_z \Big|_{i+1/2,j+1/2,k}^{n+1/2} - \mathcal{H}_z \Big|_{i+1/2,j-1/2,k}^{n+1/2}}{\kappa_y \Delta y} - \frac{\mathcal{H}_y \Big|_{i+1/2,j,k+1/2}^{n+1/2} - \mathcal{H}_y \Big|_{i+1/2,j,k-1/2}^{n+1/2}}{\kappa_z \Delta z}
$$

$$
+ \sum_{m=0}^{N-1} Z_{0_y}(m) \frac{\mathcal{H}_z \Big|_{i+1/2,j+1/2,k}^{n-m+1/2} - \mathcal{H}_z \Big|_{i+1/2,j-1/2,k}^{n-m+1/2}}{\Delta y}
$$

$$
- \sum_{m=0}^{N-1} Z_{0_z}(m) \frac{\mathcal{H}_y \Big|_{i+1/2,j,k+1/2}^{n-m+1/2} - \mathcal{H}_y \Big|_{i+1/2,j,k-1/2}^{n-m+1/2}}{\Delta z}.
$$

One should notice that the first two lines of this equation are identical to the standard FDTD update equation, except for the terms κ_y and κ_z in the denominators. In practice, the convolutions represented by the last two terms in Equation (9.93) are very computationally costly to store and implement. However, due to the simple exponential form of Z_0 from Equation (9.92), the convolutions can be performed recursively through the introduction of an auxiliary term ψ_i. With the implementation of this auxiliary variable, Equation (9.93) becomes:

$$
\epsilon_r \epsilon_0 \frac{\mathcal{E}_x \Big|_{i+1/2,j,k}^{n+1} - \mathcal{E}_x \Big|_{i+1/2,j,k}^{n}}{\Delta t} + \sigma \frac{\mathcal{E}_x \Big|_{i+1/2,j,k}^{n+1} + \mathcal{E}_x \Big|_{i+1/2,j,k}^{n}}{2} \tag{9.94}
$$

$$
= \frac{\mathcal{H}_z \Big|_{i+1/2,j+1/2,k}^{n+1/2} - \mathcal{H}_z \Big|_{i+1/2,j-1/2,k}^{n+1/2}}{\kappa_y \Delta y} - \frac{\mathcal{H}_y \Big|_{i+1/2,j,k+1/2}^{n+1/2} - \mathcal{H}_y \Big|_{i+1/2,j,k-1/2}^{n+1/2}}{\kappa_z \Delta z}
$$

$$
+ \psi_{ex,y} \Big|_{i+1/2,j,k}^{n+1/2} - \psi_{ex,z} \Big|_{i+1/2,j,k}^{n+1/2}
$$

where the ψ_i terms are given by

$$
\psi_{ex,y} \Big|_{i+1/2,j,k}^{n+1/2} = b_y \psi_{ex,y} \Big|_{i+1/2,j,k}^{n-1/2} + a_y \frac{\mathcal{H}_z \Big|_{i+1/2,j+1/2,k}^{n+1/2} - \mathcal{H}_z \Big|_{i+1/2,j-1/2,k}^{n+1/2}}{\Delta y} \tag{9.95a}
$$

$$
\psi_{ex,z} \Big|_{i+1/2,j,k}^{n+1/2} = b_z \psi_{ex,z} \Big|_{i+1/2,j,k}^{n-1/2} + a_z \frac{\mathcal{H}_y \Big|_{i+1/2,j,k+1/2}^{n+1/2} - \mathcal{H}_y \Big|_{i+1/2,j,k-1/2}^{n+1/2}}{\Delta z} \tag{9.95b}
$$

and the constants b_i are given by:

$$
b_i = e^{-(\sigma_i/\kappa_i + \alpha_i)(\Delta_t/\epsilon_0)}. \tag{9.95c}
$$

The notation $\psi_{ex,y}$ denotes that this ψ term applies to the update equation for $\mathcal{E}_x$, involving the spatial derivative in the y-direction (i.e., $\partial\mathcal{H}_z/\partial y$) and thus the constant b_y. Note that similar pairs of ψ terms must be kept for each of the six update equations in the 3D Yee cell formulation, resulting in an extra 12 "fields" to be stored in the case of PMLs on all six sides (and hence the need for many subscripts!).

Now, note that σ_i, κ_i, and α_i are scaled along their respective axes, and graded in the same way as in the Bérenger split-field formulation (except in the case of α_i, which must actually be graded in the opposite manner, i.e., α_i should be nonzero at the PML interface, fading to $\alpha_i = 0$ at the boundary). This in turn means the a_i and b_i are 1D functions, and need not be stored over the entire 3D space. Similarly, the ψ_e and ψ_h "fields" are zero outside the PML regions, and so can be stored only within the PML, saving computational time and storage space.

Finally, the reader should note that the CPML implementation here is independent of material; the implementation presented here, by simply adding ψ components to the normal update equations (and not forgetting the κ scaling in the denominators of Equation 9.94), can be applied to any material, be it dispersive, anisotropic, or nonlinear.

9.6.1   Example simulation using the CPML

Figure 9.7 shows a snapshot of the $\mathcal{H}_z$ field from an example 2D TM simulation using the CPML method to terminate the simulation space. The space is 200 m $\times$ 200 m, with $\Delta x = \Delta y = 1$ m, and the time step is 2.36 ns; this snapshot is taken at $t = 708.5$ ns. We excite a hard $\mathcal{H}_z$ source at the center of the space with an amplitude of 1 Tesla and a frequency of 15 MHz, giving a free-space wavelength of 20 m, so that our grid has 20 cells per wavelength. Figure 9.7 shows the absolute value of the $\mathcal{H}_z$ field, so the variations shown are actually half-wavelengths. The left panel shows the full simulation space, and the right panel shows a zoomed-in view of the last 20 m on the right edge of the simulation space. Note that the right panel is also scaled to -100 dB rather than -60 dB. From this panel we can see that no reflection can be discerned.

Due to the increased complexity of the UPML and CPML methods compared to the Bérenger split-field method, we provide here greater details on the implementation of this particular example. The statements shown below are written in Matlab. After specifying constants and grid parameters, we define the maximum conductivities as in Equation (9.32):

```
sxmax = -(m+1)*log(1e-6)/2/377/(10*dx);
```

where m is the order of the polynomial grading, $R(0) = 10^{-6}$, $\eta \simeq 377$, and the width of the PML is `10*dx`. The conductivities on each of the boundaries are then graded according to the fourth-order polynomial:

```
for mm = 1:10
    sy(mm+1)     = sxmax*((11-mm-0.5)/10)^4;
    sym(mm)      = sxmax*((11-mm)/10)^4;
    sy(yend+1-mm) = sxmax*((11-mm-0.5)/10)^4;
```

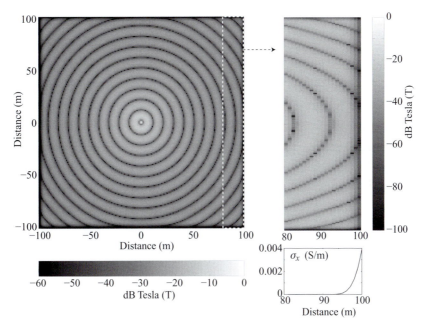

Figure 9.7 Example simulation with the CPML. The space is 200 m × 200 m and is excited with an $\mathcal{H}_z$ source at frequency 15 MHz. The panel shows a zoomed-in view of the last 20 m on the right; below, the PML conductivity profile.

```
    sym(yend+1-mm) = sxmax*((11-mm)/10)^4;
    sx(mm+1) = sxmax*((11-mm-0.5)/10)^4;
    sxm(mm) = sxmax*((11-mm)/10)^4;
    sx(xend+1-mm) = sxmax*((11-mm-0.5)/10)^4;
    sxm(xend+1-mm) = sxmax*((11-mm)/10)^4;
end
```

where xend and yend are the indices at the ends of the space, i.e., 201 in this case. Note that the conductivities are vectors of length 201. Some of the conductivity vectors must be shifted by a half-cell (hence the 0.5 terms in the equations above), but knowing which ones to shift is not at first obvious. As it turns out, the vectors are aligned with the fields that they will be applied to. For example, later $\sigma_x = $ sx will be applied to the update equation for $\mathcal{E}_y$; but $\mathcal{E}_y$ is located at grid points $(i + 1/2, j)$, so it requires conductivities σ_x at half-cell locations in x.

Next these conductivities are used to create vectors a_i and b_i from Equations (9.92) and (9.95):

```
aex = exp(-sx*dt/e0) - 1; bex = aex + 1;
aey = exp(-sy*dt/e0) - 1; bey = aey + 1;
ahx = exp(-sxm*dt/e0) - 1; bhx = ahx + 1;
ahy = exp(-sym*dt/e0) - 1; bhy = ahy + 1; .
```

This convenient feature that $b_i = a_i + 1$ is due to the fact that we have chosen κ and $\alpha = 1$; this also simplifies the form of the equations for a_i above. Next, inside a time iteration loop, we have the update equations (where xx and yy are the simulation space dimensions in number of grid cells):

```
for i = 1:xx-1,
  for j = 1:yy-1,
    Phy(i,j) = bhy(j) * Phy(i,j) + ahy(j) * (Ex(i,j+1)-Ex(i,j))/dy;
    Phx(i,j) = bhx(i) * Phx(i,j) + ahx(i) * (Ey(i+1,j)-Ey(i,j))/dx;

    Hz(i,j) = Hz(i,j) + dt/(u0*dy) * (Ex(i,j+1)-Ex(i,j)) - ...
      dt/(u0*dx) * (Ey(i+1,j)-Ey(i,j)) + dt/u0 * (Phy(i,j) - Phx(i,j));

  end
end

Hz(101,101) = sin(2*pi*f*t*dt); % update the source

for i = 2:xx-1,
  for j = 2:yy-1,
    Pey(i,j) = bey(j) * Pey(i,j) + aey(j) * (Hz(i,j)-Hz(i,j-1))/dy;
    Pex(i,j) = bex(i) * Pex(i,j) + aex(i) * (Hz(i,j)-Hz(i-1,j))/dx;

    Ex(i,j) = Ex(i,j) + dt/(dy*e0) * (Hz(i,j)-Hz(i,j-1)) + dt/e0 * Pey(i,j);
    Ey(i,j) = Ey(i,j) - dt/(dx*e0) * (Hz(i,j)-Hz(i-1,j)) - dt/e0 * Pex(i,j);
  end
end .
```

The variable Pex is $\psi_{e,x}$ from Equations (9.95), noting that we do not need the second subscript in 2D, as we have only four ψ's to deal with (rather than nine in 3D). Note that where without the PML we had only three update equations, we now have seven. In this example we have written the ψ terms as matrices with the same size as the $\mathcal{E}$ and $\mathcal{H}$ component fields; this is to simplify the for loops. Memory and computation time can be saved by defining the ψ terms only in the PML, since they will be zero elsewhere.

Note that above we have written our update equations with for loops for each of the spatial indices, simply for illustrative purposes. Matlab is particularly well suited to matrix operations, and these calculations can be made considerably faster by writing the i and j loops as matrix operations. For comparison, this simple example took 48.8 seconds to run as written above; with matrix operations, it required only 4.46 seconds. When this example is run with the split-field method, however, using the longer for loop method required 24.4 seconds, exactly half that of the CPML method. This difference can be traced to having only four update equations rather than seven (the equation $\mathcal{H}_z = \mathcal{H}_{zx} + \mathcal{H}_{zy}$ can be executed outside the loop and takes negligible time), as well as by the complexity of the $\mathcal{H}_z$ update equation in the CPML method above.

If the split-field method is twice as fast as the CPML method, then what are the advantages of using the CPML? The advantages are twofold: first, the CPML absorbs evanescent waves, which the split-field method does not, meaning the CPML boundary can be placed considerably closer to sources; and second, the CPML is independent of the medium, and so the medium can be dispersive, anisotropic, lossy, inhomogeneous, or even nonlinear; in the next chapter we will begin discussion of the FDTD method in these types of media. Because of these features as well as its simple implementation shown above, the CPML is an excellent choice for terminating almost any type of simulation.

9.7 Summary

In this chapter we have introduced the concept of a perfectly matched layer (PML) as a method for absorbing outgoing waves at the edges of an FDTD simulation. The PML is truly an "absorbing" boundary layer, while the methods of Chapter 8 are more accurately described as "radiation" boundary conditions. In a PML, a layer of finite thickness (usually 10 grid cells) is added to the boundaries of the simulation space, and this layer is made to absorb waves through non-physical electric and magnetic conductivities, causing the field amplitude to decay. Reflections at the simulation space–PML interface are eliminated by "perfect matching" of the impedances of the two media, as well as through careful grading of the PML conductivities.

The Bérenger or split-field PML method involves splitting select field components and separately applying x- and y-directed conductivities to the resulting equations. In 2D, the transverse component is split:

$$\mathcal{H}_z = \mathcal{H}_{zx} + \mathcal{H}_{zy} \qquad \text{for the TM mode}$$

or

$$\mathcal{E}_z = \mathcal{E}_{zx} + \mathcal{E}_{zy} \qquad \text{for the TE mode}$$

while in 3D, all six field components must be split. In the PML regions, then, directional conductivities such as σ_x, σ_y, $\sigma_{m,x}$, and $\sigma_{m,y}$ selectively attenuate waves propagating in their respective directions. At the $x = 0$ boundary, for example, the field components for the TM mode are analytically described by:

$$H_z = H_0\, e^{-js_x k_{1x} x + jk_{1y} y} = H_0 e^{-jk_{1x} x + jk_{1y} y}\, e^{-\sigma_x x \eta_1 \cos\theta_i}$$

$$E_x = -H_0 \eta_1 \sin\theta_i\, e^{-jk_{1x} x + jk_{1y} y}\, e^{-\sigma_x x \eta_1 \cos\theta_i}$$

$$E_y = H_0 \eta_1 \cos\theta_i\, e^{-jk_{1x} x + jk_{1y} y}\, e^{-\sigma_x x \eta_1 \cos\theta_i}.$$

The waves propagate into the PML region and attenuate according to $e^{-\sigma_x x \eta_1 \cos\theta_i}$, which is dependent on the angle of incidence. For this reason the PML is most effective at near-normal incidence.

In the split-field PML, as in the other PML types, the conductivities must be slowly increased from zero at the PML boundary to the maximum value at the edges. This maximum value is determined based on the degree of attenuation required, and the grading can be either polynomial or geometric.

A second type of PML we have introduced in this chapter is the uniaxial or UPML, which is related to the split-field PML. The UPML is a physically defined anisotropic medium with tensor permittivity and permeability, for example at the x boundaries:

$$\bar{\bar{\epsilon}}_2 = \epsilon_1 \begin{bmatrix} s_x^{-1} & 0 & 0 \\ 0 & s_x & 0 \\ 0 & 0 & s_x \end{bmatrix} \quad \text{and} \quad \bar{\bar{\mu}}_2 = \mu_1 \begin{bmatrix} s_x^{-1} & 0 & 0 \\ 0 & s_x & 0 \\ 0 & 0 & s_x \end{bmatrix}.$$

These tensor material parameters define a simple anisotropic medium, but for the UPML the medium is made both anisotropic and lossy by making the constants s_x complex. In fact, similar to the split-field PML, we use $s_x = 1 + \sigma_x/j\omega\epsilon_1$, yielding a propagation constant that includes frequency-independent attenuation (Equation 9.52).

The UPML can be implemented in the FDTD method through direct discretization of the resulting lossy equations, but this method is made more complicated by the s_x^{-1} terms in the matrices above. In those cases, intergration terms in the PDEs require infinite sums in the FDEs. Alternatively, the UPML can be implemented in a four-step method involving $\mathcal{H}$ and $\mathcal{E}$ field components as usual, but also using $\overline{\mathcal{D}}$ and $\overline{\mathcal{B}}$ fields as intermediaries. This eliminates the integral terms from the PDEs, but requires extra equations in the implementation.

The last PML method we have introduced in this chapter is the convolutional or CPML, which is an implementation of the CFS-PML method. The CPML method has the advantage that it absorbs evanescent waves in the near field of a source, and that it works equally well in dispersive, anisotropic, or nonlinear media. The CPML involves "stretching" the coordinates in the partial derivatives of Maxwell's equations:

$$\partial x \rightarrow s_x \partial x \quad \text{where} \quad s_x = \kappa_x + \frac{\sigma_x}{\alpha_x + j\omega\epsilon_0}$$

Above, s_x is nearly identical to that of the UPML, with the possibility introduced of including κ and α to experimentally increase the effectiveness of the PML. Using these stretched coordinates, the complex s_i terms appear as convolutions in the time domain, with impulse response given in Equation (9.90). These convolutions result in infinite sums in the FDE versions of Maxwell's equations, as in the UPML method. However, due to their simple exponential form, these infinite sums can be iteratively "updated" in each step of the FDTD procedure. The result is that the FDTD update equations are modified with the simple addition of ψ terms, for example in the $\mathcal{E}_x$ update equation in 3D:

$$\underbrace{\left[\frac{\partial \mathcal{E}_x}{\partial t} + \sigma \mathcal{E}_x\right]_{i+1/2,j,k}^{n+1/2} = \left[\nabla_{yz} \times \mathcal{H}_{yz}\right]_{i+1/2,j,k}^{n+1/2}}_{\text{Usual FDTD update equation}} + \psi_{ex,y}\Big|_{i+1/2,j,k}^{n+1/2} - \psi_{ex,z}\Big|_{i+1/2,j,k}^{n+1/2}$$

where the ψ_i terms are updated at each time step according to:

$$\psi_{ex,y}\Big|_{i+1/2,j,k}^{n+1/2} = b_y \psi_{ex,y}\Big|_{i+1/2,j,k}^{n-1/2} + a_y \frac{\mathcal{H}_z\Big|_{i+1/2,j+1/2,k}^{n+1/2} - \mathcal{H}_z\Big|_{i+1/2,j-1/2,k}^{n+1/2}}{\Delta y}$$

$$\psi_{ex,z}\Big|_{i+1/2,j,k}^{n+1/2} = b_z \psi_{ex,z}\Big|_{i+1/2,j,k}^{n-1/2} + a_z \frac{\mathcal{H}_y\Big|_{i+1/2,j,k+1/2}^{n+1/2} - \mathcal{H}_y\Big|_{i+1/2,j,k-1/2}^{n+1/2}}{\Delta z}$$

where b_i and a_i are parts of the impulse response of s_i, given in Equations (9.92) and (9.95). Similar pairs of ψ terms must be written for each of the six component update equations.

9.8   Problems

9.1. **Derivation of the PML for the TE case.** Recall that in Section 9.3 we did not arrive at a restriction on the parameter b_1. By conducting an analysis of the TM case, starting from the electric field version of Equation (9.40), show that we must have $b_2 = b_1^{-1}$ in order to have $\Gamma = 0$.

9.2. **PML versus second-order Mur.** Modify the 2D simulation in Problem 8.2 to use a split-field PML boundary 10 cells thick, and measure the reflections back into the simulation space as a function of θ. How does the split-field PML compare to the second-order Mur boundary?

9.3. **Comparing the PML grading.** Write a 2D FDTD simulation with the split-field PML, similar to the example in Section 9.2.3. Use the same parameters as the example, i.e., $\Delta x = \Delta y = 5$ cm in a space of 6×3 m. Excite a sinusoidal source at the center of the space with $f_0 = 600$ MHz, and grade the PML with a polynomial of order 4 and $R(0) = 10^{-6}$. Measure the reflections by comparing the field amplitude at a particular snapshot in time at a location just inside the PML.

    (a) Modify the parameters of the polynomial grading, in particular the order m, between 3 and 4 (it need not be an integer). How does this affect the reflections? What m gives the lowest reflections?

    (b) Now implement the PML with a geometric grading, choosing appropriate parameters g and $\sigma_{x,0}$ according to Equations (9.35). Compare the results to the polynomial grading. Experiment with the effect of the parameter g on the PML performance.

9.4. **UPML implementation.** Write a code to implement the UPML in a 2D TE FDTD simulation, using the same parameters as Problem 9.3 above.

    (a) Excite the simulation at a frequency of 600 MHz, and compare the propagating waves along the x and y axes as in Figure 9.5. How do the reflections here compare to those in the split-field method? How do the simulation times compare?

    (b) Excite a sinusoidal wave with $f_0 = 100$ MHz. The PML is now in the near field of the source; How well does the PML perform?

9.5. **UPML in cylindrical coordinates.** Derive an equivalent of the UPML in 2D cylindrical coordinates, for the r_{max} boundary and the $z = 0$ and z_{max} boundaries. Implement this PML in Problem 4.9, for the half-wave dipole and monopole problems. Note that for the monopole you will not need the PML at $z = 0$.

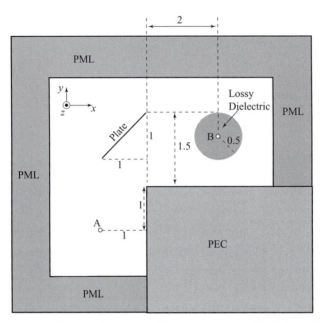

Figure 9.8 System of scattering objects to be modeled in Problem 9.9.

9.6. **UPML via D and B**. Write a code similar to Problem 9.4 using the four-step method involving **D** and **B** as intermediate fields, as described in Section 9.5. Run the simulation for 1000 time steps, and measure the computational time required (in Matlab, this is measured quite easily by enclosing the main loop with the statements `tic` and `toc`). Compare this computational time to the method in Problem 9.4. Which method is more efficient, and by how much?

9.7. **CPML implementation.** Modify the Luneberg lens simulation in Problem 8.12 to use the convolutional PML on the boundaries. Compare the results, and the computation time, to the second-order Mur boundary.

9.8. **Frequency response of UPML and CPML.** Implement the UPML and CPML separately in the simple square 2D simulation of Problem 4.5. Excite the simulations with the derivative of a Gaussian source five time steps wide. Measure the reflection coefficient as a function of frequency at locations on the PML boundary corresponding to incident angles of 0, 15, 30, and 45 degrees. Compare the UPML and CPML reflections. Which PML performs better at low frequencies, and why?

9.9. **A system of scattering objects.** Model the system shown in Figure 9.8. The dimensions are in meters. The plate is made of PEC. The sphere is made of lossy dielectric, with $\epsilon = 4.0\epsilon_0$ and $\sigma = 0.01$ S/m. The rest of the open area is free space. Use $\Delta x = \Delta y = 0.05$ m and $\Delta t = \Delta x/(\sqrt{2}\,c)$. Also, when modeling, all scatter objects and sources should be at least 10 cells away from the outer boundary. Let the PML be 10 cells thick.

(a) Drive E_z in the cell at point A with the Gaussian:

$$E_z{}^n = e^{\left[\frac{-(n-80)}{20}\right]^2}.$$

Record the E_z field over time at point B. Now place the source at point B and repeat the run, but this time record the E_z field over time at point A, and compare with the first run. Is there a relationship between the two results? Explain.

(b) Repeat part (a), but this time drive the H_y at point A using the same Gaussian function and record the H_x field over time at point B. Then as in part (a), reverse the scenario (drive H_x at point B and measure H_y at point A) and compare. Are the results as expected? Comment and explain.

9.10. **Horn antenna revisited.** Modify the 2D simulation in Problem 7.9, the horn antenna, to include a PML of your choice on the boundaries. Now, place a small conducting sphere, three grid cells in radius, 10 grid cells outside the aperture of the horn antenna, and measure the field intensity scattered back to the source.

9.11. **Radiation pattern of a monopole.** Modify your code from Problem 7.4, the monopole above a ground plane, to include a PML on five of the six boundaries. The lower boundary should still be PEC, and the monopole antenna should be modeled using the thin-wire approximation. Measure the radiation pattern of the antenna about ~3 wavelengths from the source.

9.12. **Scattering pattern in 3D.** Model a spherical PEC scatterer in a 3D Cartesian simulation, using UPML or CPML boundaries and a total-field / scattered-field formulation surrounding the scatterer.

(a) Make the scatterer exactly 1 wavelength in radius. Launch a plane wave from one side of the space at the TF/SF boundary, and measure the scattering pattern in 3D just inside the PML by monitoring a large number of points in 3D. Compare the resulting scattering pattern to the theoretical approximation from the Mie solution (see, for example, Balanis [7] or Stratton [8]).

(b) Modify the scatterer to be either one-half wavelength or two wavelengths in radius; how does the scattering pattern change?

(c) Make the scatterer a dielectric with $\epsilon_r = 4$ and compare the scattering patterns to those found for the PEC scatterer above.

References

[1] J. P. Bérenger, "A perfectly matched layer for the absorption of electromagnetic waves," *J. Comput. Phys.*, vol. 114, pp. 185–200, 1994.

[2] U. S. Inan and A. S. Inan, *Electromagnetic Waves*. Prentice-Hall, 2000.

[3] A. Taflove and S. Hagness, *Computational Electrodynamics: The Finite-Difference Time-Domain Method*, 3rd edn. Artech House, 2005.

[4] M. Kuzuoglu and R. Mittra, "Frequency dependence of the constitutive parameters of causal perfectly matched anisotropic absorbers," *IEEE Microwave and Guided Wave Lett.*, vol. 6, pp. 447–449, 1996.

[5] J. Roden and S. D. Gedney, "Convolution PML (CPML): An efficient FDTD implementation of the CFS-PML for arbitrary media," *Microwave and Opt. Tech. Lett.*, vol. 27, pp. 334–339, 2000.

[6] W. C. Chew and W. H. Weedon, "A 3D perfectly matched medium from modified Maxwell's equations with stretched coordinates," *Microwave Opt. Technol. Lett.*, vol. 7, pp. 599–604, 1994.

[7] C. A. Balanis, *Advanced Engineering Electromagnetics*. New York: Wiley, 1989.

[8] J. A. Stratton, *Electromagnetic Theory*. McGraw-Hill, 1941.

10 FDTD modeling in dispersive media

One of the reasons for the rapidly growing use of FDTD methods in numerical electro-magnetics is the relative ease with which this technique can be used to solve problems involving complex media, including dispersive materials. These are materials in which either or both of the material parameters ϵ and μ are functions of frequency.[1] In reality, all materials have frequency-dependent ϵ and μ, but many materials can be approx-imated as frequency-independent over a frequency band of interest, simplifying their analysis and simulation. In this chapter, we discuss efficient FDTD methods for mod-eling electromagnetic wave phenomena in such materials. We will focus on the much more common case of frequency-dependent permittivity, but will include discussion of magnetically dispersive materials known as ferrites at the end of the chapter.

Macroscopic electromagnetic behavior of most dielectric materials is given in terms of tabulated, plotted, or otherwise analytically specified *complex* permittivity $\epsilon_c(\omega)$. For linear materials, the constitutive relationship between the electric flux density and electric field phasors, valid at each frequency and at each point in space, is:

$$\mathbf{D}(\mathbf{r}, \omega) = \epsilon_c(\mathbf{r}, \omega) \, \mathbf{E}(\mathbf{r}, \omega) \tag{10.1}$$

where we have explicitly noted that the medium may be inhomogeneous, i.e., ϵ_c may vary in space.

Sometimes, losses within the dielectric due to bulk conductivity within the material (as otherwise would be separately represented by a conductivity σ) can be lumped into $\epsilon_c(\omega)$. Alternatively, we may account for conductive losses separately, by keeping a nonzero σ in Maxwell's equations; this conductivity itself may or may not be a function of frequency over the band of interest.

Equation (10.1) is often written as:

$$\mathbf{D}(\mathbf{r}, \omega) = \epsilon_0 \, \epsilon_\infty \, \mathbf{E}(\mathbf{r}, \omega) + \mathbf{P}(\mathbf{r}, \omega)$$

$$\mathbf{D}(\mathbf{r}, \omega) = \epsilon_0 \, \epsilon_\infty \, \mathbf{E}(\mathbf{r}, \omega) + \epsilon_0 \, \chi_e(\mathbf{r}, \omega) \, \mathbf{E}(\mathbf{r}, \omega) \tag{10.2}$$

where $\mathbf{P}$ is the electric polarization phasor, χ_e is the electric susceptibility of the material, and the quantity ϵ_∞ is the relative permittivity at the upper end of the frequency band considered in the particular application (its analog at the lower end of the frequency band

---

[1] Note that the conductivity σ may also be a function of frequency, but its effect can be rolled into the complex permittivity.

is often written as ϵ_{dc}). At extremely high frequencies, real physical materials behave like a vacuum, and so ϵ_∞ is almost always equal to one. However, in an FDTD simulation, it may be useful in certain cases to use a non-unity ϵ_∞. For example, consider again Figure 2.1; if, in our simulation, we are only interested in the first resonance in the microwave region, then ϵ_∞ will refer to the first "new" level between the first and second resonance.

Note in general that the complex permittivity is given by

$$\epsilon_c(\omega) = \epsilon_0[\epsilon_\infty + \chi_e(\omega)] \tag{10.3}$$

which implies that the susceptibility $\chi_e(\omega)$ is a complex quantity.

Due to the product of two frequency-dependent quantities, the time domain equivalent of Equation (10.2) is given by the convolution integral:

$$\overline{\mathscr{D}}(\mathbf{r}, t) = \epsilon_0 \, \epsilon_\infty \, \overline{\mathscr{E}}(\mathbf{r}, t) + \epsilon_0 \underbrace{\int_0^t \overline{\mathscr{E}}[\mathbf{r}, (t - \tau)] \, \tilde{\chi}_e(\mathbf{r}, \tau) \, d\tau}_{\overline{\mathscr{P}}(\mathbf{r},t)} \tag{10.4}$$

where $\tilde{\chi}_e(\mathbf{r}, t)$ is the electric susceptibility, i.e., the inverse Laplace transform of $\chi_e(\mathbf{r}, \omega)$. Note from Equation (10.4) that the time domain polarization vector $\overline{\mathscr{P}}(\mathbf{r}, t)$ depends on the values of $\overline{\mathscr{E}}$ at $\mathbf{r}$ not only at time t, but also at all times prior to t.

At first glance, it may appear as if all we need to do is simply to incorporate the evaluation of the convolution integral in Equation (10.4) into our FDTD formulation. However, in practice, there are two important problems which prohibit direct implementation of Equation (10.4). First, the convolution integral must be evaluated for each electric field component, at each time step and spatial location within the frequency-dependent medium. Second, and more importantly, evaluation of the convolution requires the storage of the entire time history of the electric fields.

Fortunately, an efficient FDTD method of implementing Equation (10.4) has been developed by R. Leubbers et al. [1], known as the Recursive Convolution method. This method makes use of the observation that Maxwell's equations require the evaluation of $\partial\overline{\mathscr{D}}(\mathbf{r}, t)/\partial t$ rather than $\overline{\mathscr{D}}(\mathbf{r}, t)$, and also takes advantage of appropriate analytical representations for $\epsilon_c(\omega)$. Alternatively, dispersive materials can be implemented in the FDTD method by using the $\overline{\mathscr{D}}$ and $\overline{\mathscr{B}}$ vectors, similar to the implementation of the UPML method in Chapter 9, except in this case we use the physical $\overline{\mathscr{D}}$ and $\overline{\mathscr{B}}$ fields. As we shall see, both the recursive convolution and the auxiliary differential equation (ADE) methods are very similar to the methods used for the PML implementations in Chapter 9; after all, the UPML described in Chapter 9 was represented by frequency-dependent ϵ and μ (see Section 9.4) – the UPML is, in fact, a dispersive material.

10.1   Recursive convolution method

We now consider different types of materials, categorized by the functional form of $\epsilon_c(\omega)$, using the recursive convolution method developed by Luebbers. We begin with the simplest type of dispersive material, the so-called *Debye* materials.

10.1.1 Debye materials

In general, complex permittivity results from the fact that the electric flux density $\overline{\mathscr{D}}$ in a dielectric, resulting from an applied alternating electric field $\overline{\mathscr{E}}$ has, in general, a different phase with respect to $\overline{\mathscr{E}}$. This behavior is due to the inertia of the polarization $\overline{\mathscr{P}}$ which, when the frequency becomes high enough, cannot follow the rapid variations of the field, giving rise to "relaxation" of the permittivity.[2] In other words, if we were to suddenly turn off the alternating electric field $\overline{\mathscr{E}}$, the flux density $\overline{\mathscr{D}}$ does not go to zero instantly, but instead decays to zero in some manner (for example, as a decaying exponential).

This behavior can be formally described as

$$\epsilon_c(\omega) = \epsilon_0 \left[\epsilon_\infty + \int_0^\infty \tilde{\chi}_e(t) e^{-j\omega t} dt \right] \tag{10.5}$$

where $\tilde{\chi}_e(t)$ is some sort of decay factor accounting for the lagging of the polarization behind the applied field; the form of $\tilde{\chi}_e(t)$ is what defines our different types of materials in this section. Note that we must have $\tilde{\chi}_e(t) \to 0$ as $\omega \to \infty$, since at $\omega \to \infty$ we must have $\epsilon_c(\omega) = \epsilon_0 \epsilon_\infty$. Note that $\tilde{\chi}_e(t)$ also describes the increase of polarization (and thus $\overline{\mathscr{D}}$) toward its equilibrium value when the alternating field is suddenly applied, so that it is related to the time constant of the polarization of the dielectric.

A simple exponential form for the decay factor $\tilde{\chi}_e(t)$ was proposed by P. Debye [3]:

$$\tilde{\chi}_e(t) = \tilde{\chi}_e(0) e^{-t/\tau_0} u(t) \tag{10.6}$$

where $u(t)$ is the unit step function and τ_0 is known as the Debye relaxation time constant, characteristic of the particular material; this time constant can often be a function of temperature.

Substituting Equation (10.6) into (10.5) we have:

$$\epsilon_c = \epsilon_0 \left[\epsilon_\infty + \int_0^\infty \tilde{\chi}_e(0) e^{(-j\omega - 1/\tau_0)t} dt \right]$$

$$\epsilon_c(\omega) = \epsilon_0 \left[\epsilon_\infty + \frac{\tilde{\chi}_e(0)}{\left(\dfrac{1}{\tau_0} + j\omega \right)} \right] \qquad \rightarrow \qquad \boxed{\chi_e(\omega) = \frac{\tilde{\chi}_e(0)\tau_0}{1 + j\omega\tau_0}}. \tag{10.7}$$

Denoting the value of the relative permittivity at dc ($\omega = 0$) to be ϵ_{dc}, we have from Equation (10.7):

$$\epsilon_c(0) = \epsilon_{dc}\epsilon_0 = \epsilon_0 \epsilon_\infty + \epsilon_0 \tau_0 \tilde{\chi}_e(0). \tag{10.8}$$

We then solve for $\tilde{\chi}_e(0)$ and substitute this result in Equation (10.6), to find for the time domain susceptibility:

$$\tilde{\chi}_e(t) = \frac{\epsilon_{dc} - \epsilon_\infty}{\tau_0} e^{-t/\tau_0} u(t). \tag{10.9}$$

---

[2] Probably the best description of this phenomenon can be found in Richard Feynman's famous *Lectures on Physics* [2].

Using $\tilde{\chi}_e(0)$ in Equation (10.7) gives the expression for the frequency-dependent complex permittivity:

$$\epsilon_c(\omega) = \epsilon_0 \left[\epsilon_\infty + \frac{\epsilon_{dc} - \epsilon_\infty}{1 + j\omega\tau_0} \right] = \epsilon_0 \left[\epsilon_r' - j\epsilon_r'' \right] \tag{10.10}$$

Separate expressions for the real and imaginary parts ϵ_r' and ϵ_r'' can be found by equating the real and imaginary parts of (10.10). We have:

$$\epsilon_r' = \epsilon_\infty + \frac{\epsilon_{dc} - \epsilon_\infty}{1 + \omega^2\tau_0^2} \tag{10.11a}$$

$$\epsilon_r'' = \frac{(\epsilon_{dc} - \epsilon_\infty)\omega\tau_0}{1 + \omega^2\tau_0^2} \tag{10.11b}$$

$$\tan\delta_c \equiv \frac{\epsilon_r''}{\epsilon_r'} = \frac{(\epsilon_{dc} - \epsilon_\infty)\omega\,\tau_0}{\epsilon_{dc} + \epsilon_\infty\,\omega^2\tau_0^2} \tag{10.11c}$$

where $\tan\delta_c$ is the *loss tangent*; in tabulated material parameters, one often finds the loss tangent tabulated instead of ϵ_r''.

Equations (10.11) describe the complex frequency-dependent behavior of some materials, primarily dipolar liquids, due to the so-called *orientational polarization*, arising from the rotation of molecular dipoles. Such materials, the ubiquitous material H_2O being one of them, generally carry a permanent dipole moment. With no applied electric field, the individual dipole moments point in random directions, so that the net dipole moment is zero. When an electric field is applied, it tends to line up the individual dipoles to produce an additional net dipole moment. If all the dipoles in a material were to line up, the polarization would be very large. However, at ordinary temperatures and relatively small electric fields, the collisions of the molecules in their thermal motion allow only a small fraction to line up with the field. Orientational polarization effects become important in the microwave frequency range and are important in many microwave applications, including biological applications where the materials involved are composed largely of mixtures of water with other materials.

Figure 10.1 shows the real and imaginary parts of $\chi_e(\omega)$ for the Debye, Lorentz, and Drude materials, plotted as a function of wavelength. We will come back to the latter two after they are introduced later in this chapter. The Debye-type susceptibility is given in frequency domain by the rightmost term in Equation (10.7), with a functional form of $\chi_e(\omega) \sim C/(1 + j\omega\tau_0)$ where C is some constant. In Figure 10.1a, the vertical dashed line is the wavelength corresponding to τ_0. We can see that at this characteristic wavelength, the real and imaginary parts of the susceptibility are equal, and the imaginary part is at its maximum; thus, at this wavelength the material has the greatest absorption losses. For water at room temperature, this frequency corresponds to about 20 GHz [4]. Otherwise, the real part of the susceptibility increases monotonically toward unity for increasing wavelength, and the imaginary part decreases correspondingly.

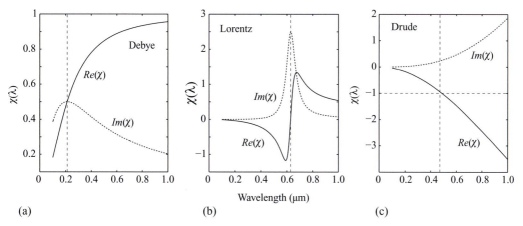

Figure 10.1 Susceptibilities of dispersive material types. The real and imaginary parts of susceptibility are plotted as a function of wavelength for fictional Debye, Lorentz, and Drude materials. The vertical dashed lines correspond to τ_0, ω_0, and ω_r in each of the respective cases.

In order to implement these Debye materials in the FDTD algorithm, we now consider time discretization of Equation (10.4):

$$\overline{\mathscr{D}}^n = \epsilon_0 \left[\epsilon_\infty \overline{\mathscr{E}}^n + \sum_{m=0}^{n-1} \overline{\mathscr{E}}^{n-m} \underbrace{\int_{m\Delta t}^{(m+1)\Delta t} \tilde{\chi}_e(\tau)\,d\tau}_{\tilde{\chi}_e^m \text{ below}} \right]. \tag{10.12a}$$

For simplicity in understanding the recursive convolution method, we also write the analogous expression for $\overline{\mathscr{D}}^{n-1}$:

$$\overline{\mathscr{D}}^{n-1} = \epsilon_0 \left[\epsilon_\infty \overline{\mathscr{E}}^{n-1} + \sum_{m=0}^{n-2} \overline{\mathscr{E}}^{n-1-m} \int_{m\Delta t}^{(m+1)\Delta t} \tilde{\chi}_e(\tau)\,d\tau \right]. \tag{10.12b}$$

Since $\overline{\mathscr{D}}$ and $\overline{\mathscr{E}}$ will be co-located on the grid, we do not need to worry about spatial discretization for now. We now introduce some compact notation that will be used to form the numerical algorithm:

$$\tilde{\chi}_e^m \equiv \int_{m\Delta t}^{(m+1)\Delta t} \tilde{\chi}_e(\tau)\,d\tau \tag{10.13a}$$

$$\Delta \tilde{\chi}_e^m \equiv \tilde{\chi}_e^m - \tilde{\chi}_e^{m+1} = \int_{m\Delta t}^{(m+1)\Delta t} \tilde{\chi}_e(\tau)\,d\tau - \int_{(m+1)\Delta t}^{(m+2)\Delta t} \tilde{\chi}_e(\tau)\,d\tau \tag{10.13b}$$

$$\Psi^{n-1} \equiv \sum_{m=0}^{n-2} \overline{\mathscr{E}}^{n-1-m} \Delta \tilde{\chi}_e^m. \tag{10.13c}$$

In Equation (10.13a), $\tilde{\chi}_e^m$ represents the integral of $\tilde{\chi}_e(t)$ over a single time step, from m to $m+1$. $\Delta \tilde{\chi}_e^m$ in Equation (10.13b) is the difference between successive $\tilde{\chi}_e^m$ terms.

Equation (10.13c) is the convolution summation, written in terms of a newly introduced accumulation variable Ψ. Using Equations (10.12) in the discrete form of Ampère's law:

$$\nabla \times \mathcal{H}\Big|^{n-1/2} = \frac{1}{\Delta t}\left(\overline{\mathcal{D}}\Big|^{n} - \overline{\mathcal{D}}\Big|^{n-1}\right)$$

we can write the update equation for the electric field as:

$$\overline{\mathcal{E}}^{n} = \frac{1}{\epsilon_\infty + \tilde{\chi}_e^0}\left[\epsilon_\infty \overline{\mathcal{E}}^{n-1} + \Psi^{n-1} + \frac{\Delta t}{\epsilon_0}\left(\nabla \times \mathcal{H}\right)^{n-1/2}\right] \qquad (10.14)$$

where, from Equation (10.13a),

$$\tilde{\chi}_e^0 \equiv \int_0^{\Delta t} \tilde{\chi}_e(\tau)\,d\tau.$$

Note that up to this point, we have not made any assumptions about the form of $\tilde{\chi}_e(t)$, and as such the derivation of the update equation is completely general and not specific to the Debye-type materials. Now, evaluation of Equation (10.14) still appears to require the storage of the time history of the electric field, through the accumulation variable Ψ; this is where we must assume an analytical approximation for $\tilde{\chi}_e(t)$. The exponential form of $\tilde{\chi}_e(t)$ in the Debye formulation greatly simplifies the problem at hand, since successive values $\Delta\tilde{\chi}_e^m$ can be recursively evaluated. Using the form of $\tilde{\chi}_e(t)$ from Equation (10.9) and evaluating the integral in Equation (10.13b), we find:

$$\Delta\tilde{\chi}_e^m = (\epsilon_{dc} - \epsilon_\infty)\left(1 - e^{-\Delta t/\tau_0}\right)^2 e^{-m\Delta t/\tau_0}$$

$$\rightarrow \quad \Delta\tilde{\chi}_e^{m+1} = e^{-\Delta t/\tau_0}\,\Delta\tilde{\chi}_e^m. \qquad (10.15)$$

Furthermore, we can write the accumulation variable Ψ^{n-1} from Equation (10.13c) as:

$$\Psi^{n-1} = \sum_{m=0}^{n-2} \overline{\mathcal{E}}^{n-1-m}\,\Delta\tilde{\chi}_e^m = \underbrace{\overline{\mathcal{E}}^{n-1}\,\Delta\tilde{\chi}_e^0}_{m=0} + \sum_{m=1}^{n-2} \overline{\mathcal{E}}^{n-1-m}\,\Delta\tilde{\chi}_e^m$$

$$= \overline{\mathcal{E}}^{n-1}\,\Delta\tilde{\chi}_e^0 + \sum_{m=0}^{n-3} \overline{\mathcal{E}}^{n-2-m}\,\Delta\tilde{\chi}_e^{m+1}$$

$$\rightarrow \quad \Psi^{n-1} = \overline{\mathcal{E}}^{n-1}\,\Delta\tilde{\chi}_e^0 + e^{-\Delta t/\tau_0}\,\Psi^{n-2} \qquad (10.16)$$

where, in the final step, we have used the fact that $\Delta\tilde{\chi}_e^{m+1} = e^{-\Delta t/\tau_0}\Delta\tilde{\chi}_e^m$ from Equation (10.15). We thus see that the convolution is reduced to a running sum, eliminating the need to store time histories of the electric field.

To summarize, implementation of the Debye material in the FDTD algorithm involves minor changes to the electric field update equations:

$$\overline{\mathcal{E}}^{n} = \frac{1}{\epsilon_\infty + \tilde{\chi}_e^0}\left[\epsilon_\infty \overline{\mathcal{E}}^{n-1} + \Psi^{n-1} + \frac{\Delta t}{\epsilon_0}\left(\nabla \times \mathcal{H}\right)^{n-1/2}\right] \qquad (10.14)$$

where $\tilde{\chi}_e^0$ is given by Equation (10.13a) with $m = 0$, and Ψ^{n-1} is a new update equation, involving a new "field" Ψ, which is updated according to Equation (10.16). Note that the

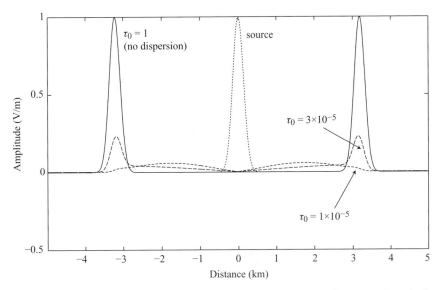

Figure 10.2 Example simulation of a Debye material in 1D. The Gaussian source is excited at the center (dotted) and allowed to propagate in both directions. The parameter τ_0 controls the amount of dispersion and dissipation in the material.

FDTD update equation for the magnetic field is unchanged from that which we normally use, namely

$$\overline{\mathcal{H}}^{n+1/2} = \overline{\mathcal{H}}^{n-1/2} - \frac{\Delta t}{\mu}\left[\nabla \times \overline{\mathcal{E}}\right]^n,\tag{10.17}$$

since we limited our attention to frequency-dependent dielectrics, assuming that the permeability μ is not dependent on frequency.

The field update Equations (10.14) and (10.17) are written in terms of total field quantities. As before, it may be advantageous to use a scattered-field formulation or a total-field/scattered-field formulation, the corresponding expressions for which are obtained from (10.14) and (10.17) as described in Chapter 7.

Example Debye simulation

Figure 10.2 shows an example of a 1D simulation involving a Debye material. This example uses $\Delta x = 20$ m, $\Delta t = 66.7$ ns (the CFL limit), and a Gaussian source with $n_\tau = 10\Delta t$. We set $\epsilon_{dc} = 10$ and $\epsilon_\infty = 1$, and vary τ_0 as shown in the figure; the snapshots are shown for $t = 200\Delta t = 13.3$ μs. For $\tau_0 = 1$ s, this time constant is so slow compared to the speed of the wave that essentially no dispersion occurs. For $\tau_0 = 3 \times 10^{-5}$ s, comparable to the time scale of the simulation, dispersion and dissipation both occur; we see that the amplitude of the pulse is reduced and low-frequency components are delayed, leaving what appears to be a trailing DC component. For $\tau_0 = 1 \times 10^{-5}$ s, the dispersion and dissipation are so strong that the original pulse is almost indistinguishable under the highly dispersed slow component.

The code for a Debye-type simulation is extremely simple. After defining the usual constants as well as eps_dc, eps_inf, and tau_0, we define the parameters from Equations (10.13):

```
dXe0 = (eps_dc - eps_inf) * (1 - exp(-dt/tau_0))^2;
Xe0 = (eps_dc - eps_inf) * (1 - exp(-dt/tau_0));
```

Then, inside the looping update equations, we update Psi and the $\mathscr{E}_z$ update equation, with i as the x-index:

```
Psi(i) = dXe0 * Ez(i) + exp_fac * Psi(i);
Ez(i) = 1/(eps_inf + Xe0) * (eps_inf * Ez(i) + Psi(i) + ...
(dt/e0/dx) * (Hy(i) - Hy(i-1)));
```

Since the $\mathscr{H}_y$ update equation is not changed, the rest of the algorithm proceeds exactly as in the usual FDTD method.

10.1.2   Lorentz materials

We mentioned above that Debye type dispersion occurs in materials, usually liquids, which exhibit orientational polarization, which is due to the re-orientation of molecules under the influence of an electric field. The other type of polarization which typically becomes important at optical frequencies is the so-called *distortional* polarization, which arises from the displacement of electrons (*electronic* polarizability) or atoms (*ionic* polarizability) due to the application of the external field.

The basic model in this case is that of charges bound elastically to an equilibrium position, so that when displaced by an applied field the charges return to their equilibrium positions by executing free oscillations about it, which die off at a rate determined by the damping forces present. This model thus leads to the possibility of resonant absorption, in which maximum power is absorbed from an alternating applied field when it is at a characteristic frequency. The charges can be regarded as behaving as linear harmonic oscillators, and a simple mathematical formulation of their behavior can thus be formulated. To see this, consider the equation of force for a harmonic oscillator, which gives the displacement x due to an applied field $\overline{\mathscr{E}}(t)$:

$$m\frac{\partial^2 x}{\partial t^2} + k_f m \frac{\partial x}{\partial t} + k_s x = q\,\overline{\mathscr{E}}(t) \tag{10.18}$$

where k_f is the coefficient of friction and k_s is a spring constant. In the frequency domain, this equation is written as:

$$-m\,\omega^2 x + j\omega\,k_f m\,x + k_s\,x = q\,\overline{\mathscr{E}}(\omega). \tag{10.19}$$

Noting that the polarization vector $\overline{\mathscr{P}}(\omega) = Nqx(\omega)$, where N is the number density of molecules, we find:

$$\overline{\mathscr{P}}(\omega) = \frac{Nq^2}{k_s + j\omega\,k_f m - m\,\omega^2}\overline{\mathscr{E}}(\omega). \tag{10.20}$$

Thus, since $\overline{\mathcal{P}}(\omega) = \chi_e(\omega)\overline{\mathcal{E}}(\omega)$, the functional form of $\chi_e(\omega)$ is given directly above. However, in the analysis of these electronic materials, this functional form is typically written as:

$$\chi_e(\omega) = \frac{\tilde{\chi}_e(0)\omega_r}{(\delta^2 + \omega_r^2) + 2j\omega\delta - \omega^2} \tag{10.21}$$

where δ is the damping constant (which has units of s^{-1}), and $\omega_0 = \sqrt{\omega_r^2 + \delta^2}$ is the resonance frequency. Note that this susceptibility is a two-pole expression, while the Debye susceptibility had a single pole.

The frequency domain expression in Equation (10.21) is plotted in Figure 10.1b. This plot uses a material with damping coefficient $\delta = 2 \times 10^{14}$ s^{-1}, and a resonance frequency $\omega_0 = \sqrt{\delta^2 + \omega_r^2} = 3 \times 10^{15}$ rad/s, which corresponds with the wavelength shown by the vertical dashed line. While this is not a real material, it is not far from the behavior of many real metals; for example, aluminum[3] can be modeled as a Lorentz-type material with a damping constant of $\delta = 4.5 \times 10^{13}$ s^{-1} and a resonance frequency of $\omega_0 = 5.37 \times 10^{14}$ rad/s; this lower resonance frequency pushes the resonance out to about 3 μm in wavelength.

We see that the imaginary part of the susceptibility – which is responsible for absorption – tends to zero at long and short wavelengths, and peaks at the resonance frequency ω_0. The real part of the susceptibility starts at zero at short wavelengths, goes through a resonance at ω_0, and then settles at a different constant value at long wavelengths. This is precisely the behavior described in Figure 2.1.

The inverse transform of Equation (10.21) gives the susceptibility in the time domain:

$$\tilde{\chi}_e(t) = \tilde{\chi}_e(0)\, e^{-t\delta}\, \cos(\omega_r t)\, u(t). \tag{10.22}$$

Once again defining ϵ_{dc} and ϵ_∞ as the values of relative permittivity at zero and infinite frequencies, we can show that

$$\tilde{\chi}_e(0) = \frac{\omega_0^2(\epsilon_\infty - \epsilon_{dc})}{\omega_r} \tag{10.23}$$

where $\omega_0 = \sqrt{\omega_r^2 + \delta^2}$. Lorentz materials are thus described by four parameters, namely ϵ_∞, ϵ_{dc}, δ, and ω_r. Using Equation (10.23), we have for the complex permittivity:

$$\epsilon(\omega) = \epsilon_0\,\epsilon_\infty + \underbrace{\frac{\epsilon_0\,(\epsilon_{dc} - \epsilon_\infty)\,\omega_0^2}{\omega_0^2 + 2j\omega\delta - \omega^2}}_{\chi_e(\omega)}. \tag{10.24}$$

Although the time domain form of Equation (10.22) does not immediately lend itself to recursive evaluation, this can be achieved by defining a complex time domain susceptibility $\tilde{\chi}_e^c(t)$, such that

$$\tilde{\chi}_e^c(t) \equiv -j\tilde{\chi}_e(0)\, e^{(-\delta + j\omega_r)t}\, u(t) \tag{10.25}$$

---

[3] In this context we have modeled aluminum with a single Lorentz pole; however, it is more realistically described by at least four such poles. The other three poles occur in visible wavelengths, but with an order of magnitude lower amplitude. See the article by W. H. P. Pernice et al. [5] for details.

so that

$$\tilde{\chi}_e(t) = \mathcal{R}e\left\{\tilde{\chi}_e^c(t)\right\}.$$

Note that it is ω_r that appears in the exponent of the time domain susceptibility, and not ω_0, although the latter is the more physically meaningful resonance frequency.

We can now implement recursive convolution with this complex time domain suscep-tibility. We can apply Equation (10.13a) to find the Lorentz material version of $\tilde{\chi}_e^m$:

$$\left[\tilde{\chi}_e^c\right]^m = \frac{-j\tilde{\chi}_e(0)}{\delta - j\omega_r} e^{(-\delta + j\omega_r)m\Delta t}\left[1 - e^{(-\delta + j\omega_r)\Delta t}\right]. \tag{10.26}$$

Evaluation of Equation (10.26) with $m = 0$ yields $\left[\tilde{\chi}_e^c\right]^0$, the real part of which is needed in the electric field update Equation (10.14). Similarly, we apply Equation (10.13b) to find the analog of $\Delta\tilde{\chi}_e^m$:

$$\left[\Delta\tilde{\chi}_e^c\right]^m = \frac{-j\tilde{\chi}_e(0)}{\delta - j\omega_r} e^{(-\delta + j\omega_r)m\Delta t}\left[1 - e^{(-\delta + j\omega_r)\Delta t}\right]^2. \tag{10.27}$$

Conveniently, this result exhibits the important property, similar to the Debye material, where:

$$\left[\Delta\tilde{\chi}_e^c\right]^{m+1} = e^{(-\delta + j\omega_r)\Delta t}\left[\Delta\tilde{\chi}_e^c\right]^m. \tag{10.28}$$

We can now introduce a complex accumulation variable Ψ_c^n such that

$$\Psi^{n-1} = \sum_{m=0}^{n-2}\overline{\mathscr{E}}^{n-1-m}\Delta\tilde{\chi}_e^m = \mathcal{R}e\left\{\Psi_c^{n-1}\right\} \tag{10.29}$$

which can be updated recursively using the same approach as shown above for the real accumulation variable for Debye materials. The new accumulation variable can be updated with an equation similar to (10.16):

$$\Psi_c^{n-1} = \overline{\mathscr{E}}^{n-1}\left[\Delta\tilde{\chi}_e^c\right]^0 + e^{(-\delta + j\omega_r)\Delta t}\Psi_c^{n-2} \tag{10.30}$$

where, similar to the Debye case,

$$\left[\Delta\tilde{\chi}_e^c\right]^0 = \frac{-j\tilde{\chi}_e(0)}{\delta - j\omega_r}\left[1 - e^{(-\delta + j\omega_r)\Delta t}\right]^2. \tag{10.31}$$

As a final note, in the implementation of the FDTD update equations for Lorentz materials, it is important to remember that we need to use the real parts of complex quantities (i.e., Equation 10.29) in the electric field update Equations (10.14).

Example Lorentz simulation

Figure 10.3 shows an example simulation in 1D of a Lorentz material. In this case, we use the same simulation parameters as the Debye example above, namely $\Delta x = 20$ m and $\Delta t = 66.7$ ns. As before, $\epsilon_{dc} = 10$ and $\epsilon_\infty = 1$, but the Gaussian source has a width of $n_\tau = 20\Delta t$ and is modulated at frequency $f = 300$ kHz. In all cases δ is kept constant relative to ω_r, such that $\delta = \omega_r/100$, which is similar to most metals [5]. Note that since

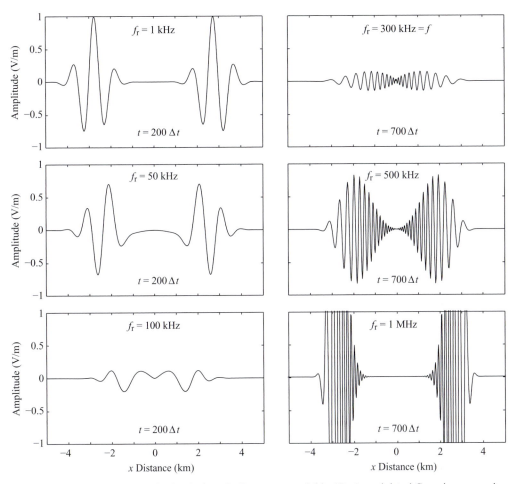

Figure 10.3 Example simulation of a Lorentz material in 1D. A modulated Gaussian source is excited at the center and allowed to propagate in both directions.

δ is much smaller than ω_r, we can state that the resonant frequency ω_0 is approximately ω_r. The snapshots shown in Figure 10.3 are taken after 200 or 700 time steps in the left and right columns, respectively.

For a resonant frequency of $f_r = 1$ kHz, the transmitted frequency of 300 kHz is so much higher that we are nowhere near the resonant absorption peak shown in Figure 10.1. Furthermore, the pulse propagates at exactly $v_p = 3 \times 10^8$ m/s, since at high frequencies the permittivity settles at $\epsilon = \epsilon_0 \epsilon_\infty = \epsilon_0$. This is the case where there is no dispersion.

When the resonant frequency f_r is increased to 50 kHz and then 100 kHz, we begin to see both dissipation (a reduction in the amplitude of the pulse) and dispersion (since the shape has changed). The dispersion is due to the changing of the Gaussian envelope rather than any effect on the carrier frequency. When the resonant frequency matches the carrier frequency of 300 kHz (top right), the wave propagates much more slowly, and seems to be trapped or confined within a small volume. As the resonant frequency is increased (so that the wavelength is longer than the resonant wavelength), the wave begins to propagate again, and has less attenuation or dissipation. However, a problem at these higher resonant frequencies is that numerical *amplification* is occurring, as evidenced by the amplitudes greater than 1 in the $f_r = 1$ MHz case. This is a numerical effect, as the stability criterion is modified in this dispersive material; however, a detailed analysis of the stability of this problem is beyond the scope of this book. The interested reader is referred to Beck and Mirotznik [6], who found that for a particular Lorentz material, the time step needed to be reduced by a factor of ~ 6 to ensure stability.

10.1.3   Drude materials

In a so-called "perfect" conductor, electrons are free to move within the metal, and so any electromagnetic wave (of any frequency) accelerates electrons perfectly in phase with the wave. These accelerated electrons re-radiate at the same frequency and in phase, causing what we interpret as a "reflection."

However, this behavior relies on the absence of collisions within the medium. In the interaction of optical frequencies with realistic (nonperfect conductor) metals, the electric field associated with the optical wave accelerates internal electrons, but their motion is damped by collisions with the metallic atoms or molecules. The resulting electron motion results in a form of dispersion that can be modeled as

$$\chi_e(\omega) = -\frac{\omega_r^2}{\omega^2 - j\omega/\tau_r} \tag{10.32}$$

where ω_r is the resonant frequency in the metal, $\tau_r = 1/\nu$ is the relaxation time, and ν is a collision frequency. This type of dispersion, known as *Drude* dispersion, has a two-pole form, similar to Lorentz materials, except that one of the poles is found at $\omega = 0$. Modeling of metals at optical frequencies in the FDTD method requires an accurate implementation of this kind of dispersion.

It is instructive to separate the susceptibility into its real and imaginary parts:

$$\chi_e(\omega) = -\frac{\omega_r^2}{\omega(\omega^2 + 1/\tau_r^2)}(\omega + j/\tau_r).$$

We see from this expression that at high frequency, both the real and imaginary parts of $\chi_e(\omega)$ tend toward zero with the denominator increasing toward infinity; and at low frequency, both the real and imaginary parts tend toward large negative values (the negative imaginary part implies absorption). At the frequency where $\omega = 1/\tau_r$, the real and imaginary parts take the same value. Finally, note that in the case where $\tau_r = \infty$ ($\nu = 0$), where there is no imaginary part and thus absorption, the susceptibility becomes $\chi_e(\omega) = -\omega_r^2/\omega^2$. Thus, where $\omega = \omega_r$, the susceptibility is -1, as shown in Figure 10.1; in turn, the permittivity $\epsilon(\omega) = 0$ and the phase velocity $v_p = \infty$! Thus, this is truly a "resonance" frequency. In reality, however, there are no materials for which the collision frequency $\nu = 0$ at the resonant frequency.

This frequency domain susceptibility is plotted in Figure 10.1a, using a relaxation time of $\tau_r = 1 \times 10^{-15}$ s^{-1}. As can be seen from the form of the equation, in addition to defining the resonance, ω_r controls the amplitude of the susceptibility (and hence the level of absorption). The imaginary part of $\chi(\omega)$ starts near 0 at very high frequency and increases monotonically, implying higher and higher absorption at longer wavelengths. Meanwhile, the real part rapidly becomes negative at long wavelengths; indeed, many metals have a negative real part of ϵ. Silver and gold are examples of metals that are well modeled as Drude materials; gold has a relative permittivity of about -40 at $\lambda = 1$ μm [7].

Taking the inverse transform of (10.32), we find the time domain susceptibility,

$$\tilde{\chi}_e(t) = \omega_r^2 \tau_r \left(1 - e^{-t/\tau_r}\right) u(t). \tag{10.33}$$

Note the similarity between this and the form of the Debye susceptibility in Equation (10.9). As for the Debye materials, we proceed by deriving the discretized variables $\tilde{\chi}_e^m$, $\Delta \tilde{\chi}_e^m$, and Ψ (because of the simple form of $\tilde{\chi}_e(t)$ above, these will be real, like the Debye materials).

Applying Equation (10.13a), we find the form of $\tilde{\chi}_e^m$:

$$\tilde{\chi}_e^m = \omega_r^2 \tau_r \left[\Delta t + \tau_r e^{-m\Delta t/\tau_r}(e^{-\Delta t/\tau_r} - 1)\right]. \tag{10.34}$$

Next, we use Equation (10.13b) to find $\Delta \tilde{\chi}_e^m$; after some simplification:

$$\Delta \tilde{\chi}_e^m = -\omega_r^2 \tau_r^2 e^{-m\Delta t/\tau_r}(e^{-\Delta t/\tau_r} - 1)^2. \tag{10.35}$$

Once again, the form of Equation (10.35), with m appearing only in the exponential, suggests the simple relationship between $\Delta \tilde{\chi}_e^m$ and $\Delta \tilde{\chi}_e^{m+1}$ as we had for the Debye and Lorentz materials, namely that given by Equation (10.15) for Debye materials. This means the update equation for Ψ^n in Equation (10.16) is valid for Drude materials as well, with τ_r in place of τ_0. Furthermore, for Drude materials, $\Delta \tilde{\chi}_e^0$ is given by:

$$\Delta \tilde{\chi}_e^0 = -\omega_r^2 \tau_r^2 (e^{-\Delta t/\tau_r} - 1)^2 \tag{10.36}$$

which is used in the update Equation (10.16) for Ψ. Similarly, $\tilde{\chi}_e^0$ is given by Equation (10.34) with $m = 0$:

$$\tilde{\chi}_e^0 = \omega_r^2 \tau_r \left[\Delta t + \tau_r(e^{-\Delta t/\tau_r} - 1)\right]. \tag{10.37}$$

We thus find that apart from the form of $\tilde{\chi}_e^0$ and $\Delta \tilde{\chi}_e^0$, the FDTD procedure for Drude materials is identical to that of Debye materials.

Example Drude simulation

Figure 10.4 shows an example 1D simulation of a Drude material, with $f_r = 100$ kHz, and varying τ_r. The simulation space is identical to the Debye example earlier, including the values of Δt and Δx, and the center of the simulation space is excited with a simple Gaussian source with a half-width of $\sigma = 20\Delta t$. Note that the implementation of the Drude material is identical to the Debye material, except for different forms of $\tilde{\chi}_e^0$ and $\Delta \tilde{\chi}_e^0$.

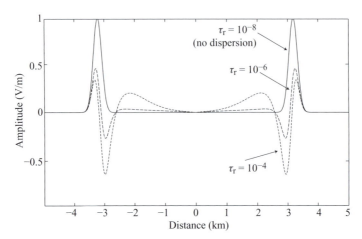

Figure 10.4 Example simulation of a Drude material in 1D. A simple Gaussian source is excited at the center and allowed to propagate in both directions.

When τ_r is very small (10^{-8}), the denominator of Equation (10.32) becomes very large, so that the susceptibility becomes very small; this behavior is seen in Figure 10.1: when the wavelength is much smaller than τ_r (or the frequency is very high), both the real and imaginary parts of χ tend toward zero, and the material behaves like free space, as we see in Figure 10.4. Now, when τ_r is increased to become comparable to $1/\omega$ and ω_r, we begin to see an increase in both the dispersion and dissipation in the material.

It should be noted, of course, that the parameter ω_r has a significant effect on the dispersion characteristics of the material. Not only does it define the resonant frequency, but its relative value compared to τ_r controls the dissipation characteristics.

10.1.4   Isotropic plasma

The final type of dispersive "dielectric" medium we consider is an isotropic plasma. A plasma is a gas in which an appreciable fraction of its constituent molecules have been ionized; as such, the plasma is a mixture of neutral atoms, free electrons, and both positive and negative ions. Strictly speaking, as plasma by definition is neutral, so the total number of electrons and negative ions equals the number of positive ions (in the simple case where each carries only one extra charge).

Due to the presence of these free electrons and ions, the plasma responds to applied electromagnetic fields, whether they are DC fields or AC fields (waves). Collisions between the excited electrons and the neutral gas molecules, however, dampen the electron motion, just as described above for Drude materials.

An *isotropic* plasma is the most basic type of plasma, in which the direction of the applied fields does not affect the plasma response. In Chapter 11 we will discuss *anisotropic* plasmas, in which an applied DC magnetic field confines the electron motion in a particular direction. The isotropic plasma here arises when either there is no applied magnetic field, or the collision frequency ν is high compared to the magnetic field.

The complex permittivity $\epsilon_c(\omega)$ of an isotropic plasma is given as [8]:

$$\epsilon_c(\omega) = \epsilon_0 \left[1 + \chi_e(\omega)\right] = \epsilon_0 \left[1 - \frac{\omega_p^2}{\omega(\omega + j\,\nu_c)}\right] \tag{10.38}$$

where ν_c in units of s^{-1} is the frequency of collisions between free electrons and other molecules, and ω_p is the so-called *plasma frequency*, related to the density of free electrons N_e as $\omega_p^2 = N_e q_e^2/(\epsilon_0 m_e)$, where q_e and m_e are the electronic charge and mass respectively. The reader should observe that the susceptibility in Equation (10.38) is almost identical to the susceptibility in Equation (10.32) for Drude materials. In fact, the plasma frequency ω_p is a resonance frequency; if the plasma is perturbed from its homogeneous state by creating a charge imbalance, this is the frequency at which charges will oscillate around the equilibrium point [9].

As in the Drude material, the susceptibility $\chi_e(\omega)$ of the isotropic plasma is seen to be negative, indicating that the permittivity of an isotropic plasma can be less than that of free space. Noting that $\chi_e(\omega)$ has two poles, at $\omega = 0$ and $\omega = j\nu_c$, it is convenient [10] to represent the effects of the pole at $\omega = 0$ in terms of a nonzero conductivity σ_{eff}, in order to avoid dealing with noncausal functions.

We can thus write the complex permittivity as

$$\epsilon_c(\omega) = \epsilon_0 \left[\epsilon_\infty - \frac{\omega_p^2/\nu_c}{\omega + j\nu_c} + \frac{\omega_p^2/\nu_c}{j\omega}\right] \tag{10.39a}$$

$$= \epsilon_0 \left[\epsilon_\infty + \underbrace{\frac{-(\omega_p^2/\nu_c)}{\omega + j\nu_c}}_{\chi_{\mathrm{eff}}(\omega)} + \underbrace{\frac{\epsilon_0(\omega_p^2/\nu_c)}{j\omega}}_{\sigma_{\mathrm{eff}}/(j\omega)}\right] \tag{10.39b}$$

$$\underbrace{\phantom{\epsilon_0 \left[\epsilon_\infty + \frac{-(\omega_p^2/\nu_c)}{\omega + j\nu_c}\right]}}_{\epsilon_{\mathrm{eff}}}$$

where we define $\sigma_{\mathrm{eff}} = \epsilon_0 \omega_p^2/\nu_c$. Maxwell's equation can be written as

$$\nabla \times \overline{\overline{\mathcal{H}}} = \sigma_{\mathrm{eff}}\,\overline{\overline{\mathcal{E}}} + j\omega\epsilon_{\mathrm{eff}}\,\overline{\overline{\mathcal{E}}} \tag{10.40a}$$

$$= j\omega \underbrace{\left(\frac{\sigma_{\mathrm{eff}}}{j\omega} + \epsilon_{\mathrm{eff}}\right)}_{\epsilon_c}\overline{\overline{\mathcal{E}}} = j\omega\,\epsilon_c\,\overline{\overline{\mathcal{E}}}. \tag{10.40b}$$

It thus appears that the isotropic plasma can be modeled as a medium with effective permittivity ϵ_{eff} and nonzero conductivity σ_{eff}. Furthermore, we note that the form of $\chi_{\mathrm{eff}}(\omega)$ is precisely the same as the complex permittivity for Debye materials as given in Equation (10.9), if we let

$$(\epsilon_{\mathrm{dc}} - \epsilon_\infty) = \frac{-\omega_p^2}{\nu_c} \quad \text{and} \quad \tau_0 = \frac{1}{\nu_c}.$$

Thus, we can use the same type of recursive convolution FDTD implementation given in Equations (10.13) and (10.14) to model electromagnetic wave propagation in an isotropic plasma, as long as we appropriately modify Equation (10.14) to include effects of a nonzero conductivity. In other words, the electric field update equation must now be a discretized version of Equation (10.40). The conductivity-modified version of Equation (10.14) is:

$$\overline{\mathscr{E}}^n = \frac{\epsilon_\infty}{\dfrac{\sigma_{\text{eff}}\Delta t}{\epsilon_\infty} + \epsilon_\infty + \tilde{\chi}_{\text{eff}}^0} \left[\overline{\mathscr{E}}^{n-1} + \frac{1}{\epsilon_\infty}\Psi^{n-1} + \frac{\Delta t}{\epsilon_0\epsilon_\infty} \left(\nabla \times \overline{\mathscr{H}} \right)^{n-1/2} \right] \tag{10.41}$$

where we also use $\tilde{\chi}_{\text{eff}}$ instead of $\tilde{\chi}_e$ in Equations (10.13), (10.15), and (10.16).

Note that just as in the Debye case, the magnetic field update Equation (10.17) is unaffected:

$$\overline{\mathscr{H}}^{n+1/2} = \overline{\mathscr{H}}^{n-1/2} - \frac{\Delta t}{\mu}\left[\nabla \times \overline{\mathscr{E}} \right]^n. \tag{10.42}$$

10.1.5   Improvement to the Debye and Lorentz formulations

The recursive convolution method presented above provides a simple, easy-to-implement method for modeling dispersive materials in the FDTD method. However, improved accuracy can be achieved with a small correction to the algorithm. The discretization of Equation (10.4) presented in Equation (10.12) involved the basic assumption that $\overline{\mathscr{E}}^{n-m}$ was approximately constant over the time step over which the integral is performed, from $m\,\Delta t$ to $(m+1)\Delta t$; as such, it could be taken out of the integral. Greater accuracy is achieved if, instead, the field is assumed to be piecewise linear over each time step. This improved method, derived by Kelley and Luebbers [11], is referred to as the *piecewise-linear* recursive convolution method. With no assumptions on the form of $\overline{\mathscr{E}}$, the correct analytical form of Equation (10.12) is:

$$\overline{\mathscr{D}}^n = \epsilon_0 \left[\epsilon_\infty \overline{\mathscr{E}}^n + \int_0^{n\Delta t} \overline{\mathscr{E}}(n\Delta t - \tau)\tilde{\chi}_e(\tau)\,d\tau \right]. \tag{10.43}$$

If $\overline{\mathscr{E}}(n\Delta t - \tau)$ is taken to be a constant over any single time step, as we did before, it can be taken out of the integral and put into a sum as presented in Equation (10.12). For the improved method, we instead use linear interpolation of $\overline{\mathscr{E}}$ over the time step to find the value of $\overline{\mathscr{E}}$ at any given instant of time:

$$\overline{\mathscr{E}}(t) = \overline{\mathscr{E}}^n + \left(\frac{\overline{\mathscr{E}}^{n+1} - \overline{\mathscr{E}}^n}{\Delta t} \right)(t - n\Delta t) \qquad \text{for} \qquad n\Delta t < t < (n+1)\Delta t.$$

In this case, the term $\overline{\mathscr{E}}(n\Delta t - \tau)$ in Equation (10.43) is presented (in m time-indexing) as

$$\overline{\mathscr{E}}(n\Delta t - \tau) = \overline{\mathscr{E}}^{n-m} + \left(\frac{\overline{\mathscr{E}}^{n-m-1} - \overline{\mathscr{E}}^{n-m}}{\Delta t} \right)(\tau - m\Delta t). \tag{10.44}$$

After substituting this expression into (10.43) and manipulating, we find a new form similar to Equation (10.12):

$$
\overline{\mathscr{D}}^n = \epsilon_0 \left[\underbrace{\epsilon_\infty \overline{\mathscr{E}}^n + \sum_{m=0}^{n-1} \overline{\mathscr{E}}^{n-m} \tilde{\chi}_e^m + \left(\overline{\mathscr{E}}^{n-m-1} - \overline{\mathscr{E}}^{n-m} \right) \tilde{\xi}_e^m}_{\text{Eq. (10.12)}} \right]
\tag{10.45}
$$

where $\tilde{\chi}_e^m$ is the same as before in Equation (10.13a), but now we must also keep track of a new variable $\tilde{\xi}_e^m$, which introduces a set of notations similar to Equations (10.13):

$$
\tilde{\xi}_e^m \equiv \frac{1}{\Delta t} \int_{m \Delta t}^{(m+1)\Delta t} (\tau - m \Delta t) \tilde{\chi}_e(\tau) \, d\tau
\tag{10.46a}
$$

$$
\Delta \tilde{\xi}_e^m \equiv \tilde{\xi}_e^m - \tilde{\xi}_e^{m+1}
\tag{10.46b}
$$

and the Ψ accumulator in Equation (10.16) is modified to take the form:

$$
\Psi^{n-1} = \overline{\mathscr{E}}^{n-1} \left(\Delta \tilde{\chi}_e^0 - \Delta \tilde{\xi}_e^0 \right) + \overline{\mathscr{E}}^{n-2} \Delta \tilde{\xi}_e^0 + e^{-\Delta t/\tau_0} \Psi^{n-2}
\tag{10.47}
$$

where we have assumed the form for Debye materials; a similar modification results for Lorentz or Drude materials, using their time domain susceptibilities. In the case of Lorentz media, Equation (10.47) will be modified to take the form:

$$
\Psi_c^{n-1} = \overline{\mathscr{E}}^{n-1} \left(\left[\Delta \tilde{\chi}_e^c \right]^0 - \left[\Delta \tilde{\xi}_e^c \right]^0 \right) + \overline{\mathscr{E}}^{n-2} \left[\Delta \tilde{\xi}_e^c \right]^0 + e^{(-\delta + j\omega_r)\Delta t} \Psi_c^{n-2}.
\tag{10.48}
$$

Note, of course, that the terms $\left[\Delta \tilde{\xi}_e^c \right]^0$ and $\left[\Delta \tilde{\chi}_e^c \right]^0$ are different (and complex) for the Lorentz case compared to Debye materials, and that in updating the electric field equation equivalent to Equation (10.14), the real part of Ψ_c^n must be taken.

10.2 Auxiliary differential equation method

For obvious reasons, the technique described in the previous section for FDTD modeling of dispersive materials is known as the recursive convolution method. While this method is quite efficient, an alternative technique, known as the auxiliary differential equation or ADE method, has also been used with success [12, 13] and is also applicable to nonlinear materials. We now briefly discuss this alternative method.

The basis for the auxiliary differential equation method is to express the relationship between $\overline{\mathscr{D}}$ and $\overline{\mathscr{E}}$ with a differential equation rather than a convolution integral. This is achieved by inverse Fourier transformation of the relationship in Equation (10.2) between $\mathbf{D}(\omega)$ and $\mathbf{E}(\omega)$ into the time domain, rather than inverse transforming the susceptibility function $\chi_e(\omega)$. In doing this, we take advantage of the particular functional form of $\chi_e(\omega)$, and transform back to the time domain by recognizing multiplication with $j\omega$ as a first-order time derivative and multiplication with $-\omega^2 = (j\omega)^2$ as a second-order time derivative. We will illustrate the method separately for Debye, Lorentz, and Drude materials.

10.2.1   Debye materials

For Debye materials, we use Equation (10.10) for the material permittivity in Equation (10.2) to write

$$\mathbf{D} = \epsilon_0 \epsilon_\infty \mathbf{E} + \frac{\epsilon_0 (\epsilon_{dc} - \epsilon_\infty)}{1 + j\omega \tau_0} \mathbf{E}$$

$$(1 + j\omega \tau_0) \mathbf{D} = \epsilon_0 \epsilon_{dc} \mathbf{E} + j\omega \tau_0 \epsilon_0 \epsilon_\infty \mathbf{E}$$

$$\mathbf{D} + j\omega \tau_0 \mathbf{D} = \epsilon_0 \epsilon_{dc} \mathbf{E} + j\omega \tau_0 \epsilon_0 \epsilon_\infty \mathbf{E}. \tag{10.49}$$

Taking the inverse Fourier transform of Equation (10.49), we find:

$$\overline{\mathcal{D}} + \tau_0 \frac{\partial \overline{\mathcal{D}}}{\partial t} = \epsilon_0 \epsilon_{dc} \overline{\mathcal{E}} + \tau_0 \epsilon_0 \epsilon_\infty \frac{\partial \overline{\mathcal{E}}}{\partial t}. \tag{10.50}$$

This differential equation can be discretized in the usual manner to find an update equation for $\overline{\mathcal{E}}$:

$$\overline{\mathcal{E}}^{n+1} = \left[\frac{\Delta t + 2\tau_0}{2\tau_0 \epsilon_0 \epsilon_\infty + \epsilon_0 \epsilon_{dc} \Delta t} \right] \overline{\mathcal{D}}^{n+1}$$

$$+ \left[\frac{\Delta t - 2\tau_0}{2\tau_0 \epsilon_0 \epsilon_\infty + \epsilon_0 \epsilon_{dc} \Delta t} \right] \overline{\mathcal{D}}^{n} + \left[\frac{2\tau_0 \epsilon_0 \epsilon_\infty - \epsilon_0 \epsilon_{dc} \Delta t}{2\tau_0 \epsilon_0 \epsilon_\infty + \epsilon_0 \epsilon_{dc} \Delta t} \right] \overline{\mathcal{E}}^{n}. \tag{10.51}$$

As usual, since there is no dispersion in the magnetic permeability,[4] we still have the same update equation for $\overline{\mathcal{H}}$:

$$\overline{\mathcal{H}}^{n+1/2} = \overline{\mathcal{H}}^{n-1/2} - \frac{\Delta t}{\mu} \left[\nabla \times \overline{\mathcal{E}} \right]^{n}. \tag{10.52}$$

Thus, at each time step, we use Equation (10.52) to find $\overline{\mathcal{H}}^{n+1/2}$, and use it to find the next value of $\overline{\mathcal{D}}$ from:

$$\frac{\partial \overline{\mathcal{D}}}{\partial t} = \nabla \times \overline{\mathcal{H}} \qquad \rightarrow \qquad \overline{\mathcal{D}}^{n+1} = \overline{\mathcal{D}}^{n} - \Delta t \left[\nabla \times \overline{\mathcal{H}} \right]^{n+1/2}. \tag{10.53}$$

Finally, we then use $\overline{\mathcal{D}}^{n+1}$ along with $\overline{\mathcal{D}}^{n}$ and $\overline{\mathcal{E}}^{n}$ in Equation (10.51) to find the next value of $\overline{\mathcal{E}}^{n+1}$. This method of FDTD modeling of dispersive materials is more straightforward than the recursive convolution method, but does require that we store values of $\overline{\mathcal{D}}$ (at time steps n and $n + 1$) in addition to $\overline{\mathcal{E}}$ and $\overline{\mathcal{H}}$.

10.2.2   Formulation for multiple Debye poles

In many realistic materials, the dispersion properties can be described by a single Debye pole or a single Lorentz pole pair (i.e., a single resonance), but only in a narrow frequency range of interest. In a wider frequency spectrum, numerous poles may exist, as shown

---

[4] Dispersion in the magnetic permeability can be easily handled by the ADE method, requiring a similar auxiliary equation relating $\mathcal{B}$ and $\mathcal{H}$, similar to the methods used for the PML in Chapter 9.

for water in Figure 2.1. In fact, some metals, such as tungsten or nickel, are well modeled by a single Drude pole in combination with as many as *four* Lorentz poles [5].

A major advantage of the auxiliary differential equation (ADE) method is its simplicity in modeling materials with multiple poles. The recursive convolution method, on the other hand, becomes cumbersome for FDTD modeling in materials with susceptibility $\chi_e(\omega)$ having multiple poles. For these multiple-pole materials, an efficient reformulation of the ADE method has been put forth [14], which we present next.

We begin by rewriting the first equation in (10.49) in the following form:

$$
\mathbf{D} = \underbrace{\epsilon_0 \epsilon_\infty \mathbf{E}}_{\text{Free-space term}} + \underbrace{\frac{\epsilon_0 (\epsilon_{\text{dc}} - \epsilon_\infty)}{1 + j\omega\tau_0} \mathbf{E}}_{\text{Polarization term}}.
\tag{10.54}
$$

Using Equation (10.54) in Ampère's law we have:

$$
\nabla \times \mathbf{H} = \underbrace{j\omega\epsilon_0\epsilon_\infty \mathbf{E}}_{\text{Free-space displacement current}} + \underbrace{j\omega\frac{\epsilon_0 (\epsilon_{\text{dc}} - \epsilon_\infty)}{1 + j\omega\tau_0} \mathbf{E}}_{\text{Polarization current } \mathbf{J}_p(\omega)}
\tag{10.55}
$$

which corresponds to the time domain equation:

$$
\nabla \times \overline{\mathcal{H}} = \epsilon_0 \epsilon_\infty \frac{\partial \overline{\mathcal{E}}(t)}{\partial t} + \overline{\mathcal{J}}_p(t).
\tag{10.56}
$$

For materials having susceptibility $\chi_e(\omega)$ with multiple (e.g., a total of M) poles, we can express the permittivity in the frequency domain as:

$$
\epsilon(\omega) = \epsilon_0 \left(\epsilon_\infty + \sum_{m=1}^{M} \frac{\Delta\epsilon_m}{1 + j\omega\tau_{0m}} \right)
\tag{10.57}
$$

where $\Delta\epsilon_m = \epsilon_{\text{dc}}^m - \epsilon_\infty^m$, representing the change in the real part of the relative permittivity in the vicinity of the mth Debye pole, specified by the relaxation time τ_{0m}. The polarization current (given by $\overline{\mathcal{J}}_p(t) = \partial\overline{\mathcal{P}}/\partial t$) corresponding to the mth pole would be given by:

$$
\mathbf{J}_{pm}(\omega) = j\omega\epsilon_0 \left(\frac{\Delta\epsilon_m}{1 + j\omega\tau_{0m}} \right) \mathbf{E}(\omega).
\tag{10.58}
$$

The corresponding Maxwell's Equation (10.56) can then be rewritten as:

$$
\nabla \times \overline{\mathcal{H}} = \epsilon_0 \epsilon_\infty \frac{\partial \overline{\mathcal{E}}(t)}{\partial t} + \sum_{m=1}^{M} \overline{\mathcal{J}}_{pm}(t)
\tag{10.59}
$$

where $\overline{\mathcal{J}}_{pm}(t)$ is the time domain counterpart of $\mathbf{J}_{pm}(\omega)$. As we update the fields with Equation (10.59), we will need the values of $\overline{\mathcal{J}}_{pm}(t)$ at $(n + 1/2)$. For this purpose, we need to transform (10.58) to the time domain. To do this, we first multiply both sides by $(1 + j\omega\tau_{0m})$ to find:

$$
\mathbf{J}_{pm}(\omega) + j\omega\tau_{0m}\mathbf{J}_{pm}(\omega) = \epsilon_0 \Delta\epsilon_m j\omega\mathbf{E}(\omega)
\tag{10.60}
$$

which transforms into the time domain as:

$$\overline{\mathcal{J}}_{pm}(t) + \tau_{0m} \frac{d\overline{\mathcal{J}}_{pm}(t)}{dt} = \epsilon_0 \Delta\epsilon_m \frac{\partial\overline{\mathcal{E}}(t)}{\partial t}. \tag{10.61}$$

This equation is now our auxiliary differential equation analogous to Equation (10.42), and can be discretized to obtain the update equation for $\overline{\mathcal{J}}_{pm}^{n+1}$:

$$\overline{\mathcal{J}}_{pm}\Big|^{n+1} = \underbrace{\left(\frac{2\tau_{0m} - \Delta t}{2\tau_{0m} + \Delta t}\right)}_{k_m}\overline{\mathcal{J}}_{pm}\Big|^{n} + \underbrace{\left(\frac{2\epsilon_0\Delta\epsilon_m\Delta t}{2\tau_{0m} + \Delta t}\right)}_{\beta_m}\left[\frac{\overline{\mathcal{E}}\Big|^{n+1} - \overline{\mathcal{E}}\Big|^{n}}{\Delta t}\right]. \tag{10.62}$$

However, in solving Equation (10.59), we shall need $\overline{\mathcal{J}}_{pm}^{n+1/2}$, which can be obtained by averaging in time:

$$\overline{\mathcal{J}}_{pm}\Big|^{n+1/2} = \frac{\overline{\mathcal{J}}_{pm}\Big|^{n+1} + \overline{\mathcal{J}}_{pm}\Big|^{n}}{2}$$

$$= \left(\frac{2\tau_{0m}}{2\tau_{0m} + \Delta t}\right)\overline{\mathcal{J}}_{pm}\Big|^{n} + \left(\frac{\epsilon_0\Delta\epsilon_m\Delta t}{2\tau_{0m} + \Delta t}\right)\left[\frac{\overline{\mathcal{E}}\Big|^{n+1} - \overline{\mathcal{E}}\Big|^{n}}{\Delta t}\right]. \tag{10.63}$$

This can now be used in Equation (10.59) to find $\overline{\mathcal{E}}^{n+1}$. We have:

$$\left[\nabla \times \overline{\mathcal{H}}\right]^{n+1/2} = \epsilon_0\epsilon_\infty\left[\frac{\overline{\mathcal{E}}^{n+1} - \overline{\mathcal{E}}^{n}}{\Delta t}\right] + \sum_{m=1}^{M}\overline{\mathcal{J}}_{pm}^{n+1/2}. \tag{10.64}$$

Note that in Equation (10.63), $\overline{\mathcal{J}}_{pm}^{n+1/2}$ is a function of $\overline{\mathcal{E}}^{n+1}$, $\overline{\mathcal{E}}^{n}$, and $\overline{\mathcal{J}}_{pm}^{n}$. Thus, we can substitute Equation (10.63) into Equation (10.64) and rearrange to find our update equation for $\overline{\mathcal{E}}^{n+1}$:

$$\overline{\mathcal{E}}^{n+1} = \overline{\mathcal{E}}^{n} + \left(\frac{2\Delta t}{2\epsilon_0\epsilon_\infty + \sum_{m=1}^{M}\beta_m}\right)\left[\left[\nabla \times \overline{\mathcal{H}}\right]^{n+1/2} - \frac{1}{2}\sum_{m=1}^{M}(1 + k_m)\overline{\mathcal{J}}_{pm}^{n}\right] \tag{10.65}$$

with the constants k_m and β_m given in Equation (10.62). Thus, we find the overall procedure for updating the fields is as follows:

$$\overline{\mathcal{E}}^{n}, \overline{\mathcal{J}}^{n}, \overline{\mathcal{H}}^{n+1/2} \xrightarrow{(10.65)} \overline{\mathcal{E}}^{n+1} \xrightarrow{(10.62)} \overline{\mathcal{J}}^{n+1} \tag{10.66}$$

after which we use Faraday's law, i.e.,

$$\left[\nabla \times \overline{\mathcal{E}}\right]^{n+1} = -\mu\left[\frac{\overline{\mathcal{H}}^{n+3/2} - \overline{\mathcal{H}}^{n+1/2}}{\Delta t}\right] \tag{10.67}$$

to find $\overline{\mathcal{H}}^{n+3/2}$ and proceed with the updating.

Note that the ADE method as described here, and in the next two subsections for Lorentz and Drude materials, can be applied to *any number* of poles. Equivalently, this means it can handle any number of polarization currents, if one considers the formulation given in Equation (10.65), with the summation over M polarization currents $\overline{\mathscr{J}}_{pm}$.

10.2.3 Lorentz materials

Lorentz materials similarly lend themselves well to the use of the auxiliary differential equation method, thanks to the simple form of the frequency domain susceptibility. Recall the form of the complex frequency domain permittivity in Lorentz materials from earlier:

$$\epsilon_c(\omega) = \epsilon_0\,\epsilon_\infty + \underbrace{\frac{\epsilon_0\,(\epsilon_{dc} - \epsilon_\infty)\,\omega_0^2}{\omega_0^2 + 2j\omega\delta - \omega^2}}_{\chi_e(\omega)}. \tag{10.24}$$

We will find that the ADE method for a single Lorentz pole-pair is simply the $M = 1$ case of the multiple-pole-pair method, and so we will generalize to M poles from the outset. For a Lorentz material with M pole-pairs, the mth pole-pair is defined by a resonant frequency ω_{0m} and a damping constant δ_m, as well as relative permittivities ϵ_{dc}^m and ϵ_∞^m. The full complex permittivity can then be written in the form:

$$\epsilon_c(\omega) = \epsilon_0\,\epsilon_\infty + \epsilon_0 \sum_{m=1}^{M} \underbrace{\frac{\Delta\epsilon_m\,\omega_{0m}^2}{\omega_{0m}^2 + 2j\omega\delta_m - \omega^2}}_{\chi_{e,m}(\omega)} \tag{10.68}$$

where again $\Delta\epsilon_m = \epsilon_{dc}^m - \epsilon_\infty^m$. Now, recall that for a single Debye pole, our auxiliary differential equation was a complex frequency domain relationship between $\mathbf{D}$ and $\mathbf{E}$, given by Equation (10.49). We then found that for multiple Debye poles, it was more practical to write a relationship between $\mathbf{E}$ and the polarization current $\mathbf{J}_p$. In the Lorentz case, we will find that it is more practical to write the relationship between $\mathbf{E}$ and $\mathbf{J}_p$ even for a single pole-pair.

Since $\overline{\mathscr{J}}_p(t) = \partial\overline{\mathscr{P}}/\partial t$ and thus $\mathbf{J}_p(\omega) = j\omega\mathbf{P}$, we can write the form of a polarization current for each pole-pair $\mathbf{J}_{pm}$ as:

$$\mathbf{J}_{pm} = \epsilon_0\Delta\epsilon_m\omega_{0m}^2\left(\frac{j\omega}{w_{0m}^2 + 2j\omega\delta_m - \omega^2}\right)\mathbf{E}. \tag{10.69}$$

Rearranging to remove the fraction:

$$w_{0m}^2\mathbf{J}_{pm} + 2j\omega\delta_m\mathbf{J}_{pm} - \omega^2\mathbf{J}_{pm} = \epsilon_0\Delta\epsilon_m\omega_{0m}^2 j\omega\mathbf{E}. \tag{10.70}$$

This equation can now be easily transformed back into the time domain by taking the inverse Fourier transform to find:

$$\omega_{0m}^2\,\overline{\mathscr{J}}_{pm} + 2\delta_m\frac{\partial\overline{\mathscr{J}}_{pm}}{\partial t} + \frac{\partial^2\overline{\mathscr{J}}_{pm}}{\partial t^2} = \epsilon_0\Delta\epsilon_m\omega_{0m}^2\frac{\partial\overline{\mathscr{E}}}{\partial t}. \tag{10.71}$$

We can now discretize using the standard second-order centered differences for both first and second derivatives, to find:

$$\omega_{0m}^2 \overline{\mathcal{J}}_{pm}^n + 2\delta_m \left(\frac{\overline{\mathcal{J}}_{pm}^{n+1} - \overline{\mathcal{J}}_{pm}^{n-1}}{2\Delta t} \right) + \left(\frac{\overline{\mathcal{J}}_{pm}^{n+1} - 2\overline{\mathcal{J}}_{pm}^n + \overline{\mathcal{J}}_{pm}^{n-1}}{\Delta t^2} \right)$$

$$= \epsilon_0 \Delta \epsilon_m \omega_{0m}^2 \left(\frac{\overline{\mathcal{E}}_m^{n+1} - \overline{\mathcal{E}}_m^{n-1}}{2\Delta t} \right). \tag{10.72}$$

Notice the use of second-order differences over two time steps, in order to use only field values at known locations (i.e., integer time steps for both $\overline{\mathcal{E}}$ and $\overline{\mathcal{J}}$). This can be solved for $\overline{\mathcal{J}}_{pm}^{n+1}$ to yield a useable update equation:

$$\overline{\mathcal{J}}_{pm}^{n+1} = A_{1m} \overline{\mathcal{J}}_{pm}^n + A_{2m} \overline{\mathcal{J}}_{pm}^{n-1} + A_{3m} \left(\frac{\overline{\mathcal{E}}_m^{n+1} - \overline{\mathcal{E}}_m^{n-1}}{2\Delta t} \right) \tag{10.73}$$

with the constants A_m given by

$$A_{1m} = \frac{2 - \omega_{0m}^2 \Delta t^2}{1 + \delta_m \Delta t} \quad ; \quad A_{2m} = \frac{\delta_m \Delta t - 1}{\delta_m \Delta t + 1} \quad ; \quad A_{3m} = \frac{\epsilon_0 \Delta \epsilon_m \omega_{0m}^2 \Delta t^2}{1 + \delta_m \Delta t}.$$

Now, once again, we are left with trying to solve (10.59), which when discretized around time step $n + 1/2$ requires the polarization currents $\overline{\mathcal{J}}_m^{n+1/2}$. As usual, these field values are found by averaging in time:

$$\overline{\mathcal{J}}_{pm}^{n+1/2} = \frac{1}{2} \left(\overline{\mathcal{J}}_{pm}^{n+1} + \overline{\mathcal{J}}_{pm}^n \right)$$

$$= \frac{1}{2} \left[(1 + A_{1m}) \overline{\mathcal{J}}_{pm}^n + A_{2m} \overline{\mathcal{J}}_{pm}^{n-1} + A_{3m} \left(\frac{\overline{\mathcal{E}}_m^{n+1} - \overline{\mathcal{E}}_m^{n-1}}{2\Delta t} \right) \right]. \tag{10.74}$$

This can now be plugged directly into Equation (10.59) to find a Lorentz equivalent to Equation (10.64), and finally manipulated into an update equation to find $\overline{\mathcal{E}}^{n+1}$:

$$\overline{\mathcal{E}}^{n+1} = C_1 \overline{\mathcal{E}}^n + C_2 \overline{\mathcal{E}}^{n-1} + C_3 \left[\left[\nabla \times \overline{\mathcal{H}} \right]^{n+1/2} - \frac{1}{2} \sum_{m=1}^{M} \left((1 + A_{1m}) \overline{\mathcal{J}}_{pm}^n + A_{2m} \overline{\mathcal{J}}_{pm}^{n-1} \right) \right]$$

$$\tag{10.75}$$

where now the C constants are given by

$$C_1 = \frac{2\epsilon_0 \epsilon_\infty}{2\epsilon_0 \epsilon_\infty + \frac{1}{2} \sum A_{3m}} \quad ; \quad C_2 = \frac{\frac{1}{2} \sum A_{3m}}{2\epsilon_0 \epsilon_\infty + \frac{1}{2} \sum A_{3m}} \quad ; \quad C_3 = \frac{2\Delta t}{2\epsilon_0 \epsilon_\infty + \frac{1}{2} \sum A_{3m}}$$

where the sum as usual runs from $m = 1$ to $m = M$, the total number of poles.

Note that in the Lorentz formulation, more field storage is required as a direct consequence of the pole-pairs (i.e., both $\overline{\mathcal{E}}^n$, $\overline{\mathcal{E}}^{n-1}$, $\overline{\mathcal{J}}^n$, and $\overline{\mathcal{J}}^{n-1}$ are required), whereas the Debye formulation with a single pole did not require these prior fields. Specifically, the updating of Ampére's law proceeds through the steps:

$$\overline{\mathscr{E}}^n, \overline{\mathscr{E}}^{n-1}, \overline{\mathscr{J}}^n, \overline{\mathscr{J}}^{n-1}, \overline{\mathscr{H}}^{n+1/2} \xrightarrow{(10.75)} \overline{\mathscr{E}}^{n+1} \xrightarrow{(10.73)} \overline{\mathscr{J}}^{n+1} \tag{10.76}$$

Again, Faraday's law proceeds as usual, given in Equation (10.67), to update the magnetic field.

10.2.4 Drude materials

Finally, at the risk of repetitiveness, we discuss the ADE method applied to Drude materials. From the previous section, the complex susceptibility of a single-pole Drude material is modeled by:

$$\chi_e(\omega) = -\frac{\omega_r^2}{\omega^2 - j\omega/\tau_r}. \tag{10.32}$$

In the case of a multiple-pole Drude material, the permittivity can be written as

$$\epsilon_c(\omega) = \epsilon_0\epsilon_\infty - \epsilon_0 \sum_{m=1}^{M} \frac{\omega_{rm}^2}{\omega^2 - j\omega/\tau_{rm}}. \tag{10.77}$$

Again, since $\mathbf{J}_p(\omega) = j\omega\mathbf{P}$, we can write the form of a polarization current for each pole $\mathbf{J}_{pm}$ as:

$$\mathbf{J}_{pm} = -j\omega\epsilon_0 \left(\frac{\omega_{rm}^2}{\omega^2 - j\omega/\tau_{rm}} \right) \mathbf{E}. \tag{10.78}$$

At this point the reader should notice the pattern in deriving these auxiliary differential equations; after writing the complex frequency domain relationship between $\mathbf{E}$ and $\mathbf{J}_p$, we rearrange to form an equation that, when transformed back into the time domain, can be easily discretized using the FDTD method. In the Drude case, we rearrange Equation (10.78) to find the frequency domain equation:

$$\omega^2\mathbf{J}_{pm} - \frac{j\omega}{\tau_{rm}}\mathbf{J}_{pm} = -j\omega\epsilon_0\omega_{rm}^2\mathbf{E} \tag{10.79}$$

which can be transformed into the time domain as before. In this case, there is a common factor of $(-j\omega)$ (recalling that $\omega^2 = (j\omega) \cdot (-j\omega)$) which can be removed from the equation before transforming to the time domain. Ultimately we find:

$$\frac{\partial \overline{\mathscr{J}}_{pm}}{\partial t} + \frac{1}{\tau_{rm}}\overline{\mathscr{J}}_{pm} = \epsilon_0\omega_{rm}^2\overline{\mathscr{E}}. \tag{10.80}$$

This equation can be discretized as usual at the $n + 1/2$ time step, using time averages to get the $\overline{\mathscr{E}}$ and $\overline{\mathscr{J}}$ field quantities at the half time step:

$$\left(\frac{\overline{\mathscr{J}}_{pm}^{n+1} - \overline{\mathscr{J}}_{pm}^n}{\Delta t} \right) + \frac{1}{\tau_{rm}} \left(\frac{\overline{\mathscr{J}}_{pm}^{n+1} + \overline{\mathscr{J}}_{pm}^n}{2} \right) = \epsilon_0\omega_{rm}^2 \left(\frac{\overline{\mathscr{E}}^{n+1} + \overline{\mathscr{E}}^n}{2} \right) \tag{10.81}$$

which yields the update equation for $\overline{\mathcal{J}}_{pm}^{n+1}$:

$$\overline{\mathcal{J}}_{pm}^{n+1} = \underbrace{\left(\frac{2 - \Delta t/\tau_{rm}}{2 + \Delta t/\tau_{rm}}\right)}_{\alpha_m} \overline{\mathcal{J}}_{pm}^{n} + \underbrace{\left(\frac{w_{rm}^2 \epsilon_0 \Delta t}{2 + \Delta t/\tau_{rm}}\right)}_{\beta_m} \left(\frac{\overline{\mathcal{E}}^{n+1} + \overline{\mathcal{E}}^{n}}{2}\right). \tag{10.82}$$

Once again, as for the Debye and Lorentz materials, we need $\overline{\mathcal{J}}_{pm}^{n+1/2}$ to use in Ampère's law (10.59), which can be found by simple averaging, as was done earlier for the Debye and Lorentz materials:

$$\overline{\mathcal{J}}_{pm}^{n+1/2} = \frac{1}{2}\left(\overline{\mathcal{J}}_{pm}^{n+1} + \overline{\mathcal{J}}_{pm}^{n}\right)$$

$$= \frac{1}{2}\left[(1 + \alpha_m)\overline{\mathcal{J}}_{pm}^{n} + \beta_m\left(\frac{\overline{\mathcal{E}}^{n+1} + \overline{\mathcal{E}}^{n}}{2}\right)\right] \tag{10.83}$$

with the constants α_m and β_m given in (10.82). This can now be substituted into Ampère's law (10.59) and manipulated to find an update equation for $\overline{\mathcal{E}}$, similar to Equation (10.65) for Debye materials or Equation (10.75) for Lorentz materials:

$$\overline{\mathcal{E}}^{n+1} = \left(\frac{2\epsilon_0\epsilon_\infty - \Delta t \sum \beta_m}{2\epsilon_0\epsilon_\infty + \Delta t \sum \beta_m}\right)\overline{\mathcal{E}}^{n}$$

$$+ \left(\frac{2\Delta t}{2\epsilon_0\epsilon_\infty + \sum \beta_m}\right)\left[\left[\nabla \times \overline{\mathcal{H}}\right]^{n+1/2} - \frac{1}{2}\sum_{m=1}^{M}(1 + \alpha_m)\overline{\mathcal{J}}_{pm}^{n}\right] \tag{10.84}$$

where, as usual, all summations run from $m = 1$ to $m = M$, the number of Drude poles. Note the similarity of this equation with Equation (10.65) for Debye materials; there is an additional scaling factor in front of $\overline{\mathcal{E}}^{n}$, but otherwise the form is identical. In a similar vein, the updating process for Debye materials follows the same steps as described in Equation (10.66).

The formulations for Debye, Lorentz, and Drude media derived in this section assumed, for simplicity, that the medium had zero conductivity σ. It is a simple extension to derive formulations that include the conductivity of the dispersive medium, and the modification to the equations follows similar lines as the modification from free space to a simple lossy medium. For example, observe the derivation above for the isotropic plasma, in which case the dispersion was separated into two parts, one of which resulted in nonzero conductivity σ_{eff}.

10.3 Summary

In this chapter we introduced FDTD modeling of *dispersive* media, which are those materials whose permittivity $\epsilon(\omega)$ and/or permeability $\mu(\omega)$ are functions of frequency ω. In reality, all materials are dispersive in a wider sense, but many materials can be treated approximately as frequency-independent over narrow bandwidths of interest.

We introduced four different types of dispersive materials. First, Debye materials are those whose electromagnetic behavior arises from *orientational* polarization of the molecules in the medium; a common example is water, which is a dipolar liquid. The susceptibility of a Debye material can be described by:

$$\chi_e(\omega) = \frac{\tilde{\chi}_e(0)\tau_0}{1 + j\omega\tau_0}.$$ (Debye dispersion)

Second, Lorentz materials are those whose behavior arises from *electronic* or *ionic* polarization, where the electrons or ions are displaced from the molecules to which they are bounded. The susceptibility of Lorentz materials is described by:

$$\chi_e(\omega) = \frac{\tilde{\chi}_e(0)\omega_r}{(\delta^2 + \omega_r^2) + 2j\omega\delta - \omega^2}.$$ (Lorentz dispersion)

Third, we discussed the Drude model of dispersion, which applies very well to most metals, whose free electrons respond easily to applied electromagnetic fields. The susceptibility of a Drude material is given by:

$$\chi_e(\omega) = -\frac{\omega_r^2}{\omega^2 - j\omega/\tau_r}.$$ (Drude dispersion)

Fourth, we introduced the isotropic plasma as a dispersive material. In an isotropic plasma, free electrons and ions respond to applied electromagnetic fields, but their movement is damped by collisions with the neutral molecules. The permittivity of an isotropic plasma is given by:

$$\epsilon_c(\omega) = \epsilon_0 \left[1 + \chi_e(\omega)\right] = \epsilon_0 \left[1 - \frac{\omega_p^2}{\omega(\omega + j\,\nu_c)}\right].$$ (Isotropic plasma)

To model these dispersive materials in the FDTD method, we discussed two approaches: namely, the recursive convolution (RC) method, and the auxiliary differential equation (ADE) method. In the RC method, we make note of the fact that the frequency domain forms of the susceptibilities above easily transform into the time domain. Then, the relationship $\overline{\mathcal{D}} = \epsilon\overline{\mathcal{E}}$ becomes a convolution problem involving the time-dependent permittivity. For example, in Debye materials:

$$\overline{\mathcal{D}}^n = \epsilon_0 \left[\epsilon_\infty \overline{\mathcal{E}}^n + \sum_{m=0}^{n-1} \overline{\mathcal{E}}^{n-m} \int_{m\Delta t}^{(m+1)\Delta t} \tilde{\chi}_e(\tau)\,d\tau\right].$$ (10.12a)

By applying this equation (and its equivalent at time $n - 1$) to Ampère's law, we find an update equation for $\overline{\mathcal{E}}^n$ that involves a convolution integral term, Ψ, which has its own update equation.

By contrast, the ADE method involves writing an equation relating $\overline{\mathcal{E}}$ and the polarization current $\overline{\mathcal{J}}_p$, and discretizing that relationship to find an update equation for $\overline{\mathcal{J}}_p$. For example, in Lorentz materials, the frequency domain relationship between the polarization current and $\overline{\mathcal{E}}$ is:

$$\mathbf{J}_p = \epsilon_0 \Delta\epsilon\omega_0^2 \left(\frac{j\omega}{w_0^2 + 2j\omega\delta - \omega^2}\right)\mathbf{E}$$ (10.69)

which can now be easily transformed back into the time domain:

$$\omega_0^2 \overline{\mathcal{I}}_p + 2\delta \frac{\partial \overline{\mathcal{I}}_p}{\partial t} + \frac{\partial^2 \overline{\mathcal{I}}_p}{\partial t^2} = \epsilon_0 \Delta \epsilon \omega_0^2 \frac{\partial \overline{\mathcal{E}}}{\partial t} \tag{10.71}$$

This equation is then easily discretized and rearranged to find an update equation for $\overline{\mathcal{I}}_p$, and the updated $\overline{\mathcal{I}}_p$ field is then used in the update equation for $\overline{\mathcal{E}}$. Note that for multiple Lorentz pole-pairs, or multiple Debye poles, there is an update equation for each $\overline{\mathcal{I}}_{pm}$.

10.4   Problems

10.1.   **A Lorentz medium with finite conductivity.** Derive the electric field update equation (analogous to Equation 10.14) for a single-pole ADE Lorentz medium with finite conductivity σ.

10.2.   **Combined Drude / Lorentz medium.** A large number of metals, including nickel, silver, and platinum [5], are modeled with a combined Drude / Lorentz model of the susceptibility. In these cases, the complex frequency domain susceptibility is given by:

$$\chi_e(\omega) = \frac{\omega_r^2}{\omega(j\nu - \omega)} + \sum_{m=1}^{M} \frac{A_m}{\omega_m^2 + 2j\omega\delta_m - \omega^2}.$$

Use this susceptibility equation to derive an FDTD ADE formulation for such a medium.

10.3.   **Platinum simulation.** Write an FDTD code to simulate a 1D plane wave incident on a slab of silver using the ADE method. Launch a sinusoidal source from the left edge of the space with a wavelength of 700 nm (i.e., a red optical wave). Choose appropriate FDTD parameters (Δx, Δt) for this wavelength. Create a slab of silver in the space that is one-half wavelength thick.
   (a)   First, treat silver as a simple Drude material with $\omega_r = 1.33 \times 10^{16}$ rad/s and $\nu = 4.94 \times 10^{13}$ s^{-1} (in the notation of Problem 10.2 above). What amplitude is transmitted through the slab? What is reflected? What are the losses inside the slab?
   (b)   In addition to the Drude pole in (a), include the single Lorentz pole-pair for silver, with $A_1 = 3.09 \times 10^{23}$ s^{-1}, $\delta_1 = 1.62 \times 10^{18}$ s^{-1}, and $w_1 = 3.14 \times 10^{16}$ rad/s.
   (c)   Vary the input optical wavelength over the range 300–1000 nm. How does the transmitted amplitude vary with wavelength?

10.4.   **An isotropic plasma.** Modify your code from Problem 10.3 to model a slab of isotropic plasma. Use a plasma frequency of 1 MHz (corresponding to an electron density of $\sim 10^4$ electrons/cm^3) and a collision frequency of $\nu_c = 0$.

Launch a Gaussian pulse at the slab that covers the frequency range up to 2 MHz. Measure the frequency response of the reflected and transmitted pulses.

10.5. **Collisional plasma.** Repeat Problem 10.4, but use a collision frequency of $v_c = 10^4$ s^{-1}. Compare the reflected and transmitted frequency responses. Repeat for $v_c = 10^3$ and 10^5 s^{-1}.

10.6. **A DNG material.** Metamaterials are materials made up of periodic structures whose macroscopic behavior can be controlled through the material's design. A double negative (DNG) material is one whose permittivity and permeability both exhibit dispersion, so that at particular frequencies both of these, and thus the index of refraction, can be negative.

(a) Model a slab of Drude material in 2D with dispersive permittivity only, with $v = 1/\tau = 4 \times 10^{-4}$ s^{-1} and $\omega_p = 3 \times 10^{11}$ rad/s. Use 20 grid cells per wavelength and make the slab 100 grid cells thick. Launch a sinusoidal wave at the slab with a wavelength of 1 cm ($f = 30$ GHz) and observe the wave phase behavior in the slab.

(b) Derive the ADE method for a DNG material. Start with the μ version of equation (10.77) and write an auxiliary equation for the magnetization current **M** related to **H**. Then, include **M** in the update equation for **H**.

(c) Now, repeat (a) but make the slab a DNG material in 2D whose permittivity ϵ and permeability μ both exhibit Drude behavior, both with the same relaxation time and resonant frequency as in part (a). Comment on the results.

10.7. **Stability of the ADE method.** Investigate the stability of the ADE method in 1D for single-pole Debye materials, using the methods in Chapter 6. Start with Equation (10.51), writing each field in the form $\mathcal{D}_i^n = C_1 e^{j(\omega n \Delta t + k i \Delta x)}$, and simplify to find a dispersion relation. Under what conditions is the method stable?

10.8. **Submarine communication.** Consider the problem of communication between two submarines submersed at some depth at sea, 1 km from each other. Seawater can be modeled as a Debye material with parameters $\epsilon_{dc} = 81$, $\epsilon_\infty = 1$, and $t_0 = 10^{-5}$ s. Analyze the problem at hand using the recursive convolution method and make recommendations as to how best to proceed.

A critical parameter to determine is the pulse rate, which determines the data rate. In other words, what is the minimum separation between successive pulses? What is the highest data rate (number of pulses per second) which can be used? Assume 2D TM$_z$ propagation between the submarines and vertical electric antennas (i.e., excite the $\mathcal{E}_z$ field at the transmitting submarine and also monitor the $\mathcal{E}_z$ field at the receiving submarine) and that the transmitted Gaussian pulses can have a peak initial electric field value of 1 V/m. Also assume that the receiving system is designed to recognize a peak signal value of 0.5 mV/m or more as a one, whereas a value of 0.05 mV/m or less is recognized as a zero.

10.9. Speed and performance of the ADE method. Repeat Problem 10.8 using the ADE method. Is the same result achieved? How do the simulation times compare?

References

[1] R. Luebbers, F. P. Hunsberger, K. S. Kunz, R. B. Standler, and M. Schneider, "A frequency-dependent finite-difference time-domain formulation formulation dispersive materials," *IEEE Trans. on Electr. Comp.*, vol. 32, no. 3, pp. 222–227, 1990.

[2] R. P. Feynman, R. B. Leighton, and M. Sands, *The Feynman Lectures on Physics*. Addison Wesley, 1964.

[3] P. Debye, *Polar Molecules, Chapter V*. New York: Chemical Catalog Co., 1929.

[4] A. Beneduci, "Which is the effective time scale of the fast Debye relaxation process in water?" *J. Molec. Liquids*, vol. 138, pp. 55–60, 2007.

[5] W. H. P. Pernice, F. P. Payne, and D. F. G. Gallagher, "An FDTD method for the simulation of dispersive and metallic structures," *Opt. and Quant. Elec.*, vol. 38, pp. 843–856, 2007.

[6] W. A. Beck and M. S. Mirotznik, "Generalized analysis of stability and numerical dispersion in the discrete-convolution FDTD method," *IEEE Trans. Ant. Prop.*, vol. 48, pp. 887–894, 2000.

[7] D. R. Lide, ed., *CRC Handbook of Chemistry and Physics; Internet Version 2010*. CRC Press, 2010.

[8] U. S. Inan and A. S. Inan, *Engineering Electromagnetics*. Addison-Wesley, 1999.

[9] U. S. Inan and A. S. Inan, *Electromagnetic Waves*. Prentice-Hall, 2000.

[10] R. Luebbers, F. Hunsberger, and K. S. Kunz, "A frequency-dependent finite-difference time-domain formulation for transient propagation in plasma," *IEEE Trans. Ant. Prop.*, vol. 39, p. 29, 1991.

[11] D. F. Kelley and R. J. Luebbers, "Piecewise linear recursive convolution for dispersive media using FDTD," *IEEE Trans. Ant. Prop.*, vol. 44, pp. 792–797, 1996.

[12] T. Kashiwa and I. Fukai, "A treatment of the dispersive characteristics associated with electronic polarization," *Microwave Opt. Technol. Lett.*, vol. 3, p. 203, 1990.

[13] R. Joseph, S. Hagness, and A. Taflove, "Direct time integration of Maxwell's equations in linear dispersive media with absorption for scattering and propagation of femtosecond electromagnetic pulses," *Optics Lett.*, vol. 16, p. 1412, 1991.

[14] M. Okoniewski, M. Mrozowski, and M. A. Stuchly, "Simple treatment of multi-term dispersion in FDTD," *IEEE Microwave and Guided Wave Lett.*, vol. 7, pp. 121–123, 1997.

11 FDTD modeling in anisotropic media

In the previous chapter, we looked at the FDTD method applied to *dispersive* media, which are those materials whose permittivity (and/or permeability) is a function of frequency. In this chapter, we discuss materials whose permittivity (and/or permeability) is a function of the vector *direction* of the wave electric and magnetic fields.[1] These are referred to an *anisotropic* materials.

Electrically anisotropic materials can be understood as having a tensor permittivity, which can be written in the general form:

$$\bar{\bar{\epsilon}} = \epsilon_0 \begin{bmatrix} \epsilon_{xx} & \epsilon_{xy} & \epsilon_{xz} \\ \epsilon_{yx} & \epsilon_{yy} & \epsilon_{yz} \\ \epsilon_{zx} & \epsilon_{zy} & \epsilon_{zz} \end{bmatrix}. \tag{11.1}$$

This tensor permittivity can be interpreted in two ways. First, mathematically, as we shall see, the $\bar{\mathscr{D}}$ field components depend on cross terms of $\bar{\mathscr{E}}$ – for instance, $\mathscr{D}_x$ may depend on $\mathscr{E}_x$, $\mathscr{E}_y$, and/or $\mathscr{E}_z$ – and physically, this means that $\bar{\mathscr{D}}$ is no longer parallel to $\bar{\mathscr{E}}$. This follows directly from applying the matrix permittivity in Equation (11.1) to the relationship $\mathbf{D}(\omega) = \epsilon(\omega)\mathbf{E}(\omega)$. Second, and possibly more intuitively, it means that the index of refraction in the medium will be a function of the wave propagation direction and polarization of the wave. This interpretation will become more intuitive when we discuss modeling of liquid crystals.

In this chapter we will discuss the FDTD method applied to general anisotropic media; then, we will discuss particular examples of a magnetized plasma, a liquid crystal medium, and a ferrite material.

11.1 FDTD method in arbitrary anisotropic media

The following discussion is adapted from the work of Dou and Sebak [1]. This very simple method involves a four-step updating process involving the four fields **E**, **D**, **H**, and **B**, similar to the ADE method for dispersive materials. In the most general sense, an anisotropic medium may have a tensor permittivity (electric anisotropy) and/or a tensor

---

[1] Note that we define anisotropy as the dependence on the direction of the field vectors, and not on the wave propagation direction. In this latter case, the anisotropy can be thought of as a combination of the wave propagation direction and the wave polarization.

permeability (magnetic anisotropy), each with nine unique matrix entries:

$$\bar{\bar{\epsilon}} = \epsilon_0 \begin{bmatrix} \epsilon_{xx} & \epsilon_{xy} & \epsilon_{xz} \\ \epsilon_{yx} & \epsilon_{yy} & \epsilon_{yz} \\ \epsilon_{zx} & \epsilon_{zy} & \epsilon_{zz} \end{bmatrix} \quad \text{and} \quad \bar{\bar{\mu}} = \mu_0 \begin{bmatrix} \mu_{xx} & \mu_{xy} & \mu_{xz} \\ \mu_{yx} & \mu_{yy} & \mu_{yz} \\ \mu_{zx} & \mu_{zy} & \mu_{zz} \end{bmatrix}. \tag{11.2}$$

Working from the medium-independent versions of Maxwell's equations, we can arrange the following set of four equations:

$$\frac{\partial \overline{\mathcal{D}}}{\partial t} = \nabla \times \overline{\mathcal{H}} \tag{11.3a}$$

$$\overline{\mathcal{D}} = \bar{\bar{\epsilon}} \, \overline{\mathcal{E}} \quad \rightarrow \quad \overline{\mathcal{E}} = [\bar{\bar{\epsilon}}]^{-1} \overline{\mathcal{D}} \tag{11.3b}$$

$$\frac{\partial \overline{\mathcal{B}}}{\partial t} = -\nabla \times \overline{\mathcal{E}} \tag{11.3c}$$

$$\overline{\mathcal{B}} = \bar{\bar{\mu}} \, \overline{\mathcal{H}} \quad \rightarrow \quad \overline{\mathcal{H}} = [\bar{\bar{\mu}}]^{-1} \overline{\mathcal{B}} \tag{11.3d}$$

where we have assumed that the entries of $\bar{\bar{\epsilon}}$ and $\bar{\bar{\mu}}$ are not frequency dependent; otherwise, the conversion from the true linear relationship $\mathbf{D}(\omega) = \epsilon(\omega)\mathbf{E}(\omega)$ would require convolutions. We will see the effect of a frequency-dependent (i.e., dispersive) anisotropic medium when we discuss the magnetized plasma. Note that the four Equations (11.3) are written in order: beginning with the $\overline{\mathcal{H}}$ field, one would follow this sequence of equations in order to update all four fields.

The inverses of the tensor permittivity and permeability that appear in Equations (11.3) can be written as

$$[\bar{\bar{\epsilon}}]^{-1} = \frac{1}{\Lambda_e} \begin{bmatrix} \epsilon'_{xx} & \epsilon'_{xy} & \epsilon'_{xz} \\ \epsilon'_{yx} & \epsilon'_{yy} & \epsilon'_{yz} \\ \epsilon'_{zx} & \epsilon'_{zy} & \epsilon'_{zz} \end{bmatrix} \quad \text{and} \quad [\bar{\bar{\mu}}]^{-1} = \frac{1}{\Lambda_h} \begin{bmatrix} \mu'_{xx} & \mu'_{xy} & \mu'_{xz} \\ \mu'_{yx} & \mu'_{yy} & \mu'_{yz} \\ \mu'_{zx} & \mu'_{zy} & \mu'_{zz} \end{bmatrix} \tag{11.4}$$

where Λ_e and Λ_h represent the determinants of ϵ and μ respectively; for example,

$$\Lambda_e = \epsilon_0(\epsilon_{xx}\epsilon_{yy}\epsilon_{zz} + \epsilon_{xy}\epsilon_{yz}\epsilon_{zx} + \epsilon_{xz}\epsilon_{yx}\epsilon_{zy} - \epsilon_{xz}\epsilon_{yy}\epsilon_{zx} - \epsilon_{xy}\epsilon_{yx}\epsilon_{zz} - \epsilon_{xx}\epsilon_{yz}\epsilon_{zy}).$$

The primed entries such as ϵ'_{xx}, it should be recognized, are different from the given ϵ_{xx}, and are determined from the matrix inversion process. Now, applying these inverse tensors to Equation (11.3b), we find update equations for $\overline{\mathcal{D}} \rightarrow \overline{\mathcal{E}}$, explicitly written at the time step n and the appropriate discretized spatial coordinates:

$$\mathcal{E}_x\Big|^n_{i+1/2,j,k} = \frac{\epsilon'_{xx}}{\Lambda_e} \mathcal{D}_x\Big|^n_{i+1/2,j,k} + \frac{\epsilon'_{xy}}{\Lambda_e} \mathcal{D}_y\Big|^n_{i+1/2,j,k} + \frac{\epsilon'_{xz}}{\Lambda_e} \mathcal{D}_z\Big|^n_{i+1/2,j,k} \tag{11.5a}$$

$$\mathcal{E}_y\Big|^n_{i,j+1/2,k} = \frac{\epsilon'_{yx}}{\Lambda_e} \mathcal{D}_x\Big|^n_{i,j+1/2,k} + \frac{\epsilon'_{yy}}{\Lambda_e} \mathcal{D}_y\Big|^n_{i,j+1/2,k} + \frac{\epsilon'_{yz}}{\Lambda_e} \mathcal{D}_z\Big|^n_{i,j+1/2,k} \tag{11.5b}$$

$$\mathcal{E}_z\Big|^n_{i,j,k+1/2} = \frac{\epsilon'_{zx}}{\Lambda_e} \mathcal{D}_x\Big|^n_{i,j,k+1/2} + \frac{\epsilon'_{zy}}{\Lambda_e} \mathcal{D}_y\Big|^n_{i,j,k+1/2} + \frac{\epsilon'_{zz}}{\Lambda_e} \mathcal{D}_z\Big|^n_{i,j,k+1/2}. \tag{11.5c}$$

The problem with this set of equations, of course, is that $\overline{\mathcal{D}}$ field components are asked for at locations where they are not defined. For example, in Equation (11.5a) above, the

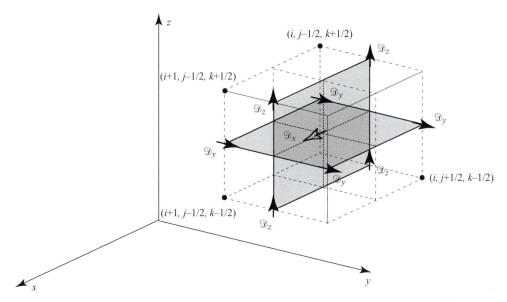

Figure 11.1 Locations of $\mathcal{D}_x$, $\mathcal{D}_y$, and $\mathcal{D}_z$ fields in the 3D grid. The solid arrows show the components of $\mathcal{D}_y$ and $\mathcal{D}_z$ that are used to find their values at the grid location $(i + 1/2, j, k)$ at the center of this cell.

$\mathcal{D}_x$ component is fine, because it is co-located with $\mathcal{E}_x$ at grid location $(i + 1/2, j, k)$. However, the $\mathcal{D}_y$ and $\mathcal{D}_z$ components are asked for at this location, but are known at grid points $(i, j + 1/2, k)$ and $(i, j, k + 1/2)$ respectively. The usual quick fix is applied: we take the average values across known spatial points to find the values of fields at desired locations. In Equation (11.5a), the $\mathcal{D}_y$ and $\mathcal{D}_z$ values are found by taking the average over four points surrounding the desired location, as shown in Figure 11.1:

$$\mathcal{D}_y\Big|_{i+1/2,j,k} = \frac{1}{4}\left(\mathcal{D}_y\Big|_{i,j-1/2,k} + \mathcal{D}_y\Big|_{i,j+1/2,k} + \mathcal{D}_y\Big|_{i+1,j-1/2,k} + \mathcal{D}_y\Big|_{i+1,j+1/2,k}\right)$$
$$(11.6a)$$

$$\mathcal{D}_z\Big|_{i+1/2,j,k} = \frac{1}{4}\left(\mathcal{D}_z\Big|_{i,j,k-1/2} + \mathcal{D}_z\Big|_{i,j,k+1/2} + \mathcal{D}_z\Big|_{i+1,j,k-1/2} + \mathcal{D}_z\Big|_{i+1,j,k+1/2}\right).$$
$$(11.6b)$$

In Figure 11.1, the location of interest $(i + 1/2, j, k)$ is at the center of the grid, where the hollow $\mathcal{D}_x$ arrow is located. The solid arrows surrounding it are the field values of $\mathcal{D}_y$ and $\mathcal{D}_z$ that are used to find their values at the location $(i + 1/2, j, k)$. Pairs of equations similar to Equations (11.6) apply to the terms in Equations (11.5b) and (11.5c).

Next, having found $\overline{\mathcal{E}}$, the updating proceeds as usual for Equation (11.3c), taking us from $\overline{\mathcal{E}}$ to $\overline{\mathcal{B}}$. Since this equation is not material-specific, it can be applied using the straightforward differencing equations that we have used throughout this book. Having found $\overline{\mathcal{B}}$ (which is co-located with $\overline{\mathcal{H}}$), we discretize Equation (11.3d) similarly for the anisotropic permeability, yielding the equations for $\overline{\mathcal{B}} \to \overline{\mathcal{H}}$:

$$
\mathcal{H}_x \Big|_{i,j+1/2,k+1/2}^{n+1/2} = \frac{\mu'_{xx}}{\Lambda_h} \mathcal{B}_x \Big|_{i,j+1/2,k+1/2}^{n+1/2} + \frac{\mu'_{xy}}{\Lambda_h} \mathcal{B}_y \Big|_{i,j+1/2,k+1/2}^{n+1/2} + \frac{\mu'_{xz}}{\Lambda_h} \mathcal{B}_z \Big|_{i,j+1/2,k+1/2}^{n+1/2}
$$
(11.7a)

$$
\mathcal{H}_y \Big|_{i+1/2,j,k+1/2}^{n+1/2} = \frac{\mu'_{yx}}{\Lambda_h} \mathcal{B}_x \Big|_{i+1/2,j,k+1/2}^{n+1/2} + \frac{\mu'_{yy}}{\Lambda_h} \mathcal{B}_y \Big|_{i+1/2,j,k+1/2}^{n+1/2} + \frac{\mu'_{yz}}{\Lambda_h} \mathcal{B}_z \Big|_{i+1/2,j,k+1/2}^{n+1/2}
$$
(11.7b)

$$
\mathcal{H}_z \Big|_{i+1/2,j+1/2,k}^{n+1/2} = \frac{\mu'_{zx}}{\Lambda_h} \mathcal{B}_x \Big|_{i+1/2,j+1/2,k}^{n+1/2} + \frac{\mu'_{zy}}{\Lambda_h} \mathcal{B}_y \Big|_{i+1/2,j+1/2,k}^{n+1/2} + \frac{\mu'_{zz}}{\Lambda_h} \mathcal{B}_z \Big|_{i+1/2,j+1/2,k}^{n+1/2}.
$$
(11.7c)

Again, we are asked for field values of $\overline{\mathcal{B}}$ at locations where they are not defined, and again we take four-point spatial averages. For the two terms in Equation (11.7a), the following two averaging equations are required:

$$
\mathcal{B}_y \Big|_{i,j+1/2,k+1/2} = \frac{1}{4} \Big(\mathcal{B}_y \Big|_{i+1/2,j+1,k+1/2} + \mathcal{B}_y \Big|_{i+1/2,j,k+1/2}
$$

$$
+ \mathcal{B}_y \Big|_{i-1/2,j+1,k+1/2} + \mathcal{B}_y \Big|_{i-1/2,j,k+1/2} \Big)
$$
(11.8a)

$$
\mathcal{B}_z \Big|_{i,j+1/2,k+1/2} = \frac{1}{4} \Big(\mathcal{B}_z \Big|_{i+1/2,j+1/2,k+1} + \mathcal{B}_z \Big|_{i+1/2,j+1/2,k}
$$

$$
+ \mathcal{B}_z \Big|_{i-1/2,j+1/2,k+1} + \mathcal{B}_z \Big|_{i-1/2,j+1/2,k} \Big).
$$
(11.8b)

Two similar averaging equations are required to go along with Equations (11.7b) and (11.7c). Finally, the updating scheme is completed by applying Equation (11.3a), which again is material independent and proceeds with straightforward difference equations.

For the general, nondispersive anisotropic medium, that's it! The updating process becomes a four-step procedure, involving Ampère's law and Faraday's law as usual, but also requiring the material constitutive relations, used as auxiliary equations.[2]

11.2   FDTD in liquid crystals

While the method described in Section 11.1 above is applicable in general to any tensor permeability and/or permittivity with simple (i.e., frequency-independent) entries, specific types of media lend themselves to simplified and/or alternative formulations. In this section we will discuss the FDTD analysis of liquid crystals, which, under certain circumstances, can be even simpler than the general formulation above. In the next section we will discuss FDTD in a magnetized plasma, which is both anisotropic and dispersive, and whose formulation is somewhat more complex.

---

[2] Note, however, these are not auxiliary *differential* equations.

We begin with some background on waves in crystals[3] in order to understand the basis of the anisotropy in crystals.

A general crystal has a tensor permittivity as in the general formulation above, but when aligned along the principal axes, this permittivity can have only diagonal components ϵ_{xx}, ϵ_{yy}, and ϵ_{zz}, with all off-diagonal components equal to zero. In this general crystal, the frequency domain wave equation can be written as:

$$\mathbf{k} \times (\mathbf{k} \times \mathbf{E}) + \omega^2 \mu_0 \bar{\bar{\epsilon}} \, \mathbf{E} = 0 \tag{11.9}$$

where now $\bar{\bar{\epsilon}}$ is a diagonal matrix; note that our entries ϵ_{xx}, etc., include ϵ_0 multiplied by the relative permittivities. This wave equation can be rewritten in matrix form, and the dispersion relation for the medium is found by setting the determinant of the multiplying matrix A to zero. We thus find the dispersion relation for the crystal:

$$\begin{vmatrix} \omega^2 \mu_0 \epsilon_{xx} - k_y^2 - k_z^2 & k_x k_y & k_x k_z \\ k_y k_x & \omega^2 \mu_0 \epsilon_{yy} - k_x^2 - k_z^2 & k_y k_z \\ k_z k_x & k_z k_y & \omega^2 \mu_0 \epsilon_{zz} - k_x^2 - k_y^2 \end{vmatrix} = 0. \tag{11.10}$$

This dispersion relation can be interpreted most easily when it is represented as a surface in 3D $\mathbf{k}$-space; this surface is known as the *wave-normal surface*, and consists of two shells which overlap. Lines drawn through the origin and the points where the shells intersect define the *optic axes* (there are two of them). In general, except along the optic axes, any direction of propagation will intersect the surface at two points (due to the two shells), meaning the material can support two phase velocities at any direction of propagation.

Now, $\hat{\mathbf{k}}$ is the direction of phase propagation, but in an anisotropic medium, where $\mathbf{D}$ and $\mathbf{E}$ are not parallel (as mentioned in the introduction to this chapter), the direction of *energy* propagation can be different.[4] The direction of energy propagation is given by $\hat{\mathbf{s}}$ (or equivalently, the group velocity vector $\mathbf{v}_g$), and $\mathbf{E}$ and $\mathbf{H}$ are always perpendicular to the direction of $\mathbf{s}$.

Using the plane wave relation $\mathbf{k} = (\omega/c)n \, \mathbf{s}$, where n is the index of refraction, we can rewrite (11.10) as

$$\frac{s_x^2}{n^2 - \epsilon_{xx}/\epsilon_0} + \frac{s_y^2}{n^2 - \epsilon_{yy}/\epsilon_0} + \frac{s_z^2}{n^2 - \epsilon_{zz}/\epsilon_0} = \frac{1}{n^2}. \tag{11.11}$$

This equation is known as Fresnel's equation of wave normals, and the solution (i.e., the eigenvalues) gives the "eigenindices" of refraction. These eigenindices form an *index ellipsoid*, illustrated in Figure 11.2, which when oriented with the Cartesian axes is

$$\frac{x^2}{n_x^2} + \frac{y^2}{n_y^2} + \frac{z^2}{n_z^2} = 1. \tag{11.12}$$

In general, crystals can be categorized as *isotropic*, *uniaxial*, or *biaxial*. An isotropic crystal is the trivial case where the permittivity (and hence index of refraction) is the

---

[3] For a thorough background of waves in crystals, see, for example, Yariv and Yeh [2].

[4] In particular, note that in any medium, the direction of energy propagation $\hat{\mathbf{s}}$ is given by the Poynting vector $\mathbf{E} \times \mathbf{H}$, whereas the direction of phase propagation $\hat{\mathbf{k}}$ is defined by $\mathbf{D} \times \mathbf{B}$.

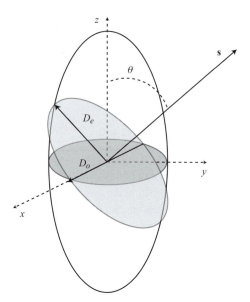

Figure 11.2 Index ellipsoid for a uniaxial crystal. The large ellipsoid has major axes of length $2n_x$, $2n_y$, and $2n_z$. The vectors in the angled plane (perpendicular to **s**) that intersect the main ellipsoid have lengths n_o and n_e, giving the supported indices of refraction in that propagation direction.

same in all directions, and the permittivity is given by a scalar quantity. In a uniaxial crystal, there is only one optic axis, and there exist both *ordinary* and *extraordinary* indices of refraction n_o and n_e, leading to an index ellipsoid that is symmetric in the x–y plane but elongated (or shortened) in the z-direction; Figure 11.2 shows an example of a uniaxial crystal index ellipsoid. When the z-axis is aligned along the "optical axis" as described, the permittivity is given by the tensor:

$$\bar{\bar{\epsilon}} = \epsilon_0 \begin{bmatrix} n_o^2 & 0 & 0 \\ 0 & n_o^2 & 0 \\ 0 & 0 & n_e^2 \end{bmatrix}. \tag{11.13}$$

In a biaxial crystal, all three indices may differ, leading to a complicated index ellipsoid and a more general permittivity tensor

$$\bar{\bar{\epsilon}} = \epsilon_0 \begin{bmatrix} n_x^2 & 0 & 0 \\ 0 & n_y^2 & 0 \\ 0 & 0 & n_z^2 \end{bmatrix}. \tag{11.14}$$

In general, the principal coordinate axis in a biaxial crystal is defined such that $n_x < n_y < n_z$.

From here on we will restrict our discussion to uniaxial crystals for simplicity. In the uniaxial crystal, the index ellipsoid takes the form

$$\frac{x^2}{n_o^2} + \frac{y^2}{n_o^2} + \frac{z^2}{n_e^2} = 1. \tag{11.15}$$

Now, in Figure 11.2, a wave propagating in the direction **s** at angle θ to the optic axis has two indices of refraction. A wave with its **D** vector in the x–y plane will see the ordinary index of refraction n_o, regardless of angle θ (in the figure it has been aligned with the x-axis). The wave with the perpendicular **D** sees a different index of refraction, found from:

$$\frac{1}{n_e^2(\theta)} = \frac{\cos^2 \theta}{n_o^2} + \frac{\sin^2 \theta}{n_e^2}. \tag{11.16}$$

The fact that the medium supports two different phase velocities at the same propagation angle, and that these velocities depend on the polarization of the incident wave, means the medium will cause a *rotation* of the polarization of the wave as it traverses the medium, similar to *Faraday rotation*, which is caused by a polarization rotation along a steady magnetic field.

11.2.1 FDTD formulation

With this background, we now proceed to set up the FDTD formulation for a uniaxial crystal. For our discussion we rely on the work of Kriezis and Elston [3]. The formulation begins, as it did with the general anisotropic medium, with the medium-independent Maxwell's equations and the constitutive relations:

$$\frac{\partial \overline{\mathcal{D}}}{\partial t} = \nabla \times \overline{\mathcal{H}} \tag{11.17a}$$

$$\frac{\partial \overline{\mathcal{B}}}{\partial t} = -\nabla \times \overline{\mathcal{E}} \tag{11.17b}$$

$$\overline{\mathcal{D}} = \overline{\overline{\epsilon}}\, \overline{\mathcal{E}} \tag{11.17c}$$

$$\overline{\mathcal{B}} = \mu_0 \overline{\mathcal{H}}. \tag{11.17d}$$

The treatment herein assumes that the crystal properties vary in the x–z plane, as does the permittivity. In general, the crystal optic axis can be aligned with respect to the x, y, z coordinates with a *tilt angle* θ and a *twist angle* ϕ, as shown in Figure 11.3. In this case, the simple tensor permittivity given in Equation (11.13) must be rotated through the angles θ and ϕ, yielding the rotated tensor permittivity

$$\overline{\overline{\epsilon}}(x, z) = \epsilon_0 \begin{bmatrix} n_o^2 + \Delta n^2 \cos^2 \theta \cos^2 \phi & \Delta n^2 \cos^2 \theta \sin \phi \cos \phi & \Delta n^2 \sin \theta \cos \theta \cos \phi \\ \Delta n^2 \cos^2 \theta \sin \phi \cos \phi & n_o^2 + \Delta n^2 \cos^2 \theta \sin^2 \phi & \Delta n^2 \sin \theta \cos \theta \sin \phi \\ \Delta n^2 \sin \theta \cos \theta \cos \phi & \Delta n^2 \sin \theta \cos \theta \sin \phi & n_o^2 + \Delta n^2 \sin^2 \theta \end{bmatrix}$$

$$\tag{11.18}$$

where $\Delta n^2 = n_e^2 - n_o^2$ is the difference between the two relative permittivities.

The model formulation described here uses all six field components, despite being only a 2D simulation, and as a result the Yee cell in 2D is modified as shown in Figure 11.3. The presence of the tilt and twist angles and thus the full 3×3 tensor permittivity in (11.18) means that all field components are coupled and are thus necessary in the model, despite symmetry in the y-direction.

Equation (11.17d) can be skipped, since we are not dealing with tensor permeability, and μ_0 can be folded into Equation (11.17a). Thanks to the use of material-independent

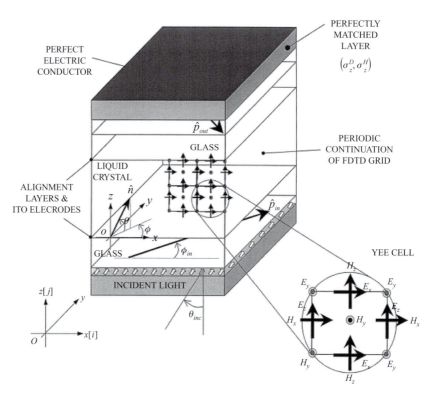

Figure 11.3 Yee cell for the liquid crystal display problem. Taken from Kriezis and Elston [3]. Note that in this 2D model, all three components of each field are required and, as a result, some field components (e.g., $\mathscr{E}_z$ and $\mathscr{H}_x$) are co-located (inset at bottom right). The FDTD model presented by Kriezis also includes a PML and periodic boundaries, discussed in Chapter 12.

Maxwell's equations, Equations (11.17a) and (11.17b) can be readily discretized using the usual algorithm; for example, the y components of $\overline{\mathscr{D}}$ and $\overline{\mathscr{B}}$ are given by

$$\mathscr{D}_y\Big|_{i,k}^{n+1} = \mathscr{D}_y\Big|_{i,k}^{n} + \Delta t \left(\frac{\mathscr{H}_x\Big|_{i,k+1/2}^{n+1/2} - \mathscr{H}_x\Big|_{i,k-1/2}^{n+1/2}}{\Delta z} - \frac{\mathscr{H}_z\Big|_{i+1/2,k}^{n+1/2} - \mathscr{H}_z\Big|_{i-1/2,k}^{n+1/2}}{\Delta x} \right)$$

(11.19a)

$$\mathscr{B}_y\Big|_{i+1/2,k+1/2}^{n+1/2} = \mathscr{B}_y\Big|_{i+1/2,k+1/2}^{n-1/2}$$

$$+ \frac{\Delta t}{\mu_0} \left(\frac{\mathscr{E}_z\Big|_{i+1,k+1/2}^{n} - \mathscr{E}_z\Big|_{i,k+1/2}^{n}}{\Delta x} - \frac{\mathscr{E}_x\Big|_{i+1/2,k+1}^{n} - \mathscr{E}_x\Big|_{i+1/2,k}^{n}}{\Delta z} \right).$$

(11.19b)

These two equations have not made any use of the inherent anisotropy of the medium; that comes about through the use of the constitutive relation relating $\overline{\mathscr{E}}$ to $\overline{\mathscr{D}}$,

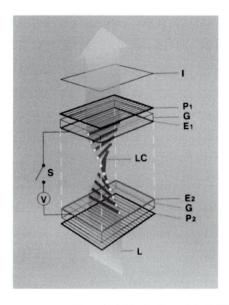 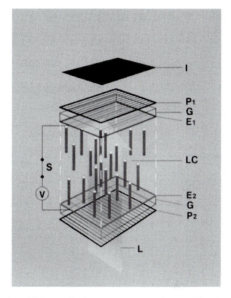

Figure 11.4 Illustration of the LCD cell operation. This particular arrangement is considered "active dark," meaning the ON state is such that no light is emitted. "Active light" devices can be made with parallel polarizers. On the left, with no applied voltage at the electrodes E_1 and E_2, the crystals are arranged in such a way that the light polarization rotates 90 degrees and is passed through the polarizers P_1 and P_2 to form image I. When voltage is applied, the crystals line up so that no polarization rotation occurs, and P_1 blocks any outgoing light.

Equation (11.17c), implemented as

$$\overline{\mathscr{E}} = [\overline{\overline{\epsilon}}]^{-1}\overline{\mathscr{D}} \qquad (11.20)$$

with $\overline{\overline{\epsilon}}$ given in Equation (11.18). This inverse can be taken exactly as described in Equation (11.4) and applied using the same algorithm outlined in Equations (11.5) and (11.6), where averages need to be taken to find field values at the required locations.

In an actual LCD cell as used in real displays, the pixel is set up as a *twisted nematic pixel*, as shown in Figure 11.4. The lower electrode is treated with a coating that forces the crystal molecules to align with a particular twist angle ϕ. Typically, on the lower electrode the coating is applied such that $\phi = 0$, and on the upper electrode $\phi = 90$ degrees (θ can be set to any number, but is usually a small angle such as $\theta = 2$ degrees).

The application of the two constraints on the electrodes causes the crystal molecules throughout the cell to align themselves with a twist as shown on the left side of Figure 11.4. Now, external to the electrodes, two polarizers are placed around the cell, allowing only light of linear polarization to pass. As shown, the two polarizers are perpendicular. The light passing through the first polarizer is thus linearly polarized, and due to the birefringence of the liquid crystal medium, the polarization of the incident light is caused to rotate, as described earlier. If the second polarizer is perpendicular

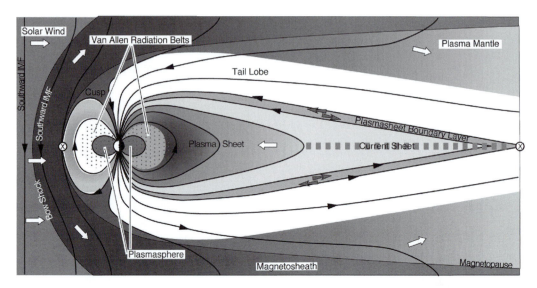

Figure 11.5 The Earth's magnetosphere. The Earth's plasmasphere is an example of a cold plasma, and the radiation belts are a hot plasma. Both of these plasma regions, as well as the enlongated tail region, are magnetized by the Earth's static magnetic field. Figure courtesy P. H. Reiff, Rice University.

to the first and the cell is arranged such that the light twists through 90 degrees of polarization, the light is allowed to pass through; this is an "active dark" OFF state – the presence of light at the output signifies that the cell has NOT been activated.

Now, the application of an electric field to the cell causes the crystal molecules to align with the field, due to their inherent polarity, as shown on the right of Figure 11.4. The polarization of the light is not affected as it only sees the ordinary refractive index, and is blocked by the second polarizer. Note that by switching the second polarizer to be parallel with the first, we turn the cell into an "active light" state.

These twisted-nematic LCD cells were developed in the late 1960s and 1970s, but suffered from a number of drawbacks, including the need for two polarizers, and a poor visibility with viewing angle. However, it should be noted that they are only one class of LCD, and one should note the variety of methods that has grown in recent years. FDTD methods such as the one described above have been used extensively to model the birefringence in liquid crystals, enabling modeling of the transmission of light at arbitrary incidence angles.

11.3   FDTD in a magnetized plasma

As an introduction to magnetized plasmas, Figure 11.5 shows the region around the Earth known as the *magnetosphere*. The Earth's magnetosphere is the region of space where the Earth's magnetic field, generated by currents deep within the Earth, is dominant over other extraterrestrial magnetic fields such as that of the Sun (the interplanetary magnetic field, or IMF). The upper regions of the Earth's atmosphere are ionized by solar radiation and are known as the ionosphere, and its upper regions are also influenced by the Earth's magnetic field. Similarly, the plasmasphere and radiation belts contain electrons and protons but very few neutral atoms or molecules; these regions are *fully ionized* magnetized plasmas. The "tail" of the magnetosphere, as illustrated in Figure 11.5, is elongated due to the pressure exerted by the solar wind.

Other planets, such as Jupiter and Saturn, also have magnetospheres, although other planets' magnetospheres may arise for different reasons.[5] Magnetospheres provide an excellent laboratory for studying plasma physics and wave-particle interactions on scales that could never be created in a man-made laboratory. For further discussion of the Earth's magnetosphere, see, for example, the texts by Kivelson and Russell [4] and Gurnett and Bhattacharjee [5].

Before deriving the FDTD algorithm for a magnetized plasma, we must first examine the functional form of the permittivity. In Chapter 10 we saw how a collisional plasma with electron density N_e and collision frequency v_c leads to a complex permittivity given by:

$$\epsilon_c(\omega) = \epsilon_0 \left[1 + \chi_e(\omega) \right] = \epsilon_0 \left[1 - \frac{\omega_p^2}{\omega(\omega + jv_c)} \right] \tag{10.38}$$

where the electron density is related to the plasma frequency $\omega_p = \sqrt{N_e q_e^2/\epsilon_0 m_e}$. This leads to a modification of Ampère's law in the frequency domain:

$$\nabla \times \mathbf{H} = j\omega\epsilon_0 \left(1 - \frac{\omega_p^2/\omega^2}{1 - jv_c/\omega} \right) \mathbf{E}. \tag{11.21}$$

It should be recognized that the expression in parentheses is the square of the index of refraction, n, in the plasma medium, and that, in addition to being complex, in general it can be less than unity.

The addition of a static, background magnetic field $\mathbf{B}_0$ adds a degree of complexity to the analysis of the medium, and leads to *anisotropy*, where the permittivity is a tensor.[6] To see this, we can write an equation of force ($F = m a = m\, dv/dt$) on an electron in the magnetized plasma medium, considering the forces of the electromagnetic wave ($\overline{\mathscr{E}}$ and $\overline{\mathscr{B}}$) and the static magnetic field $\mathbf{B}_0$:

$$m_e \frac{d\overline{v}}{dt} = q_e\overline{\mathscr{E}} + q_e\overline{v} \times (\overline{\mathscr{B}} + \mathbf{B}_0) \tag{11.22}$$

where $\overline{v}$ is the vector velocity of the electron, and m_e and q_e are its mass and charge. Note that we have neglected collisions v above; as such this is considered a cold,[7] *collisionless*

---

[5] Mars also has a weak magnetosphere, which is create by magnetic currents in the crust of the planet, rather than large currents in the core.

[6] Note that the presence of a magnetic field $\mathbf{B}_0$ alone will not affect the propagation of waves; the plasma must also be present. The effect of the magnetic field on wave propagation occurs through the electron and ion motion in the plasma.

[7] A plasma can be cold, warm, or hot, depending on the electron energies; for warm and hot plasmas, more terms are required in the force Equation (11.22). For details, see for example the plasma physics text by Bittencourt [6].

magnetized plasma, as can be found in the outer magnetosphere, where there are very few neutral particles.

To simplify the analysis, we will take the external magnetic field $\mathbf{B}_0$ to be in the z-direction so $\mathbf{B}_0 = \hat{z} B_0$, and we consider only small signal oscillations (i.e., $|\overline{\mathscr{B}}| \ll B_0$), so the $\overline{v} \times \overline{\mathscr{B}}$ term drops out (as the product of two small-signal terms). In the frequency domain, we can write the resulting force equation as

$$j\omega \mathbf{v} \simeq \frac{q_e}{m_e}(\mathbf{E} + \mathbf{v} \times \mathbf{B}_0). \tag{11.23}$$

With $\mathbf{B}_0$ aligned with the z-axis, writing the individual components of this vector equation is greatly simplified:

$$j\omega v_x = \frac{q_e}{m_e} E_x - \frac{q_e}{m_e} B_0 v_y \tag{11.24a}$$

$$j\omega v_y = \frac{q_e}{m_e} E_y - \frac{q_e}{m_e} B_0 v_x \tag{11.24b}$$

$$j\omega v_z = \frac{q_e}{m_e} E_z. \tag{11.24c}$$

Note that due to the $\mathbf{v} \times \mathbf{B}_0$ force, the z-component of velocity is not affected by B_0. The above set of equations can be solved together to find the individual components of velocity:

$$v_x = \frac{-j\omega(q_e/m_e)E_x + (q_e/m_e)\omega_c E_y}{\omega_c^2 - \omega^2} \tag{11.25a}$$

$$v_y = \frac{-(q_e/m_e)\omega_c E_x - j\omega(q_e/m_e)E_y}{\omega_c^2 - \omega^2} \tag{11.25b}$$

$$v_z = \frac{jq_e/m_e}{\omega} E_z \tag{11.25c}$$

where $\omega_c = q_e B_0/m_e$ is the *electron cyclotron frequency* or *electron gyrofrequency*, which is the frequency at which electrons orbit around static magnetic field lines. This frequency is independent of the applied field $\overline{\mathscr{E}}$ or the electron energy. Note that in this formulation we have ignored positive ions and large negative ions: in general, the masses of these ions are so much larger than that of the electrons that their motion can be ignored at all but the lowest frequencies.[8]

Now, recalling that the motion of a density of N_e electrons per unit volume with velocity $\mathbf{v}_e$ results in a current density of $\mathbf{J} = N_e q_e \mathbf{v}_e$, we see that the above velocity components form a current density $\mathbf{J}$ in the plasma. We can write the relationship between $\mathbf{D}$ and $\mathbf{E}$ in the plasma medium as [7]:

$$j\omega \mathbf{D} = j\omega\epsilon_0 \mathbf{E} + \mathbf{J} = j\omega\epsilon_{\text{eff}} \mathbf{E}. \tag{11.26}$$

---

[8] In plasmas where ion motion is important, there are also *ion* cyclotron frequencies, where q_i and m_i define their specific frequencies. Note that the sign of the charge is important, so that positive ions gyrate around field lines in the opposite direction of negative ions and electrons.

In the plasma medium with no static magnetic field, we had a scalar effective permittivity $\epsilon_{\text{eff}} = \epsilon_0(1 - \omega_p^2/\omega^2)$. Now, we will end up with a tensor effective permittivity after substituting in the current components above:

$$j\omega \begin{bmatrix} D_x \\ D_y \\ D_z \end{bmatrix} = j\omega\epsilon_0 \begin{bmatrix} E_x \\ E_y \\ E_z \end{bmatrix} + \begin{bmatrix} J_x \\ J_y \\ J_z \end{bmatrix}$$

$$= j\omega\epsilon_0 \begin{bmatrix} \epsilon_{\text{eff}}^{11} & \epsilon_{\text{eff}}^{12} & 0 \\ \epsilon_{\text{eff}}^{12*} & \epsilon_{\text{eff}}^{11} & 0 \\ 0 & 0 & \epsilon_{\text{eff}}^{33} \end{bmatrix} \begin{bmatrix} E_x \\ E_y \\ E_z \end{bmatrix} \qquad (11.27)$$

where the entries of the equivalent permittivity tensor are given by:

$$\epsilon_{\text{eff}}^{11} = \left[1 + \frac{\omega_p^2}{\omega_c^2 - \omega^2} \right] \qquad (11.28a)$$

$$\epsilon_{\text{eff}}^{12} = \frac{j\omega_p^2(\omega_c/\omega)}{\omega_c^2 - \omega^2} \qquad (11.28b)$$

$$\epsilon_{\text{eff}}^{33} = \left[1 - \frac{\omega_p^2}{\omega^2} \right] \qquad (11.28c)$$

and $\epsilon_{\text{eff}}^{12*}$ is the complex conjugate of $\epsilon_{\text{eff}}^{12}$. The reader should be satisfied to note that by setting $B_0 = 0$ (and thus $\omega_c = 0$), we arrive at the effective permittivity for the isotropic plasma from Chapter 10.

Note how our magnetized plasma now resembles an anisotropic medium, where $\mathbf{D}(\omega) = \epsilon_0\epsilon_{\text{eff}}(\omega)\mathbf{E}(\omega)$; this of course has the added complexity that the entries of the permittivity tensor are frequency dependent, so the medium is both anisotropic and dispersive.

11.3.1 Implementation in FDTD

With this simple background, we can now start to see how a magnetized plasma can be modeled in FDTD. For this discussion we rely on the method described by Lee and Kalluri [8].

This formulation has a few variations compared to the discussion presented above. First, collisions between neutrals and electrons are included, as they were for our discussion of the unmagnetized plasma as a dispersive medium in Chapter 10. Collisions form another "force" on the right-hand side of Equation (11.23), equal to $F_v = -\nu v$; it is a damping force proportional to the electron velocity. Second, note that the $\mathbf{v} \times \mathbf{B}_0$ term in (11.23) can be written as $\omega_c \times \mathbf{J}$ with a constant scaling factor, since ω_c is proportional to $\mathbf{B}_0$ and $\mathbf{J}$ is proportional to $\mathbf{v}$; specifically we have:

$$\overline{\omega}_c \times \overline{\mathbf{J}} \quad \rightarrow \quad \frac{q_e}{m_e}\mathbf{B}_0 \times N_e q_e \mathbf{v} \quad \rightarrow \quad -\epsilon_0\omega_p^2 \mathbf{v} \times \mathbf{B}_0$$

Third, the magnetic field $\mathbf{B}_0$ is considered to point in some arbitrary direction, not necessarily aligned with any axis; this results in a permittivity matrix that is not simple and symmetric as given in (11.26).

With these modifications, the Lorentz force Equation (11.22) can be written in time domain as:[9]

$$\frac{\partial \overline{\mathcal{J}}}{\partial t} + v\overline{\mathcal{J}} = \epsilon_0 \omega_p^2(\mathbf{r}, t)\overline{\mathcal{E}} - \overline{\omega}_c(\mathbf{r}, t) \times \overline{\mathcal{J}}. \tag{11.29}$$

Note that we have explicitly written ω_p and ω_c as functions of space and time; in most applications the electron density and magnetic field vary throughout space, and in some cases can change with time. Note also that we have written the cyclotron frequency ω_c as a vector, such that

$$\overline{\omega}_c = \omega_{cx}\hat{x} + \omega_{cy}\hat{y} + \omega_{cz}\hat{z}$$

and each component is proportional to its respective component of static magnetic field, i.e., $\omega_{cx} = q_e B_{0x}/m_e$. Solving for each component of $\overline{\mathcal{J}}$ and including Ampère's law and Faraday's law in our discussion, the x-components of our set of three equations to solve become

$$\frac{\partial \mathcal{H}_x}{\partial t} = -\frac{1}{\mu_0}\left(\frac{\partial \mathcal{E}_z}{\partial y} - \frac{\partial \mathcal{E}_y}{\partial z}\right) \tag{11.30a}$$

$$\frac{\partial \mathcal{E}_x}{\partial t} = \frac{1}{\epsilon_0}\left(\frac{\partial \mathcal{H}_z}{\partial y} - \frac{\partial \mathcal{H}_y}{\partial z} - \mathcal{J}_x\right) \tag{11.30b}$$

$$\begin{bmatrix} \partial \mathcal{J}_x/\partial t \\ \partial \mathcal{J}_y/\partial t \\ \partial \mathcal{J}_z/\partial t \end{bmatrix} = \Omega \begin{bmatrix} \mathcal{J}_x \\ \mathcal{J}_y \\ \mathcal{J}_z \end{bmatrix} + \epsilon_0 \omega_p^2(\mathbf{r}, t) \begin{bmatrix} \mathcal{E}_x \\ \mathcal{E}_y \\ \mathcal{E}_z \end{bmatrix} \tag{11.30c}$$

where the matrix Ω is given by:

$$\Omega = \begin{bmatrix} -v & -\omega_{cz} & \omega_{cy} \\ \omega_{cz} & -v & -\omega_{cx} \\ -\omega_{cy} & \omega_{cx} & -v \end{bmatrix}. \tag{11.31}$$

The other component equations of Faraday's law and Ampère's law follow a similar form to Equations (11.30a) and (11.30b).

We now proceed to derive an FDTD algorithm in order to update these equations using the grid shown in Figure 11.6. The $\overline{\mathcal{E}}$ and $\overline{\mathcal{H}}$ field components are placed at the same locations as usual in the Yee cell, but rather than placing the current components of $\overline{\mathcal{J}}$ co-located with $\overline{\mathcal{E}}$, they are placed at the integer cell locations (i, j, k), which were otherwise unoccupied; further, the current components are updated at time step $(n + 1/2)$ along with the $\overline{\mathcal{H}}$ components. Then, the updating of Equations (11.30a) and

---

[9] Note that in Lee and Kalluri [8], there is a plus sign before the last term, since they have taken the minus sign out from q_e; in our notation here, $q_e = -1.6 \times 10^{-19}$ C and so ω_c is negative.

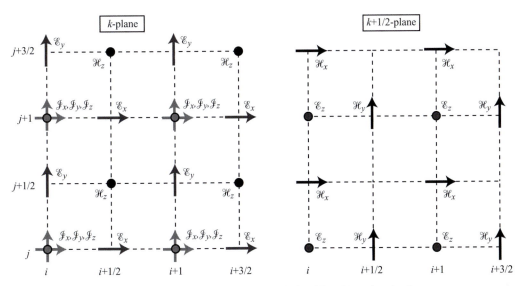

Figure 11.6 Yee cell adapted to Lee and Kalluri algorithm. Note that the $\mathcal{I}$ components are co-located at integer grid points; otherwise, the Yee cell is the same we are used to.

(11.30b) are straightforward and follow the basic differencing scheme on the Yee cell:

$$\mathcal{E}_x\Big|_{i+1/2,j,k}^{n+1} = \mathcal{E}_x\Big|_{i+1/2,j,k}^{n} \tag{11.32a}$$

$$+ \frac{\Delta t}{\epsilon_0}\left(\frac{\partial \mathcal{H}_z}{\partial y}\Big|_{i+1/2,j,k}^{n+1/2} - \frac{\partial \mathcal{H}_y}{\partial z}\Big|_{i+1/2,j,k}^{n+1/2} - \frac{\mathcal{I}_x\Big|_{i+1,j,k}^{n+1/2} + \mathcal{I}_x\Big|_{i,j,k}^{n+1/2}}{2}\right)$$

$$\mathcal{H}_x\Big|_{i,j+1/2,k+1/2}^{n+1/2} = \mathcal{H}_x\Big|_{i,j+1/2,k+1/2}^{n-1/2} + \frac{\Delta t}{\mu_0}\left(\frac{\partial \mathcal{E}_z}{\partial y}\Big|_{i,j+1/2,k+1/2}^{n} - \frac{\partial \mathcal{E}_y}{\partial z}\Big|_{i,j+1/2,k+1/2}^{n}\right).$$

$$\tag{11.32b}$$

Notice that the current term $\mathcal{I}_x$ is averaged at its integer cell locations in order to find the value at $(i+1/2, j, k)$.[10]

The updating of the $\overline{\mathcal{I}}$ Equation (11.30c) is considerably more complex. Generally, one would perform the standard time differencing of the terms on the right, and spatial averaging of the $\overline{\mathcal{I}}$ terms on the left, and rearrange to find equations of the form $\mathcal{I}_i^{n+1/2} = f_1(\mathcal{I}_i^{n-1/2}) + f_2(\overline{\mathcal{E}})$. However, the analysis is made much simpler through the use of Laplace transforms. Recalling that the Laplace transform of a time derivative is

---

[10] A brief note on averaging: the Yee cell was specifically designed so that averaging of fields is not required, at least in the most basic, source-free isotropic medium. Any time spatial averaging is required, it introduces an artificial diffusive component, which can range from negligible to problematic depending on how the averaging is conducted. Details of this problem for the magnetized plasma can be found in [9]. In general, averaging should only be used when absolutely necessary.

$\mathcal{L}\left[\partial\overline{\mathcal{G}}(t)/\partial t\right] = s\mathbf{G}(s)$, similar to Fourier transforms, we can take the Laplace transform of (11.30c) to find:

$$s\mathbf{J}(s) - \mathbf{J}_0 = \Omega\mathbf{J}(s) + \epsilon_0\omega_p^2\mathbf{E}$$

where ω_p, ω_c, ν, and $\mathbf{E}$ are assumed to be a constant over the time step. Solving for $\mathbf{J}(s)$ we find

$$\mathbf{J}(s) = (s\mathbf{I} - \Omega)^{-1}\mathbf{J}_0 + \epsilon_0\omega_p^2\frac{1}{s}(s\mathbf{I} - \Omega)^{-1}\mathbf{E} \tag{11.33}$$

where $\mathbf{I}$ is the identity matrix in three dimensions, and $\mathbf{J}_0$, from Laplace transform theory, is the "initial value" of $\mathbf{J}$; below, when we apply this algorithm to a single time step, $\mathbf{J}_0$ will actually be the previous time step value of $\mathbf{J}$. Note that the $1/s$ multiplier on the last term comes from Laplace transform theory, in which $\overline{\mathcal{E}}$ is actually taken to be $\overline{\mathcal{E}} \cdot u(t)$. Now taking the inverse Laplace transform we arrive back at an expression for $\overline{\mathcal{J}}$ in time domain:

$$\overline{\mathcal{J}}(t) = \mathbf{A}(t)\mathbf{J}_0 + \epsilon_0\omega_p^2\mathbf{K}(t)\overline{\mathcal{E}}(t) \tag{11.34}$$

where the expressions $\mathbf{A}(t)$ and $\mathbf{K}(t)$ come from inverse Laplace transform theory and are given by

$$\mathbf{A}(t) = \exp(\Omega t) \tag{11.35a}$$

$$\mathbf{K}(t) = \Omega^{-1}[\exp(\Omega t) - \mathbf{I}]. \tag{11.35b}$$

When fully expanded, these expressions take the following form, provided for completeness:

$$\mathbf{A}(t) = \exp(\Omega t) \tag{11.36a}$$

$$= e^{-\nu t}\begin{bmatrix} C_1\omega_{cx}^2 + \cos(\omega_c t) & C_1\omega_{cx}\omega_{cy} - S_1\omega_{cz} & C_1\omega_{cx}\omega_{cz} + S_1\omega_{cy} \\ C_1\omega_{cy}\omega_{cx} + S_1\omega_{cz} & C_1\omega_{cy}^2 + \cos(\omega_c t) & C_1\omega_{cy}\omega_{cz} + S_1\omega_{cx} \\ C_1\omega_{cz}\omega_{cx} + S_1\omega_{cy} & C_1\omega_{cz}\omega_{cy} + S_1\omega_{cx} & C_1\omega_{cz}^2 + \cos(\omega_c t) \end{bmatrix}$$

$$\mathbf{K}(t) = \Omega^{-1}[\exp(\Omega t) - \mathbf{I}] \tag{11.36b}$$

$$= \frac{e^{-\nu t}}{\omega_c^2 + \nu^2}\begin{bmatrix} C_2\omega_{cx}^2 + C_3 & C_2\omega_{cx}\omega_{cy} - C_4\omega_{cz} & C_2\omega_{cx}\omega_{cz} + C_4\omega_{cy} \\ C_2\omega_{cy}\omega_{cx} + C_4\omega_{cz} & C_2\omega_{cy}^2 + C_3 & C_2\omega_{cy}\omega_{cz} - C_4\omega_{cx} \\ C_2\omega_{cz}\omega_{cx} - C_4\omega_{cy} & C_2\omega_{cz}\omega_{cy} + C_4\omega_{cx} & C_2\omega_{cz}^2 + C_3 \end{bmatrix}.$$

The parameters S_1 and C_i are time dependent and are given by:

$$S_1 = \sin(\omega_c t)/\omega_c$$
$$C_1 = (1 - \cos(\omega_c t))/\omega_c^2$$
$$C_2 = (e^{\nu t} - 1)/\nu - \nu C_1 - S_1$$
$$C_3 = \nu(e^{\nu t} - \cos(\omega_c t)) + \omega_c\sin(\omega_c t)$$
$$C_4 = e^{\nu t} - \cos(\omega_c t) - \nu S_1$$

with $\omega_c^2 = \omega_{cx}^2 + \omega_{cy}^2 + \omega_{cz}^2$. Given these expressions, we can proceed to discretize the current Equation (11.30c) to find the system of update equations:

$$
\begin{bmatrix} \mathcal{J}_x \big|_{i,j,k}^{n+1/2} \\ \mathcal{J}_y \big|_{i,j,k}^{n+1/2} \\ \mathcal{J}_z \big|_{i,j,k}^{n+1/2} \end{bmatrix} = \mathbf{A}(\Delta t) \begin{bmatrix} \mathcal{J}_x \big|_{i,j,k}^{n-1/2} \\ \mathcal{J}_y \big|_{i,j,k}^{n-1/2} \\ \mathcal{J}_z \big|_{i,j,k}^{n-1/2} \end{bmatrix} + \frac{\epsilon_0}{2} \omega_p^2 \big|_{i,j,k}^{n} \mathbf{K}(\Delta t) \begin{bmatrix} \mathcal{E}_x \big|_{i+1/2,j,k}^{n} + \mathcal{E}_x \big|_{i-1/2,j,k}^{n} \\ \mathcal{E}_y \big|_{i,j+1/2,k}^{n} + \mathcal{E}_y \big|_{i,j-1/2,k}^{n} \\ \mathcal{E}_z \big|_{i,j,k+1/2}^{n} + \mathcal{E}_z \big|_{i,j,k+1/2}^{n} \end{bmatrix}.
$$

$$(11.37)$$

This is, of course, a set of three independent update equations, written in matrix form for simplicity. Notice that the $\overline{\mathcal{E}}$ components have been spatially averaged to find the value of the field at the integer grid locations, rather than spatially averaging $\overline{\mathcal{J}}$ components as we have done in the past; this is a direct result of locating $\overline{\mathcal{J}}$ components at integer grid locations (i, j, k).

The algorithm presented above, then, is all that is required to simulate a magnetized plasma in the FDTD method. Maxwell's equations, as given in equations (11.30a) and (11.30b), are updated in the straightforward FDTD algorithm presented in Equations (11.32). Note, however, that $\overline{\mathcal{J}}^{n+1/2}$ must be updated using Equations (11.37) *before* updating the electric field in Equation (11.32a), since it requires $\overline{\mathcal{J}}$ at time step $n + 1/2$. The parameters S and C, and in turn the matrices $\mathbf{A}$ and $\mathbf{K}$, can be pre-calculated and stored in memory. Note that these matrices are used as $\mathbf{A}(\Delta t)$ and $\mathbf{K}(\Delta t)$ in Equation (11.37), i.e., they are functions of Δt.

11.4   FDTD in ferrites

As our third and final example of anisotropic materials, we now investigate wave propagation and the FDTD method in materials known as *ferrites*. In this section we will see an alternative formulation for the FDTD method in anisotropic materials, but both methods could be applied to either of these materials.

Ferrites are so named because of the presence of iron; their general composition can be given by the chemical formula MO Fe_2O_3, where M is a divalent (i.e., having two electrons in its outer shell) metal such as manganese, zinc, nickel, or iron. Ferrite materials differ from ferromagnetic metals in that their conductivities can be considerably lower; typical ferromagnetics such as iron, cobalt, and nickel (and their alloys) have such high conductivities that their magnetic properties are irrelevant in terms of their effects on electromagnetic fields. Ferrites typically have conductivities of about 10^{-4}–10^0 S-m^{-1}, compared to $\sim 10^7$ for iron; their relative permeabilities μ_r are on the order of a few thousand, compared to tens of thousands for ferromagnetics.

The origin of ferromagnetism is illustrated in Figure 11.7. In all materials, spinning electrons have a dipole moment $\mathbf{m}$ and angular momentum $\overline{\mathcal{L}}$, but in non-magnetic materials, the random arrangement of atoms results in cancellation of the electron spin

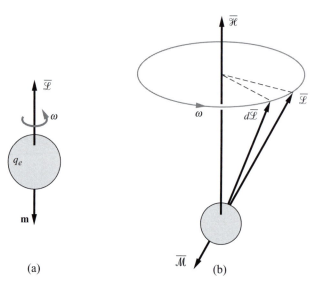

Figure 11.7 The dipole moment and angular momentum in ferrites. (a) a spinning electron has a dipole moment **m** and angular momentum $\overline{\mathscr{L}}$ which are anti-parallel. (b) The magnetization vector $\overline{\mathscr{M}}$ precesses around the field vector $\overline{\mathscr{H}}$ in the presence of a magnetic field or electromagnetic wave.

macroscopically. In ferrites, however, a net resulting spin exists, and a dipole moment is established. A rotating charge has dipole moment **m** and angular momentum $\overline{\mathscr{L}}$ as shown in Figure 11.7a; for a single electron these have quantifiable values, $|\overline{\mathscr{L}}| = h/4\pi = 0.527 \times 10^{-34}$ J-m and $|\mathbf{m}| = 9.27 \times 10^{-24}$ A-m^2. The ratio $\gamma_m = |\mathbf{m}|/|\overline{\mathscr{L}}|$ is known as the *gyromagnetic ratio*; this ratio is also given by $\gamma_m = q_e/m_e$, which also appears in the plasma gyrofrequency described in the previous section.

Now, if we place this electron in a magnetic field $\overline{\mathscr{H}}$, as in Figure 11.7b, the magnetic field exerts a torque $\overline{\mathscr{T}}$ on the dipole moment **m**, following the equation of angular motion:

$$\overline{\mathscr{T}} = \frac{d\overline{\mathscr{L}}}{dt} \tag{11.38}$$

which tends to rotate the dipole moment into alignment with $\overline{\mathscr{H}}$; however, since it has angular momentum and we have not introduced any damping force, the spin does not change angle but precesses gyroscopically about $\overline{\mathscr{H}}$. The torque acting on **m** due to the magnetic field $\overline{\mathscr{H}}$ is related to $\overline{\mathscr{H}}$ and the angle between it and **m**:

$$\overline{\mathscr{T}} = \mu_0 \mathbf{m} \times \overline{\mathscr{H}} \tag{11.39}$$

and since $\mathbf{m} = \gamma_m \overline{\mathscr{L}}$, we have:

$$\frac{d\mathbf{m}}{dt} = -\gamma_m \mu_0 \mathbf{m} \times \overline{\mathscr{H}}.$$

With N_e electrons per unit volume, the magnetization vector $\overline{\mathcal{M}} = N_e\mathbf{m}$, so

$$\frac{d\overline{\mathcal{M}}}{dt} = -\gamma_m\mu_0\overline{\mathcal{M}} \times \overline{\mathcal{H}}. \tag{11.40}$$

Now, our goal in this analysis is to see how the anisotropic permeability arises from the electron motion. As such, we seek the relationship between $\overline{\mathcal{B}}$ and $\overline{\mathcal{H}}$ such that $\overline{\mathcal{B}} = \mu_0(\overline{\mathcal{H}} + \overline{\mathcal{M}}) = \mu_0\mu_r\overline{\mathcal{H}}$ in order to find the anisotropic (and, it turns out, dispersive) μ_r. In the frequency domain we can write Equation (11.40) as:

$$j\omega\mathbf{M} = -\gamma_m\mu_0\mathbf{M} \times \mathbf{H} \tag{11.41}$$

Let us assume the magnetic field $\mathbf{H}$ consists of a large polarizing field $\mathbf{H}_0$ and a small time-varying wave field $\mathbf{H}_1$; these will give rise, in turn, to a large constant magnetization $\mathbf{M}_0$ parallel to $\mathbf{H}_0$ and a small time-varying component $\mathbf{M}_1$. The large magnetization $\mathbf{M}_0$ is related to the applied magnetic field as $\mathbf{M}_0 = \mu_0(\mu_r - 1)\mathbf{H}_0$. If we assume $\mathbf{H}_0 = \hat{z}H_0$, i.e., it is aligned with the z-direction, the wave fields are:

$$\mathbf{H} = \hat{x}H_{1x} + \hat{y}H_{1y} + \hat{z}H_0 \tag{11.42a}$$

$$\mathbf{M} = \hat{x}M_{1x} + \hat{y}M_{1y} + \hat{z}M_0 \tag{11.42b}$$

where we have ignored the time-varying components in the z-direction, since they are negligible compared to H_0 and M_0. Substituting these into Equation (11.41) and solving for components of M_1, we have:

$$M_{1x} = \frac{\omega_0\omega_M}{\omega_0^2 - \omega^2}H_{1x} + j\frac{\omega\omega_M}{\omega_0^2 - \omega^2}H_{1y} \tag{11.43a}$$

$$M_{1y} = -j\frac{\omega\omega_M}{\omega_0^2 - \omega^2}H_{1x} + \frac{\omega_0\omega_M}{\omega_0^2 - \omega^2}H_{1y} \tag{11.43b}$$

$$M_{1z} = M_0 \tag{11.43c}$$

where $\omega_0 = -\mu_0\gamma_m H_0$ is the so-called *Larmor frequency* (which, as it turns out, is equal to the electron gyrofrequency in a plasma), and $\omega_M = -\mu_0\gamma_m M_0$ is the magnetization frequency. Equations (11.43) define a matrix relationship between $\mathbf{H}_1$ and $\mathbf{M}_1$; we can use this to write the relationship between the wave fields $\mathbf{B}_1$ and $\mathbf{H}_1$ as:

$$\mathbf{B}_1 = \begin{bmatrix} B_x \\ B_y \\ B_z \end{bmatrix} = \mu_0(\mathbf{M}_1 + \mathbf{H}_1) = \mu_0 \begin{bmatrix} \mu_{\text{eff}}^{11} & \mu_{\text{eff}}^{12} & 0 \\ \mu_{\text{eff}}^{21} & \mu_{\text{eff}}^{22} & 0 \\ 0 & 0 & 1 \end{bmatrix} \begin{bmatrix} H_x \\ H_y \\ H_z \end{bmatrix} \tag{11.44}$$

where

$$\mu_{\text{eff}}^{11} = \mu_{\text{eff}}^{22} = 1 + \frac{\omega_0\omega_M}{\omega_0^2 - \omega^2} \tag{11.45a}$$

$$\mu_{\text{eff}}^{12} = \left(\mu_{\text{eff}}^{21}\right)^* = j\frac{\omega_M\omega}{\omega_0^2 - \omega^2}. \tag{11.45b}$$

Note the similarity between the permeability tensor for ferrites here, and the permittivity tensor for magnetized plasmas in the previous section. It should come as no surprise, then, that similar analysis methods can be used for the FDTD implementation. However, to make things more complicated, we can introduce a damping factor in ferrites which will change the form of the permeability tensor somewhat. In a general magnetic material, Equation (11.40) above is replaced by the so-called *Landau-Lifshitz-Gilbert* equation, which can be written in the form:

$$\frac{d\overline{\mathcal{M}}}{dt} = -\gamma_m \mu_0 \overline{\mathcal{M}} \times \overline{\mathcal{H}} - \alpha \overline{\mathcal{M}} \times \frac{d\overline{\mathcal{M}}}{dt} \tag{11.46}$$

where α is a loss coefficient known as the Landau-Lifshitz damping parameter. Notice now that $d\overline{\mathcal{M}}/dt$ appears on both sides of the equation. We leave the derivation of the constitutive relation $\overline{\mathcal{B}} = \mu_0 \mu_r \overline{\mathcal{H}}$ to the problems at the end of this chapter. It turns out that the permeability tensor entries are modified by substituting $\omega_0 \to \omega_0 + j\omega\alpha$, in which case they take the form:

$$\mu_{\text{eff}}^{11} = \mu_{\text{eff}}^{22} = 1 + \frac{\omega_M(\omega_0 + j\omega\alpha)}{(\omega_0 + j\omega\alpha)^2 - \omega^2} \tag{11.47a}$$

$$\mu_{\text{eff}}^{12} = \left(\mu_{\text{eff}}^{21}\right)^* = j\frac{\omega_M\omega}{(\omega_0 + j\omega\alpha)^2 - \omega^2}. \tag{11.47b}$$

11.4.1   Implementation in FDTD

For our analysis of ferrites in the FDTD algorithm, we rely on the implementation derived by M. Alsunaidi [10], which includes the general case with damping parameter α. The implementation follows a method similar to the auxiliary differential equation method that we discussed in Chapter 10 for dispersive materials; in fact, since these ferrite materials are dispersive, the use of the ADE method is natural. The catch, of course, is that they are also anisotropic, and the tensor permeability will need to be dealt with.

We start by multiplying out Equations (11.47) to find:

$$\mu_{\text{eff}}^{11} = \mu_{\text{eff}}^{22} = 1 + \frac{\omega_M\omega_0 + j\omega\alpha\omega_M}{\omega_0^2 + j\omega\,2\alpha\omega_0 + (j\omega)^2(1 + \alpha^2)} \tag{11.48a}$$

$$\mu_{\text{eff}}^{12} = \left(\mu_{\text{eff}}^{21}\right)^* = \frac{j\omega\omega_M}{\omega_0^2 + j\omega\,2\alpha\omega_0 + (j\omega)^2(1 + \alpha^2)}. \tag{11.48b}$$

If we then use the above effective permeabilities in the matrix Equation (11.44) and multiply out the terms, we find the following equations relating B_x and B_y to H_x and H_y (note that $B_z = \mu_0 H_z$):

$$\left[\omega_0^2 + j\omega\,2\alpha\omega_0 + (j\omega)^2(1 + \alpha^2)\right] B_x = \mu_0 \left[\omega_0^2 + j\omega\,2\alpha\omega_0 + (j\omega)^2(1 + \alpha^2)\right] H_x$$
$$+ \mu_0 \left[\omega_M\omega_0 + j\omega\,\alpha\omega_M\right] H_x$$
$$+ j\omega\mu_0\omega_M H_y \tag{11.49a}$$

$$\left[\omega_0^2 + j\omega\, 2\alpha\omega_0 + (j\omega)^2(1+\alpha^2)\right] B_y = \mu_0 \left[\omega_0^2 + j\omega\, 2\alpha\omega_0 + (j\omega)^2(1+\alpha^2)\right] H_y$$
$$+ \mu_0 \left[\omega_M\omega_0 + j\omega\,\alpha\omega_M\right] H_y$$
$$- j\omega\mu_0\omega_M H_x. \tag{11.49b}$$

Notice the symmetry between these two equations; the only difference is the minus sign in front of the last term, due to the complex conjugate in the permeability matrix. Now, each of the terms in Equations (11.49) is proportional to either $j\omega$, $(j\omega)^2$, or a constant, so they can easily be transformed back to the time domain. Equation (11.49a) in particular becomes:

$$\omega_0^2 \mathcal{B}_x + 2\alpha\omega_0 \frac{\partial \mathcal{B}_x}{\partial t} + (1+\alpha^2)\frac{\partial^2 \mathcal{B}_x}{\partial t^2} = \mu_0(\omega_0^2 + \omega_0\omega_M)\mathcal{H}_x + \mu_0\alpha(2\omega_0 + \omega_M)\frac{\partial \mathcal{H}_x}{\partial t}$$
$$+ \mu_0(1+\alpha^2)\frac{\partial^2 \mathcal{H}_x}{\partial t^2} + \mu_0\omega_M\frac{\partial \mathcal{H}_y}{\partial t}. \tag{11.50}$$

In order to find our update equation for $\mathcal{H}^{n+1/2}$, this equation can now be discretized using second-order centered differences, as we have done throughout this book. Note that the second derivatives in time will require three-point differencing; as such, our first-order time derivatives will be spaced across two time points, spanning from $(n + 1/2)$ to $(n - 3/2)$, so that all discretized fields are at known locations in time. We find:

$$\frac{\omega_0^2}{2}\left(\mathcal{B}_x\big|^{n+1/2}+\mathcal{B}_x\big|^{n-3/2}\right) + \frac{2\alpha\omega_0}{2\Delta t}\left(\mathcal{B}_x\big|^{n+1/2}-\mathcal{B}_x\big|^{n-3/2}\right) \tag{11.51}$$

$$+ \frac{1+\alpha^2}{\Delta t^2}\left(\mathcal{B}_x\big|^{n+1/2}-2\mathcal{B}_x\big|^{n-1/2}+\mathcal{B}_x\big|^{n-3/2}\right)$$

$$= \frac{\mu_0\omega_0(\omega_0+\omega_M)}{2}\left(\mathcal{H}_x\big|^{n+1/2}+\mathcal{H}_x\big|^{n-3/2}\right) + \frac{\mu_0\alpha(2\omega_0+\omega_M)}{2\Delta t}\left(\mathcal{H}_x\big|^{n+1/2}-\mathcal{H}_x\big|^{n-3/2}\right)$$

$$+ \frac{\mu_0(1+\alpha^2)}{\Delta t^2}\left(\mathcal{H}_x\big|^{n+1/2}-2\mathcal{H}_x\big|^{n-1/2}+\mathcal{H}_x\big|^{n-3/2}\right) + \frac{\mu_0\omega_M}{2\Delta t}\left(\mathcal{H}_y\big|^{n+1/2}-\mathcal{H}_y\big|^{n-3/2}\right)$$

where it should be noted that we have used temporal averaging of the $\mathcal{B}_x$ and $\mathcal{H}_x$ fields at time steps $(n + 1/2)$ and $(n - 3/2)$ to find the fields at time step $(n - 1/2)$; this is not necessary, but it provides symmetry in the update equation coefficients, as we shall see shortly, without affecting stability: this algorithm can be applied at the CFL limit. This equation can be solved for $\mathcal{H}^{n+1/2}$ to find:

$$c_1 \mathcal{H}_x\big|^{n+1/2} = c_2 \mathcal{B}_x\big|^{n+1/2} + c_3 \mathcal{B}_x\big|^{n-3/2} + c_4 \mathcal{B}_x\big|^{n-1/2} \tag{11.52}$$

$$+ c_5 \mathcal{H}_x\big|^{n-1/2} + c_6 \mathcal{H}_x\big|^{n-3/2} + c_7 \mathcal{H}_y\big|^{n-3/2} - c_8 \mathcal{H}_y\big|^{n+1/2}$$

where the coefficients $c_1, \ldots, c_8$ are given as functions of ω_0, ω_M, α, and Δt:

$$c_1 = \frac{\mu_0 \omega_0 (\omega_0 + \omega_M)}{2} + \frac{\mu_0 \alpha (2\omega_0 + \omega_m)}{2\Delta t} + \frac{\mu_0 (1 + \alpha^2)}{\Delta t^2} \tag{11.53a}$$

$$c_2 = \frac{\omega_0^2}{2} + \frac{\alpha \omega_0}{\Delta t} + \frac{1 + \alpha^2}{\Delta t^2} \tag{11.53b}$$

$$c_3 = \frac{\omega_0^2}{2} - \frac{\alpha \omega_0}{\Delta t} + \frac{1 + \alpha^2}{\Delta t^2} \tag{11.53c}$$

$$c_4 = -\frac{2(1 + \alpha^2)}{\Delta t^2} \tag{11.53d}$$

$$c_5 = \frac{2\mu_0 (1 + \alpha^2)}{\Delta t^2} \tag{11.53e}$$

$$c_6 = -\frac{\mu_0 \omega_0 (\omega_0 + \omega_M)}{2} + \frac{\mu_0 \alpha (2\omega_0 + \omega_m)}{2\Delta t} - \frac{\mu_0 (1 + \alpha^2)}{\Delta t^2} \tag{11.53f}$$

$$c_7 = c_8 = \frac{\mu_0 \omega_M}{2\Delta t}. \tag{11.53g}$$

Using the same coefficients, Equation (11.49b) can be used to write an update equation for $\mathcal{H}_y^{n+1/2}$:

$$c_1 \mathcal{H}_y \big|^{n+1/2} = c_2 \mathcal{B}_y \big|^{n+1/2} + c_3 \mathcal{B}_y \big|^{n-3/2} + c_4 \mathcal{B}_y \big|^{n-1/2} \tag{11.54}$$

$$+ c_5 \mathcal{H}_y \big|^{n-1/2} + c_6 \mathcal{H}_y \big|^{n-3/2} - c_7 \mathcal{H}_x \big|^{n-3/2} + c_8 \mathcal{H}_x \big|^{n+1/2}.$$

Note that Equations (11.52) and (11.54) are symmetric except for the signs in front of c_7 and c_8. Now, we are faced with the problem that Equation (11.52) requires $\mathcal{H}_y^{n+1/2}$ in order to update $\mathcal{H}_x^{n+1/2}$, and vice versa for (11.54); i.e., the two equations are coupled due to the anisotropy of the medium. The solution to this problem is to substitute Equation (11.54) into (11.52). Solving for $\mathcal{H}_x^{n+1/2}$ we find:

$$\mathcal{H}_x \big|^{n+1/2} = c_1' \mathcal{B}_x \big|^{n+1/2} + c_2' \mathcal{B}_x \big|^{n-3/2} + c_3' \mathcal{B}_x \big|^{n-1/2} \tag{11.55}$$

$$+ c_4' \mathcal{H}_x \big|^{n-1/2} + c_5' \mathcal{H}_x \big|^{n-3/2} + c_6' \mathcal{H}_y \big|^{n-3/2} - c_7' \mathcal{B}_y \big|^{n+1/2} - c_8' \mathcal{B}_y \big|^{n-3/2}$$

$$- c_9' \mathcal{B}_y \big|^{n-1/2} - c_{10}' \mathcal{H}_y \big|^{n-1/2}$$

where the new prime coefficients are given in terms of the original c's:

$$c_1' = \frac{c_2}{g} \qquad\qquad c_5' = \frac{c_6 + c_7 c_8 / c_1}{g} \qquad\qquad c_7' = \frac{c_2 c_8 / c_1}{g}$$

$$c_2' = \frac{c_3}{g} \qquad\qquad c_6' = \frac{c_7 - c_6 c_8 / c_1}{g} \qquad\qquad c_8' = \frac{c_3 c_8 / c_1}{g}$$

$$c_3' = \frac{c_4}{g} \qquad\qquad\qquad\qquad\qquad\qquad\qquad\quad c_9' = \frac{c_4 c_8 / c_1}{g}$$

$$c_4' = \frac{c_5}{g} \qquad\qquad\qquad\qquad\qquad\qquad\qquad\quad c_{10}' = \frac{c_5 c_8 / c_1}{g}$$

where $g = c_1 + c_8^2 / c_1$.

An equation similar to Equation (11.55) is required to update $\mathcal{H}_y^{n+1/2}$ (the derivation of which is trivial starting from Equation 11.54), but for the z-component (parallel to H_0) we have simply $\mathcal{H}_z = \mu_0 \mathcal{B}_z$. Thus, our auxiliary differential equation relating $\overline{\mathcal{B}}$ to $\overline{\mathcal{H}}$ is complete. Note that this algorithm requires storage of fields two time steps back, in addition to requiring the auxiliary $\overline{\mathcal{B}}$ field.

For the other FDTD update equations, we can simply use Maxwell's equations as we have in the past, except that we use $\overline{\mathcal{E}}$ to update $\overline{\mathcal{B}}$ rather than $\overline{\mathcal{H}}$. The complete procedure takes the form (in the notation of Equation 10.66):

$$\overline{\mathcal{E}}^{n-1}, \overline{\mathcal{H}}^{n-1/2} \xrightarrow{(11.3a)} \overline{\mathcal{E}}^n \xrightarrow{(11.3c)} \overline{\mathcal{B}}^{n+1/2} \xrightarrow{(11.55)} \overline{\mathcal{H}}^{n+1/2} \quad (11.56)$$

where, of course, the storage of $\overline{\mathcal{B}}$ and $\overline{\mathcal{H}}$ at time steps $(n-1/2)$ and $(n-3/2)$ is required for the last step.

As a final note, the reader should observe that the magnetized plasma problem in Section 11.3 can also be formulated in the ADE-like method described here, and the ferrite problem described here can also be formulated in a method similar to the one presented in Section 11.3. We will explore these possibilities in the problems at the end of the chapter.

11.5 Summary

In this chapter we introduced methods for simulating anisotropic materials in the FDTD method. A general, nondispersive anisotropic material has tensor permittivity and permeability:

$$\overline{\overline{\epsilon}} = \epsilon_0 \begin{bmatrix} \epsilon_{xx} & \epsilon_{xy} & \epsilon_{xz} \\ \epsilon_{yx} & \epsilon_{yy} & \epsilon_{yz} \\ \epsilon_{zx} & \epsilon_{zy} & \epsilon_{zz} \end{bmatrix} \quad \text{and} \quad \overline{\overline{\mu}} = \mu_0 \begin{bmatrix} \mu_{xx} & \mu_{xy} & \mu_{xz} \\ \mu_{yx} & \mu_{yy} & \mu_{yz} \\ \mu_{zx} & \mu_{zy} & \mu_{zz} \end{bmatrix}.$$

The simulation of these anisotropic materials in the FDTD method requires the use of material-independent Maxwell's equations, along with the two constitutive relations:

$$\overline{\mathcal{E}} = [\overline{\overline{\epsilon}}]^{-1} \overline{\mathcal{D}}$$

$$\overline{\mathcal{H}} = [\overline{\overline{\mu}}]^{-1} \overline{\mathcal{B}}.$$

These two matrix equations can then each be written as a set of three equations, in which each of the $\overline{\mathcal{E}}$ and $\overline{\mathcal{H}}$ components are functions of all three $\overline{\mathcal{D}}$ or $\overline{\mathcal{B}}$ components. Four-point spatial averaging is used to find field values at required locations.

We introduced three examples of anisotropic materials, namely the liquid crystal, the magnetized plasma, and the ferrite. In crystals in general, an *index ellipsoid* defines the indices of refraction for waves propagating in all directions through the crystal. The FDTD formulation for crystals, and for liquid crystals, follows the formulation for general anisotropic materials, after defining the tensor permittivity.

Next, we introduced the magnetized plasma, which is both anisotropic and dispersive. The analysis of a magnetized plasma begins with the statement of the equation of force on an electron in the presence of a static magnetic field:

$$m_e \frac{\partial \overline{v}}{\partial t} = q_e \overline{\mathscr{E}} + q_e \overline{v} \times (\overline{\mathscr{B}} + \mathbf{B}_0) - \nu\, \overline{v}.$$

In the FDTD formulation of Lee and Kalluri [8], we write this equation of force as an equation for the current, since $\overline{\mathscr{J}} = q_e N_e \overline{v}$:

$$\begin{bmatrix} \partial \mathscr{J}_x / \partial t \\ \partial \mathscr{J}_y / \partial t \\ \partial \mathscr{J}_z / \partial t \end{bmatrix} = \begin{bmatrix} -\nu & -\omega_{cz} & \omega_{cy} \\ \omega_{cz} & -\nu & -\omega_{cx} \\ -\omega_{cy} & \omega_{cx} & -\nu \end{bmatrix} \begin{bmatrix} \mathscr{J}_x \\ \mathscr{J}_y \\ \mathscr{J}_z \end{bmatrix} + \epsilon_0 \omega_p^2(\mathbf{r}, t) \begin{bmatrix} \mathscr{E}_x \\ \mathscr{E}_y \\ \mathscr{E}_z \end{bmatrix}.$$

Finally, the solution of this equation in the FDTD method involves writing the equation in the Laplace domain, rearranging, and taking the inverse Laplace transform, which results in the algorithm given in Equations (11.36) and (11.37).

Similar to the magnetized plasma, ferrites, which are low-loss materials exhibiting ferromagnetism, are dispersive anisotropic materials, but in this case with anisotropic permeability. The tensor permeability takes the form:

$$\mu_{\text{eff}}(\omega) = \mu_0 \begin{bmatrix} \mu_{\text{eff}}^{11} & \mu_{\text{eff}}^{12} & 0 \\ \mu_{\text{eff}}^{21} & \mu_{\text{eff}}^{22} & 0 \\ 0 & 0 & 1 \end{bmatrix}$$

where, in the general lossy case with damping factor α, the coefficients are given by:

$$\mu_{\text{eff}}^{11} = \mu_{\text{eff}}^{22} = 1 + \frac{\omega_M(\omega_0 + j\omega\alpha)}{(\omega_0 + j\omega\alpha)^2 - \omega^2}$$

$$\mu_{\text{eff}}^{12} = \left(\mu_{\text{eff}}^{21}\right)^* = j\frac{\omega_M \omega}{(\omega_0 + j\omega\alpha)^2 - \omega^2}.$$

To derive the FDTD update equation, we use a method similar to the ADE method of Chapter 10. We use this tensor permeability to write a frequency-dependent equation of the form $\mathbf{B} = \mu_{\text{eff}}\mathbf{H}$, which we then transform into the time domain, resulting in first- and second-time derivatives. This is then discretized in the usual manner, after analytically solving the coupled update equations for $\mathscr{H}_x$ and $\mathscr{H}_y$.

11.6   Problems

11.1. **Anisotropy in 1D.** Write a 1D simulation for a generic nondispersive anisotropic medium, as described in Section 11.1. Note that even in 1D, you will need to include all three components of each field $\overline{\mathscr{E}}$, $\overline{\mathscr{D}}$, $\overline{\mathscr{H}}$, and $\overline{\mathscr{B}}$, for 12 components total. Use $\Delta x = 0.5$ cm and make the space 200 cm long, and launch a sinusoidal plane wave with a wavelength of 10 cm. Create a slab of anisotropic material 20 cm long in the middle of the space, and truncate both ends with a Mur second-order boundary condition.

    (a)   Let $\epsilon_{xx} = \epsilon_{yy} = \epsilon_{zz} = 3$, and set all off-diagonal entries to one (i.e., $\epsilon_{ij} = 1$ if $i \neq j$). Launch a plane wave with $\mathscr{E}_z$ polarization; what is the polarization

at the output of the slab? Measure the amplitude of the reflected and transmitted waves.

(b) Repeat (a) for an $\mathcal{E}_y$ polarized wave.

(c) Modify your code to include anisotropic permeability, and set $\mu_{xx} = 1$, $\mu_{yy} = 2$, $\mu_{zz} = 3$; repeat (a) and (b) and compare the results.

11.2. **Uniaxial crystal in 2D**. Write a 2D TM mode FDTD simulation in the x–y plane for a uniaxial crystal with $n_o = 1.5$ and $n_e = 2$.

(a) Create a slab of this crystal material that spans y and covers two wavelengths in x, and align the optic axis with y. Launch an $\mathcal{E}_z$ plane wave from $x = 0$ incident on the slab at frequency 1 THz and use 20 grid cells per wavelength. Observe the polarization and amplitude of the transmitted and reflected waves.

(b) Repeat (a), but rotate the crystal in the y–z plane through angles of 15, 30, 45, 60, and 90 degrees, using the rotation matrix in Equation (11.18). How do the transmitted polarization and amplitude change?

11.3. **Ferrite permittivity.** Derive Equations (11.47) from Equation (11.46), by first deriving the constitutive relation between $\overline{\mathcal{B}}$ and $\overline{\mathcal{H}}$ in this complex medium. Hint: Write out the component Equations in (11.47) in the frequency domain and follow the steps beginning from Equation (11.41).

11.4. **Magnetized plasma method.** Derive the matrices $\mathbf{A}$ and $\mathbf{K}$ in the Lee and Kalluri method, Equations (11.36), using symbolic matrix inversion.

11.5. **Stability of the Lee and Kalluri method.** Derive the stability of the Lee and Kalluri method in Equations (11.37), using the methods of Chapter 6. If you are having trouble arriving at a closed-form stability condition, make plots of the amplification matrix eigenvalues with variations in Δt and Δx until you arrive at a stability condition semi-empirically.

11.6. **Magnetized plasma with an ADE method.** The method used in this chapter for ferrites is a type of auxiliary differential equation method. It turns out that one can also write an ADE-type method for the magnetized plasma. Derive such a method, using the frequency-domain version of Equation (11.29) and following the logic in the ferrite section.

11.7. **Ferrites with a Laplace method.** Similarly, derive an FDTD method for ferrites, using the Laplace transform derivation described in this chapter for the magnetized plasma. Assume $\alpha = 0$ for simplicity.

11.8. **Magnetized plasma in 2D.** Write a 2D simulation for a magnetized plasma using the Lee and Kalluri method described in this chapter. Create a slab of plasma, as in Problem 10.5, and use the same plasma parameters therein.

(a) Use a background magnetic field $B_0 = 1$ μT and $v_c = 10^5$ s^{-1}. Orient the magnetic field perpendicular to the slab; i.e., if the slab spans y, orient B_0 in the x-direction. Launch an $\mathcal{E}_z$-polarized wave incident on the slab from

$x = 0$. How small does Δt need to be to ensure stability? Does this agree with the stability analysis in Problem 11.5? How does this stability criterion relate to the plasma frequency, and what does this mean physically?

(b) Repeat (a), but with B_0 oriented in the y-direction.

(c) Repeat (a), but with B_0 oriented at 45 degrees in the x–y plane. Compare with transmitted and reflected waves and polarization in these three cases.

References

[1] L. Dou and A. R. Sebak, "3D FDTD method for arbitrary anisotropic materials," *IEEE Microwave Opt. Technol. Lett.*, vol. 48, pp. 2083–2090, 2006.

[2] A. Yariv and P. Yeh, *Optical Waves in Crystals*. Wiley, 1984.

[3] E. E. Kriezis and S. J. Elston, "Light wave propagation in liquid crystal displays by the 2-D finite-difference time domain method," *Optics Comm.*, vol. 177, pp. 69–77, 2000.

[4] M. G. Kivelson and C. T. Russell, *Introduction to Space Physics*. Cambridge University Press, 1995.

[5] D. A. Gurnett and A. Bhattacharjee, *Introduction to Plasma Physics: With Space and Laboratory Applications*. Cambridge University Press, 2005.

[6] J. A. Bittencourt, *Fundamentals of Plasma Physics*. Springer-Verlag, 2004.

[7] U. S. Inan and A. S. Inan, *Electromagnetic Waves*. Prentice-Hall, 2000.

[8] J. H. Lee and D. Kalluri, "Three-dimensional FDTD simulation of electromagnetic wave transformation in a dynamic inhomogeneous magnetized plasma," *IEEE Trans. Ant. Prop.*, vol. 47, no. 7, pp. 1146–1151, 1999.

[9] T. W. Chevalier, U. S. Inan, and T. F. Bell, "Terminal impedance and antenna current distribution of a VLF electric dipole in the inner magnetosphere," *IEEE Trans. Ant. Prop.*, vol. 56, no. 8, pp. 2454–2468, 2008.

[10] M. A. Alsunaidi, "FDTD analysis of microstrip patch antenna on ferrite substrate," *IEEE Microwave Opt. Technol. Lett.*, vol. 50, pp. 1848–1851, 2008.

12 Some advanced topics

At this point, the reader should have all the necessary tools to create a useful FDTD model, in any number of dimensions, in almost any material.[1] In the remainder of this book we cover a number of topics of general interest that have been found to be very useful in a number of FDTD applications. In this chapter, we will discuss modeling of periodic structures, modeling of physical features smaller than a grid cell size, the method known as Bodies of Revolution (BOR) for modeling cylindrical structures in 3D, and finally, the near-to-far field transformation, which is used to extrapolate the far-field radiation or scattering pattern from a confined FDTD simulation.

12.1 Modeling periodic structures

Quite a number of problems in numerical modeling involve structures that are periodic in one or more dimensions. Examples include photonic bandgap structures, which involve periodic arrays of dielectric structures; or arrays of antennas, in cases where the array is large enough that it can be analyzed as a periodic structure. These problems can be reduced to the modeling of a single period of the structure through the methods described in this section. Figure 12.1 shows a simple example of a periodic structure, consisting of rows of circular dielectric "balls" that repeat in the y-direction.

The main issue of complexity in modeling periodic structures, and which requires some ingenuity in the modeling effort, arises from the fact that *future values* of fields are involved in the FDTD update equations. To see this, we write the phasor **E** and **H** fields in 2D at the boundaries of unit cells, as shown in Figure 12.1:[2]

$$H_x(x, y = y_p + \Delta y/2) = H_x(x, y = \Delta y/2) \cdot e^{-jk_y y_p} \qquad (12.1a)$$

$$E_z(x, y = 0) = E_z(x, y = y_p) \cdot e^{jk_y y_p} \qquad (12.1b)$$

---

[1] In the previous two chapters we covered dispersive and anisotropic materials, and inhomogeneous materials are simply handled through spatially dependent ϵ, μ, and σ. We did not, however, cover nonlinear materials, whose material parameters can be functions of the electric field intensity $\overline{\mathscr{E}}$. For a good background in nonlinear materials and their implementation in FDTD, we refer the reader to papers by R. W. Ziolkowski and J. B. Judkins [1, 2] and F. L. Teixeira [3], and the book by Taflove and Hagness [4]. Nonlinear materials are most often also dispersive, and so both effects must be treated together.

[2] Note that we do not have a problem with the $\mathscr{H}_y$ field component, since its update equation will not require any fields that cross the boundary.

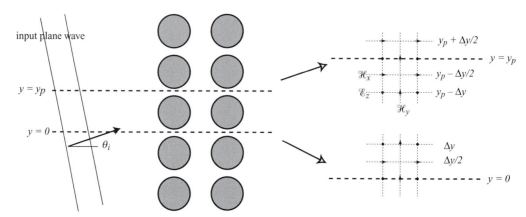

Figure 12.1 Periodic structures. The material is periodic in the y-dimension with period y_p. We wish to reduce the modeling requirement in the y-dimension to include only the range from 0 to y_p.

where y_p is the width of the unit cell, and $k_y = k_0 \sin \theta_i$ is the y-component of the k-vector for a plane wave incident at angle θ_i. The equations above show that the field values at the top boundary must be the same as those at the bottom boundary, with a phase shift determined by the incident angle. In the time domain, these become:

$$\mathcal{H}_x(x, y = y_p + \Delta y/2, t) = \mathcal{H}_x(x, y = \Delta y/2, t - y_p \sin \theta_i) \tag{12.2a}$$

$$\mathcal{E}_z(x, y = 0, t) = \mathcal{E}_z(x, y = y_p, t + y_p \sin \theta_i). \tag{12.2b}$$

Note that the $\mathcal{E}_z$ field at $y = 0$ and time t is equal to the same field at $y = y_p$ but at time $t + y_p \sin \theta_i$, a *future value*. This is where periodic structures become tricky. Now, there are various techniques to try to circumvent this problem, which can be categorized as *direct-field* methods and *field-transformation* methods. The former work directly with Maxwell's equations; in the latter, Maxwell's equations are transformed into nonphysical intermediate fields (similar to an auxiliary equation approach).

12.1.1   Direct-field methods

Normal incidence

The case of a plane wave at normal incidence is rather trivial; in this case $\sin \theta_i$ is equal to zero, and there is no problem with future values. We can then simply set the field values on the boundaries equal:

$$\mathcal{H}_x(x, y = y_p + \Delta y/2, t) = \mathcal{H}_x(x, y = \Delta y/2, t)$$

$$\mathcal{E}_z(x, y = 0, t) = \mathcal{E}_z(x, y = y_p, t). \tag{12.3}$$

In the FDTD algorithm, we require the value of $\mathcal{H}_x$ just above the boundary, at $y = y_p + \Delta y/2$ in Figure 12.1, in order to update the electric field $\mathcal{E}_z$ at $y = y_p$. For normal incidence, this value of $\mathcal{H}_x$ is the same as its counterpart at $y = \Delta y/2$ just above the lower boundary of our periodic structure, and so the field value there can be used

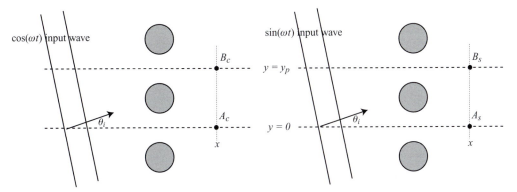

Figure 12.2 Sine-cosine method of modeling periodic structures. Fields at the known boundaries ($\mathscr{E}_z$ at B and $\mathscr{H}_x$ at A) are used to find their unknown counterparts.

directly. This is similar to the correction that was made in polar coordinates in Chapter 4 to close the 2π discontinuity.

Sine-cosine method

At any non-normal angle of incidence, we have the problem of requiring future values. The sine-cosine method is a successful method of attacking this problem, albeit for a single-frequency source only.[3] Consider a point on each of the boundaries at $(x, 0)$ and (x, y_p). We will need to run the simulation twice; in one simulation, we launch an incident wave with $\cos(\omega t)$ dependence, and label those two locations on the periodic boundary A_c and B_c, as shown in Figure 12.2. In a second simulation, which can be run in parallel, we launch a wave with $\sin(\omega t)$ dependence; the boundary points are labeled A_s and B_s. The fields should then be 90 degrees out of phase from each other. The instantaneous fields on the boundaries are combined to form a complex number of the form $\mathscr{E}_z(A_c) + j\mathscr{E}_z(A_s)$, where $k_y = k \sin \theta_i$ takes into account the incidence angle. Phase shifts of the form $e^{jk_y y_p}$ are used to find the fields at the unknown locations (A for $\mathscr{E}_z$ and B for $\mathscr{H}_x$). The final field values are found as:

$$\mathscr{E}_z(A_c) = \mathfrak{Re}\{[\mathscr{E}_z(B_c) + j\mathscr{E}_z(B_s)] \cdot e^{jk_y y_p}\} \tag{12.4a}$$

$$\mathscr{E}_z(A_s) = \mathfrak{Im}\{[\mathscr{E}_z(B_c) + j\mathscr{E}_z(B_s)] \cdot e^{jk_y y_p}\} \tag{12.4b}$$

$$\mathscr{H}_x(B_c) = \mathfrak{Re}\{[\mathscr{H}_x(A_c) + j\mathscr{H}_x(A_s)] \cdot e^{-jk_y y_p}\} \tag{12.4c}$$

$$\mathscr{H}_x(B_s) = \mathfrak{Im}\{[\mathscr{H}_x(A_c) + j\mathscr{H}_x(A_s)] \cdot e^{-jk_y y_p}\}. \tag{12.4d}$$

Note that the fields as they are denoted above are merely shorthand: for example, $\mathscr{H}_x(A_c)$ actually refers to the value of $\mathscr{H}_x$ in the cosine simulation that is just below the boundary, as shown in Figure 12.1. As mentioned above, this method of solving periodic boundaries is simple and effective for a single-frequency simulation, but loses the inherent broadband nature of the FDTD algorithm that is its primary strength.

---

[3] The loss of broadband capability in any FDTD method is not insignificant; in fact, this problem can be solved in the frequency domain (FDFD) method, which we will introduce in Chapter 14, quite simply.

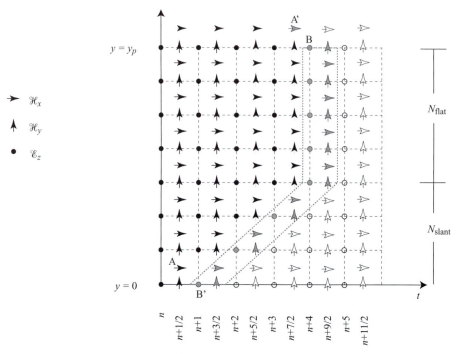

Figure 12.3 Angled-update method of modeling periodic structures. The fields in gray are about to be updated at the same time in the FDTD algorithm, even though they are not defined at the same absolute time.

Angled-update method

The last direct-field method we will discuss is the angled-update method [4]. The angled-update method utilitizes the observation that updating particular locations on the boundary can be *delayed* until the time when the corresponding field on the other boundary is useful. This means that at any particular "time step," field values on the grid are not all defined at the same absolute time.

Consider Figure 12.3. What appears as the x-axis in this figure is actually time progression; operations performed at a particular location in y–t space are thus performed over all x. Notice that $\mathscr{E}_z$ fields are located at integer locations in both y and t; $\mathscr{H}_x$ and $\mathscr{H}_y$ fields are located at half-integer time steps. All of the field components in black are known, and those in gray (inside the dashed box) are about to be updated.

We wish to drive the simulation with a plane wave at incident angle θ_i. We begin by breaking up the space as shown in Figure 12.3 into a slanted section with N_{slant} spatial steps, and a flat section with N_{flat} spatial steps. As the FDTD algorithm is iterated in time, the field components shown in gray will be updated at the same time, even though they do not exist at the same absolute time. Given a spatial step Δy, we can see that $N_y = N_{\text{flat}} + N_{\text{slant}} = y_p/\Delta y$. The number of steps required in the slanted section is given by

$$N_{\text{slant}} = \text{ceil}\left[(N_y - 1)\sqrt{2}\sin\theta_i\right] \tag{12.5}$$

where the ceiling function ceil means the result in brackets is rounded up to the nearest integer. Now, the updating procedure can be conducted as follows:

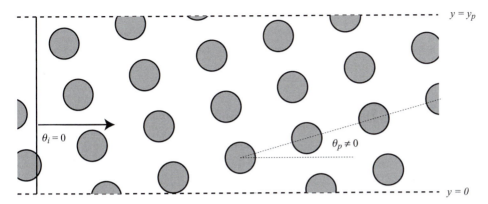

Figure 12.4 Tilting the structure in a periodic boundary problem. The incident plane wave is launched normally, but the structure itself is angled at some nonzero angle θ_p. A new structure must be designed for each angle of incidence.

1. Use the field value of $\mathcal{H}_x$ at A (at time $n + 1/2$) to update $\mathcal{H}_x$ at A' (at some later time), since these two fields are equal according to the boundary condition in Equation (12.2a).
2. Use the FDTD algorithm to update $\mathcal{E}_z$ at B, which requires $\mathcal{H}_x$ at A'.
3. Use this value of $\mathcal{E}_z$ at B to update $\mathcal{E}_z$ at B', which are equal according to Equation (12.2b).
4. Update each of the fields along the slant boundary (in gray) sequentially, starting from the earliest in time (i.e., from the lower left to the upper right), using the FDTD algorithm.
5. Update each of the fields along the flat boundary (in gray) with the usual FDTD algorithm.

The steps above form a single iteration of the updating procedure, and must then be repeated. Note that the method is limited to incident angles of less than 45 degrees.

This method has the distinct advantage over the sine-cosine method in that it is broadband, and not confined to a single frequency. Furthermore, the stability criterion is the same as for the usual FDTD method, and no extra fields need to be stored for earlier time steps. All that is required is careful planning in setting up the update equations, and recognition that for a given snapshot of the simulation results, not all of the field values are defined at the same absolute time!

The angled periodic structure

As a final comment on direct-field methods, we note that it is often possible, and more straightforward, to simply angle the periodic structure in question rather than the incident wave. In this case, the normal-incidence direct-field method above can be used, and the somewhat complex methods described in the next section can be avoided. This method, where the periodic structure is angled, is illustrated in Figure 12.4. The plane wave is launched at normal incidence, but the structure itself is tilted at some nonzero angle θ_p.

This simple solution to the angled incidence problem can be easier to implement in many cases, but has the drawback that the periodic structure must be redefined for each angle of incidence. Furthermore, in cases where the structures fit well with the orthogonal FDTD grid (i.e., if they are small squares or cubes), these rotated structures will now result in staircasing along the edges.

12.1.2   Field-transformation methods

The second group of methods for modeling periodic structures are the so-called field-transformation methods, introduced by Veysoglu et al. [5]. These methods use a transformation from $\overline{\mathcal{E}}$ and $\overline{\mathcal{H}}$ to a pair of auxiliary fields, which are nonphysical, to remove the time gradient across the grid. To understand the concept, it must be understood that the time gradient corresponds to a simple phase shift in frequency domain; this is the feature that was taken advantage of by the sine-cosine method. As such, Maxwell's equations in frequency domain are substituted with the following fields, for the case of the 2D TM mode:

$$P_z = E_z \cdot e^{jk_y y}$$
$$Q_x = \eta_0 H_x \cdot e^{jk_y y} \qquad (12.6)$$
$$Q_y = \eta_0 H_y \cdot e^{jk_y y}$$

where, again, $k_y = k_0 \sin \theta_i$ and θ_i is the incidence angle of a plane wave. In essence, we have substituted the usual fields $\mathbf{E}$ and $\mathbf{H}$ with nonphysical fields $\mathbf{P}$ and $\mathbf{Q}$, which are phase shifted with respect to the real fields. Multiplication by η_0 in the Q components serves as a convenience and removes ϵ_0 and μ_0 from the update equations that follow.

With the periodic boundary conditions $P_z(x, y=0) = P_z(x, y=y_p)$ and $Q_z(x, y=0) = Q_z(x, y=y_p)$, these yield the following transformed Maxwell's equations in frequency and time domain:

$$\frac{j\omega\mu_r}{c} Q_x = -\frac{\partial P_z}{\partial y} + jk_y P_z \qquad \rightarrow \qquad \frac{\mu_r}{c}\frac{\partial \mathcal{Q}_x}{\partial t} = -\frac{\partial \mathcal{P}_z}{\partial y} + \frac{\sin\theta_i}{c}\frac{\partial \mathcal{P}_z}{\partial t} \qquad (12.7a)$$

$$\frac{j\omega\mu_r}{c} Q_y = \frac{\partial P_z}{\partial x} \qquad \rightarrow \qquad \frac{\mu_r}{c}\frac{\partial \mathcal{Q}_y}{\partial t} = \frac{\partial \mathcal{P}_z}{\partial x} \qquad (12.7b)$$

$$\frac{j\omega\epsilon_r}{c} P_z = \frac{\partial Q_y}{\partial x} - \frac{\partial Q_x}{\partial y} + jk_y Q_x \qquad \rightarrow \qquad \frac{\epsilon_r}{c}\frac{\partial \mathcal{P}_z}{\partial t} = \frac{\partial \mathcal{Q}_y}{\partial x} - \frac{\partial \mathcal{Q}_x}{\partial y} + \frac{\sin\theta_i}{c}\frac{\partial \mathcal{Q}_x}{\partial t} \qquad (12.7c)$$

where, as usual, script notation $\mathcal{P}$ and $\mathcal{Q}$ imply time domain quantities. Time derivatives can be collected on the left-hand side by combining Equations (12.7a) and (12.7b):

$$\left(\frac{\epsilon_r \mu_r - \sin^2\theta_i}{c}\right)\frac{\partial \mathcal{Q}_x}{\partial t} = -\epsilon_r \frac{\partial \mathcal{P}_z}{\partial y} + \left(\frac{\partial \mathcal{Q}_y}{\partial x} - \frac{\partial \mathcal{Q}_x}{\partial y}\right)\sin\theta_i \qquad (12.8a)$$

$$\frac{\mu_r}{c}\frac{\partial \mathcal{Q}_y}{\partial t} = \frac{\partial \mathcal{P}_z}{\partial x} \qquad (12.8b)$$

$$\left(\frac{\epsilon_r \mu_r - \sin^2\theta_i}{c}\right)\frac{\partial \mathcal{P}_z}{\partial t} = -\frac{\partial \mathcal{P}_z}{\partial y}\sin\theta_i + \mu_r\left(\frac{\partial \mathcal{Q}_y}{\partial x} - \frac{\partial \mathcal{Q}_x}{\partial y}\right). \qquad (12.8c)$$

Now, we have a set of coupled partial differential equations, like Maxwell's equations, that we wish to solve on an FDTD grid. However, an added complexity arises because, unlike in Maxwell's equations, there are time derivatives and spatial derivatives of the same fields in the same equations (12.8a and 12.8c above). We will also find that stability must be re-analyzed compared to the normal FDTD update equations, in part because of the time and space fields but also because of the factor multiplying the time derivatives in each of the Equations (12.8). As a result, the stability criterion becomes angle-dependent and thus a function of θ_i.

It can be shown [4] that for this method, the analytical dispersion relation is

$$\frac{v_p}{c} = \frac{\sin\alpha \sin\theta_i + \sqrt{(\sin\alpha \sin\theta_i)^2 + \cos^2\theta_i}}{\cos^2\theta_i} \tag{12.9}$$

where α is the propagation angle of a scattered field, defined such that $k_{x,\text{scat}} = (\omega/v_p)\cos\alpha$ and $k_{y,\text{scat}} = (\omega/v_p)\sin\alpha$. Note here that multiple α's are possible for a single incidence angle due to different scattering components. It is also of interest that for $\theta_i = 0$ we find $v_p = c$ for any α; there is no dispersion at normal incidence. This relation, Equation (12.9), will be useful to compare the two methods for discretizing the update equations that are described below.

As mentioned above, the stability of the field-transformation methods is modified compared to the usual FDTD method. It has been shown that a good approximation, which turns out to be conservative, is given by:

$$\frac{v_p \Delta t}{\Delta x} \leq \frac{1 - \sin\theta_i}{\sqrt{D}} \tag{12.10}$$

where D is the dimensionality of the problem. At normal incidence, this reduces to the usual CFL condition; at, say, 45-degree incidence, the stability criterion is now more strict by a factor of $\sqrt{2}$.

Now, we are still left with the problem of mixed time and spatial derivatives in our coupled PDEs; these are dealt with by a number of methods, which are briefly discussed next.

Multiple-grid method

The first approach for discretizing Equations (12.8) involves using multiple grids, with the two grids having $\mathcal{P}$ and $\mathcal{Q}$ fields defined at different spatial and temporal locations. The grids are defined as shown in Figure 12.5. The second grid is merely shifted one-half spatial cell in the direction of periodicity (y) and one-half time step compared to the first grid.

The two grids in Figure 12.5 are solved simultaneously. The hollow-arrowed fields ($\mathcal{P}_{z1}$, $\mathcal{Q}_{x2}$, and $\mathcal{Q}_{y2}$) are solved at time step n, while the complementary fields are solved at time step $n + 1/2$. There are thus six update equations for the 2D case. We begin by

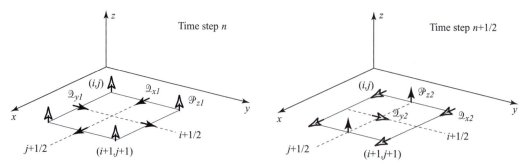

Figure 12.5 Multi-grid placement of field components. Note that neither cell is the same as the standard Yee cell, but field components are located such that the difference equations are symmetric.

updating the $n + 1/2$ fields using their previous values and known fields at time n:

$$\mathcal{D}_{x1}\Big|_{i,j}^{n+1/2} = \mathcal{D}_{x1}\Big|_{i,j}^{n-1/2} - \frac{\epsilon_r c \Delta t}{\Delta y} A \left(\mathcal{P}_{z1}\Big|_{i,j}^{n} - \mathcal{P}_{z1}\Big|_{i,j-1}^{n} \right) \qquad (12.11a)$$

$$+ \frac{c \Delta t}{\Delta x} A \sin \theta_i \left(\mathcal{D}_{y2}\Big|_{i,j-1}^{n} - \mathcal{D}_{y2}\Big|_{i-1,j-1}^{n} \right)$$

$$- \frac{c \Delta t}{\Delta y} A \sin \theta_i \left(\mathcal{D}_{x2}\Big|_{i,j}^{n} - \mathcal{D}_{x2}\Big|_{i,j-1}^{n} \right)$$

$$\mathcal{D}_{y1}\Big|_{i,j}^{n+1/2} = \mathcal{D}_{y1}\Big|_{i,j}^{n-1/2} - \frac{c \Delta t}{\mu_r \Delta x} \left(\mathcal{P}_{z1}\Big|_{i+1,j}^{n} - \mathcal{P}_{z1}\Big|_{i,j}^{n} \right) \qquad (12.11b)$$

$$\mathcal{P}_{z2}\Big|_{i,j}^{n+1/2} = \mathcal{P}_{z2}\Big|_{i,j}^{n-1/2} - \frac{c \Delta t}{\Delta y} A \sin \theta_i \left(\mathcal{P}_{z1}\Big|_{i,j}^{n} - \mathcal{P}_{z1}\Big|_{i,j-1}^{n} \right) \qquad (12.11c)$$

$$+ \frac{\mu_r c \Delta t}{\Delta x} A \left(\mathcal{D}_{y2}\Big|_{i,j-1}^{n} - \mathcal{D}_{y2}\Big|_{i-1,j-1}^{n} \right) - \frac{\mu_r c \Delta t}{\Delta y} A \left(\mathcal{D}_{x2}\Big|_{i,j}^{n} - \mathcal{D}_{x2}\Big|_{i,j-1}^{n} \right)$$

where $A = 1/(\mu_r \epsilon_r - \sin^2 \theta_i)$ and, as usual, $c = (\mu_0 \epsilon_0)^{-1/2}$. We then update the second grid using the fields from the first grid:

$$\mathcal{D}_{x2}\Big|_{i,j}^{n+1} = \mathcal{D}_{x2}\Big|_{i,j}^{n} - \frac{\epsilon_r c \Delta t}{\Delta y} A \left(\mathcal{P}_{z2}\Big|_{i,j}^{n+1/2} - \mathcal{P}_{z2}\Big|_{i,j-1}^{n+1/2} \right) \qquad (12.12a)$$

$$+ \frac{c \Delta t}{\Delta x} A \sin \theta_i \left(\mathcal{D}_{y1}\Big|_{i,j-1}^{n+1/2} - \mathcal{D}_{y1}\Big|_{i-1,j-1}^{n+1/2} \right)$$

$$- \frac{c \Delta t}{\Delta y} A \sin \theta_i \left(\mathcal{D}_{x1}\Big|_{i,j}^{n+1/2} - \mathcal{D}_{x1}\Big|_{i,j-1}^{n+1/2} \right)$$

$$\mathcal{D}_{y2}\Big|_{i,j}^{n+1} = \mathcal{D}_{y2}\Big|_{i,j}^{n} - \frac{c \Delta t}{\mu_r \Delta x} \left(\mathcal{P}_{z2}\Big|_{i+1,j}^{n+1/2} - \mathcal{P}_{z2}\Big|_{i,j}^{n+1/2} \right) \qquad (12.12b)$$

$$\left.\mathcal{P}_{z1}\right|_{i,j}^{n+1} = \left.\mathcal{P}_{z1}\right|_{i,j}^{n} - \frac{c\Delta t}{\Delta y} A \sin\theta_i \left(\left.\mathcal{P}_{z2}\right|_{i,j}^{n+1/2} - \left.\mathcal{P}_{z2}\right|_{i,j-1}^{n+1/2}\right) \tag{12.12c}$$

$$+ \frac{\mu_r c\Delta t}{\Delta x} A \left(\left.\mathcal{D}_{y1}\right|_{i,j-1}^{n+1/2} - \left.\mathcal{D}_{y1}\right|_{i-1,j-1}^{n+1/2}\right) - \frac{\mu_r c\Delta t}{\Delta y} A \left(\left.\mathcal{D}_{x1}\right|_{i,j}^{n+1/2} - \left.\mathcal{D}_{x1}\right|_{i,j-1}^{n+1/2}\right).$$

Lossy materials can be incorporated into this method quite simply [4], as was done for the standard FDTD equations. This involves terms of the form $\sigma\eta_0\mathcal{P}_z$ and $\sigma^m/\eta_0\mathcal{D}_{x,y}$ (for magnetic losses) added to the left-hand side of the transformed Maxwell's equations, which can then be updated by including spatial and temporal averaging where appropriate.

This multiple-grid method has a number of relatively minor disadvantages, the first being the fact that we now have six update equations rather than only three. In three dimensions, the number of equations increases to 24: each of the usual six field components becomes four components, and four grids are required. However, this is usually more than compensated for by the fact that the space has been reduced in both x and y dimensions to a single period of the structure.

The second major disadvantage is that the two grids (in the 2D case) can become decoupled from one another, similar to how the TE and TM modes are decoupled. It is possible, in fact, to excite only one of the grids, or to have scattering objects which affect only one of the grids.

Split-field method

The second field-transformation method for modeling periodic structures is the split-field method. This method, as the name implies, requires that the $\overline{\mathcal{P}}$ and $\overline{\mathcal{D}}$ fields are "split" into two component fields, but which are placed on the same grid; these split fields are operated on separately and then later added back together. This scheme parallels the split-field PML that was introduced in Chapter 9.

Recall that as with the multiple-grid method, the goal is to create a set of equations which decouple the spatial and temporal partial derivatives. We begin with the time domain versions of Equations (12.7) rather than (12.8), and rather than solving for the time derivatives on the left-hand side, we instead eliminate the time-derivative terms from the right-hand side by the introduction of new fields:

$$\mathcal{D}_x = \mathcal{D}_{xa} + \frac{\sin\theta_i}{\mu_r}\mathcal{P}_z \tag{12.13a}$$

$$\mathcal{P}_z = \mathcal{P}_{za} + \frac{\sin\theta_i}{\epsilon_r}\mathcal{D}_x. \tag{12.13b}$$

These are then substituted into Equations (12.7) to find:

$$\frac{\mu_r}{c}\frac{\partial\mathcal{D}_{xa}}{\partial t} = -\frac{\partial\mathcal{P}_z}{\partial y} \tag{12.14a}$$

$$\frac{\mu_r}{c}\frac{\partial\mathcal{D}_y}{\partial t} = \frac{\partial\mathcal{P}_z}{\partial x} \tag{12.14b}$$

$$\frac{\epsilon_r}{c}\frac{\partial\mathcal{P}_{za}}{\partial t} = \frac{\partial\mathcal{D}_y}{\partial x} - \frac{\partial\mathcal{D}_x}{\partial y}. \tag{12.14c}$$

This set of equations has successfully decoupled the spatial and temporal derivatives, as desired, so that each equation has only temporal derivatives of $\mathcal{P}$ or $\mathcal{Q}$ and only spatial derivatives of the other field. The method, now, is to use Equations (12.14) to determine the partial (subscripted "xa" and "za") fields from the total fields, and then to use Equations (12.13) to find the total fields from the partial fields. The problem remains that Equations (12.13) have total fields on the right-hand sides. Two similar, but not equivalent, approaches are available to circumvent this problem, both of which use Equations (12.13) together to simplify each other:

1. Substitute Equation (12.13b) into Equation (12.13a) to eliminate $\mathcal{P}_z$. This yields the two successive equations

$$\mathcal{Q}_x \left(1 - \frac{\sin^2 \theta_i}{\mu_r \epsilon_r}\right) = \mathcal{Q}_{xa} + \frac{\sin \theta_i}{\mu_r} \mathcal{P}_{za} \tag{12.15a}$$

$$\epsilon_r \mathcal{P}_z = \epsilon_r \mathcal{P}_{za} + \mathcal{Q}_x \sin \theta_i. \tag{12.15b}$$

2. Substitute Equation (12.13a) into Equation (12.13b) to eliminate $\mathcal{Q}_x$. This yields two alternative successive equations

$$\mathcal{P}_z \left(1 - \frac{\sin^2 \theta_i}{\mu_r \epsilon_r}\right) = \mathcal{P}_{za} + \frac{\sin \theta_i}{\epsilon_r} \mathcal{Q}_{xa} \tag{12.16a}$$

$$\mu_r \mathcal{Q}_x = \mu_r \mathcal{Q}_{xa} + \mathcal{P}_z \sin \theta_i. \tag{12.16b}$$

The first strategy, as it turns out, has a slightly more relaxed stability limit; also, the two methods have different dispersion characteristics. Using the first strategy, we can discretize Equations (12.14) and (12.15) to find our sequence of five update equations which, in the order they would be updated, are:

$$\mathcal{Q}_{xa}\Big|_{i,j+1/2}^{n+1/2} = \mathcal{Q}_{xa}\Big|_{i,j+1/2}^{n-1/2} - \frac{c\Delta t}{\mu_r \Delta y}\left(\mathcal{P}_z\Big|_{i,j+1}^{n} - \mathcal{P}_z\Big|_{i,j}^{n}\right) \tag{12.17a}$$

$$\mathcal{Q}_y\Big|_{i+1/2,j}^{n+1/2} = \mathcal{Q}_y\Big|_{i+1/2,j}^{n-1/2} - \frac{c\Delta t}{\mu_r \Delta x}\left(\mathcal{P}_z\Big|_{i+1,j}^{n} - \mathcal{P}_z\Big|_{i,j}^{n}\right) \tag{12.17b}$$

$$\mathcal{P}_{za}\Big|_{i,j}^{n+1/2} = \mathcal{P}_{za}\Big|_{i,j}^{n-1/2} + \frac{c\Delta t}{\epsilon_r \Delta x}\left(\mathcal{Q}_y\Big|_{i+1/2,j}^{n} - \mathcal{Q}_y\Big|_{i-1/2,j}^{n}\right)$$
$$- \frac{c\Delta t}{\epsilon_r \Delta y}\left(\mathcal{Q}_x\Big|_{i,j+1/2}^{n} - \mathcal{Q}_x\Big|_{i,j-1/2}^{n}\right) \tag{12.17c}$$

$$\mathcal{Q}_x\Big|_{i,j+1/2}^{n+1/2} = \frac{\mu_r \epsilon_r}{\mu_r \epsilon_r - \sin^2 \theta_i} \mathcal{Q}_{xa}\Big|_{i,j+1/2}^{n+1/2}$$
$$+ \frac{\epsilon_r \sin \theta_i}{2(\mu_r \epsilon_r - \sin^2 \theta_i)}\left(\mathcal{P}_{za}\Big|_{i,j+1}^{n+1/2} + \mathcal{P}_{za}\Big|_{i,j}^{n+1/2}\right) \tag{12.17d}$$

$$\mathcal{P}_z\Big|_{i,j}^{n+1/2} = \mathcal{P}_{za}\Big|_{i,j}^{n+1/2} + \frac{\sin \theta_i}{2\epsilon_r}\left(\mathcal{Q}_x\Big|_{i,j+1/2}^{n+1/2} + \mathcal{Q}_x\Big|_{i,j-1/2}^{n+1/2}\right). \tag{12.17e}$$

Note the spatial averaging of fields in the last two equations. This implementation introduces additional computation in that all fields $\mathcal{P}_z$, $\mathcal{Q}_x$, and $\mathcal{Q}_y$ must be updated every half time step, and storage of the fields is required two half time steps back; i.e., the fields must be known at times n and $(n - 1/2)$ in order to update time step $(n + 1/2)$.

The stability criterion for this method can be shown to be (in 2D):

$$\frac{v_p \Delta t}{\Delta y} \le \frac{\cos^2 \theta_i}{\sqrt{1 + \cos^2 \theta_i}}. \tag{12.18}$$

This stability criterion is less stringent than that of the multiple-grid method described above. The requirement shows that $v_p \Delta t / \Delta y \le 1/\sqrt{2}$ for $\theta_i = 0$, i.e., at normal incidence, but is then more stringent for increasing θ_i.

Again, one can very simply introduce lossy materials into this problem by including real conductivities applied to the $\mathcal{P}$ and $\mathcal{Q}$ fields. In this case the time domain differential equations have the form:

$$\frac{\mu_r}{c} \frac{\partial \mathcal{Q}_{xa}}{\partial t} + \frac{\sigma^m}{\eta_0} \mathcal{Q}_{xa} = -\frac{\partial \mathcal{P}_z}{\partial y} - \frac{\sigma^m \sin \theta_i}{\eta_0 \mu_r} \mathcal{P}_z \tag{12.19a}$$

$$\frac{\mu_r}{c} \frac{\partial \mathcal{Q}_y}{\partial t} + \frac{\sigma^m}{\eta_0} \mathcal{Q}_y = \frac{\partial \mathcal{P}_z}{\partial x} \tag{12.19b}$$

$$\frac{\epsilon_r}{c} \frac{\partial \mathcal{P}_{za}}{\partial t} + \sigma \eta_0 \mathcal{P}_{za} = \frac{\partial \mathcal{Q}_y}{\partial x} - \frac{\partial \mathcal{Q}_x}{\partial y} + \frac{\sigma \eta_0 \sin \theta_i}{\epsilon_r} \mathcal{Q}_x \tag{12.19c}$$

where σ and σ^m are electric and magnetic conductivities, respectively. Unfortunately, in this method, simple temporal averaging of the extra fields on the left-hand side, as was done for the lossy simple FDTD equations, leads to instability at incident angles greater than 30 degrees. A more general temporal averaging has been formulated that provides stability:

$$\frac{\sigma^m}{\eta_0} Q_{xa} \to \frac{\sigma^m}{\eta_0} \left[\beta Q_{xa} \Big|_{i,j}^{n+1} + (1 - 2\beta) Q_{xa} \Big|_{i,j}^{n+1/2} + \beta Q_{xa} \Big|_{i,j}^{n} \right] \tag{12.20}$$

where β is a parameter that is dependent on the incident angle, and has been optimized empirically:

$$\beta_{opt} = \frac{1}{2 \cos^2 \theta_i}.$$

This provides a brief introduction to the methods used to analyze periodic structures in the FDTD method. For further reading and example simulations, we refer the reader to papers by Roden et al. [6] and Harms et al. [7], as well as the book by Taflove and Hagness [4].

12.2   Modeling fine geometrical features

Often one finds that the structures we wish to model in FDTD have scale sizes smaller than the local cell size. These might be thin wires (as described in Chapter 7) or thin plates, or curved structures that are poorly modeled in Cartesian or other coordinate systems. It is possible to model these structures more accurately in a number of ways, the most obvious of which are nonuniform grids and nonorthogonal grids. Nonuniform orthogonal grids have variation of Δx, Δy, and/or Δz along the respective direction, but maintain orthogonal cell shapes (while not necessarily cubes). Nonorthogonal grids use mesh spacings that are irregular and do not fit the standard coordinate systems. These methods have been implemented with great success in a variety of applications; we will briefly discuss nonuniform grids in more detail along with other types of grids in Chapter 15. The downside of nonuniform grids is that the number of grid cells will increase, although depending on the structure, this can be a relatively small increase; however, the time step is determined by the smallest grid cell size, so the total simulation time may increase considerably.

In this section, we discuss a different method for modeling small-scale structures that allows us to remain within the uniform, orthogonal grid space. The methods use *local subcell gridding* around the structures of interest, but maintain the normal grid elsewhere. These can be very valuable in modeling curved structures, thin wires, and thin sheets, and these are the applications we will discuss in this section.

The general approach of this method is to use Ampère's law and/or Faraday's law in integral form, integrated on an array of orthogonal contours rather than cells; this is similar to the representation in Figure 2.3 in Chapter 2. Consider the extension of Figure 2.3 into three dimensions, yielding a 3D array of interlinked loops. The presence of irregular structures is accounted for by modifying the integral results on the relevant contours. Furthermore, contours can be more easily deformed to conform with the surface of structures. In the following subsections, we will see this contour method applied to a number of interesting structures.

12.2.1   Diagonal split-cell model

As an introduction to the concepts of subcell modeling, the simplest subcell model is an application for perfect electric conductor (PEC) structures where the PEC boundary cuts across the cell at a diagonal. Rather than simply applying $\mathscr{E} = 0$ to the nearest cell faces, we can model the PEC boundary along the diagonal of the cell; this is illustrated for the 2D TE case in Figure 12.6 as the thin solid line, where the PEC is the shaded region in the lower half-space. The inset figure should be ignored for now, as it will be used in the average properties model in the next section. The usual staircase pattern does a poor job of modeling the curved conductive surface; the diagonal model is a first-order improvement. If we focus on the $\mathscr{H}_z$ field component at the center of the enclosed cell and apply Faraday's law to the cell, assuming zero field in the PEC,

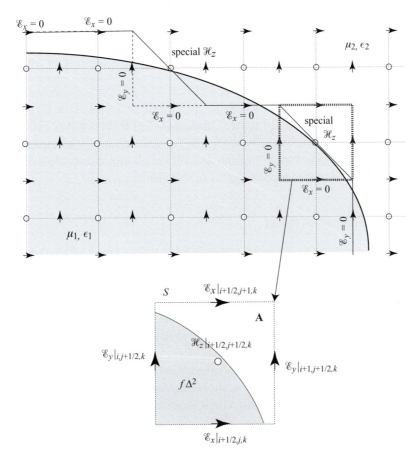

Figure 12.6 Simple contour-path models. The physical boundary we wish to approximate is shown with the thick, curved line; the usual staircase pattern is shown with a dashed line, while the diagonal split cell is shown with a solid line. The enlarged cell shows how the average properies model is applied to a cell.

we find

$$
\int \mu \frac{\partial \overline{\mathcal{H}}}{\partial t} \cdot d\mathbf{A} = \int \overline{\mathcal{E}} \cdot dS \longrightarrow \mu_0 \left(\frac{\mathcal{H}_z \Big|_{i+1/2,j+1/2,k}^{n+1/2} - \mathcal{H}_z \Big|_{i+1/2,j+1/2,k}^{n-1/2}}{\Delta t} \right) \cdot \underbrace{\frac{\Delta^2}{2}}_{\text{1/2 area of cell}}
$$

$$
= \left(\mathcal{E}_x \Big|_{i+1/2,j+1,k}^{n} - \underbrace{\mathcal{E}_x \Big|_{i+1/2,j,k}^{n}}_{=0 \text{ in PEC}} + \underbrace{\mathcal{E}_y \Big|_{i,j+1/2,k}^{n}}_{=0 \text{ in PEC}} - \mathcal{E}_y \Big|_{i+1,j+1/2,k}^{n} \right) \Delta
$$

which can then be solved for the update equation,

$$
\mathcal{H}_z\Big|_{i+1/2,j+1/2,k}^{n+1/2} = \mathcal{H}_z\Big|_{i+1/2,j+1/2,k}^{n-1/2} + \frac{2\Delta t}{\mu_0\Delta} \left(\mathcal{E}_x\Big|_{i+1/2,j+1,k}^{n} - \mathcal{E}_y\Big|_{i+1,j+1/2,k}^{n} \right) \tag{12.21}
$$

where $\Delta = \Delta x = \Delta y$ is the spatial grid size. In these equations we have included the k index to show that it can denote the 2D TE case or a slice of the 3D case. Thus, we simply invoke the modified update Equation (12.21) at the center of cells that are split diagonally. Note that this derivation for the TE case can easily be adapted to the TM case that is found more often in this book, but that the TE case is more physically satisfying, since we are setting tangential electric fields to zero.

12.2.2   Average properties model

In the more general case where the material boundary is not a PEC, a similar method can be used to more accurately model the structure. This model uses the fractions of the cell area that are covered by each material, which one might expect to give better accuracy than the diagonal split-cell model, in which it is assumed that half of the split cells lie within the PEC. Note that a diagonal split cell with non-PEC material can be applied if it is simply assumed that half of the cell lies on either side of the boundary.

   With ϵ_1 and μ_1 in the lower, shaded space in Figure 12.6, below the solid curved line, and ϵ_2 and μ_2 in the upper space, we use Faraday's law with the contour S around the area $\mathbf{A}$ to form the following discretized equation at the center of the cell enlarged at the right of the figure:

$$
\int \mu \frac{\partial \overline{\mathcal{H}}}{\partial t} \cdot d\mathbf{A} = \int \overline{\mathcal{E}} \cdot d\mathbf{S} \longrightarrow
$$

$$
\left(\frac{\mathcal{H}_z\Big|_{i+1/2,j+1/2,k}^{n+1/2} - \mathcal{H}_z\Big|_{i+1/2,j+1/2,k}^{n-1/2}}{\Delta t} \right) \left[\mu_1 f \Delta^2 + \mu_2(1-f)\Delta^2 \right] \tag{12.22}
$$

$$
= \left(\mathcal{E}_x\Big|_{i+1/2,j+1,k}^{n} - \mathcal{E}_x\Big|_{i+1/2,j,k}^{n} + \mathcal{E}_y\Big|_{i,j+1/2,k}^{n} - \mathcal{E}_y\Big|_{i+1,j+1/2,k}^{n} \right) \Delta
$$

where f is the fraction of the cell that lies in region 1. Note that the factor in square brackets on the left-hand side replaces $\mu_0\Delta^2$, the permeability times the area of the cell, with the sum of two such "areas," where to each is applied the appropriate value of μ. The update equation for this method becomes:

$$
\mathcal{H}_z\Big|_{i+1/2,j+1/2,k}^{n+1/2} = \mathcal{H}_z\Big|_{i+1/2,j+1/2,k}^{n-1/2} + \frac{\Delta t}{[\mu_1 f + \mu_2(1-f)]\Delta} \tag{12.23}
$$

$$
\times \left(\mathcal{E}_x\Big|_{i+1/2,j+1,k}^{n} - \mathcal{E}_x\Big|_{i+1/2,j,k}^{n} + \mathcal{E}_y\Big|_{i,j+1/2,k}^{n} - \mathcal{E}_y\Big|_{i+1,j+1/2,k}^{n} \right).
$$

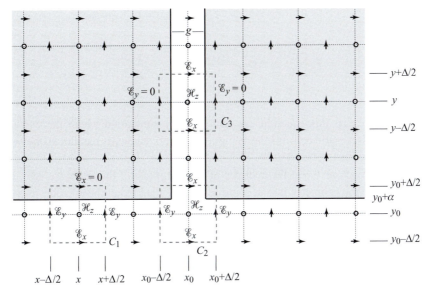

Figure 12.7 Narrow slot model. Faraday's law contours are drawn around the locations shown in order to derive update equations at those locations. Note that each of the three contours shown must be treated with special consideration.

Similar modifications can be made to the other update equations; in each, μ or ϵ is simply replaced by the "average" μ or ϵ in the cells along the boundary, where the parameters are averaged by their relative contributions in area.

12.2.3   The narrow slot

In Chapter 7, we introduced the contour method for simulating a thin wire, where the structure is narrower than the smallest grid cell size. A method that is nearly identical can be applied to the opposite case, known as the "narrow slot," where we have a slot through a conducting medium that is narrower than the grid cell size, through which we wish to propagate an electromagnetic wave.

The two-dimensional slot in Figure 12.7 shows how one would implement such a subcell structure. A narrow gap of width $g < \Delta$ is introduced, and for generality the slot may begin at a height $y_0 + \alpha$ that is not coincident with a grid cell edge. There are three distinct regions in this problem that require special attention, with contours marked C_1, C_2, and C_3. Inside the shaded PEC region, the electric fields are zero; we use this information to form our modified update equations.

1. On contour C_1, which is away from the slot region but borders on the conductor, we apply Faraday's law as before, where now the area **A** has dimensions $\Delta(\Delta/2 + \alpha)$, and the contour S has edges of length $\Delta/2 + \alpha$ along the y-direction. Noting that

$E_x|_{x,y_0+\Delta/2} = 0$ along the top edge of the grid cell, we find:

$$\mathcal{H}_z\Big|_{x,y_0}^{n+1/2} = \mathcal{H}_z\Big|_{x,y_0}^{n-1/2} + \Delta t \frac{\left(\mathcal{E}_y\Big|_{x-\Delta/2,y_0}^{n} - \mathcal{E}_x\Big|_{x+\Delta/2,y_0}^{n}\right) \cdot (\Delta/2 + \alpha) - \mathcal{E}_x\Big|_{x,y_0-\Delta/2}^{n} \Delta}{\mu_0 \Delta(\Delta/2 + \alpha)}.$$

(12.24)

Note that this equation can apply to any simulation where the conductor border crosses a grid cell, as long as the conductor boundary is parallel to one of the cardinal directions.

2. On contour C_2, at the aperture of the slot, none of the fields are zero, but we must take the geometry into account as we integrate along the contour, using the correct segment lengths and area to apply to S and $\mathbf{A}$. This yields the update equation:

$$\mathcal{H}_z\Big|_{x_0,y_0}^{n+1/2} = \mathcal{H}_z\Big|_{x_0,y_0}^{n-1/2} + \frac{\Delta t}{\mu_0\left[(\Delta/2 + \alpha)\Delta + (\Delta/2 - \alpha)g\right]}$$

(12.25)

$$\times \left[\mathcal{E}_x\Big|_{x_0,y_0+\Delta/2}^{n} g - \mathcal{E}_x\Big|_{x_0,y_0-\Delta/2}^{n} \Delta + \left(\mathcal{E}_y\Big|_{x_0-\Delta/2,y_0}^{n} - \mathcal{E}_y\Big|_{x_0+\Delta/2,y_0}^{n}\right) \cdot (\Delta/2 + \alpha)\right]$$

where the denominator in square brackets is the area of the region outside the PEC.

3. On contour C_3, within the slot, the integral of the contour leaves only the $\mathcal{E}_x$ components along segments of S with length g, so the update equation becomes:

$$\mathcal{H}_z\Big|_{x_0,y}^{n+1/2} = \mathcal{H}_z\Big|_{x_0,y}^{n-1/2} + \Delta t \frac{\left(\mathcal{E}_x\Big|_{x_0,y+\Delta/2}^{n} - \mathcal{E}_x\Big|_{x_0,y-\Delta/2}^{n}\right) g}{\mu_0 g \Delta}.$$

(12.26)

Note that the Equations (12.24)–(12.26) above have not been fully simplified; terms can cancel such as the g terms in Equation (12.26). However, they have been left this way to emphasize the Faraday's law application. In the numerator on the right-hand side of each equation is the line integral around the respective contours; this simplifies as field values multiplied by the respective line lengths that are outside the PEC. In the denominators are the areas within the contours; more accurately these would be multiplied on the left-hand side, as the left-hand side represents the surface integral of $\partial\mathcal{H}/\partial t$ within the contour. We have moved them to the right-hand side denominator in order to yield usable update equations.

Note that similar procedures can be applied in the 2D TM mode or in 3D by following similar procedures, and similarly, a technique can be derived where the material is not PEC but a real dielectric. Some of these ideas will be explored in the problems at the end of this chapter.

12.2.4　Dey-Mittra techniques

We wish now to find a technique that can accurately model curved structures in a method analogous to the narrow slot above. Dey and Mittra [8, 9] introduced conformal modeling techniques for PEC and material structures that have since evolved to a very high degree of stability and accuracy. In this scenario, depicted in Figure 12.8, the PEC boundary of a

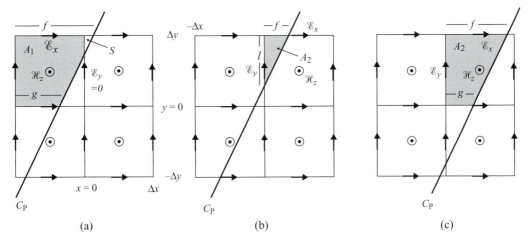

Figure 12.8 The three scenarios for the Dey-Mittra PEC cell. Depending on where the PEC boundary crosses the grid cell, these are the possible configurations which must each be treated separately.

curved surface in 3D cuts across grid cells at arbitrary angles, and in each cell, these cuts are approximated as straight edges. Doing so simplifies our mathematical description of the area $\mathbf{A}$ and contour S of the resulting polygons.

As shown in Figure 12.8, there are three different types of cell cuts that each require unique treatment. In this figure, the PEC is to the right of C_P, and the shaded region is the part of the cell of interest that is outside the PEC. Note that we are looking at the TM mode in 2D, or, equivalently, a TE_z slice of a 3D space; similar constructions are necessary for the complementary update equations.

1. Consider Figure 12.8a. We wish to find the corrected update equation for the top left cell, whose area in free space is given by A_1. If we apply Faraday's law and integrate the contour, noting that $\mathscr{E}_y|_{0,\Delta y/2} = 0$ since it is inside the PEC, we find

$$\mathscr{H}_z\Big|_{-\Delta x/2,\Delta y/2}^{n+1/2} = \mathscr{H}_z\Big|_{-\Delta x/2,\Delta y/2}^{n-1/2} \tag{12.27}$$

$$+ \frac{\Delta t}{\mu_0 A_1}\left(\mathscr{E}_y\Big|_{-\Delta x,\Delta y/2}^{n}\Delta y + \mathscr{E}_x\Big|_{-\Delta x/2,\Delta y}^{n} f - \mathscr{E}_x\Big|_{-\Delta x/2,0}^{n} g\right)$$

where A_1 is the area of the shaded region, which upon inspection of the geometry is equal to

$$A_1 = g\Delta y + (f-g)\Delta y/2 = (f+g)\Delta y/2. \tag{12.28}$$

Note that this integration has neglected the small area S in the top right cell. It has been shown that for stability, this region must be small, and the CFL stability condition is affected by the size of this region; for details pertaining to this stability requirement, see [10].

2. In Figure 12.8b, we have a different geometry, where the cut through the cell yields the small triangle A_2 in the top right cell to be accounted for. For the top left cell, all four of the electric field components are outside the PEC, so it can be updated in the usual manner, after adjusting for the area of the region. Turning our attention to the small shaded triangular cell, we need to update $\mathscr{H}_z$ in the center of the adjacent cell on the right, despite the fact that it is inside the PEC. Only two sides of the triangle are in free space (outside the PEC), so the line integral yields

$$\mathscr{H}_z\Big|_{\Delta x/2,\Delta y/2}^{n+1/2} = \mathscr{H}_z\Big|_{\Delta x/2,\Delta y/2}^{n-1/2} + \frac{\Delta t}{\mu_0 A_2}\left(\mathscr{E}_y\Big|_{0,\Delta y/2}^{n} l + \mathscr{E}_x\Big|_{\Delta x/2,\Delta y}^{n} f\right) \qquad (12.29)$$

where the area $A_2 = fl/2$. It is important to note that in this case, we are updating a field $\mathscr{H}_z$ that is *inside the PEC*. Also, the $\mathscr{E}_x$ component used may be zero if it lies within the PEC, as in the case shown, but in other geometries might be outside C_P.

3. In Figure 12.8c, we are interested in A_2 of the top right cell again, but this time we have three sides of the polygon to deal with. The only difference between this and the first case is the small region S in the first case, which, as mentioned earlier, affects stability. Here, we have the same update equation (adjusted for the field locations shown):

$$\mathscr{H}_z\Big|_{\Delta x/2,\Delta y/2}^{n+1/2} = \mathscr{H}_z\Big|_{\Delta x/2,\Delta y/2}^{n-1/2} + \frac{\Delta t}{\mu_0 A_1}\left(\mathscr{E}_y\Big|_{0,\Delta y/2}^{n} \Delta y + \mathscr{E}_x\Big|_{\Delta x/2,\Delta y}^{n} f - \mathscr{E}_x\Big|_{\Delta x/2,0}^{n} g\right).$$

$$(12.30)$$

Similar to Figure 12.8b, the $\mathscr{H}_z$ component of interest may or may not be inside the PEC, but it must be updated regardless. The surrounding $\overline{\mathscr{E}}$ components, on the other hand, will be zero if they are inside the PEC and updated normally if they are not.

Material structures

The Dey-Mittra technique is also easily adapted to material structures other than PEC. The method here simply involves computing a local *effective permittivity* or permeability near the boundary of the structure, similar to the average properties model described earlier. Simply put, in a particular 3D cell located at i, j, k that is sliced by a material boundary, yielding volume V_1 in the region with ϵ_1, and V_2 in the region with ϵ_2, the "effective permittivity" of this cell is

$$\epsilon_{\text{eff}} = \frac{\epsilon_1 V_1 + \epsilon_2 V_2}{V_1 + V_2} \qquad (12.31)$$

where $V_1 + V_2 = \Delta x \Delta y \Delta z$, the volume of the cell. This effective permittivity is then used in the appropriate update equations for $\overline{\mathscr{E}}$ components in the cell. A similar expression is used when the material boundary separates two materials with different permeabilities μ_1 and μ_2.

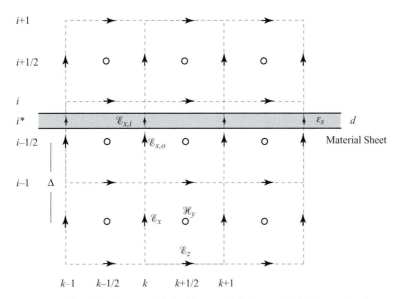

Figure 12.9 The thin-sheet model. A thin material sheet with thickness d and parameters ϵ_s and σ_s cuts through the FDTD cell at some arbitrary location. Update equations must be modified inside the sheet and in the cell surrounding the sheet.

12.2.5   Thin material sheets

Figure 12.9 shows the setup for a problem involving a sheet of some material (perfect conductor or otherwise) that has a thickness less than the cell width. This technique is similar to the thin-wire approximation that was discussed in Chapter 7, only now our conducting wire is a 2D plane with finite thickness, and we consider the case where it is not a perfect conductor. A method for the solution of this type of this material sheet was developed by Maloney and Smith [11]. Note that this figure illustrates a 3D problem in which we are showing a TE slice, but a similar analysis could be applied as a 2D TM problem.

The thin sheet has permittivity ϵ_s and conductivity σ_s (it could also have permeability μ_s), and thickness $d < \Delta$, where Δ is the grid cell size. In this setup it is necessary to split the $\mathcal{E}_x$ field nearest to the material sheet into two parts, $\mathcal{E}_{x,o}$ outside the sheet and $\mathcal{E}_{x,i}$ inside the sheet, and to then implement the contour integral in three regions: above, within, and below the sheet. In this problem we will consider a material with $\mu = \mu_0$, but with permittivity ϵ_s and conductivity σ_s. Note that we have drawn the general case where the thin sheet cuts through the cell at an arbitrary location, but it could easily be aligned with a cell edge to simplify the problem.

$\mathcal{H}_x$, $\mathcal{H}_y$, and $\mathcal{E}_z$ do not need to be split in this case, because $\mathcal{H}_y$ and $\mathcal{E}_z$ are tangential and thus continuous across the boundary, and $\mathcal{H}_x$ is continuous due to the fact that $\mu = \mu_0$. Note that in other setups of this problem, other field components need to be split if the contour S or area $\mathbf{A}$ used in the respective integral equations is crossed by the material sheet.

The equations for $\mathcal{E}_{x,o}$ and $\mathcal{E}_{x,i}$ look exactly as they would normally:

$$\mathcal{E}_{x,o}\Big|_{i-1/2,j,k}^{n+1} = \mathcal{E}_{x,o}\Big|_{i-1/2,j,k}^{n} + \frac{\Delta t}{\epsilon_0}[\nabla \times \mathcal{H}]_{yz}\Big|_{i-1/2,j,k}^{n+1/2} \tag{12.32a}$$

$$\mathcal{E}_{x,i}\Big|_{i^*,j,k}^{n+1} = \left(\frac{2\epsilon_s - \sigma_s \Delta t}{2\epsilon_s + \sigma_s \Delta t}\right)\mathcal{E}_{x,i}\Big|_{i^*,j,k}^{n} + \left(\frac{2\Delta t \sigma_s}{2\epsilon_s + \sigma_s \Delta t}\right)[\nabla \times \mathcal{H}]_{yz}\Big|_{i-1/2,j,k}^{n+1/2}. \tag{12.32b}$$

Note the curl operator is applied in the y–z plane, implying differences across z and y. Also, since i^* corresponds the closest with $(i - 1/2)$ in the grid, the curl operator applies at the $(i - 1/2)$ level in x. We next use Ampère's law contours that cross the sheet to find the update equations for $\mathcal{E}_y$ and $\mathcal{E}_z$ near the sheet:

$$\mathcal{E}_y\Big|_{i,j+1/2,k}^{n+1} = \left(\frac{2\epsilon_{av} - \sigma_{av}\Delta t}{2\epsilon_{av} + \sigma_{av}\Delta t}\right)\mathcal{E}_y\Big|_{i,j+1/2,k}^{n} + \left(\frac{2\Delta t \sigma_{av}}{2\epsilon_{av} + \sigma_{av}\Delta t}\right)[\nabla \times \mathcal{H}]_{xz}\Big|_{i,j+1/2,k}^{n+1/2} \tag{12.33a}$$

$$\mathcal{E}_z\Big|_{i,j,k+1/2}^{n+1} = \left(\frac{2\epsilon_{av} - \sigma_{av}\Delta t}{2\epsilon_{av} + \sigma_{av}\Delta t}\right)\mathcal{E}_z\Big|_{i,j,k+1/2}^{n} + \left(\frac{2\Delta t \sigma_{av}}{2\epsilon_{av} + \sigma_{av}\Delta t}\right)[\nabla \times \mathcal{H}]_{xz}\Big|_{i,j,k+1/2}^{n+1/2} \tag{12.33b}$$

where, as in the Dey-Mittra model, ϵ_{av} and σ_{av} are defined as averaged quantities, taking into account the thickness of the sheet:

$$\epsilon_{av} = \left(1 - \frac{d}{\Delta x}\right)\epsilon_0 + \frac{d}{\Delta x}\epsilon_s \quad \text{and} \quad \sigma_{av} = \frac{d}{\Delta x}\sigma_s. \tag{12.34}$$

Now, the $\mathcal{H}_x$ components do not require any modification from the usual free-space FDTD equations, because a contour circling these components will lie parallel to the sheet and below it. The $\mathcal{H}_y$ and $\mathcal{H}_z$ components, however, are tangential to the sheet (and thus have contours perpendicular to the sheet), and those contours include both $\mathcal{E}_{x,o}$ and $\mathcal{E}_{x,i}$ terms. The update equations become:

$$\mathcal{H}_z\Big|_{i-1/2,j+1/2,k}^{n+1/2} = \mathcal{H}_z\Big|_{i-1/2,j+1/2,k}^{n-1/2} + \frac{\Delta t}{\mu_0}\left[\left(\frac{\Delta x - d}{\Delta x}\right)\frac{\partial \mathcal{E}_{x,o}}{\partial y}\Big|_{i-1/2,j+1/2,k}^{n}\right. \tag{12.35a}$$

$$\left. + \frac{d}{\Delta x}\frac{\partial \mathcal{E}_{x,i}}{\partial y}\Big|_{i^*,j+1/2,k}^{n} - \frac{\partial \mathcal{E}_y}{\partial x}\Big|_{i-1/2,j+1/2,k}^{n}\right]$$

$$\mathcal{H}_y\Big|_{i-1/2,j,k+1/2}^{n+1/2} = \mathcal{H}_y\Big|_{i-1/2,j,k+1/2}^{n-1/2} + \frac{\Delta t}{\mu_0}\left[\left(\frac{\Delta x - d}{\Delta x}\right)\frac{\partial \mathcal{E}_{x,o}}{\partial z}\Big|_{i-1/2,j,k+1/2}^{n}\right. \tag{12.35b}$$

$$\left. + \frac{d}{\Delta x}\frac{\partial \mathcal{E}_{x,i}}{\partial z}\Big|_{i^*,j,k+1/2}^{n} - \frac{\partial \mathcal{E}_z}{\partial x}\Big|_{i-1/2,j,k+1/2}^{n}\right]$$

where the partial derivatives are implemented with the usual centered differencing at the relevant location.

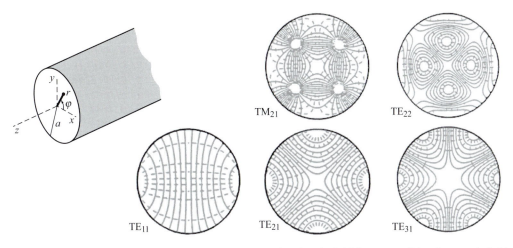

Figure 12.10 Modes in a cylindrical waveguide. Electric field lines are solid and magnetic field lines are dashed. The modes shown are not symmetric in either radius r or azimuth ϕ; but the ϕ-dependence can be written analytically. Adapted from [12].

In this section, we have introduced a number of examples of modeling subcell structures. In each case, we have applied Faraday's law and/or Ampère's law to a grid cell in which a material boundary is present, and found modified versions of the FDTD algorithm for the appropriate field components. We must point out, however, that these methods do not preserve the stability criterion of the FDTD algorithm; in fact, their stability is not well known analytically, and typically the stability is established through numerical trial-and-error.

12.3   Bodies of revolution

In Section 4.6 we introduced the FDTD algorithm for cylindrical and spherical coordinates. Among the choices for cylindrical coordinates in 2D is the case where the simulation is symmetric in azimuth, and we can then reduce the simulation space to an r–z plane. On the other hand, without cylindrical symmetry, we are forced to use the full 3D cylindrical FDTD method.

However, in some cases, the simulation space may not be azimuthally symmetric, but a simulation can be designed so that the full 3D method is not required. The *Bodies of Revolution* (BOR) method is used in modeling systems in cylindrical coordinates where the objects have cylindrical symmetry, but the electromagnetic fields resulting may not. For example, Figure 12.10 shows a number of mode patterns inside a cylindrical waveguide (i.e., an optical fiber); one can observe that none of these patterns are azimuthally symmetric, but their azimuthal dependence can be written as a sine or cosine. By writing the azimuthal ϕ variation analytically as a Fourier series, we can reduce the numerical scheme to the equivalent of a 2D simulation. This algorithm is sometimes labeled "2.5-dimensional"; each simulation run yields a solution for a single Fourier

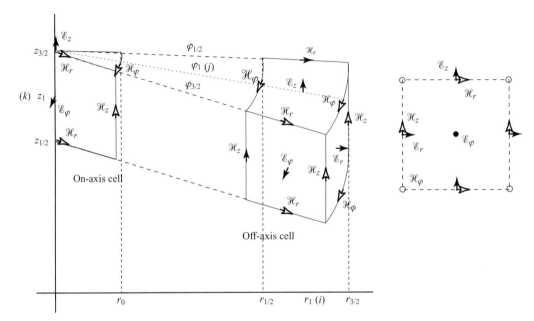

Figure 12.11 The cylindrical coordinate FDTD cell used for bodies of rotation methods. Note the placement of fields on-axis for the unique on-axis cells. The 3D cylindrical coordinate Yee cell collapses to the 2D cell shown on the right when the azimuthal dependence is removed.

mode of azimuthal variation, and so a sequence of runs must be computed until the desired number of Fourier modes are included.

Derivation of the bodies of revolution method begins with Ampère's and Faraday's laws in integral form,

$$\oint_C \overline{\mathcal{H}} \cdot dl = \iint_S \sigma \overline{\mathcal{E}} \cdot dS + \frac{\partial}{\partial t} \iint_S \overline{\mathcal{D}} \cdot dS \tag{12.36a}$$

$$\oint_C \overline{\mathcal{E}} \cdot dl = -\frac{\partial}{\partial t} \iint_S \overline{\mathcal{B}} \cdot dS. \tag{12.36b}$$

Next, we write the electric and magnetic fields in terms of Fourier series with an arbitrary number of sine and cosine expansions:

$$\overline{\mathcal{E}} = \sum_{m=0}^{\infty} \left[\overline{\mathcal{E}}_u \cos(m\phi) + \overline{\mathcal{E}}_v \sin(m\phi) \right] \tag{12.37a}$$

$$\overline{\mathcal{H}} = \sum_{m=0}^{\infty} \left[\overline{\mathcal{H}}_u \cos(m\phi) + \overline{\mathcal{H}}_v \sin(m\phi) \right] \tag{12.37b}$$

where the u and v subscripts are simply to keep track of the cosine and sine terms of the Fourier expansion.

The unit cell that we will use is shown in Figure 12.11. Similar to our analysis of the FDTD cylindrical method in Section 4.6, the cells on-axis (at $r = 0$) will require special consideration, so we will leave those for last. For the off-axis cells, we compute each of the fields by substituting Equations (12.37) into (12.36). Starting with the $\mathscr{E}_r$ fields, we integrate along the edge of the rightmost face in Figure 12.11 (i.e., the face surrounding $\mathscr{E}_r$ at $r = r_{3/2}$).[4] We have:

$$\epsilon \frac{\partial}{\partial t} \int_{z_{1/2}}^{z_{3/2}} \int_{\phi_{1/2}}^{\phi_{3/2}} \left[\mathscr{E}_{r,u}(\phi_1, z_1, t) \cos m\phi + \mathscr{E}_{r,v}(\phi_1, z_1, t) \sin m\phi \right] r_{3/2} d\phi dz \quad (12.38)$$

$$= \int_{z_{1/2}}^{z_{3/2}} \left[\mathscr{H}_{z,u}(\phi_{3/2}, z_1, t) \cos m\phi_{3/2} + \mathscr{H}_{z,v}(\phi_{3/2}, z_1, t) \sin m\phi_2 \right] dz$$

$$+ \int_{\phi_{3/2}}^{\phi_{1/2}} \left[\mathscr{H}_{\phi,u}(\phi_1, z_{3/2}, t) \cos m\phi + \mathscr{H}_{\phi,v}(\phi_1, z_{3/2}, t) \sin m\phi \right] r_{3/2} d\phi$$

$$+ \int_{z_{3/2}}^{z_{1/2}} \left[\mathscr{H}_{z,u}(\phi_{1/2}, z_1, t) \cos m\phi_{1/2} + \mathscr{H}_{z,v}(\phi_{1/2}, z_1, t) \sin m\phi_1 \right] dz$$

$$+ \int_{\phi_{1/2}}^{\phi_{3/2}} \left[\mathscr{H}_{\phi,u}(\phi_1, z_{1/2}, t) \cos m\phi + \mathscr{H}_{\phi,v}(\phi_1, z_{1/2}, t) \sin m\phi \right] r_{3/2} d\phi$$

where it is understood that each of the field components lies at $r = r_{3/2}$. We can go ahead and integrate this equation, assuming the fields to be constant along their respective edges. The resulting equation has terms in $\sin m\phi_{1/2}$, $\sin m\phi_{3/2}$, $\cos m\phi_{1/2}$, and $\cos m\phi_{3/2}$, which can be grouped into separate equations, yielding the following four equations:

$$\left[-\Delta z \mathscr{H}_{z,u}(\phi_{1/2}, z_1, t) - \frac{r_{3/2}}{m} \mathscr{H}_{\phi,v}(\phi_1, z_{3/2}, t) + \frac{r_{3/2}}{m} \mathscr{H}_{\phi,v}(\phi_1, z_{1/2}, t) \right] \cos m\phi_{1/2}$$

$$= \left[\frac{r_{3/2} \epsilon \Delta z}{m} \frac{\partial}{\partial t} \mathscr{E}_{r,v}(\phi_1, z_1, t) \right] \cos m\phi_{1/2} \quad (12.39a)$$

$$\left[\Delta z \mathscr{H}_{z,u}(\phi_{3/2}, z_1, t) + \frac{r_{3/2}}{m} \mathscr{H}_{\phi,v}(\phi_1, z_{3/2}, t) - \frac{r_{3/2}}{m} \mathscr{H}_{\phi,v}(\phi_1, z_{1/2}, t) \right] \cos m\phi_{3/2}$$

$$= \left[-\frac{r_{3/2} \epsilon \Delta z}{m} \frac{\partial}{\partial t} \mathscr{E}_{r,v}(\phi_1, z_1, t) \right] \cos m\phi_{3/2} \quad (12.39b)$$

$$\left[-\Delta z \mathscr{H}_{z,v}(\phi_{1/2}, z_1, t) - \frac{r_{3/2}}{m} \mathscr{H}_{\phi,u}(\phi_1, z_{3/2}, t) + \frac{r_{3/2}}{m} \mathscr{H}_{\phi,u}(\phi_1, z_{1/2}, t) \right] \sin m\phi_{1/2}$$

$$= \left[\frac{r_{3/2} \epsilon \Delta z}{m} \frac{\partial}{\partial t} \mathscr{E}_{r,u}(\phi_1, z_1, t) \right] \sin m\phi_{1/2} \quad (12.39c)$$

$$\left[\Delta z \mathscr{H}_{z,v}(\phi_{3/2}, z_1, t) + \frac{r_{3/2}}{m} \mathscr{H}_{\phi,u}(\phi_1, z_{3/2}, t) - \frac{r_{3/2}}{m} \mathscr{H}_{\phi,u}(\phi_1, z_{1/2}, t) \right] \cos m\phi_{3/2}$$

$$= \left[-\frac{r_{3/2} \epsilon \Delta z}{m} \frac{\partial}{\partial t} \mathscr{E}_{r,u}(\phi_1, z_1, t) \right] \cos m\phi_{3/2}. \quad (12.39d)$$

Now, we can remove the ϕ dependence of each of the field components when we collapse the problem to the analysis of a single 2D Fourier mode. This is shown in

---

[4] In the grid locations, we have left off the i, j, k labels for simplicity; in reality, the locations are mapped such that $r_1 \to r_i$, $r_{3/2} \to r_{i+1/2}$, etc., as shown.

Figure 12.11: the 3D cell is reduced to the 2D cell on the right, and ϕ is no longer defined. After canceling the $\sin m\phi$ and $\cos m\phi$ terms from these equations, we can see that the first two equations are identical, as are the last two, so we really have two independent equations for solving $\mathcal{E}_r$. Similar equations can be derived for the $\mathcal{E}_\phi$ component using the nearest face in Figure 12.11 (the face at $\phi_{3/2}$), and for the $\mathcal{E}_z$ component using the topmost face (at $z_{3/2}$). The complete set of equations is given in [4]. Ultimately, we are left with two sets of six equations, the first of which is shown below (note that we have now reverted back to the usual i, j, k indexing):

$$\frac{\partial}{\partial t}\mathcal{E}_{r,v}\Big|_{i+1/2,k} = \frac{1}{\epsilon}\left[\frac{\mathcal{H}_{\phi,v}\big|_{i+1/2,k-1/2}-\mathcal{H}_{\phi,v}\big|_{i+1/2,k+1/2}}{\Delta z}\right] - \frac{m}{\epsilon r_{i+1/2}}\mathcal{H}_{z,u}\Big|_{i+1/2,k}$$

$$\text{(12.40a)}$$

$$\frac{\partial}{\partial t}\mathcal{E}_{\phi,u}\Big|_{i,k} = \frac{1}{\epsilon}\left[\frac{\mathcal{H}_{z,u}\big|_{i-1/2,k}-\mathcal{H}_{z,u}\big|_{i+1/2,k}}{\Delta r}+\frac{\mathcal{H}_{r,u}\big|_{i,k+1/2}-\mathcal{H}_{r,u}\big|_{i,k-1/2}}{\Delta z}\right] \quad \text{(12.40b)}$$

$$\frac{\partial}{\partial t}\mathcal{E}_{z,v}\Big|_{i,k+1/2} = \frac{2}{\epsilon(r_{i+1/2}^2 - r_{i-1/2}^2)}\Big[r_{i+1/2}\mathcal{H}_{\phi,v}\big|_{i+1/2,k+1/2}$$

$$-r_{i-1/2}\mathcal{H}_{\phi,v}\big|_{i-1/2,k+1/2}+m\,\Delta r\,\mathcal{H}_{r,u}\big|_{i,k+1/2}\Big] \quad \text{(12.40c)}$$

$$\frac{\partial}{\partial t}\mathcal{H}_{r,u}\Big|_{i,k+1/2} = \frac{1}{\mu}\left[\frac{\mathcal{E}_{\phi,u}\big|_{i,k+1}-\mathcal{E}_{\phi,u}\big|_{i,k}}{\Delta z}\right] - \frac{m}{\mu r}\mathcal{E}_{z,v}\big|_{i,k+1/2} \quad \text{(12.40d)}$$

$$\frac{\partial}{\partial t}\mathcal{H}_{\phi,v}\Big|_{i+1/2,k+1/2} = \frac{1}{\mu}\left[\frac{\mathcal{E}_{z,v}\big|_{i+1,k+1/2}-\mathcal{E}_{z,v}\big|_{i,k+1/2}}{\Delta r}+\frac{\mathcal{E}_{r,v}\big|_{i+1/2,k}-\mathcal{E}_{r,v}\big|_{i+1/2,k+1}}{\Delta z}\right]$$

$$\text{(12.40e)}$$

$$\frac{\partial}{\partial t}\mathcal{H}_{z,u}\Big|_{i+1/2,k} = \frac{2}{\mu(r_{i+1/2}^2 - r_{i-1/2}^2)}$$

$$\times\left[r_{i+1/2}\mathcal{E}_{\phi,u}\big|_{i,k}-r_{i-1/2}\mathcal{E}_{\phi,u}\big|_{i+1,k}+m\,\Delta r\,\mathcal{E}_{r,v}\big|_{i+1/2,k}\right]. \quad \text{(12.40f)}$$

Notice in the above set of six equations that we have left off the ϕ index in the subscripts; thanks to the fact that we have eliminated the ϕ-dependence by using Fourier series, there are no longer any ϕ indices! The 3D cell at the center of Figure 12.11 has now been "collapsed" in the ϕ-direction to the 2D cell shown at the right. In this 2D cell, some of the fields are co-located, as shown.

A second set of six equations results with the reciprocal terms used, i.e., exchanging $\mathcal{E}_{r,u}$ anywhere it is found by $\mathcal{E}_{r,v}$, and so forth. This second set of six equations is simply rotated 90 degrees from the first set, and is independent in isotropic media. We can choose the first set to difference without loss of generality. By applying simple time-differencing on the left-hand side, we have our discrete update equations; the terms on the right-hand side are already differenced.

On-axis cells

To deal with the on-axis field components at the left of Figure 12.11, we use the contours shown in the figure. Note that $\mathcal{E}_\phi$ is defined on the axis and not at the center of the "cell" as one might expect. To compute $\mathcal{E}_z$ at top left, we integrate the three-sided contour on the top, which involves two $\mathcal{H}_r$ terms and one $\mathcal{H}_\phi$ term. After integrating and collecting terms, we are left with four equations, two of which are redundant:

$$\left[mr_0 \mathcal{H}_{r,u}(0, z_{3/2}, t) + r_0 \mathcal{H}_{\phi,v}(r_0, z_{3/2}, t) \right] \cos m\phi_{1/2}$$

$$= \left[\frac{\epsilon r_0^2}{2} \frac{\partial}{\partial t} \mathcal{E}_{z,v}(0, z_{3/2}, t) \right] \cos m\phi_{1/2} \tag{12.41a}$$

$$\left[mr_0 \mathcal{H}_{r,v}(0, z_{3/2}, t) - r_0 \mathcal{H}_{\phi,u}(r_0, z_{3/2}, t) \right] \sin m\phi_{1/2}$$

$$= \left[-\frac{\epsilon r_0^2}{2} \frac{\partial}{\partial t} \mathcal{E}_{z,u}(0, z_{3/2}, t) \right] \sin m\phi_{1/2} \tag{12.41b}$$

where $r_0 = \Delta r/2$ is the radius at the location of the first $\mathcal{H}_z$ component. Since we are on axis, we would expect in this case that there is no azimuthal dependence, i.e. $m = 0$; this eliminates the $\mathcal{H}_r$ contribution from the equations above, and the two equations become identical apart from the u and v fields. We are thus left with the single difference equation for $\mathcal{E}_z$:

$$\frac{\partial}{\partial t} \mathcal{E}_{z,v}(r = 0, z_{3/2}, t) = \frac{2}{\epsilon r_0} \mathcal{H}_{\phi,v}(r_0, z_{3/2}, t) \tag{12.42}$$

which yields the update equation,

$$\mathcal{E}_z \Big|_{0,k+1/2}^{n+1} = \mathcal{E}_z \Big|_{0,k+1/2}^{n} + \frac{4\Delta t}{\epsilon \Delta r} \mathcal{H}_{\phi,v} \Big|_{i=1/2,k+1/2}^{n+1/2} \tag{12.43}$$

where again $\Delta r = 2r_0$ to remain consistent with the off-axis cells. Note that this equation is the same as one would use in a standard 2D cylindrical coordinate FDTD simulation for the $r = 0$ update, as derived in Section 4.6.

We also need to find $\mathcal{E}_\phi$ on the axis, and this uses the rectangular contour at the left of Figure 12.11. This component, it turns out, is zero for $m \neq 1$; notice that in Figure 12.10, of the modes shown, only the TE$_{11}$ mode has nonzero electric fields at the origin in the r–ϕ plane shown. This contour again involves three terms, two $\mathcal{H}_r$ terms and one $\mathcal{H}_z$ term, since $\mathcal{H}_z$ is zero on the axis for $m = 1$. Following the same technique as above, and assuming the $\mathcal{E}_\phi$ field is uniform over the area of the cell from $r = 0$ to $r_0 = \Delta r/2$, we find the differential equation as:

$$\epsilon \Delta z \frac{\Delta r}{2} \frac{\partial}{\partial t} \mathcal{E}_{\phi,v}(0, z_1, t) = -\Delta z \mathcal{H}_{z,v}(r_0, z_1, t) + \frac{\Delta r}{2} \left[\mathcal{H}_{r,v}(0, z_{3/2}, t) - \mathcal{H}_{r,v}(0, z_{1/2}, t) \right] \tag{12.44}$$

which can be manipulated to find the update equation:

$$\mathcal{E}_\phi \Big|_{0,k}^{n+1} = \mathcal{E}_\phi \Big|_{0,k}^{n} - \frac{2\Delta t}{\epsilon \Delta r} \mathcal{H}_z \Big|_{1/2,k}^{n+1/2} + \frac{\Delta t}{\epsilon \Delta z} \left(\mathcal{H}_r \Big|_{0,k+1/2}^{n+1/2} - \mathcal{H}_r \Big|_{0,k-1/2}^{n+1/2} \right). \tag{12.45}$$

Finally, we have the $\mathcal{H}_r$ term on the axis to deal with, which is also zero for $m \neq 1$. Consider the $\mathcal{H}_r$ component at the very top left corner of the cell shown in Figure 12.11. We start from Faraday's law and substitute the Fourier expanded fields with $m = 1$ to find:

$$-\frac{\partial}{\partial t}\mu\mathcal{H}_u = \nabla \times \mathscr{E}_u + \frac{1}{r}\hat{\phi} \times \mathscr{E}_v \tag{12.46a}$$

$$-\frac{\partial}{\partial t}\mu\mathcal{H}_v = \nabla \times \mathscr{E}_v - \frac{1}{r}\hat{\phi} \times \mathscr{E}_u. \tag{12.46b}$$

Using the first equation above for consistency with our earlier analysis, and appling differencing techniques, we find the update equation:

$$\mathcal{H}_r\Big|_{0,k+1/2}^{n+1/2} = \mathcal{H}_r\Big|_{0,k+1/2}^{n-1/2} - \frac{\Delta t}{\mu\Delta r}\mathscr{E}_z\Big|_{1,k+1/2}^{n} + \frac{\Delta t}{\mu\Delta z}\left(\mathscr{E}_\phi\Big|_{0,k+1}^{n} - \mathscr{E}_\phi\Big|_{0,k}^{n}\right). \tag{12.47}$$

Note that in the equation above, we have used the $\mathscr{E}_z$ component at index $i = 1$, or at $r = \Delta r$, rather than at $r = 0$. This is because Faraday's law, as applied in Equations (12.46), gives a measure of the derivative of $\mathscr{E}_z$ in the ϕ direction, although the resulting equation looks like simply the value of $\mathscr{E}_z$. For $m > 0$ we found that $\mathscr{E}_z$ on the axis is equal to zero, so for this approximation we use the nearest $\mathscr{E}_z$ component, namely at $r = \Delta r$.

As a final note, remember that these solutions apply to a single mode m, and that the simulation must be run once for each mode of interest; the variable m appears in Equations (12.40), so we should expect different solutions for each mode. However, this can be considerably faster than a 3D simulation, when one considers that the same number of equations are involved (six), but the space is reduced in size by however many steps one would use in the azimuthal direction. For, say, 100 steps in azimuth, one would need to simulate 100 modes (more than necessary) to reach the same computational time. However, there is a drawback in that the stability criteria are more stringent for higher order modes. An empirical formulation finds that:

$$\frac{v_p \Delta t}{\Delta x} \leq \begin{cases} 1/(m+1) & m > 0, \\ 1/\sqrt{2} & m = 0. \end{cases} \tag{12.48}$$

Thus, for higher-order modes $m > 0$, the stability criterion becomes more stringent than even the 3D FDTD algorithm.

12.4   Near-to-far field transformation

The fourth and final technique that we will introduce in this chapter applies to using the confined space of an FDTD simulation to find fields at distances far beyond the space boundaries.

Any FDTD simulation involving scattering from various structures, or the simulation of antenna radiation, necessarily yields the near-field scattering or radiation pattern, due to the limitations of the size of the computational space. For example, assume we have a

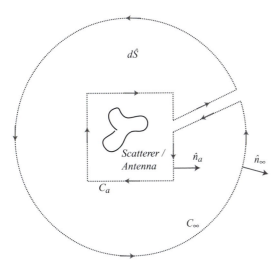

Figure 12.12 Near-to-far field transformation. Schematic diagram showing the integration contour for the near-to-far field transformation. The complete, closed contour integral around C_a and C_∞ is used to map fields on C_a to the far field at C_∞.

100-cell space in one dimension, with a coarse discretization of 10 cells per wavelength. If we place a radiating antenna at the center of the space, the boundary of the FDTD space is only 50 cells or five wavelengths from the source. Typically, the far field is defined as a minimum of $2D^2/\lambda$ from the source, where D is the largest dimension of the radiator [13]. Thus, we cannot resolve the far-field pattern with this small space. To circumvent this problem, we can either (a) increase the size of the space to reach greater than 10 wavelengths, which will increase the computation time considerably; or (b) apply a numerical technique known as the near-to-far field transformation. This technique maps the near-field results of the FDTD simulation analytically into the far field, thus yielding an estimate of the far-field radiation or scattering pattern of the object of interest.

12.4.1   Frequency domain formulation

The frequency domain formulation of the near-to-far field transformation (and, for that matter, the time domain formulation) relies on Green's theorem.[5] Referring to Figure 12.12, we draw an almost-closed contour C_a around the scattering/radiating object and around a boundary at a distance C_∞ that are connected, creating a single open surface dS. The Green's function $G(r, r')$ maps the phasor field $E(r)$ from a location on

---

[5] For an introduction to Green's theorem, we refer the reader to the classic book on electromagnetics by Jackson [14].

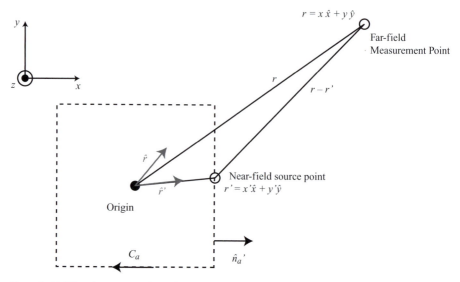

Figure 12.13 Visual representation of the near-to-far field transformation in 2D. The near-field points on the contour C_a are located at r' and mapped to a distant point r. The vectors $\hat{r}$ and $\hat{r}'$ are the unit vectors for the respective directions. Note that *every* point on C_a contributes to the field at r.

the inner boundary r' to a distant location r on the outer boundary:

$$\int_S \left[E_z(r')(\nabla^2)'G(r,r') - G(r,r')(\nabla^2)'E_z(r') \right] ds'$$

$$= \oint_{C_\infty} \left[E_z(r')\frac{\partial G(r,r')}{\partial r'} - G(r,r')\frac{\partial E_z(r')}{\partial r'} \right] dC'$$

$$- \oint_{C_a} \left[E_z(r')\frac{\partial G(r,r')}{\partial n'_a} - G(r,r')\frac{\partial E_z(r')}{\partial n'_a} \right] dC' \tag{12.49}$$

where we are considering the 2D TM mode for simplicity, thus deriving the equations for E_z. Here n'_a is the normal to the contour C_a (see Figure 12.13 for the diagrammatic representation). This shows that the complete surface integral is written as the sum of two line integrals, with a minus sign because the inner contour is in the opposite direction.

We wish to simplify this equation in order to find a relationship between the fields on the inner boundary (which we will calculate in the FDTD algorithm) and the fields on the outer boundary (the far field, which we desire). We can begin by simplifying the left-hand side of Equation (12.21) using the definition of Green's functions,

$$(\nabla^2)'G(r,r') = \delta(r-r') - k^2 G(r,r') \tag{12.50}$$

and the Helmholtz equation,

$$(\nabla^2)'E_z(r') = -k^2 E_z(r'). \tag{12.51}$$

Application of these two leaves simply $E_z(r')$ on the left-hand side of the equation. Furthermore, it is simple to show[6] that the first integral on the right of Equation (12.21) (the integral over C_∞) goes to zero as $r' \to \infty$, so that Equation (12.49) reduces to:

$$E_z(r') = \oint_{C_a} \left[G(r, r')n'_a \cdot \nabla' E_z(r') - E_z(r')n'_a \nabla' G(r, r') \right] dC'. \qquad (12.52)$$

Now we must take the far-field limit of this equation to discern the form of $E_z(r')$. The two-dimensional Green's function is given by the Hankel function,

$$G(r, r') = \frac{j}{4} H_0^{(2)}(k|r - r'|)$$

where $H_0^{(2)}$ is the Hankel function of the second kind or order 0.[7] In the far field, we assume r' is many wavelengths away, so that $k|r - r'|$ is very large; then the limit of the Green's function is:

$$\lim_{k|r-r'| \to \infty} G(r, r') = \frac{j^{3/2}}{\sqrt{8\pi k}} \frac{e^{-jk|r-r'|}}{|r - r'|^{1/2}} \quad \to \quad \frac{j^{3/2}}{\sqrt{8\pi kr}} e^{-jkr} e^{jk\hat{r}\cdot r'} \qquad (12.53)$$

and the gradient is given by:

$$\lim_{k|r-r'| \to \infty} \nabla' G(r, r') = jk\hat{r} \frac{j^{3/2}}{\sqrt{8\pi k}} \frac{e^{-jk|r-r'|}}{|r - r'|^{1/2}} \quad \to \quad (jk\hat{r}) \frac{j^{3/2}}{\sqrt{8\pi kr}} e^{-jkr} e^{jk\hat{r}\cdot r'}$$

$$(12.54)$$

where the final forms of the above two equations were found using binomial expansions[8] of $|r - r'|^{1/2}$ and $|r - r'|$. Substituting Equations (12.53) and (12.54) back into Equation (12.52), we find:

$$\lim_{k|r-r'| \to \infty} E_z(r) = \frac{e^{j3\pi/4}}{\sqrt{8\pi kr}} e^{-jkr} \oint_{C_a} \left[n'_a \cdot \nabla' E_z(r') - jk E_z(r')n'_a \cdot \hat{r} \right] e^{jk\hat{r}\cdot r'} dC'. $$

$$(12.55)$$

---

[6] In 2D, both $E_z(r')$ and $G(r, r')$ decay as $(r')^{-1/2}$, so their derivatives with respect to r' decay as $(r')^{-3/2}$; thus the integral term over C_∞ decays as

$$\int_{C_\infty} \sim 2\pi r' \left((r')^{-1/2} \right) \left((r')^{-3/2} \right) \sim (r')^{-1}$$

which goes to zero for large r'.

[7] The Hankel function of the second kind is given by

$$H_n^{(2)}(z) = J_n(z) - i Y_n(z)$$

where J_n and Y_n are the Bessel function of the first and second kind, respectively, which are common solutions to cylindrical problems.

[8] In particular, $|r - r'|^{1/2} \to r^{1/2}$, and $|r - r'| \to \hat{r} \cdot r'$.

Now, this equation is not yet in a usable form, since it requires the gradient of $E_z(r')$. To convert it into its standard form, we write the gradient as:

$$\nabla' E_z(r') = \hat{x}' \frac{\partial E_z}{\partial x'} + \hat{y}' \frac{\partial E_z}{\partial y'}$$

$$= \hat{x}' \left(-j\omega\mu_0 H_y \right) + \hat{y}' \left(j\omega\mu_0 H_x \right)$$

$$= j\omega\mu_0 \hat{z}' \times \vec{H}(r').$$

Furthermore, a vector identity allows us to write the second term as

$$E_z(r')n_a' \cdot \hat{r} = \left[\hat{z}' \times \left(n_a' \times \vec{E}(r') \right) \right] \cdot \hat{r}.$$

Substituting these two relationships into Equation (12.55), we find the standard form for the near-to-far field transformation in Cartesian coordinates:

$$\lim_{k|r-r'|\to\infty} E_z(r) = \frac{e^{-jkr}}{\sqrt{r}} \frac{e^{j\pi/4}}{\sqrt{8\pi k}} \oint_{C_a} \left[\omega\mu_0\hat{z}' \cdot J_{eq}(r') - k\hat{z}' \times M_{eq}(r') \cdot \hat{r} \right] e^{jk\hat{r}\cdot r'} dC'$$

$$(12.56)$$

where $J_{eq} = \hat{n}_a \times H$ and $M_{eq} = -\hat{n}_a \times E$ are the equivalent electric and magnetic currents observed on the contour C_a. Equation (12.56) thus represents the frequency domain transformation of E_z from the contour C_a to a distant location r.

This can be implemented in a straightforward manner in any FDTD space. One must record fields at each grid location around a boundary C_a over all time, and use these fields to find $\mathcal{J}_{eq}$ and $\mathcal{M}_{eq}$. After the simulation is complete, these values can be transformed to the frequency domain to yield J_{eq} and M_{eq}, which are then used in (12.56) above. In this way one can find the far-field values for any distant location. Note that a full integration of Equation (12.56), including every point on the contour C_a is required for *each* far-field point of interest; hence, if a full radiation pattern is sought, this will require a large number of integrations.

The numerical choice of r, the far-field point of interest, depends only on the structures involved in the simulation. As mentioned earlier, the far field is generally defined as $2D^2/\lambda$ from the source, where D is the largest dimension of the source. Hence, one would presumably choose far-field points at a minimum of this distance.

12.4.2 Time domain implementation

The above formulation can also be implemented in real time, as the FDTD simulation is progressing, by using a recursive discrete fourier transform (DFT) to find the time domain fields at distant points. This method is described in [4]. Alternatively, one can implement a simple time domain algorithm for the near-to-far field transformation, as derived by Luebbers et al. [15].

We begin with the frequency domain equivalent currents $J_{eq} = \hat{n}_a \times H$ and $M_{eq} = -\hat{n}_a \times E$ as above. These are used to define the vectors W and U:

$$\vec{W} = \frac{je^{-jkr}}{2\lambda r} \int_{C_a} J_{eq} e^{jk\hat{r}\cdot r'} dS' \tag{12.57a}$$

$$\vec{U} = \frac{je^{-jkr}}{2\lambda r} \int_{C_a} M_{eq} e^{jk\hat{r}\cdot r'} dS'. \tag{12.57b}$$

Note that C_a encloses the scattering or radiating object; in 2D, C_a is a closed contour, while in 3D, C_a is a closed surface. The phasors W and U are easily inverse-transformed back into the time domain:

$$\vec{\mathcal{W}}(\mathbf{r}, t) = \frac{1}{4\pi rc} \frac{\partial}{\partial t} \left\{ \int_{C_a} \overline{\mathcal{J}}_{eq} \left[t + (r' \cdot \hat{r})/c - r/c \right] dS' \right\} \tag{12.58a}$$

$$\vec{\mathcal{U}}(\mathbf{r}, t) = \frac{1}{4\pi rc} \frac{\partial}{\partial t} \left\{ \int_{C_a} \overline{\mathcal{M}}_{eq} \left[t + (r' \cdot \hat{r})/c - r/c \right] dS' \right\}. \tag{12.58b}$$

Note that this is a surface integral over the closed surface C_a in 3D; in 2D it is a contour integral. In either case, *every* point on the contour or surface contributes to the $\mathcal{U}$ and $\mathcal{W}$ fields at **r**.

Using Equations (12.58), the time domain electric field components in spherical coordinates are:

$$\begin{aligned} \mathcal{E}_\theta &= -\eta_0 \mathcal{W}_\theta - \mathcal{U}_\phi \\ \mathcal{E}_\phi &= -\eta_0 \mathcal{W}_\phi + \mathcal{U}_\theta. \end{aligned} \tag{12.59}$$

These equations show the simplicity of the time domain method: one simply calculates $\vec{\mathcal{W}}(\mathbf{r}, t)$ and $\vec{\mathcal{U}}(\mathbf{r}, t)$ using Equations (12.58) from the currents on the contour C_a, and then transforms them back into $\mathcal{E}$ components using Equations (12.59).

We now show the implementation of this algorithm into the standard FDTD algorithm. Along the surface or contour C_a we can evaluate the equivalent currents $\mathcal{M}_{eq}$ and $\mathcal{J}_{eq}$ as the normal components of the $\mathcal{H}$ and $\mathcal{E}$ fields on each of the faces of C_a. For instance, on a face with outward component $\hat{y}$, $\mathcal{E}_x$ contributes to $\mathcal{M}_{eq}$ and thus $\mathcal{U}_z$ only. For a cubic surface C_a in 3D, components on all six faces must be included.

We must also keep track of the time delay for each location of the surface integral; this is taken care of through the r' term in Equations (12.58). The vector r' is

$$r' = (i - i_c)\Delta x \,\hat{x} + (j + 1/2 - j_c)\Delta y \,\hat{y} + (k - k_c)\Delta z \,\hat{z}$$

where i, j, k are the indices of the location r', and i_c, j_c, k_c are the indices of a central reference location. The extra 1/2 represents a location in the y-direction that is a 1/2-cell from the cell center. To represent $\mathcal{U}$ in Equation (12.58b), we need a centered difference time derivative. As an example, consider the contribution of $\mathcal{E}_x$ (at some point on C_a)

to $\mathcal{U}_z$:

$$\mathcal{U}_z\left[(n+1/2)\Delta t - (r' \cdot r)/c\right] = \frac{\Delta x \, \Delta z}{4\pi c \Delta t}(\mathcal{E}_x^{n+1} - \mathcal{E}_x^n) \qquad (12.60a)$$

$$\mathcal{U}_z\left[(n-1/2)\Delta t - (r' \cdot r)/c\right] = \frac{\Delta x \, \Delta z}{4\pi c \Delta t}(\mathcal{E}_x^n - \mathcal{E}_x^{n-1}) \qquad (12.60b)$$

where it should be understood that the true $\mathcal{U}_z$ field involves the sum of the contributions of all the $\overline{\overline{\mathcal{E}}}$ components on C_a. The square brackets on the left represent the actual time at which the contribution on the right applies to $\mathcal{U}_z$. The $\mathcal{E}_x$ fields fall on integer time steps, but will not contribute at integer time steps to the $\mathcal{U}$ and $\mathcal{W}$ fields because of the variation in distance. The time step index m closest to the time period over which $\mathcal{E}_x$ contributes to $\mathcal{U}_z$ is

$$m = \text{floor}(t_c/\Delta t + 1/2)$$

where

$$t_c = n\Delta t - (r' \cdot r)/c + R_0/c$$

and where the floor() function rounds down to the nearest integer, and R_0 is the distance from the origin to the reference surface C_a in the direction of the far-field measurement point. We then find the contribution of $\mathcal{E}_x^n$ at integer time steps to $\mathcal{U}_z$ by linear interpolation:

$$\mathcal{U}_z\left[(m+1)\Delta t\right] = \frac{\Delta x \, \Delta z}{4\pi c \Delta t}\left[1/2 - (t_c/\Delta t) + m\right]\mathcal{E}_x^n \qquad (12.61a)$$

$$\mathcal{U}_z\left[(m)\Delta t\right] = \frac{\Delta x \, \Delta z}{4\pi c \Delta t}\left[2(t_c/\Delta t - m)\right]\mathcal{E}_x^n \qquad (12.61b)$$

$$\mathcal{U}_z\left[(m-1)\Delta t\right] = \frac{\Delta x \, \Delta z}{4\pi c \Delta t}\left[1/2 + (t_c/\Delta t) - m\right]\mathcal{E}_x^n. \qquad (12.61c)$$

Note that the factors in square brackets on the right are fractions between 0 and 1, and are thus "weights" giving the contribution of $\mathcal{E}_x$ to three time steps of $\mathcal{U}_z$.

This gives the contribution of $\mathcal{E}_x$ at time n and at location (i, j, k) to $\mathcal{U}_z$. The contributions then need to be summed at this time step over the surface C_a, taking into account the different tangential field components.

In total, after running the simulation for N time steps, the vector $\mathcal{U}_z$ will have approximately $N + 2R_0$ entries. We will need to store similar vectors for $\mathcal{U}_x$, $\mathcal{U}_y$, and the corresponding $\mathcal{W}$ fields, for six extra components in total. However, it is more prudent to store the spherical coordinate fields using a coordinate transformation in real time:

$$\mathcal{U}_\theta(r, t) = \mathcal{U}_x(r, t)\cos\theta\cos\phi + \mathcal{U}_y(r, t)\cos\theta\sin\phi - \mathcal{U}_z(r, t)\sin\theta \qquad (12.62a)$$

$$\mathcal{U}_\phi(r, t) = -\mathcal{U}_x(r, t)\sin\phi + \mathcal{U}_y(r, t)\cos\phi \qquad (12.62b)$$

$$\mathcal{W}_\theta(r, t) = \mathcal{W}_x(r, t)\cos\theta\cos\phi + \mathcal{W}_y(r, t)\cos\theta\sin\phi - \mathcal{W}_z(r, t)\sin\theta \qquad (12.62c)$$

$$\mathcal{W}_\phi(r, t) = -\mathcal{W}_x(r, t)\sin\phi + \mathcal{W}_y(r, t)\cos\phi. \qquad (12.62d)$$

This now reduces the number of fields to four. Next, also in real time, these fields can be transformed into $\mathcal{E}_\theta$ and $\mathcal{E}_\phi$ using Equation (12.59), and only the two components need to be stored.

This algorithm must be executed as described above for *each* location of interest in the far field. If, for example, a complete radiation pattern for an antenna is desired, the algorithm must be implemented for a grid of points covering 4π steradians in the far-field space.

12.5   Summary

In this chapter we have introduced a number of more advanced topics that are of interest in a variety of FDTD problems. First, we introduced modeling of periodic structures. The methods presented in Section 12.1 allow us to limit our attention to a single period of a structure that is periodic in one dimension. For normal incidence, the problem is trivial: the fields just outside the periodic boundary are updated with the fields on the opposite boundary. For oblique incidence, however, phase-matching requires future values of the fields, and so more advanced methods are required.

We introduced direct-field methods, which allow us to work directly with the $\overline{\overline{\mathcal{E}}}$ and $\overline{\overline{\mathcal{H}}}$ field components. However, the field-transformation methods are more powerful. In these methods, our $\overline{\overline{\mathcal{E}}}$ and $\overline{\overline{\mathcal{H}}}$ fields are transformed into auxiliary $\overline{\overline{\mathcal{P}}}$ and $\overline{\overline{\mathcal{Q}}}$ fields according to (in 2D TM mode):

$$P_z = E_z \cdot e^{jk_y y} \qquad Q_x = \eta_0 H_x \cdot e^{jk_y y} \qquad Q_y = \eta_0 H_y \cdot e^{jk_y y}.$$

These fields then have their own versions of Maxwell's equations, given by Equations (12.8), which unfortunately have temporal and spatial derivatives of the same fields in each equation, precluding the use of the FDTD algorithm directly. To circumvent this problem, we present two approaches: in the multiple-grid approach, each of these field components is split into two parts and staggered on a Yee-like grid cell; we then have a set of six update equations rather than three. In the split-field method, $\mathcal{Q}_x$ and $\mathcal{P}_z$ are split into two parts, but the parts are co-located on the grid. However, the differential equations are now split into parts that can be solved in the FDTD algorithm.

In Section 12.2, we introduced a number of methods for modeling interactions with structures that are smaller than a grid cell. These methods apply to PEC or material (ϵ_r, μ_r) structures, and in each case, the approach is to use Ampère's law and/or Faraday's law, applied to a single grid cell, and incorporating the appropriate areas and contour segments. These result in modified update equations for grid cells which are intersected by material boundaries.

In Section 12.3, we introduced a method for modeling cylindrical structures, such as optical fibers, whose field patterns can be assumed to have periodicity in the azimuthal (ϕ) direction. In such cases, the problem can be reduced from 3D to 2D, where a full 2D simulation is required for each Fourier component of the azimuthal dependence. For example, if the fields are known to have $\sin\phi$ dependence, only one Fourier component is required.

The method involves expanding the $\overline{\mathcal{E}}$ and $\overline{\mathcal{H}}$ fields into a Fourier series:

$$\overline{\mathcal{E}} = \sum_{m=0}^{\infty} \left[\overline{\mathcal{E}}_u \cos(m\phi) + \overline{\mathcal{E}}_v \sin(m\phi) \right]$$

and similarly for $\overline{\mathcal{H}}$. These equations are then substituted into Maxwell's equations in cylindrical coordinates, and by applying the tricks of Section 12.3, the fields are "collapsed" in ϕ into a 2D simulation for each mode m.

Finally, Section 12.4 introduced the near-to-far field transformation, which allows us to calculate far fields from a compact FDTD simulation. Typically, the FDTD simulation of an antenna or scatterer is confined to the near field; but, by applying Green's functions to a near-field contour, those fields can be transformed to the far field, yielding a radiation or scattering pattern for the object of interest. In Section 12.4 we introduced both frequency domain and time domain methods, both of which can be computed in real time with the time-stepping of the FDTD algorithm.

12.6   Problems

12.1. **Basic periodic boundaries.** Repeat Problem 7.1, the parallel-plate waveguide with a scatterer, but replace the PEC walls on the top and bottom with normal-incidence periodic boundary conditions. In this way, the simulation is of a 1D periodic array of scatterers spaced every 20 mm. Launch a plane wave from the left that is constant in amplitude in x, rather than a waveguide mode. Measure the field pattern that is transmitted and reflected; don't forget to include an absorbing boundary and a total-field / scattered-field boundary.

12.2. **Periodic boundaries at angled incidence.** Modify Problem 12.1 to launch an incident plane wave at some angle θ_i. Use either the multiple-grid or split-field method. Measure the scattering pattern of the periodic structures in Figure 7.7 as a function of incident angle.

12.3. **A narrow slot through a dielectric.** Derive a narrow slot method for a slot through a material with real ϵ_r and μ_r, rather than a simple PEC material.

12.4. **The narrow slot in TM mode.** Derive the update equations for the narrow slot problem of Section 12.2.3, only for the 2D TM mode. Start by drawing a similar diagram to that of Figure 12.7, but with $\mathcal{H}_x$, $\mathcal{H}_y$, and $\mathcal{E}_z$ components.

(a)   Write an FDTD simulation of a narrow slot for the 2D TM mode. Create a slot whose width can be varied from $0.1\Delta x$ to Δx, and a material whose length in y is $10\Delta y$. Launch a sinusoidal plane wave from the $y = 0$ plane toward the slot, with a wavelength equal to $20\Delta x$. What is the amplitude of the wave that passes through the slot? How does it vary with the slot width?

(b) Repeat (a), but this time hold the slot width to $0.5\Delta x$ and vary the slot length from Δy to $10\Delta y$. How does the transmitted amplitude vary with the slot length?

12.5. Thin sheet of conducting material. Create a 2D TM mode simulation with a thin sheet of conducting material that is less than a grid cell in thickness. In this thin sheet, use $\epsilon_r = 2$ and $\sigma_s = 10$ S/m. Launch a Gaussian pulse from the left of the sheet, and observe the amplitude that is transmitted through the sheet. How thick does it need to be to block 99% of the incident wave amplitude (i.e., the transmitted amplitude is only 1%)?

12.6. Cylindrical scatterer. Repeat Problem 7.2, the cylindrical scatterer in 2D, including a TF/SF boundary and an absorbing boundary. This time, however, modify the updates on the surface of the cylinder to more accurately conform to the cylindrical surface, using the diagonal split-cell method described in this chapter. Compare the scattering pattern to that found in Problem 8.11.

12.7. Dielectric scatterer. Repeat Problem 12.6 above, but make the cylinder a dielectric material with $\epsilon_r = 3$. Use the average properties method to accurately model the cylindrical surface. Measure the scattering pattern with and without the average properties model and compare.

12.8. Bodies of revolution. Write a 2D simulation to utilize the bodies of revolution method for a cylindrical waveguide, similar to Figure 12.10. Use the same waveguide parameters as in Problem 4.10. Include an absorbing boundary at the far end of the waveguide. Excite the waveguide with an $\mathscr{E}_r$ source at the edge of the waveguide (near the PEC boundary).
(a) Allow for only one expansion in ϕ ($m = 1$) and solve the update Equations (12.40). Why does $m = 0$ not work?
(b) Repeat (a), but add expansions one at a time, up to $m = 5$. In each case, plot the field pattern 5 cm down the waveguide and compare.
(c) Explore different methods for exciting the waveguide, i.e., different field components at different locations at the end of the waveguide. How does the excitation affect the relative amplitudes in each "mode"?

12.9. Far-field radiation pattern. In this problem we will use the code from Problem 4.7, modeling a half-wave dipole antenna. Modify this code to include a PML as in Problem 9.11. Now, include a near-to-far field transformation boundary around the antenna using the time domain implementation of Luebbers et al. presented in this chapter.
(a) Measure the far-field pattern ~ 100 wavelengths from the antenna, at 100 locations equally spaced in θ. Since the antenna is symmetric in ϕ, there is no reason to measure the radiation pattern in more than one ϕ direction. Compare this far-field pattern to the near-field pattern measured in Problem 9.11.

(b) Repeat (a), but include the thin-wire approximation around the antenna. Does this affect the far-field pattern? Why or why not?

References

[1] R. W. Ziolkowski and J. B. Judkins, "Applications of the nonlinear finite difference time domain (NL-FDTD) method to pulse propagation in nonlinear media: Self-focusing and linear-nonlinear interfaces," *Radio Science*, vol. 28, pp. 901–911, 1993.

[2] R. W. Ziolkowski and J. B. Judkins, "Nonlinear finite-difference time-domain modeling of linear and nonlinear corrugated waveguides," *J. Opt. Soc. Am.*, vol. 11, pp. 1565–1575, 1994.

[3] F. L. Teixeira, "Time-domain finite-difference and finite-element methods for Maxwell equations in complex media," *IEEE Trans. Ant. Prop.*, vol. 56, pp. 2150–2166, 2008.

[4] A. Taflove and S. Hagness, *Computational Electrodynamics: The Finite-Difference Time-Domain Method*, 3rd edn. Artech House, 2005.

[5] M. E. Veysoglu, R. T. Shin, and J. A. Kong, "A finite-difference time-domain analysis of wave scattering from periodic surfaces: oblique incidence case," *J. Electro. Waves and Appl.*, vol. 7, pp. 1595–1607, 1993.

[6] J. A. Roden, S. D. Gedney, M. P. Kesler, J. G. Maloney, and P. H. Harms, "Time-domain analysis of periodic structures at oblique incidence: Orthogonal and nonorthogonal FDTD implementations," *IEEE Trans. Microwave Theory Techqs.*, vol. 46, pp. 420–427, 1998.

[7] P. H. Harms, J. A. Roden, J. G. Maloney, M. P. Kesler, E. J. Kuster, and S. D. Gedney, "Numerical analysis of periodic structures using the split-field algorithm," in *Proc. 13th Annual Review of Progress in Applied Computational Electromagnetics*, March 1997.

[8] S. Dey and R. Mittra, "A locally conformal finite-difference time-domain algorithm for modeling three-dimensional perfectly conducting objects," *IEEE Microwave Guided Wave Lett.*, vol. 7, pp. 273–275, 1997.

[9] S. Dey and R. Mittra, "A modified locally conformal finite-difference time-domain algorithm for modeling three-dimensional perfectly conducting objects," *IEEE Microwave Opt. Techqs. Lett.*, vol. 17, pp. 349–352, 1998.

[10] C. J. Railton and J. B. Schneider, "An analytical and numerical analysis of several locally conformal FDTD schemes," *IEEE Trans. Microwave Theory Techqs.*, vol. 47, pp. 56–66, 1999.

[11] J. G. Maloney and G. S. Smith, "The efficient modeling of thin material sheets in the finite-difference time-domain FDTD method," *IEEE Trans. Ant. Prop.*, vol. 40, pp. 323–330, 1992.

[12] U. S. Inan and A. S. Inan, *Electromagnetic Waves*. Prentice-Hall, 2000.

[13] C. A. Balanis, *Advanced Engineering Electromagnetics*. Wiley, 1989.

[14] J. D. Jackson, *Classical Electrodynamics*, 2nd edn. Wiley, 1975.

[15] R. Luebbers, F. Hunsberger, and K. S. Kunz, "A frequency-dependent finite-difference time-domain formulation for transient propagation in plasma," *IEEE Trans. Ant. Prop.*, vol. 39, p. 29, 1991.

13 Unconditionally stable implicit FDTD methods

Until now, we have discussed the standard FDTD algorithm, based on the interleaved second-order centered difference evaluation of the time and space derivatives, as originally introduced by Yee in 1966 [1]. This method is an *explicit* method, in which the finite difference approximations of the partial derivatives in the PDE are evaluated at the time level $(n + 1/2)$, so that the solution at the next time level $(n + 1)$ can be expressed *explicitly* in terms of the known values of the quantities at time index n.

Explicit finite difference methods have many advantages. For the hyperbolic wave equation, an important advantage is the fact that numerical solutions obtained with the explicit method exhibit finite numerical propagation speed, closely matching the physical propagation speed inherent in the actual PDE. However, an important disadvantage of explicit methods is the fact that they are only *conditionally stable*. As a result, the allowable time step (vis-à-vis the CFL condition) is generally quite small, making certain types of problems prohibitive simply due to the enormous computation times required. In such cases, which include some important electromagnetic applications as discussed in [2], it is highly desirable to avoid the time step limitation.

Implicit finite difference methods provide the means with which this time step limitation can be avoided. With an implicit method, the finite difference approximations of the partial derivatives in the PDE are evaluated at the as-yet unknown time level $(n + 1)$. These methods turn out to be *unconditionally stable*, so that there is no limit on the time step allowed for a numerically stable solution. Practical limits are of course required on the time step so that the truncation errors can be maintained within reasonable limits (as defined by the application), but as a matter of numerical *accuracy* rather than numerical stability.

The major disadvantage of implicit methods is the fact that the solution (i.e., the value of a physical quantity) at time level $(n + 1)$ depends on the values of field quantities at the same time level $(n + 1)$. As a result, the desired solution cannot be explicitly determined in terms of past values, and, in general, a system of simultaneous equations must be solved to obtain the values of the quantities at each time level.

In terms of electromagnetic applications, the standard explicit FDTD methods become prohibitive (due to the time-step requirements) in tackling some important classes of problems, including low-frequency bioelectromagnetics and modeling of VLSI circuits, as discussed in [2]. For these problems, a new alternating direction implicit (ADI)-FDTD method has been put forth that has been quite successful [3, 4, 5]; the purpose of this chapter is to introduce this method. However, in view of the fact that until now we have only discussed explicit FDTD methods, we first provide a brief general discussion of implicit methods.

13.1   Implicit versus explicit finite difference methods

When would one need to use implicit methods rather than the explicit methods we have discussed so far? First, there are scenarios when extremely small-scale structures must be simulated on a much larger grid. The CFL stability limitation requires that these problems will not only need to have very small grid sizes, but in turn very small time steps.

Second, vastly different time scales may exist in some media, including plasmas; in these media, wide ranges of relaxation times and thus wide ranges of frequencies can exist, and the standard explicit FDTD method requires that we model for the highest frequency. For example, in a plasma, frequencies can be excited simultaneously at kHz and MHz frequencies. This means the smallest wavelength will be on the order of meters, requiring $\sim$0.1 m grid cell size; but the longest wavelengths will be on the order of km, requiring over 10,000 grid cells (in each dimension) to cover one wavelength. In turn, thanks to the CFL limit, the simulation will need to run for $\sim$10,000 time steps to reach the edge of the simulation space.

Implicit methods will not help us with the problem above of requiring 10,000 grid cells, but it can circumvent the CFL limit and reduce the number of time steps. We shall introduce implicit methods by using the so-called backward-time centered space method, which is the counterpart of the explicit forward-time centered space method which was studied in Section 3.4.1. We first briefly review the forward-time centered space method.

13.1.1   The forward-time centered space method

As a quick refresher, recall the derivation of the discretized version of the forward-time centered space method applied to the convection equation, from Section 3.4.1 of Chapter 3:

$$\frac{\partial V}{\partial t} + v_p \frac{\partial V}{\partial x} = 0 \qquad \rightarrow \qquad \frac{V_i^{n+1} - V_i^n}{\Delta t} + v_p \frac{V_{i+1}^n - V_{i-1}^n}{2\Delta x} = 0. \qquad (13.1)$$

Solving Equation (13.1) for V_i^{n+1} we find the update equation:

Forward-time centered space

$$V_i^{n+1} = V_i^n - \left(\frac{v_p \Delta t}{2\Delta x}\right) \left[V_{i+1}^n - V_{i-1}^n\right]. \qquad (13.2)$$

However, we further recall from Section 5.1 of Chapter 5 that this method is uncon-ditionally *unstable* for the convection equation, since we found the error amplification factor $q > 1$ regardless of the value of Δt and the numerical wavenumber k. Hence, the forward-time centered space method cannot be used to solve the convection equation. Recall that in Chapter 3, this led us to the leapfrog method (which is centered in time and space) as a stable alternative.

13.1.2　The backward-time centered space method

Another alternative to the forward time centered space method is the *backward*-time centered space (BTCS) method. This is a fully *implicit* method, the finite difference equation (FDE) for which approximates the partial derivatives of the original PDE with first-order backward-difference approximation:

$$\frac{\partial \mathcal{V}}{\partial t}\bigg|_i^{n+1} = \frac{\mathcal{V}_i^{n+1} - \mathcal{V}_i^n}{\Delta t}. \tag{13.3}$$

The discretization of the partial differential equation at time step $(n + 1)$ is thus given as:

$$\frac{\partial \mathcal{V}}{\partial t} + v_p \frac{\partial \mathcal{V}}{\partial x} = 0 \qquad \rightarrow \qquad \frac{\mathcal{V}_i^{n+1} - \mathcal{V}_i^n}{\Delta t} + v_p \frac{\mathcal{V}_{i+1}^{n+1} - \mathcal{V}_{i-1}^{n+1}}{2\Delta x} = 0. \tag{13.4}$$

Equation (13.4) can be rearranged as:

$$\boxed{-\left(\frac{v_p \Delta t}{2\Delta x}\right) \mathcal{V}_{i-1}^{n+1} + \mathcal{V}_i^{n+1} + \left(\frac{v_p \Delta t}{2\Delta x}\right) \mathcal{V}_{i+1}^{n+1} = \mathcal{V}_i^n}. \tag{13.5}$$

It is evident that Equation (13.5) cannot be solved explicitly for $\mathcal{V}_i^{n+1}$, since the two extra values of $\mathcal{V}$ at time $n + 1$ (i.e., $\mathcal{V}_{i\pm1}^{n+1}$) also appear in the same equation; these field components are illustrated in Figure 13.1.

Before addressing the solution to this problem, let us first look at the stability of this method. We can conduct a stability analysis similar to that carried out in Chapter 5 for the forward-time centered space method. Using the relationships $\mathcal{V}_i^{n+1} = q\mathcal{V}_i^n$ and $\mathcal{V}_{i\pm1}^n = \mathcal{V}_i^n e^{\pm jk\Delta x}$, we can write Equation (13.5) as:

$$-\left(\frac{v_p \Delta t}{2\Delta x}\right) q\, e^{-jk\Delta x}\, \mathcal{V}_i^n + q\, \mathcal{V}_i^n + \left(\frac{v_p \Delta t}{2\Delta x}\right) q\, e^{jk\Delta x}\, \mathcal{V}_i^n = \mathcal{V}_i^n$$

which, when solved for q, yields

$$q = \frac{1}{1 + j\left(\dfrac{v_p \Delta t}{\Delta x}\right) \sin(k\Delta x)}. \tag{13.6}$$

Now, since

$$\left|1 + j\left(\frac{v_p \Delta t}{\Delta x}\right) \sin(k\Delta x)\right| > 1, \tag{13.7}$$

we have $q < 1$ for all nonzero values of Δt, Δx, and k, and the backward-time centered space method is thus *unconditionally stable* for the convection equation.

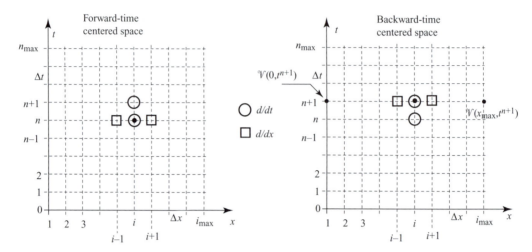

Figure 13.1 Mesh diagram for the backward-time centered space method. The grid points involved in the evaluation of the time derivative are indicated as open circles, whereas those involved in calculating the space derivative are indicated as squares.

To numerically solve the convection equation using the backward-time centered space method, we rely on boundary conditions[1] at $i = 1$ and $i = i_{max}$ as indicated in Figure 13.1. Noting that Equation (13.5) is applicable at points $i = 3$ to $i = i_{max} - 2$, and modifying this equation at $i = 2$ and $i = i_{max} - 1$ by transferring the boundary values $\mathcal{V}(0, t^{n+1})$ and $\mathcal{V}(i_{max}, t^{n+1})$ to the right-hand side, we can obtain the following simultaneous linear equations:

$$\mathcal{V}_2^{n+1} + \left(\frac{v_p \Delta t}{2\Delta x}\right)\mathcal{V}_3^{n+1} = \mathcal{V}_2^n + \left(\frac{v_p \Delta t}{2\Delta x}\right) \overbrace{\mathcal{V}(0, t^{n+1})}^{\text{left boundary condition}}$$

$$-\left(\frac{v_p \Delta t}{2\Delta x}\right)\mathcal{V}_2^{n+1} + \mathcal{V}_3^{n+1} + \left(\frac{v_p \Delta t}{2\Delta x}\right)\mathcal{V}_4^{n+1} = \mathcal{V}_3^n$$

$$-\left(\frac{v_p \Delta t}{2\Delta x}\right)\mathcal{V}_3^{n+1} + \mathcal{V}_4^{n+1} + \left(\frac{v_p \Delta t}{2\Delta x}\right)\mathcal{V}_5^{n+1} = \mathcal{V}_4^n \qquad (13.8)$$

$$\vdots$$

$$-\left(\frac{v_p \Delta t}{2\Delta x}\right)\mathcal{V}_{i_{max}-2}^{n+1} + \mathcal{V}_{i_{max}-1}^{n+1} = \mathcal{V}_{i_{max}-1}^n - \left(\frac{v_p \Delta t}{2\Delta x}\right)\underbrace{\mathcal{V}(i_{max}, t^{n+1})}_{\text{right boundary condition}} .$$

---

[1] It should be noted here that the requirement of two boundary conditions is not natural for the convection equation, which being a first-order differential equation in space, should require only one boundary condition. Physically, the convection equation propagates any initial disturbance in only one direction, so that the solution domain must be open (i.e., not closed with a boundary condition) in that direction. In practice, this open boundary can be represented in an implicit finite difference method by placing the open boundary at a large distance from the region of interest and by applying the initial (in time) condition at that location as the boundary condition.

Equations (13.8) constitute a tridiagonal[2] system of equations, which can be written in matrix form as:

$$\mathbf{A}\,\mathbf{V}^{n+1} = \mathbf{b} \tag{13.9}$$

where $\mathbf{A}$ is an $(i_{max} - 2) \times (i_{max} - 2)$ tridiagonal coefficient matrix, $\mathbf{V}^{n+1}$ is a column vector having entries $\mathcal{V}_1^{n+1}, \ldots, \mathcal{V}_{i_{max}-2}^{n+1}$, and $\mathbf{b}$ is a column vector having as its entries the right-hand sides of Equations (13.8).

The matrix Equation (13.9) can be solved by many different algorithms, although a method known as the Thomas algorithm [see, e.g., 6, Sec. 1.5] is particularly applicable for tridiagonal systems.

The discussion above provides a brief introduction to the idea behind implicit methods; however, the backward-time centered space method is only first-order accurate, and is not nearly accurate enough to be used for FDTD problems of real interest. More advanced methods for implicit time integration, resulting in unconditional stability and a very high degree of accuracy, have been developed from known mathematical methods for solving ODEs and PDEs. In essence, the solution of Maxwell's equations in an implicit scheme is determined by the time integration used. The space is discretized in the usual scheme, leaving us with differential equations of the form

$$\frac{\partial \overline{\mathcal{E}}}{\partial t} = \left[\nabla \times \overline{\mathcal{H}} \right] \Big|_{i,j,k}$$

where it now remains to numerically integrate the equation in time, and determine a time-stepping algorithm. One such algorithm is briefly introduced here.

13.2   Crank-Nicolson methods

The Crank-Nicolson methods [7, 8], which are used often in diffusion and other problems, actually take the average of a forward difference and a backward difference in space, but result in unconditional stability as desired. As one might expect and as we will see below, this averaging of the forward- and backward-time methods yields the second-order accuracy we desire.

For the 1D voltage problem discussed above, we write the spatial derivative in Equation (13.1) as the average of two spatial derivatives at different time steps, so

---

[2] A tridiagonal matrix is a square matrix in which all of the elements not on the major diagonal, or on the two diagonals on both sides of the major diagonal, are zero. The elements of the three diagonals may or may not be zero. As an example, a 5×5 tridiagonal matrix is given by:

$$\mathbf{A} = \begin{bmatrix} a_{11} & a_{12} & 0 & 0 & 0 \\ a_{21} & a_{22} & a_{23} & 0 & 0 \\ 0 & a_{32} & a_{33} & a_{34} & 0 \\ 0 & 0 & a_{43} & a_{44} & a_{45} \\ 0 & 0 & 0 & a_{54} & a_{55} \end{bmatrix}.$$

Large tridiagonal matrices arise quite often in physical problems, especially involving implicit numerical solutions of differential equations.

that Equation (13.1) becomes:

$$\frac{\partial \mathcal{V}}{\partial t}\Big|_i^{n+1/2} + \frac{v_p}{2}\left(\frac{\partial \mathcal{V}}{\partial x}\Big|_i^{n+1} + \frac{\partial \mathcal{V}}{\partial x}\Big|_i^{n}\right) = 0 \tag{13.10}$$

where the first derivative on the right is the backward-time difference, the second derivative is the forward-time difference, and we take an average of the two approximations to the spatial derivative. When discretized, this takes the form

$$\frac{\mathcal{V}_i^{n+1} - \mathcal{V}_i^n}{\Delta t} + \frac{v_p}{2}\left(\frac{\mathcal{V}_{i+1}^{n+1} - \mathcal{V}_{i-1}^{n+1}}{2\Delta x} + \frac{\mathcal{V}_{i+1}^n - \mathcal{V}_{i-1}^n}{2\Delta x}\right) = 0. \tag{13.11}$$

Rearranging terms, we can arrive at a usable update equation:

$$K\mathcal{V}_{i+1}^{n+1} + \mathcal{V}_i^{n+1} - K\mathcal{V}_{i-1}^{n+1} = -K\mathcal{V}_{i+1}^n + \mathcal{V}_i^n + K\mathcal{V}_{i-1}^n \tag{13.12}$$

where we have introduced the shorthand $K = v_p\Delta t/4\Delta x$. Once again, this can take the form of a system of equations and thus a matrix equation of the form

$$\mathbf{M}_1 \mathbf{V}^{n+1} = \mathbf{M}_2 \mathbf{V}^n$$

and thus

$$\mathbf{V}^{n+1} = \mathbf{M}_1^{-1}\mathbf{M}_2 \mathbf{V}^n.$$

Note that both matrices $\mathbf{M}_1$ and $\mathbf{M}_2$ will be tridiagonal.

We can analyze the stability of this algorithm in the usual way by substituting $V_{i\pm1}^n = V_i^n e^{\pm jk\Delta x}$ to find, with K defined as above,

$$KV_i^{n+1}e^{jk\Delta x} + V_i^{n+1} - KV_i^{n+1}e^{-jk\Delta x} = V_i^n - KV_i^n e^{jk\Delta x} + KV_i^n e^{-jk\Delta x}$$

$$(1 + 2jK\sin k\Delta x)\mathcal{V}_i^{n+1} = (1 - 2jK\sin k\Delta x)\mathcal{V}_i^n$$

$$\frac{\mathcal{V}_i^{n+1}}{\mathcal{V}_i^n} = q = \frac{(1 - 2jK\sin k\Delta x)}{(1 + 2jK\sin k\Delta x)} \tag{13.13}$$

and, since the numerator and denominator have identical imaginary and real component magnitudes, we have $|q| = 1$ exactly, and the method is stable for any values of k, Δt, and Δx.

Next, we investigate this method applied to Maxwell's equations, the set of coupled PDEs. First, however, we must note that in this method, the time stepping is actually achieved without interleaving, for reasons which will become clear momentarily; this means that both $\overline{\mathcal{E}}$ and $\overline{\mathcal{H}}$ components are stored at integer time steps n. Recall Maxwell's equations in 1D from Chapter 4:

$$\frac{\partial \mathcal{H}_y}{\partial t} = \frac{1}{\mu}\frac{\partial \mathcal{E}_z}{\partial x} \tag{4.6a}$$

$$\frac{\partial \mathcal{E}_z}{\partial t} = \frac{1}{\epsilon}\frac{\partial \mathcal{H}_y}{\partial x}. \tag{4.6b}$$

By discretizing these equations at time step $(n + 1/2)$ and applying the average of two spatial derivatives as above, the finite difference equation version of Equation (4.6b) becomes:

$$\frac{\mathcal{E}_z|_i^{n+1} - \mathcal{E}_z|_i^n}{\Delta t} = \frac{1}{2\epsilon} \left(\frac{\mathcal{H}_y|_{i+1/2}^{n+1} - \mathcal{H}_y|_{i-1/2}^{n+1}}{\Delta x} + \frac{\mathcal{H}_y|_{i+1/2}^n - \mathcal{H}_y|_{i-1/2}^n}{\Delta x} \right). \quad (13.14a)$$

Similarly, the FDE for Equation (4.6a) becomes:

$$\frac{\mathcal{H}_y|_{i+1/2}^{n+1} - \mathcal{H}_y|_{i+1/2}^n}{\Delta t} = \frac{1}{2\mu} \left(\frac{\mathcal{E}_z|_{i+1}^{n+1} - \mathcal{E}_z|_i^{n+1}}{\Delta x} + \frac{\mathcal{E}_z|_{i+1}^n - \mathcal{E}_z|_i^n}{\Delta x} \right). \quad (13.14b)$$

Notice that in both of these equations, the first fractional term on the right is the backward difference and the second term is the forward difference, and the average of these two approximations to the spatial derivative is then taken, giving the approximate value of the spatial derivative at time step $(n + 1/2)$. Overall, the scheme is implicit because unknown field values are present on the right-hand side of the equations, and hence a matrix inversion problem exists. In the case of coupled PDEs such as Maxwell's equations, these can be solved simultaneously by (a) moving all fields at time step $(n + 1)$ to the left-hand side; (b) creating a matrix of all $\mathcal{E}_z$ and $\mathcal{H}_y$ unknowns; and then (c) filling out the coefficient matrix relating the vector of fields at $(n + 1)$ to the known field values at time n. This becomes somewhat tricky when the equations are coupled, as they are in Equations (13.14); the alternating direction implicit (ADI) method, described next, demonstrates a similar matrix formation.

13.3 Alternating direction implicit (ADI) method

One implicit method which is particularly useful for multi-dimensional (2D and 3D) systems is the so-called alternating direction implicit (ADI) method, developed by Zheng et al. [3, 4, 5]. This method involves the solution of each partial differential equation using a two-step scheme in time. In the first *half*-time step, the spatial derivatives in one direction (e.g., $\partial/\partial x$) are evaluated at the known time step n while the other spatial derivatives (e.g., $\partial/\partial y$) are evaluated at time step $(n + 1)$. In the second half-time step, the process is reversed. To illustrate, we begin with the 3D Maxwell's equations, written in component form:

$$\frac{\partial \mathcal{E}_x}{\partial t} = \frac{1}{\epsilon} \left(\frac{\partial \mathcal{H}_z}{\partial y} - \frac{\partial \mathcal{H}_y}{\partial z} \right) \qquad \frac{\partial \mathcal{H}_x}{\partial t} = \frac{1}{\epsilon} \left(\frac{\partial \mathcal{E}_y}{\partial z} - \frac{\partial \mathcal{E}_z}{\partial y} \right) \quad (13.15a)$$

$$\frac{\partial \mathcal{E}_y}{\partial t} = \frac{1}{\epsilon} \left(\frac{\partial \mathcal{H}_x}{\partial z} - \frac{\partial \mathcal{H}_z}{\partial x} \right) \qquad \frac{\partial \mathcal{H}_y}{\partial t} = \frac{1}{\epsilon} \left(\frac{\partial \mathcal{E}_z}{\partial x} - \frac{\partial \mathcal{E}_x}{\partial z} \right) \quad (13.15b)$$

$$\frac{\partial \mathcal{E}_z}{\partial t} = \frac{1}{\epsilon} \left(\frac{\partial \mathcal{H}_y}{\partial x} - \frac{\partial \mathcal{H}_x}{\partial y} \right) \qquad \frac{\partial \mathcal{H}_z}{\partial t} = \frac{1}{\epsilon} \left(\frac{\partial \mathcal{E}_x}{\partial y} - \frac{\partial \mathcal{E}_y}{\partial x} \right). \quad (13.15c)$$

Working with Equation (13.15a) for $\partial \mathcal{E}_x / \partial t$, we can difference it in the usual manner, except in this case we will only increment a half-time step, using the $\mathcal{H}_z$ fields at time step $(n + 1/2)$ and the $\mathcal{H}_y$ fields at time step n:

$$
\frac{\mathcal{E}_x \Big|_{i+1/2,j,k}^{n+1/2} - \mathcal{E}_x \Big|_{i+1/2,j,k}^{n}}{\Delta t / 2} = \frac{1}{\epsilon} \left[\frac{\mathcal{H}_z \Big|_{i+1/2,j+1/2,k}^{n+1/2} - \mathcal{H}_z \Big|_{i+1/2,j-1/2,k}^{n+1/2}}{\Delta y} \right.
$$
$$
\left. - \frac{\mathcal{H}_y \Big|_{i+1/2,j,k+1/2}^{n} - \mathcal{H}_y \Big|_{i+1/2,j,k-1/2}^{n}}{\Delta z} \right]. \qquad (13.16a)
$$

For the second half-time step (to advance from time step $n + 1/2$ to time step $n + 1$), we apply the same equation except we use the $\mathcal{H}_y$ fields at time step $n + 1$:

$$
\frac{\mathcal{E}_x \Big|_{i+1/2,j,k}^{n+1} - \mathcal{E}_x \Big|_{i+1/2,j,k}^{n+1/2}}{\Delta t / 2} = \frac{1}{\epsilon} \left[\frac{\mathcal{H}_z \Big|_{i+1/2,j+1/2,k}^{n+1/2} - \mathcal{H}_z \Big|_{i+1/2,j-1/2,k}^{n+1/2}}{\Delta y} \right.
$$
$$
\left. - \frac{\mathcal{H}_y \Big|_{i+1/2,j,k+1/2}^{n+1} - \mathcal{H}_y \Big|_{i+1/2,j,k-1/2}^{n+1}}{\Delta z} \right]. \qquad (13.16b)
$$

This second equation can be considered the equivalent of the first, but applied in the *opposite direction* (hence the name of the method); relative to the electric field time difference on the left side of the equation, the $\partial \mathcal{H}_z / \partial y$ term is applied at the *later* time step in the first equation, and at the *earlier* time step in the second equation, and vice versa for the $\partial \mathcal{H}_y / \partial z$ term. Note that since we are differencing over only a half-time step, $\Delta t / 2$ appears in the denominator on the left-hand side.

Similar equations can be easily written down for the other five component Equations (13.15). Notice that all of the field components maintain their usual positions in the Yee cell; however, they must also be known at every half-time step (n and $n + 1/2$). This will, of course, double the computation time compared to the standard explicit FDTD method, but will not increase the memory requirement.

It should also be noted that these equations are *implicit*: each requires future, unknown values of the $\overline{\mathcal{H}}$ fields. This will lead to a matrix inversion process, typical of all implicit methods. This process is simplified by substituting the six equations into each other to arrive at simplified matrix equations.[3] To see this, let us try to solve for $\mathcal{E}_x^{n+1/2}$ in

---

[3] The formulation of the matrix equations can also be derived quickly using Kronecker products, which will be discussed in Chapter 14.

Equation (13.16a) above. Here is the first-half-step update equation for the $\mathcal{H}_z$ component:

$$\frac{\mathcal{H}_z\Big|_{i+1/2,j+1/2,k}^{n+1/2} - \mathcal{H}_z\Big|_{i+1/2,j+1/2,k}^{n}}{\Delta t/2} = \frac{1}{\mu}\left[\frac{\mathcal{E}_x\Big|_{i+1/2,j+1,k}^{n+1/2} - \mathcal{E}_x\Big|_{i+1/2,j,k}^{n+1/2}}{\Delta y}\right.$$

$$\left. - \frac{\mathcal{E}_y\Big|_{i+1,j+1/2,k}^{n} - \mathcal{E}_z\Big|_{i,j+1/2,k}^{n}}{\Delta x}\right]. \quad (13.17)$$

This equation is used because Equation (13.16a) has $\mathcal{H}_z^{n+1/2}$ as its future, unknown field. We can solve this equation for $\mathcal{H}_z^{n+1/2}$ at the indices required in Equation (13.16a) (noting that two values of $\mathcal{H}_z^{n+1/2}$ are required at different spatial locations) and substitute, yielding the following update equation for $\mathcal{E}_x$:

$$-\left(\frac{\Delta t^2}{4\mu\epsilon\,\Delta y^2}\right)\mathcal{E}_x\Big|_{i+1/2,j+1,k}^{n+1/2} + \left(1 + \frac{\Delta t^2}{2\mu\epsilon\,\Delta y^2}\right)\mathcal{E}_x\Big|_{i+1/2,j,k}^{n+1/2} - \left(\frac{\Delta t^2}{4\mu\epsilon\,\Delta y^2}\right)\mathcal{E}_x\Big|_{i+1/2,j-1,k}^{n+1/2}$$

$$= \mathcal{E}_x\Big|_{i+1/2,j,k}^{n}$$

$$+ \frac{\Delta t}{2\epsilon}\left(\frac{\mathcal{H}_z\Big|_{i+1/2,j+1/2,k}^{n} - \mathcal{H}_z\Big|_{i+1/2,j-1/2,k}^{n}}{\Delta y} - \frac{\mathcal{H}_y\Big|_{i+1/2,j,k+1/2}^{n} - \mathcal{H}_y\Big|_{i+1/2,j,k-1/2}^{n}}{\Delta z}\right)$$

$$- \frac{\Delta t^2}{4\mu\epsilon\,\Delta y\,\Delta x}\left(\mathcal{E}_y\Big|_{i+1,j+1/2,k}^{n} - \mathcal{E}_y\Big|_{i,j+1/2,k}^{n} - \mathcal{E}_y\Big|_{i+1,j-1/2,k}^{n} + \mathcal{E}_y\Big|_{i,j-1/2,k}^{n}\right). \quad (13.18)$$

Note that this equation has only $\mathcal{E}_x$ terms on the left-hand side at time step $n + 1/2$, and only field components at time step n on the right-hand side. Five other equations for the other field components follow, as well as another set of six equations for the second half-time step. These form a system of six matrix equations that can be solved using numerical techniques and packages that are readily available (such as the Thomas algorithm [6] mentioned earlier). Each of these equations has the general form:

$$M_1 X^{n+1/2} = P_1 X^n \qquad \text{For the first half-time step} \qquad (13.19a)$$

$$M_2 X^{n+1} = P_2 X^{n+1/2} \qquad \text{For the second half-time step} \qquad (13.19b)$$

where $M_{1,2}$ and $P_{1,2}$ are matrices of coefficients from the update equations, and X is a vector composed of all six field components at each spatial location in the grid – hence, quite a long vector! The coefficient matrices are both sparse, and the two Equations (13.19) can be combined into a single time step equation:

$$\left.\begin{array}{l} X^{n+1/2} = M_1^{-1}P_1 X^n \\ X^{n+1} = M_2^{-1}P_2 X^{n+1/2} \end{array}\right\} \rightarrow \quad \begin{array}{l} X^{n+1} = M_2^{-1}P_2 M_1^{-1}P_1 X^n \\ = \Lambda X^n \end{array} \quad (13.20)$$

with $\Lambda = M_2^{-1}P_2 M_1^{-1}P_1$.

We will now look into the stability of this algorithm, to show that it is, in fact, unconditionally stable. Because of the large size of the matrices and vectors involved, we will provide only an overview of the stability analysis here; details can be found in the work by Zheng et al. [4]. In that work, the authors conduct stability analysis on the two-step method (Equations 13.19) rather than the combined method (13.20), and so we will do the same. We apply the usual stability analysis as in Chapter 5 by substituting phasor (i.e., frequency domain) terms such as

$$E_x\big|_{i+1/2,j,k}^{n} = E_x^n e^{-j(k_x(i+1/2)\Delta x + k_y j \Delta y + k_z k \Delta z)}$$

and creating a vector of six field components of the form:

$$X^n = \begin{bmatrix} E_x^n & E_y^n & E_z^n & H_x^n & H_y^n & H_z^n \end{bmatrix}^T.$$

We can form the stability equations for the two-step procedure:

$$X^{n+1/2} = \Lambda_1 \cdot X^n \tag{13.21a}$$

$$X^{n+1} = \Lambda_2 \cdot X^{n+1/2} \tag{13.21b}$$

where the matrices Λ_1 and Λ_2 are given by

$$\Lambda_1 = \begin{bmatrix} \frac{1}{Q_y} & \frac{W_x W_y}{\mu\epsilon Q_y} & 0 & 0 & \frac{jW_z}{\epsilon Q_y} & \frac{-jW_y}{\epsilon Q_y} \\ 0 & \frac{1}{Q_z} & \frac{W_z W_y}{\mu\epsilon Q_z} & \frac{-jW_z}{\epsilon Q_z} & 0 & \frac{jW_x}{\epsilon Q_z} \\ \frac{W_x W_z}{\mu\epsilon Q_x} & 0 & \frac{1}{Q_x} & \frac{jW_y}{\epsilon Q_x} & \frac{-jW_x}{\epsilon Q_z} & 0 \\ 0 & \frac{-jW_z}{\mu Q_z} & \frac{jW_z}{\mu Q_z} & \frac{1}{Q_z} & 0 & \frac{W_x W_z}{\mu\epsilon Q_z} \\ \frac{jW_z}{\mu Q_x} & 0 & \frac{-jW_x}{\mu Q_x} & \frac{W_x W_y}{\mu\epsilon Q_x} & \frac{1}{Q_x} & 0 \\ \frac{-jW_y}{\mu Q_y} & \frac{jW_x}{\mu Q_y} & 0 & 0 & \frac{W_z W_y}{\mu\epsilon Q_y} & \frac{1}{Q_y} \end{bmatrix} \tag{13.22a}$$

$$\Lambda_2 = \begin{bmatrix} \frac{1}{Q_z} & 0 & \frac{W_z W_x}{\mu\epsilon Q_z} & 0 & \frac{jW_z}{\epsilon Q_z} & \frac{-jW_y}{\epsilon Q_z} \\ \frac{W_x W_y}{\mu\epsilon Q_x} & \frac{1}{Q_x} & 0 & \frac{-jW_z}{\epsilon Q_x} & 0 & \frac{jW_x}{\epsilon Q_x} \\ 0 & \frac{W_y W_z}{\mu\epsilon Q_y} & \frac{1}{Q_y} & \frac{jW_y}{\epsilon Q_y} & \frac{-jW_x}{\epsilon Q_y} & 0 \\ 0 & \frac{-jW_z}{\mu Q_y} & \frac{jW_y}{\mu Q_y} & \frac{1}{Q_y} & 0 & \frac{W_z W_y}{\mu\epsilon Q_y} \\ \frac{jW_z}{\mu Q_z} & 0 & \frac{-jW_x}{\mu Q_z} & 0 & \frac{1}{Q_z} & \frac{W_z W_y}{\mu\epsilon Q_z} \\ \frac{-jW_y}{\mu Q_x} & \frac{jW_x}{\mu Q_x} & 0 & 0 & \frac{W_x W_z}{\mu\epsilon Q_x} & \frac{1}{Q_x} \end{bmatrix}. \tag{13.22b}$$

The W_i and Q_i factors are given, for $i = x, y, z$ by:

$$W_i = \frac{\Delta t}{\Delta i} \sin\left(\frac{k_i \Delta i}{2}\right) \tag{13.23a}$$

$$Q_i = 1 + v_p^2 W_i^2 \tag{13.23b}$$

where, as usual, $v_p = (\mu\epsilon)^{-1/2}$. These two matrices can be multiplied to find $\Lambda = \Lambda_1 \Lambda_2$, the complete matrix for the update from time step n to time step $(n + 1)$. Stability of the algorithm is determined by finding the eigenvalues of this matrix Λ and determining under what conditions the magnitudes of the eigenvalues are less than or equal to one. Zheng et al. used the mathematical package *Maple* to find the six eigenvalues, which turn out to be:

$$\lambda_1 = \lambda_2 = 1 \qquad\qquad (13.24)$$

$$\lambda_3 = \lambda_5 = \frac{\sqrt{R^2 - S^2} + jS}{R}$$

$$\lambda_4 = \lambda_6 = \lambda_3^* = \frac{\sqrt{R^2 - S^2} - jS}{R}$$

where R and S are given by:

$$R = (\mu\epsilon + W_x^2) \cdot (\mu\epsilon + W_y^2) \cdot (\mu\epsilon + W_z^2)$$

$$S = \sqrt{4\mu\epsilon(\mu\epsilon W_x^2 + \mu\epsilon W_y^2 + \mu\epsilon W_z^2 + W_x^2 W_y^2 + W_x^2 W_z^2 + W_y^2 W_z^2) \cdot (\mu^3\epsilon^3 W_x^2 W_y^2 W_z^2)}.$$

One can readily see that the magnitudes of $\lambda_{3:6}$ are all equal to one by evaluating $|\lambda_{3:6}|^2 = \lambda_{3:6}\lambda_{3:6}^*$; for instance,

$$|\lambda_3|^2 = \left(\frac{\sqrt{R^2 - S^2} + jS}{R}\right) \cdot \left(\frac{\sqrt{R^2 - S^2} - jS}{R}\right)$$

$$= \frac{(R^2 - S^2) - (jS)^2}{R^2} = \frac{R^2 - S^2 + S^2}{R^2} = \frac{R^2}{R^2} = 1.$$

Hence, for *any* value of Δt, grid spacing, and wavelength, all of the eigenvalues are identically equal to one, and this algorithm is *unconditionally stable*.

13.3.1 Accuracy of the ADI method

We have shown so far in this chapter that the implicit methods are unconditionally stable. However, stability for any value of Δt does not imply that a given method is necessarily *accurate* for any value of Δt, and so the numerical dispersion characteristics become even more important. Similar to the methods discussed in Chapter 6, the numerical dispersion relation can be found by evaluating the determinant of the matrix $(e^{j\omega\Delta t}\mathbf{I} - \Lambda)$, where Λ is the product of the two matrices above. For the ADI method, this analysis results in:

$$\sin^2(\omega\Delta t) = \frac{4\mu\epsilon\left(\mu\epsilon W_x^2 + \mu\epsilon W_y^2 + \mu\epsilon W_z^2 + W_x^2 W_y^2 + W_x^2 W_z^2 + W_y^2 W_z^2\right)}{\left(\mu\epsilon + W_x^2\right) \cdot \left(\mu\epsilon + W_y^2\right) \cdot \left(\mu\epsilon + W_z^2\right)}.$$

$$\qquad\qquad (13.25)$$

This relation is plotted in Figure 13.2 for variations with Δt and with incident angle θ; for more details see [5]. In this plot we have assumed propagation in the x–y plane, so that $\phi = 0$, but have varied θ, where $k_x = k\cos\theta$ and $k_y = k\sin\theta$. The left plot shows the variation in numerical phase velocity $\tilde{v}_p$ (normalized by c) with Δx; the different plots correspond to different values of Δt, where $C = v_p\Delta t/\sqrt{3}\Delta x$ is the CFL limit in 3D. Note that $k\Delta x \simeq 0.3$ corresponds to $\Delta x = \lambda/20$. We can see that for time steps resulting in greater than $3C$, the numerical dispersion is significant.

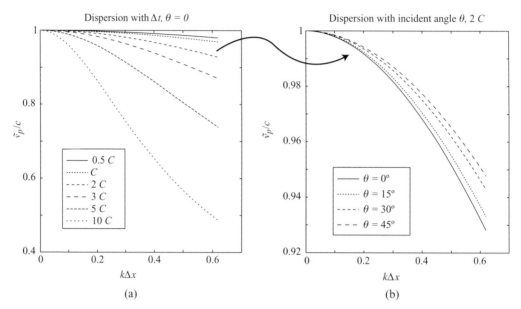

Figure 13.2 Numerical dispersion in the ADI method. (a) Numerical phase velocity versus $k\Delta x$ for variations in Δt, where $C = v_p \Delta t / \sqrt{3}\Delta x$. (b) Numerical phase velocity versus $k\Delta x$ for variations in incident angle θ, with Δt constant at $2C$.

The right plot shows the variations in $\tilde{v}_p$ at different incident angles. Note that the variations in dispersion are small for variation in incident angle, and the angular dispersion is worst along the principal axes. However, the angular dispersion is not nearly as significant as the numerical dispersion.

A couple of conclusions about the accuracy of this method can be ascertained from the dispersion Equation (13.25) and the associated plot:

- *Propagation direction*: Noting that W_x, W_y, and W_z are related to k_x, k_y, and k_z, we can see that the method results in different phase velocities that depend on direction of propagation. Generally, numerical experiments have shown that for time steps of $\Delta t < 3C$, where $C = \Delta x / v_p \sqrt{3}$ is the CFL limit, the phase velocity varies from about 0.975 to 1 times the actual phase velocity c (for $\Delta x = \lambda/20$); hence, for a threefold increase in the time step compared to the CFL limit, we achieve accuracy to within $\sim 2.5\%$.
- *Large time steps*: the method results in increased dispersion (i.e., lower phase velocity) for large time steps above the CFL limit. In fact, with time steps resulting in $v_p \Delta t / \Delta x = 5/\sqrt{3}$ (5C), with spatial steps of $\lambda/10$, the phase velocity is only $0.73c$.

In addition, analysis of this method has shown that there are truncation errors which grow as $(\Delta t)^2$ (i.e., the method is second-order accurate in time), and it has been experimentally shown that the method is really only sufficiently accurate for time steps up to 5–10 times the CFL limit. This improvement over the standard FDTD method is really only competitive in 1D; for larger problems, the matrix inversions, which may require anywhere from 10–1,000 iterations to converge, will slow down the simulation considerably. The real power of these implicit methods comes from their ability to allow high-frequency oscillations in dispersive materials, which would normally create instabilities in the explicit FDTD method.

13.4 Summary

In this chapter we have introduced *implicit* FDTD methods. These are methods in which the update equations have future, unknown values on the right-hand side of the equation, and so those future values can only be implicitly determined. These methods have the considerable advantage of avoiding the CFL stability condition, and can be stable for *any* value of Δt.

We first introduced the backward-time, centered space (BTCS) method, the implicit analog of the forward-time centered space method from Chapter 3 for solving the convection equation. The BTCS method results in an update equation of the form:

$$-\left(\frac{v_p \Delta t}{2\Delta z}\right)\mathcal{V}_{i-1}^{n+1} + \mathcal{V}_i^{n+1} + \left(\frac{v_p \Delta t}{2\Delta z}\right)\mathcal{V}_{i+1}^{n+1} = \mathcal{V}_i^n.$$

By writing out each of the equations for all i, one can form a set of matrix equations, which are ultimately written in the form

$$\mathbf{A}\,\mathbf{V}^{n+1} = \mathbf{b}$$

where $\mathbf{b}$ is a function of the values of $\mathcal{V}$ at the current time step n.

We next introduced a more efficient and accurate method for solving Maxwell's equations implicitly; namely, the Crank-Nicolson method, in which the spatial derivatives in Maxwell's equations are averaged at two time steps. The update equation in 1D for the $\mathcal{H}_y$ term in this method becomes:

$$\frac{\mathcal{H}_y\big|_{i+1/2}^{n+1} - \mathcal{H}_y\big|_{i+1/2}^n}{\Delta t} = \frac{1}{2\mu}\left(\frac{\mathcal{E}_z\big|_{i+1}^{n+1} - \mathcal{E}_z\big|_i^{n+1}}{\Delta x} + \frac{\mathcal{E}_z\big|_{i+1}^n - \mathcal{E}_z\big|_i^n}{\Delta x}\right).$$

Lastly, we introduced the alternating direction implicit (ADI)-FDTD method developed by Zheng et al. In this method, time-stepping is advanced at half-time steps, so all fields are known at every half-step. Then, the spatial derivatives are advanced in an alternating method, before and after the time step of that particular equation. Equations (13.16) clarify this process. This method is shown to be unconditionally stable, and very accurate for time steps up to 5–10 times the CFL limit.

13.5 Problems

13.1. Backward Euler in 1D. Write a 1D simulation utilizing the Backward-Euler method in Equation (13.5). Launch a Gaussian pulse from the left edge of the simulation space with a temporal duration of 10 time steps, and use a first-order

Mur boundary at the far end. Use $\Delta x = 1$ m and make the space 200 m long. Monitor the pulse as it passes a point 150 m from the source, and measure the frequency response of the system.

Vary Δt up to and beyond the CFL condition, and carefully monitor the performance of the simulation. How large can you make Δt before the simulation degrades in accuracy by more than 10%?

For this and the following problems where matrix inversions are required, the Matlab functions `spalloc`, `spones`, and the backslash \ (for matrix division) will be useful.

13.2. **Crank-Nicolson method.** Repeat Problem 13.1 using the Crank-Nicolson method. Compare the accuracy of this method as Δt is increased.

13.3. **Crank-Nicolson for Maxwell's equations.** Use the Crank-Nicolson method to solve Maxwell's equations in 1D. Launch a Guassian pulse toward a slab of dielectric material, as in Problem 4.4. How large can you make Δt and maintain accuracy in this simulation? At this largest value of Δt, how does the simulation time compare to the FDTD algorithm in Problem 4.4?

13.4. **ADI method in 2D.** Repeat Problem 8.2 using the Alternating Direction Implicit method, including the second-order Mur boundary.
   (a) Use 10 grid cells per wavelength. How large can you make Δt before the solution differs from the explicit method by more than 5%? How does the computation time compare at this time step?
   (b) Repeat (a) for 20 grid cells per wavelength.

13.5. **Luneberg lens in ADI.** Repeat Problem 8.12, the Luneberg lens, using the ADI method. Include the second-order Mur absorbing boundary, or include a PML if you wish. Compare the speed and accuracy of the ADI method for this problem to the explicit FDTD method.

References

[1] K. Yee, "Numerical solution of initial boundary value problems involving Maxwell's equations in isotropic media," *IEEE Trans. Ant. Prop.*, vol. 14, no. 3, pp. 302–307, 1966.

[2] A. Taflove and S. Hagness, *Computational Electrodynamics: The Finite-Difference Time-Domain Method*, 3rd edn. Artech House, 2005.

[3] F. Zheng, Z. Chen, and J. Zhang, "A finite-difference time-domain method without the courant stability conditions," *IEEE Microwave Guided Wave Lett.*, vol. 9, pp. 441–443, 1999.

[4] F. Zheng, Z. Chen, and J. Zhang, "Toward the development of a three-dimensional unconditionally stable finite-difference time-domain method," *IEEE Trans. on Microwave Theory Tech.*, vol. 48, pp. 1550–1558, 2000.

[5] F. Zheng and Z. Chen, "Numerical dispersion analysis of the unconditionally stable 3-D ADI-FDTD method," *IEEE Trans. Microwave Theory and Techqs.*, vol. 49, pp. 1006–1009, 2000.

[6] J. D. Hoffman, *Numerical Methods for Engineers and Scientists*. McGraw-Hill, 1992.

[7] J. Crank and P. Nicolson, "A practical method for numerical evaluation of solutions of partial differential equations of the heat conduction type," *Proc. Camb. Phil. Soc.*, vol. 43, pp. 50–67, 1947.

[8] Y. Yang, R. S. Chen, and E. K. N. Yung, "The unconditionally stable Crank Nicolson FDTD method for three-dimensional Maxwell's equations," *IEEE Microwave Opt. Technol. Lett.*, vol. 48, pp. 1619–1622, 2006.

14 Finite difference frequency domain

The FDTD method is widely applicable to many problems in science and engineering, and is gaining in popularity in recent years due in no small part to the fast increases in computer memory and speed. However, there are many situations when the frequency domain method is useful.

The finite difference frequency domain (FDFD) method utilizes the frequency domain representation of Maxwell's equations. In practice, however, rather than Maxwell's equations, the frequency domain wave equation is more often used, as it is a more compact form, requires only a single field, and does not require interleaving. In this chapter we will focus on the latter method, and give a brief overview of the Maxwell's equations formulation toward the end of the chapter. Because the method uses the frequency domain equations, the results yield only a single-frequency, steady-state solution. This is often the most desirable solution to many problems, as many engineering problems involve quasi–steady state fields; and, if a single-frequency solution is sought, the FDFD is much faster than the transient broadband analysis afforded by FDTD.

14.1 FDFD via the wave equation

The simplest way to set up an FDFD simulation is via the wave equation. This method is often most desirable as well; since it does not involve coupled equations, it does not require interleaving or a Yee cell, and all fields (E, H, and J) are co-located on the grid. This simplifies both indexing into arrays as well as secondary calculations such as the Poynting flux or energy dissipation, as no spatial averages are necessary. Furthermore, in the frequency domain all auxiliary variables such as J and M can be put in terms of the desired field quantity (E or H), and thus the matrix solution involves only E or H, and the components of this variable (E_x, E_y, and E_z) are co-located in space. Note also that only one wave equation is necessary; after solving for $\mathbf{E}$, for example, $\mathbf{H}$ follows from the frequency domain Maxwell's equations.

The wave equations can be written in a very general form as [e.g., 1]:

$$\nabla^2 \overline{\mathcal{E}} = \nabla \times \overline{\mathcal{M}} + \mu \frac{\partial \overline{\mathcal{J}}}{\partial t} + \frac{1}{\epsilon} \nabla \tilde{\rho}_e + \mu\sigma \frac{\partial \overline{\mathcal{E}}}{\partial t} + \mu\epsilon \frac{\partial^2 \overline{\mathcal{E}}}{\partial t^2} \tag{14.1a}$$

$$\nabla^2 \overline{\mathcal{H}} = -\nabla \times \overline{\mathcal{J}} + \epsilon \frac{\partial \overline{\mathcal{M}}}{\partial t} + \frac{1}{\mu} \nabla \tilde{\rho}_m + \mu\sigma \frac{\partial \overline{\mathcal{H}}}{\partial t} + \mu\epsilon \frac{\partial^2 \overline{\mathcal{H}}}{\partial t^2} + \sigma \overline{\mathcal{M}} \tag{14.1b}$$

where we have included both electric and magnetic currents $\overline{\mathscr{J}}$ and $\overline{\mathscr{M}}$, and electric and magnetic volume charge densities $\tilde{\rho}_e$ and $\tilde{\rho}_m$, the latter of which is unphysical, but could be included in a simulation for various purposes. To convert these to the frequency domain, we replace $\partial/\partial t$ with $j\omega$, resulting in:

$$\nabla^2 \mathbf{E} = \nabla \times \mathbf{M} + j\omega\mu\mathbf{J} + \frac{1}{\epsilon}\nabla\rho_e + j\omega\mu\sigma\mathbf{E} - \omega^2\mu\epsilon\mathbf{E} \tag{14.2a}$$

$$\nabla^2 \mathbf{H} = -\nabla \times \mathbf{J} + j\omega\epsilon\mathbf{M} + \frac{1}{\mu}\nabla\rho_m + j\omega\mu\sigma\mathbf{H} - \omega^2\mu\epsilon\mathbf{H} + \sigma\mathbf{M} \tag{14.2b}$$

In source-free regions, $\mathbf{J} = \mathbf{M} = \rho_e = \rho_m = 0$, and these equations simplify dramatically to:

$$\nabla^2 \mathbf{E} = j\omega\mu\sigma\mathbf{E} - \omega^2\mu\epsilon\mathbf{E} = \gamma^2\mathbf{E} \tag{14.3a}$$

$$\nabla^2 \mathbf{H} = j\omega\mu\sigma\mathbf{H} - \omega^2\mu\epsilon\mathbf{H} = \gamma^2\mathbf{H} \tag{14.3b}$$

where γ is the propagation constant:

$$\gamma^2 = j\omega\mu\sigma - \omega^2\mu\epsilon = j\omega\mu(\sigma + j\omega\epsilon).$$

Here, γ can be written as $\gamma = \alpha + j\beta$, where α is the attenuation constant (in np/m) and β is the phase constant (in rad/m). To illustrate how these equations are solved in FDFD, we will consider the 1D version of the electric field wave equation,[1] including a source term J_z for generality (which implies inclusion of the $j\omega\mu J_z$ term in Equation 14.2a):

$$\frac{\partial^2 E_z}{\partial x^2} - \gamma^2 E_z = j\omega\mu J_z. \tag{14.4}$$

We now proceed to discretize Equation (14.4). Note that we need only discretize in space, as there is no time dependence. Notice also that in simple media, where ϵ and μ are scalar, γ is purely a complex constant[2] at any given location in space, since ω is a constant. We discretize in space using a second-order centered difference:

$$\left.\frac{\partial^2 E_z}{\partial x^2}\right|_i \simeq \frac{E_z|_{i+1} - 2E_z|_i + E_z|_{i-1}}{(\Delta x)^2} - \gamma^2 E_z|_i = j\omega\mu J_z|_i \tag{14.5}$$

where we have used Equation (3.59), without the superscript n, since there is no time indexing. Multiplying by $(\Delta x)^2$, we have the following update equation:

$$E_z|_{i+1} - \left[2 + \gamma^2(\Delta x)^2\right]E_z|_i + E_z|_{i-1} = j\omega\mu(\Delta x)^2 J_z|_i.$$

---

[1] Note that in the wave equation FDFD formulation, only one of Equations (14.3) needs to be solved. Once the E-field is found, for instance, the H-field follows directly from Maxwell's equations.

[2] In anisotropic materials, γ may be a tensor rather than a simple constant; however, the 3D wave equation can then be written as three component equations with cross terms, each of which has scalar multiplying factors.

This is, in general, a tridiagonal matrix, with $A = -[2 + \gamma^2(\Delta x)^2]$:

$$\underbrace{\begin{bmatrix} A & 1 & 0 & 0 & \cdots & & \\ 1 & A & 1 & 0 & \cdots & & \\ 0 & 1 & A & 1 & \cdots & & \\ & \vdots & & & & & \\ \cdots & 0 & 1 & A & 1 \\ \cdots & 0 & 0 & 1 & A \end{bmatrix}}_{M} \underbrace{\begin{bmatrix} E_z|_{i=0} \\ E_z|_1 \\ E_z|_2 \\ \vdots \\ E_z|_{m-1} \\ E_z|_m \end{bmatrix}}_{[E_z]} = j\omega\mu(\Delta x)^2 \underbrace{\begin{bmatrix} J_z|_{i=0} \\ J_z|_1 \\ J_z|_2 \\ \vdots \\ J_z|_{m-1} \\ J_z|_m \end{bmatrix}}_{[J_z]} \tag{14.6}$$

where the total number of spatial cells in our 1D grid is $m + 1$. The solution of this equation can be found by simple matrix inversion, resulting in

$$[E_z] = [M]^{-1} j\omega\mu(\Delta x)^2 [J_z].$$

However, the formulation specified in Equation (13.6) is not particularly interesting, since it is a 1D simulation of a homogeneous medium, where γ is the same everywhere. Note, however, that incorporation of real media, both inhomogeneous and frequency-dependent with $\sigma(x, \omega)$, $\mu(x, \omega)$ and $\epsilon(x, \omega)$ is very simple:

$$\underbrace{\begin{bmatrix} A_{i=0} & 1 & 0 & 0 & \cdots & & \\ 1 & A_1 & 1 & 0 & \cdots & & \\ 0 & 1 & A_2 & 1 & \cdots & & \\ & \vdots & & & & & \\ \cdots & 0 & 1 & A_{m-1} & 1 \\ \cdots & 0 & 0 & 1 & A_m \end{bmatrix}}_{M} \underbrace{\begin{bmatrix} E_z|_{i=0} \\ E_z|_1 \\ E_z|_2 \\ \vdots \\ E_z|_{m-1} \\ E_z|_m \end{bmatrix}}_{[E_z]} = j\omega(\Delta x)^2 \underbrace{\begin{bmatrix} \mu_{i=0} J_z|_{i=0} \\ \mu_1 J_z|_1 \\ \mu_2 J_z|_2 \\ \vdots \\ \mu_{m-1} J_z|_{m-1} \\ \mu_m J_z|_m \end{bmatrix}}_{[J_z]} \tag{14.7}$$

where the A_i are functions of σ_i, μ_i, ϵ_i, and ω at the particular grid locations. Notice that we have moved μ into the vector on the right-hand side, since it is spatially dependent.

In the 1D case, solutions can usually be found quite simply by taking an inverse of M, since the inverse of a tri-diagonal matrix is very simple to calculate. However, in higher dimensions, more advanced linear algebra techniques are required as the matrix M can become very large. For example, if E is defined in 2D on a 100×100 grid of cells, the vector $[E_z]$ will have 10,000 elements, and the matrix M will be $10,000 \times 10,000$. However, this large matrix will be sparse (and it is always square), and many techniques have been developed for the particular matrix geometries that result.

Note the resemblance here to the *implicit* methods of Chapter 13. However, frequency domain solutions are not implicit or explicit, as such nomenclature refers to time-dependent ODEs and PDEs only. The solution process for steady-state equations, on the other hand, is either convergent or divergent, depending on how well conditioned the matrix solution is. Later in this chapter we will come back to discuss the degree of accuracy of the method.

14.2   Laplace matrix and Kronecker product

As mentioned above, in two dimensions the matrix M can become very large. Before delving into the details, it will be prudent to review a few mathematical concepts which will aid in the matrix representations. First, consider Equation (14.6); if we remove γ from A, we can write this equation as follows:

$$\underbrace{\begin{bmatrix} -2 & 1 & 0 & 0 & \cdots \\ 1 & -2 & 1 & 0 & \cdots \\ 0 & 1 & -2 & 1 & \cdots \\ \vdots & & & & \\ \cdots & 0 & 1 & -2 & 1 \\ \cdots & 0 & 0 & 1 & -2 \end{bmatrix}}_{L} [E_z] - \gamma^2(\Delta x)^2[E_z] = j\omega\mu(\Delta x)^2[J_z]. \qquad (14.8)$$

The matrix L is called the Laplace matrix, as it appears often in the solution of Laplace's equation. With this notation, Equation (14.8) can be written as:

$$\left(L - \gamma^2(\Delta x)^2 I\right)[E_z] = j\omega(\Delta x)^2[\mu J_z] \qquad (14.9)$$

where I is the identity matrix, and we have explicitly included μ inside the vector on the right-hand side, so that this solution can apply to inhomogeneous media. The solution to this equation is then quite simply:

$$[E_z] = jw(\Delta x)^2 \left(L - \gamma^2(\Delta x)^2 I\right)^{-1}[\mu J_z]. \qquad (14.10)$$

As mentioned earlier, the matrices become very large in 2D or 3D; a 100×100 grid will produce a $10{,}000 \times 10{,}000$ matrix A or L. This matrix arises out of the Kronecker product of two 100×100 matrices. In general, if the matrix A is $m \times n$ and the matrix B is $p \times q$, the Kronecker product $A \otimes B$ is a $(mp) \times (nq)$ matrix defined as:

$$A \otimes B = \begin{bmatrix} a_{11}B & a_{12}B & \cdots & a_{1n}B \\ a_{21}B & a_{22}B & \cdots & a_{2n}B \\ \vdots & & \ddots & \vdots \\ a_{m1}B & a_{m2}B & \cdots & a_{mn}B \end{bmatrix}$$

where a_{ij} are the entries of the matrix A. The Kronecker product obeys associativity, $(A \otimes B) \otimes C = A \otimes (B \otimes C)$ but not commutativity, that is, in general $A \otimes B \neq B \otimes A$.

Next we will see how the Kronecker product arises and involves the matrix L. To demonstrate how these matrices arise, let us consider the numerical solution of the 2D Poisson's equation:

$$\nabla^2 \phi = \frac{\partial^2 \phi}{\partial x^2} + \frac{\partial^2 \phi}{\partial y^2} = -\frac{\tilde{\rho}(x, y)}{\epsilon}$$

(14.11)

where $\tilde{\rho}$ is the charge density, ϕ is the electric potential, and the medium is assumed to be homogeneous (so we could move ϵ to the right-hand side). As usual, we subdivide our solution space into a 2D grid, and then construct a vector of unknowns by converting the 2D space of dimensions $n_x \times n_y$ into a new vector $\tilde{\phi}$:

$$\tilde{\phi} = \text{vec } \phi = \text{vec} \left(\left[\begin{bmatrix} \phi_{1,1} \\ \phi_{2,1} \\ \vdots \\ \phi_{n_x,1} \end{bmatrix} \begin{bmatrix} \phi_{1,2} \\ \phi_{2,2} \\ \vdots \\ \phi_{n_x,2} \end{bmatrix} \cdots \begin{bmatrix} \phi_{1,n_y} \\ \phi_{2,n_y} \\ \vdots \\ \phi_{n_x,n_y} \end{bmatrix} \right] \right) = \begin{bmatrix} \phi_{1,1} \\ \phi_{2,1} \\ \vdots \\ \phi_{n_x,1} \\ \phi_{1,2} \\ \phi_{2,2} \\ \vdots \\ \phi_{n_x,2} \\ \vdots \\ \phi_{1,n_y} \\ \phi_{2,n_y} \\ \vdots \\ \phi_{n_x,n_y} \end{bmatrix}.$$

(14.12)

The function **vec** X refers to the vectorization of the matrix X as shown, by stacking columns of the matrix on top of one another. Consider the x component of the Laplacian on the right-hand side of Equation (14.11). As before, we approximate the derivative using central differences, yielding:

$$\frac{\partial^2 \phi}{\partial x^2} \approx \frac{\phi_{i-1,j} - 2\phi_{i,j} + \phi_{i+1,j}}{(\Delta x)^2}.$$

(14.13)

Note the structure of the vector $\tilde{\phi}$ in Equation (14.12). We have n_x entries along the x direction, from $(x_1, y_1), (x_2, y_1), \ldots, (x_{n_x}, y_1)$ followed by another n_x entries at (x_1, y_2), $(x_2, y_2), \ldots, (x_{n_x}, y_2)$ and so on. As before, we construct a matrix which we call $L_{x,2D}$ that operates on ϕ and returns an estimate of its second derivative in the x direction. Since each of these 1D slices along the x direction is the same as the L matrix we constructed in the previous section, we have the following block matrix:

$$
L_{x,2D} = \begin{bmatrix} L & 0 & 0 & \cdots & 0 & 0 \\ 0 & L & 0 & \cdots & 0 & 0 \\ \vdots & & & \ddots & & \vdots \\ 0 & 0 & 0 & \cdots & 0 & L \end{bmatrix} \tag{14.14}
$$

where 0 is a $n_x \times n_x$ matrix of zeros. To illustrate, consider the case of a small $n_x \times n_y = 4 \times 3$ space. We have:

$$
L_{x,2D} = \left[\begin{array}{cccc|cccc|cccc}
-2 & 1 & 0 & 0 & 0 & 0 & 0 & 0 & 0 & 0 & 0 & 0 \\
1 & -2 & 1 & 0 & 0 & 0 & 0 & 0 & 0 & 0 & 0 & 0 \\
0 & 1 & -2 & 1 & 0 & 0 & 0 & 0 & 0 & 0 & 0 & 0 \\
0 & 0 & 1 & -2 & 0 & 0 & 0 & 0 & 0 & 0 & 0 & 0 \\
\hline
0 & 0 & 0 & 0 & -2 & 1 & 0 & 0 & 0 & 0 & 0 & 0 \\
0 & 0 & 0 & 0 & 1 & -2 & 1 & 0 & 0 & 0 & 0 & 0 \\
0 & 0 & 0 & 0 & 0 & 1 & -2 & 1 & 0 & 0 & 0 & 0 \\
0 & 0 & 0 & 0 & 0 & 0 & 1 & -2 & 0 & 0 & 0 & 0 \\
\hline
0 & 0 & 0 & 0 & 0 & 0 & 0 & 0 & -2 & 1 & 0 & 0 \\
0 & 0 & 0 & 0 & 0 & 0 & 0 & 0 & 1 & -2 & 1 & 0 \\
0 & 0 & 0 & 0 & 0 & 0 & 0 & 0 & 0 & 1 & -2 & 1 \\
0 & 0 & 0 & 0 & 0 & 0 & 0 & 0 & 0 & 0 & 1 & -2
\end{array}\right].
$$

$$\tag{14.15}$$

We can rewrite the above block matrix in a more compact form using the Kronecker product, where I_{n_y} is the $n_y \times n_y$ identity matrix and L_{n_x} is the $n_x \times n_x$ discrete 1D Laplacian matrix defined in Equation (14.8):

$$
L_{x,2D} = I_{n_y} \otimes L_{n_x}.
$$

Now consider the y component of the Laplacian:

$$
\frac{\partial^2 \phi}{\partial y^2} \approx \frac{\phi_{i,j-1} - 2\phi_{i,j} + \phi_{i,j+1}}{(\Delta y)^2}. \tag{14.16}
$$

We can see that the finite difference in the y direction resembles our original matrix L but now with block structure. That is,

$$
L_{y,2D} = \begin{bmatrix} -2I & I & 0 & \cdots & 0 & 0 & 0 \\ I & -2I & I & & 0 & 0 & 0 \\ \vdots & & & \ddots & & & \vdots \\ 0 & 0 & 0 & & I & -2I & I \\ 0 & 0 & 0 & \cdots & 0 & I & -2I \end{bmatrix} \tag{14.17}
$$

where now I is an $n_x \times n_x$ identity matrix and there are n_y matrices along the diagonal. Again we illustrate for the specific case of a 4×3 system:

$$
L_{y,2D} =
\left[
\begin{array}{cccc|cccc|cccc}
-2 & 0 & 0 & 0 & 1 & 0 & 0 & 0 & 0 & 0 & 0 & 0 \\
0 & -2 & 0 & 0 & 0 & 1 & 0 & 0 & 0 & 0 & 0 & 0 \\
0 & 0 & -2 & 0 & 0 & 0 & 1 & 0 & 0 & 0 & 0 & 0 \\
0 & 0 & 0 & -2 & 0 & 0 & 0 & 1 & 0 & 0 & 0 & 0 \\
\hline
1 & 0 & 0 & 0 & -2 & 0 & 0 & 0 & 1 & 0 & 0 & 0 \\
0 & 1 & 0 & 0 & 0 & -2 & 0 & 0 & 0 & 1 & 0 & 0 \\
0 & 0 & 1 & 0 & 0 & 0 & -2 & 0 & 0 & 0 & 1 & 0 \\
0 & 0 & 0 & 1 & 0 & 0 & 0 & -2 & 0 & 0 & 0 & 1 \\
\hline
0 & 0 & 0 & 0 & 1 & 0 & 0 & 0 & -2 & 0 & 0 & 0 \\
0 & 0 & 0 & 0 & 0 & 1 & 0 & 0 & 0 & -2 & 0 & 0 \\
0 & 0 & 0 & 0 & 0 & 0 & 1 & 0 & 0 & 0 & -2 & 0 \\
0 & 0 & 0 & 0 & 0 & 0 & 0 & 1 & 0 & 0 & 0 & -2
\end{array}
\right].
$$

$$\tag{14.18}$$

Written in terms of Kronecker products, the matrix is:

$$
L_{y,2D} = L_{n_y} \otimes I_{n_x}.
$$

Therefore, we can write the system of equations required to solve Poisson's equation in 2D in terms of Kronecker products as:

$$
\left(\frac{1}{(\Delta x)^2} L_{x,2D} + \frac{1}{(\Delta y)^2} L_{y,2D} \right) \Phi = -\frac{1}{\epsilon} \tilde{\rho}.
\tag{14.19}
$$

Note that the utility of the Kronecker product in solving equations of this type is no accident. Kronecker products were developed for and arise naturally in the field of *multilinear algebra*, where instead of vectors and vector spaces, we deal with tensors and tensor spaces and solutions of systems of equations involving them. For example, consider the matrix equation $AXB^T = C$, where A,B,C, and X are all matrices (that is, rank-2 tensors) and we wish to find a solution X to this system of equations. We can restate this problem as a plain linear algebra problem using the **vec** operation and the Kronecker product:

$$
(B \otimes A)(\mathbf{vec}\, X) = \mathbf{vec}\, C.
$$

For the simple case of $AX = C$, we have a result resembling our Poisson matrix L_x:

$$
(I \otimes A)(\mathbf{vec}\, X) = \mathbf{vec}\, C.
$$

And for the case of $XB^T = C$, the result resembles L_y:

$$
(B \otimes I)(\mathbf{vec}\, X) = \mathbf{vec}\, C.
$$

14.3 Wave equation in 2D

We are now ready to tackle the wave equation in 2D. To see how the method is constructed, we begin with the 2D electric field wave equation. In the 2D x–y plane we have two possible sets of equations for E, either the TM mode:

$$\frac{\partial^2 E_z}{\partial x^2} + \frac{\partial^2 E_z}{\partial y^2} - \gamma^2 E_z = j\omega\mu J_z \tag{14.20a}$$

or the TE mode:

$$\frac{\partial^2 E_x}{\partial x^2} + \frac{\partial^2 E_x}{\partial y^2} - \gamma^2 E_x = j\omega\mu J_x \tag{14.20b}$$

$$\frac{\partial^2 E_y}{\partial x^2} + \frac{\partial^2 E_y}{\partial y^2} - \gamma^2 E_y = j\omega\mu J_y. \tag{14.20c}$$

Here we will consider the TE mode. We can discretize these equations using second-order centered differences, and co-locate the E_x and E_y components at integer (i, j) locations, since we do not need to separately solve the magnetic field wave equation. Discretization in this way yields:

$$\frac{E_x|_{i+1,j} - 2E_x|_{i,j} + E_x|_{i-1,j}}{(\Delta x)^2} + \frac{E_x|_{i,j+1} - 2E_x|_{i,j} + E_x|_{i,j-1}}{(\Delta y)^2} - \gamma^2 E_x|_{i,j} = j\omega\mu J_x \tag{14.21a}$$

$$\frac{E_y|_{i+1,j} - 2E_y|_{i,j} + E_y|_{i-1,j}}{(\Delta x)^2} + \frac{E_y|_{i,j+1} - 2E_y|_{i,j} + E_y|_{i,j-1}}{(\Delta y)^2} - \gamma^2 E_y|_{i,j} = j\omega\mu J_y. \tag{14.21b}$$

Recalling the 2D discrete Laplacians defined in the previous section:

$$L_{x,2D} = I_{n_y} \otimes L_{n_x}$$

$$L_{y,2D} = L_{n_y} \otimes I_{n_x}.$$

We can again stack the unknowns E_x and E_y into vector form, and rewrite our system of equations as:

$$\left(\frac{1}{(\Delta x)^2} L_{x,2D} + \frac{1}{(\Delta y)^2} L_{y,2D} - \gamma^2 I\right) E_x = j\omega\mu J_x \tag{14.22a}$$

$$\left(\frac{1}{(\Delta x)^2} L_{x,2D} + \frac{1}{(\Delta y)^2} L_{y,2D} - \gamma^2 I\right) E_y = j\omega\mu J_y. \tag{14.22b}$$

Note the strong resemblance to Equation (14.19). However, the careful reader will notice a problem. Suppose we want a y-directed infinitesimal dipole current source J_y at a single point in the domain, as we might have done in the FDTD method. The wave equations above, however, are completely *uncoupled*, implying that a current source J_y will not excite any x-directed component E_x, which is an unphysical result. The detail we are missing is that in formulating the wave equation, we assumed the divergence of $\overline{\mathscr{E}}$ and $\overline{\mathscr{H}}$ to both equal zero, that is:

$$\nabla \times \left(\nabla \times \overline{\mathscr{E}} \right) = -\nabla^2 \overline{\mathscr{E}} + \nabla \left(\nabla \cdot \overline{\mathscr{E}} \right) = -\nabla^2 \overline{\mathscr{E}}$$

$$\nabla \times \left(\nabla \times \overline{\mathscr{H}} \right) = -\nabla^2 \overline{\mathscr{H}} + \nabla \left(\nabla \cdot \overline{\mathscr{H}} \right) = -\nabla^2 \overline{\mathscr{H}}.$$

This condition $\nabla \cdot \overline{\mathscr{E}} = \nabla \cdot \overline{\mathscr{H}} = 0$, however, does not hold at the source, as we know from Gauss's law, Equation (2.2).

There are some different approaches we might use to solve this problem: First, we could assume that the divergence *might* be nonzero and solve the system of equations that results. Unfortunately, the matrix that results is no longer symmetric (and thus is much more difficult to solve numerically) and the result will typically have a large unwanted divergence even in source-free regions. Alternatively, we might use some other method to solve for the fields in a small region around the source, e.g., analytically or with FDTD, and then use the wave equation as above to find the solution in the rest of the domain.

Fortunately, in 2D we have a simple alternative. Consider the TE mode with a source term, where we now include the H_z component wave equation:

$$\frac{\partial^2 E_x}{\partial x^2} + \frac{\partial^2 E_x}{\partial y^2} - \gamma^2 E_x = j\omega\mu J_x \tag{14.23a}$$

$$\frac{\partial^2 E_y}{\partial x^2} + \frac{\partial^2 E_y}{\partial y^2} - \gamma^2 E_y = j\omega\mu J_y \tag{14.23b}$$

$$\frac{\partial^2 H_z}{\partial x^2} + \frac{\partial^2 H_z}{\partial y^2} - \gamma^2 H_z = -\underbrace{\left(\frac{\partial J_y}{\partial x} - \frac{\partial J_x}{\partial y} \right)}_{(\nabla \times \mathbf{J})_z} \tag{14.23c}$$

where the curl of $\mathbf{J}$ term in the last equation can be traced to Equation (14.2b). Notice that because we have not included the nonzero divergence, the wave equation fails to couple E_x and E_y at the source. However, the H_z component has no such issue; H_z only depends on the $\hat{z}$ component of the curl of the source current J. Essentially, rather than assuming $\nabla \cdot \overline{\mathscr{E}} = 0$ in the formulation of the wave equation as we did earlier, we are assuming $\nabla \cdot \overline{\mathscr{H}} = 0$, which is perfectly valid even at the source.

Suppose we have a y-directed current source $J_y|_{i,j}$. Approximating the derivatives at nearby points using centered differences, we have:

$$(\nabla \times J)_z|_{i+1,j} \approx \frac{J_y|_{i+2,j} - J_y|_{i,j}}{2\Delta x} = -\frac{J_y|_{i,j}}{2\Delta x} \tag{14.24a}$$

$$(\nabla \times J)_z|_{i,j} \approx \frac{J_y|_{i+1,j} - J_y|_{i-1,j}}{2\Delta x} = 0 \tag{14.24b}$$

$$(\nabla \times J)_z|_{i-1,j} \approx \frac{J_y|_{i,j} - J_y|_{i-2,j}}{2\Delta x} = \frac{J_y|_{i,j}}{2\Delta x} \tag{14.24c}$$

since $J_y|_{i,j}$ is our only nonzero current component. So, we can replace a vertical J_y source with a pair of sources in the H_z equations, and solve for H_z using the wave equation formulation described above in Equation (14.23c). Then, if the $\vec{E}$ components are needed, we can use Maxwell's equations to solve for E_x and E_y from the H_z field.

14.4 Wave equation in 3D

The discretized FDFD equations in 3D follow straightforwardly from the previous sections. We present them here without derivation:

$$\frac{\partial^2 E_x}{\partial x^2} + \frac{\partial^2 E_x}{\partial y^2} + \frac{\partial^2 E_x}{\partial z^2} - \gamma^2 E_x = j\omega\mu J_x \tag{14.25a}$$

$$\frac{\partial^2 E_y}{\partial x^2} + \frac{\partial^2 E_y}{\partial y^2} + \frac{\partial^2 E_y}{\partial z^2} - \gamma^2 E_y = j\omega\mu J_y \tag{14.25b}$$

$$\frac{\partial^2 E_z}{\partial x^2} + \frac{\partial^2 E_z}{\partial y^2} + \frac{\partial^2 E_z}{\partial z^2} - \gamma^2 E_z = j\omega\mu J_z. \tag{14.25c}$$

After discretizing using centered differences as before, and following the patterns of the 2D method, we have:

$$\left(\frac{1}{(\Delta x)^2}L_{x,3D} + \frac{1}{(\Delta y)^2}L_{y,3D} + \frac{1}{(\Delta z)^2}L_{z,3D} - \gamma^2 I\right)E_x = j\omega\mu J_x \tag{14.26a}$$

$$\left(\frac{1}{(\Delta x)^2}L_{x,3D} + \frac{1}{(\Delta y)^2}L_{y,3D} + \frac{1}{(\Delta z)^2}L_{z,3D} - \gamma^2 I\right)E_y = j\omega\mu J_y \tag{14.26b}$$

$$\left(\frac{1}{(\Delta x)^2}L_{x,3D} + \frac{1}{(\Delta y)^2}L_{y,3D} + \frac{1}{(\Delta z)^2}L_{z,3D} - \gamma^2 I\right)E_z = j\omega\mu J_z \tag{14.26c}$$

where the 3D Laplace matrices are defined in terms of Kronecker products:

$$L_{x,3D} = I_{n_z} \otimes I_{n_y} \otimes L_{n_x} \tag{14.27a}$$

$$L_{y,3D} = I_{n_z} \otimes L_{n_y} \otimes I_{n_x} \tag{14.27b}$$

$$L_{z,3D} = L_{n_z} \otimes I_{n_y} \otimes I_{n_x}. \tag{14.27c}$$

Note that the 3D method suffers from the same problem at the source. For this reason, it is more common to use the magnetic field wave equation and include $\nabla \times \mathbf{J}$ on the right-hand side of the component equations (the magnetic field equivalents of Equations 14.26).

Wave equation caveats

The wave equation method is an efficient method for solving Maxwell's equations, provided that the system of equations remains symmetric. The inclusion of sources, however, must be handled extremely carefully due to the lack of implicit coupling between the orthogonal field components, as we found above in the 2D formulation.

Typically this means that the wave equation method is a good choice for certain types of boundary value problems. For example, problems which have a plane wave source at one end of a domain scattering off some structure are well-suited to this approach. In other scenarios, we must use some other method to find a solution to the fields at the boundaries of a small section within the domain, either analytically or with another method such as FDTD.

14.5   FDFD from Maxwell's equations

It is entirely possible to formulate the FDFD method starting with Maxwell's equations rather than from the wave equation. We present the method here for completeness and to show how the process differs. We begin with Ampère's law and Faraday's law in differential form in the frequency domain:

$$j\omega\mu\mathbf{H} = -\nabla \times \mathbf{E} - \mathbf{M}_{\text{source}} - \sigma_m\mathbf{H} \tag{14.28a}$$

$$j\omega\epsilon\mathbf{E} = \nabla \times \mathbf{H} - \mathbf{J}_{\text{source}} - \sigma\mathbf{E}. \tag{14.28b}$$

We will again consider here the 1D case without sources. For x-directed propagation and $\mathbf{E}$ polarized along y, we have

$$j\omega\mu H_z = -\frac{\partial E_y}{\partial x} - \sigma_m H_z \tag{14.29a}$$

$$j\omega\epsilon E_y = -\frac{\partial H_z}{\partial x} - \sigma E_y. \tag{14.29b}$$

Discretizing using the leapfrog method, we have the finite difference equations:

$$(j\omega\mu + \sigma_m)H_z\big|_{i+1/2} = -\frac{E_y\big|_{i+1} - E_y\big|_i}{\Delta x} \tag{14.30a}$$

$$(j\omega\epsilon + \sigma)E_y\big|_i = -\frac{H_z\big|_{i+1/2} - H_z\big|_{i-1/2}}{\Delta x}. \tag{14.30b}$$

These can be written as a simple linear system:

$$A_h H_z\big|_{i+1/2} + E_y\big|_{i+1} - E_y\big|_i = 0 \tag{14.30c}$$

$$A_e E_y\big|_i + H_z\big|_{i+1/2} - H_z\big|_{i-1/2} = 0 \tag{14.30d}$$

where $A_h = (j\omega\mu + \sigma_m)\Delta x$ and $A_e = (j\omega\epsilon + \sigma)\Delta x$. To formulate the matrix for this system, we create a column vector that includes each of the E components, followed by each of the H components; i.e., $[\mathbf{F}] = [E_{y|0}; E_{y|1}; E_{y|2}; \ldots E_{y|n}; H_{z|1/2}; H_{z|3/2}; \ldots H_{z|n-1/2}]$ (where the boundaries at 0 and n are taken to be E components).

We then fill in the appropriate matrix components, resulting in:

$$
\underbrace{
\begin{bmatrix}
-1 & 1 & 0 & 0 & \cdots & A_h & 0 \\
1 & -1 & 1 & 0 & \cdots & 0 & A_h \\
0 & 1 & -1 & 1 & \cdots & & \\
\vdots & & & & & & \\
A_e & 0 & \cdots & 0 & 1 & -1 & 1 \\
0 & A_e & \cdots & 0 & 0 & 1 & -1
\end{bmatrix}
}_{M}
\underbrace{
\begin{bmatrix}
E_0 \\
E_1 \\
\vdots \\
E_n \\
H_{1/2} \\
\vdots \\
H_{n-1/2}
\end{bmatrix}
}_{F}
= 0. \tag{14.31}
$$

The inclusion of sources in this method follows directly as in the wave equation method. For a hard source, say $E_0 = 1$ V/m, the first row of the matrix M is zeroed except for the first entry. This value is then moved to the right-hand side of the equation to create a "source" vector on the right side. Source currents can be added by including the J_{source} term in Equation (14.28b), which ends up on the right-hand side of the matrix equation.

There are a number of features which make the FDFD method from Maxwell's equations less attractive than the direct method from the wave equation. First, notice that the matrix M, which must now be inverted, is no longer tridiagonal (though it is still sparse); this means the inversion process will be somewhat more complicated. Second, the vector of field values has doubled in length, since we are simultaneously solving for E and H. In the wave equation method, we solve for E only, and H is found very simply by applying one or the other of Maxwell's equations. The doubling of the field vector results in a *quadrupling* of the matrix M, increasing computational cost considerably. Finally, as mentioned before, while it is advantageous for accuracy in the FDTD method, having the fields interleaved in space on the Yee cell in FDFD makes calculations of Poynting fluxes, losses through $J \cdot E$, and so forth less accurate and adds the necessity of spatial averaging.

14.6   Summary

In this chapter we have introduced the finite difference frequency domain (FDFD) method. The FDTD method is useful when a single-frequency solution to a problem is sought, or when the source wave can be approximated by a single frequency. In such cases, the frequency domain versions of Maxwell's equations or the wave equation are discretized. In the 2D TM mode, the wave equation can be written as:

$$
\frac{\partial^2 E_z}{\partial x^2} + \frac{\partial^2 E_z}{\partial y^2} - \gamma^2 E_z = j\omega\mu J_z
$$

where the complex propagation constant is given by:

$$
\gamma^2 = j\omega\mu\sigma - \omega^2\mu\epsilon = j\omega\mu(\sigma + j\omega\epsilon).
$$

Discretizing this equation using second-order centered differences, we find a matrix equation of the form:

$$\left(\frac{1}{(\Delta z)^2} L_{x,2D} + \frac{1}{(\Delta y)^2} L_{y,2D} - \gamma^2 I \right) E_z = j\omega\mu J_z$$

where $L_{x,2D}$ and $L_{y,2D}$ are the 2D Laplace matrices. Similar equations are derived for the 2D TM mode and for the 3D algorithm. Note that the FDTD method can also be implemented using Maxwell's equations, but such an algorithm presents a number of disadvantages, so the wave equation method is far more commonly used.

14.7   Problems

14.1. **A PML in FDFD.** Because the FDTD method yields a steady-state solution, and it cannot be excited by a Gaussian source, or have sources turned off at convenient times, the PML is even more important than it is in the FDTD method. Derive the stretched-coordinate PML in the FDFD method using the wave equation, using the same s_x, etc., stretching parameters as in the UPML. Hint: start from Equations (9.57) and combine them to form a wave equation, then discretize. For more details on the PML in the FDTD method, see the article by Rappaport et al. [2].

14.2. **TF/SF method in FDFD.** Similarly, the total-field / scattered-field formulation is of great importance in the FDFD method, since it is the only way to separate scattered fields from incident fields. Derive a TF/SF method in FDFD. Start by using the wave equation update, Equation (14.5), and then make the correct modifications near the TF/SF boundary, referring to Figure 7.3. For a single TF/SF boundary, how does this modify the matrix M?

14.3. **FDFD method in 1D.** Write a simulation using the FDFD method in 1D from the wave equation. Incorporate a PML boundary at both ends, and excite the source using the TF/SF method $\sim$10 grid cells from the end. Model the dielectric slab as in Problem 4.4, launch an incident wave with wavelength 0.3 m, and measure the steady-state amplitude in the simulation space. Does this agree with the results of Problem 4.4?

14.4. **FDFD method in 2D.** Repeat Problem 9.9, the system of scatterers with a PML boundary, in the FDFD method using the wave equation for $\mathcal{E}_z$. Derive the simulation with a sinusoidal source with a wavelength of 0.5 m. Measure the field distribution throughout the space; evaluate the reciprocity of the simulation by alternating points A and B as source and receiver.

14.5. **FDFD in cylindrical coordinates.** Derive an FDFD algorithm in 2D cylindrical coordinates, starting from the 2D wave equation, where $\mathcal{H}_\phi$ is the singular component. Include the possibility of a current source at a given point in the simulation space, and a PML on the boundaries.

14.6. **Radiation pattern of half-wave dipole.** Use the FDFD method in 2D cylindrical coordinates to model a half-wave dipole antenna, excited by an $\mathcal{E}_z$ field in a small gap, similar to Problem 12.9. Excite the antenna at the appropriate frequency and measure the radiation pattern some distance from the antenna. How does this compare to the FDTD analysis of this antenna?

References

[1] J. Van Bladel, *Electromagnetic Fields*, 2nd edn. Wiley Interscience, 2007.
[2] C. M. Rappaport, M. Kilmer, and E. Miller, "Accuracy considerations in using the PML ABC with FDFD Helmholtz equation computation," *Int. J. Numer. Model.*, vol. 13, pp. 471–482, 2000.

15 Finite volume and finite element methods

In this chapter we introduce the framework for two alternative methods, the finite volume (FV) and finite element (FE) methods, as they apply to time domain electromagnetic problems. These two methods are particularly useful in cases where the geometry of the problem involves complicated, curved surfaces, the scattering from which we would like to model with high accuracy. In such scenarios, the FDTD method's inherent "staircasing" of the grid cells means that those curved surfaces are not well described in the model.

In Section 15.2 we will introduce the finite volume method, and in Section 15.3 we will discuss the finite element method. We conclude the chapter with a brief introduction to the discontinuous Galerkin method, using features of both finite volume and finite element methods. Our intention here is not to provide a full treatment of these methods, but simply to introduce them to the student, in order to show their relationship to the FDTD method, and to clarify the conditions under which these alternative methods might be used. As such we provide only brief introductions to the FV and FE methods and refer the readers to other materials for a full treatment. Before describing these methods, we provide an introduction to advanced grid techniques, which will be required for both of these methods.

15.1 Irregular grids

The FDTD method as we have described it so far involves a regular, orthogonal grid of identical cells with dimensions Δx, Δy, and Δz, in Cartesian coordinates. We also derived the FDTD method in Chapter 4 for cylindrical and spherical coordinates, which can be used when the simulation space (and/or scattering structures) lend themselves more accurately to those coordinates. The resulting grids in cylindrical and spherical coordinates are orthogonal, but *nonuniform*, since the sizes of the grid cells varied with ϕ and θ. In addition, grids can be *nonorthogonal* or even *unstructured*.

Many of these grid techniques can be used in the FDTD method; for example, an orthogonal, nonuniform grid, as described in the next subsection, can be easily implemented in FDTD. However, the more advanced techniques of nonorthogonal or unstructured grids become cumbersome in FDTD, and it is in those circumstances that finite volume or finite element methods are a better choice. In this section, we provide

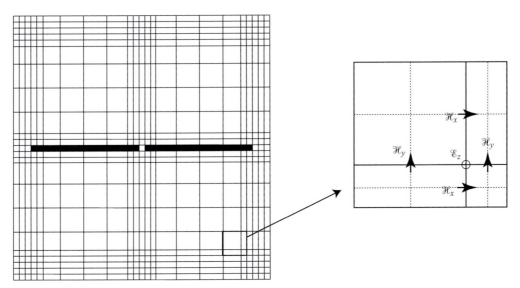

Figure 15.1 Nonuniform orthogonal grid surrounding a dipole antenna. Left: This grid retains all of the same corners as the regular FDTD grid, so that each cell simply has its own Δx and Δy. Right: This grid has square cells which grow with distance from the antenna; this method introduces "nodes" where two cells meet one larger cell, and these nodes require special treatment.

brief introductions to grid techniques, and refer the reader to other sources for more detail.

15.1.1 Nonuniform, orthogonal grids

The FDTD algorithm lends itself to orthogonal grids, as the Yee difference equations are simplest to describe in the standard orthogonal coordinate systems. However, the grids can be made *nonuniform* quite easily. For example, while the Cartesian coordinate system is uniform, the cylindrical and spherical systems are nonuniform, as the grid cell sizes in ϕ and θ vary with distance r.

Other types of nonuniform, orthogonal grids are possible. The simplest example is a grid in Cartesian coordinates where the grid sizes Δx, Δy, and/or Δz vary in their respective dimensions; an example is shown in Figure 15.1. In this example, each spatial grid size varies along its respective direction, but each of the cells is still abutting rectangles. Now, however, the usual centered difference approximations to the spatial derivatives in Maxwell's equations are no longer valid. To see this, notice that in the expanded cell in Figure 15.1, the $\mathcal{E}_z$ component at the cell corner is no longer equidistant from the pairs of $\mathcal{H}_x$ and $\mathcal{H}_y$ components, so the "centered" nature of the difference equation is lost.

These types of cells require special handling through a new derivation of the difference equations, which takes into account the changing cell size and assigns Δx_i and Δy_j to each location in the grid. We refer the reader to work by P. Monk [1, 2] for full details.

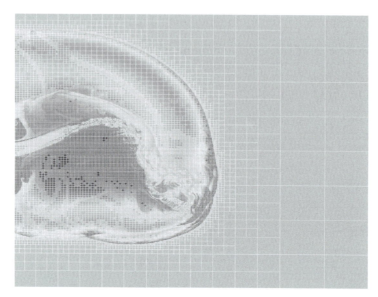

Figure 15.2 Adaptive mesh refinement. The mesh has smaller grid cells at the boundary of a propagating shock, where the gradient of the fields is highest. This example shows a fluid shock, but the concept applies equally to electromagnetic wavefronts.

We must caution that this method reduces the simulation to first-order accuracy in these transition regions; however, Monk [1, 2] has shown that second-order accuracy is maintained *globally*.

Other types of nonuniform grids can be imagined; for example, one might use a grid of square cells, but as the distance grows from the scattering object, larger square cells can be made up of four smaller cells. This is the scenario shown in Figure 15.2, which will be discussed next.

The orthogonal nature of these grids, as well as the uniform FDTD grid, makes them appealing in FDTD. The FDTD algorithm relies on contour interpretations of Maxwell's equations, as described in Chapter 12, and these contours are made trivial if the field components of interest are parallel to the contour C and normal to its enclosed surface S. When the fields are not perpendicular, we must evaluate the fields normal to the surface and parallel to the contour, which becomes quite cumbersome. We will begin to see the complexity of nonorthogonal grids when we discuss the finite volume method in the next section.

Adaptive mesh refinement

One highly advanced version of a nonuniform but orthogonal grid can be described through adaptive mesh refinement (AMR). In this method, the grid cell sizes are adapted as the fields propagate, so that where high gradients in the field (i.e., high frequencies) exist, smaller grid cells are used, but where the fields are zero or have low gradients (low frequencies), larger grid cells can be used. This is illustrated for a fluid

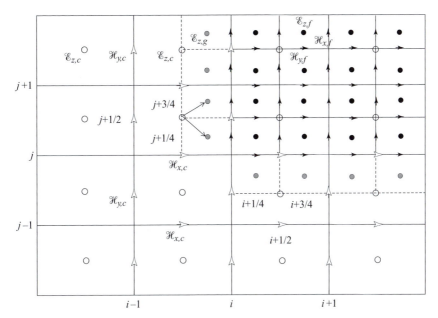

Figure 15.3 Grid-level boundaries in AMR. Interpolation and averaging of fields are required at the interface of different grid levels. In AMR, these must also occur when a grid cell or region is refined or unrefined between time steps. Adapted from [3].

shock in Figure 15.2. In this method, our grid is now *dynamic*, where until now we have discussed only *static* grids.

In either the static, irregular grid or the dynamic AMR grid, one must treat the interfaces between coarse and fine grid cells carefully, as shown in Figure 15.3. A number of different methods for interface interpolation have been used with success in different AMR algorithms. We refer to Figure 15.3, which demonstrates the method of Zakharian et al. [3] for a grid refinement factor of 2, i.e., the cells have been split in half along each dimension. We assume that the grid has been established at the given time step, and that $\mathcal{E}_z$ is known on both the coarse ($\mathcal{E}_{z,c}$) and fine ($\mathcal{E}_{z,f}$) grids at time step n; the corresponding magnetic field components are known at time step $(n - 1/2)$ on the coarse grid, and at time step $(n - 1/4)$ on the fine grid. We then update the fields as follows:

1. On the coarse grid, $\mathcal{H}_{x,c}$ and $\mathcal{H}_{y,c}$ are updated to time step $n + 1/2$ in the usual FDTD algorithm, and then the electric field $\mathcal{E}_{z,c}$ is advanced to time step $(n + 1)$.
2. The "ghost" values of $\mathcal{E}_{z,g}$ along the boundary between the coarse and fine grids are determined at time step n through interpolation of the known $\mathcal{E}_z$ fields, using both coarse and fine grids. This can be done in a number of ways, as outlined by [3], and the different methods result in different degrees of accuracy and stability.
3. The $\mathcal{H}_{x,f}$ and $\mathcal{H}_{y,f}$ fields are advanced a *half*-time step from $(n - 1/4)$ to $(n + 1/4)$ using the values of $\mathcal{E}_{z,f}$ at time step n. $\mathcal{E}_{z,f}$ is advanced to time step $(n + 1/2)$.
4. The "ghost" values of $\mathcal{E}_{z,g}$ are interpolated at time step $(n + 1/2)$ in the same way as step 2 above.

5. The fields on the fine grid are advanced to time step $(n + 3/4)$ for the $\overline{\overline{\mathcal{H}}}$ components and $(n + 1)$ for the $\mathcal{E}_{z,f}$ component, as in step 3 above.
6. The magnetic fields on the interface boundary, i.e., $\mathcal{H}_{x,c}|_{i+1/2,j}$, are evaluated by averaging (in time and space) the neighboring fine-grid fields. For example,

$$\mathcal{H}_{x,c}\Big|_{i+1/2,j}^{n+1/2} = \frac{1}{4}\left(\mathcal{H}_{x,c}\Big|_{i+1/4,j}^{n+1/4} + \mathcal{H}_{x,c}\Big|_{i+1/4,j}^{n+3/4} + \mathcal{H}_{x,c}\Big|_{i+3/4,j}^{n+1/4} + \mathcal{H}_{x,c}\Big|_{i+3/4,j}^{n+3/4}\right).$$

7. The $\mathcal{E}_{z,c}$ fields are updated to time step $(n + 1)$ along the interface boundary (i.e., at $(i + 1/2, j - 1/2)$ using the usual FDTD algorithm).
8. In the regions where the fine grid overlaps the coarse grid, i.e., the $\mathcal{E}_{z,c}$ circles within the fine grid, $\mathcal{E}_{z,c}$ is replaced by a volume average of $\mathcal{E}_{z,f}$.

Note that steps 2–3 and steps 4–5 constitute a single updating of a fractional time step on the fine grid. For a refinement factor of N, these pairs of steps must be repeated N times. The refined regions are thus updated N times for every update on the coarse grid. The method is a variant of the commonly used Berger-Oliger algorithm [4], in which block regions of the grid are refined at each time step using a "tree" structure, allowing finer and finer grids to be nested within one another.

The algorithm above is just one of a handful of possible ways to deal with the refinement interface. Now, this algorithm can work with both static and dynamic grids. In the dynamic AMR grid, the need for refinement is usually determined by gradients in the propagating fields. Once a region has been marked for refinement, a new subgrid is defined by interpolation of the known fields to the new, refined grid locations. This interpolation must be conducted very carefully, as it has a strong effect on the overall accuracy and stability of the simulation. For more details on AMR methods, we refer the reader to the papers cited above and the book by Sarris [5].

15.1.2   Nonorthogonal structured grids

One can easily devise a grid that is structured, but not orthogonal; furthermore, the grid does not need to be regular, in that adjacent cells are not the same shape. For example, the grid cells might be arbitrary parallelepipeds, the coordinates of which are the unit vectors along the edges of the cell. Such a grid structure can be easily conformed to structures within the simulation space. This type of grid structure leads to the finite volume formulation described in the next section. These types of grids are still *structured*, however, in that the vertices of the grid cells can be described by ordered arrays $x(i, j, k)$, $y(i, j, k)$, and $z(i, j, k)$. A detailed discussion of the mathematical basis of these grids is beyond the scope of this book; the reader is referred to [6] for details.

15.1.3   Unstructured grids

Lastly, for the sake of completeness, we briefly turn our attention to *unstructured grids*. In any discretized grid, consider the nodes of the grid to have locations (i, j, k). Fundamentally, a grid is considered *structured* if the nodes can be ordered; for example, (i, j, k)

and $(i, j, k + 1)$ are neighbors. This means that only the node locations themselves need to be defined, and it is assumed that each is connected to its nearest neighbors.

In an *unstructured* grid, however, this assumption is not valid. As such, not only must the node locations be defined, but information about how they are connected must be provided. Unstructured grids provide complete freedom in conforming to material surfaces, and as such they provide the most powerful method for simulating detailed structures. Typically, the grids themselves are generated using commercial software packages; the extreme complexity of these grids lies in defining the mathematical relationships between fields on the nodes or faces of the grid elements. An example of an unstructured grid is shown in Figure 15.7 in Section 15.3.

Entire books have been written on grid generation techniques; the reader is referred to the book by Thompson et al. [7] for a thorough and detailed discussion of grid generation.

15.2   The finite volume method

In this section, we introduce a class of methods known as finite volume methods. These methods are particularly well suited to solutions of hyperbolic partial differential equations such as Maxwell's equations. Finite volume methods were originally developed for applications in computational fluid dynamics (CFD), and can be very intuitive for partial differential equations such as the Navier-Stokes equation. For coupled PDEs such as Maxwell's equations, recent developments have allowed for the interleaved nature of the fields, similar to the FDTD method.

Before describing this interleaved method, we will first discuss the basic finite volume formulation. This formulation relies on conservation laws; for example, the equation of conservation of charge:

$$\frac{\partial \tilde{\rho}}{\partial t} + \nabla \cdot \bar{\mathcal{J}} = 0. \tag{15.1}$$

In general, conservation laws can be written in the form:

$$\frac{\partial \vec{u}}{\partial t} + \nabla \cdot \bar{\bar{\mathcal{F}}}(\vec{u}) = \overline{\mathcal{G}}(\vec{u}) \tag{15.2a}$$

or, in component form,

$$\frac{\partial \vec{u}}{\partial t} + \frac{\partial \vec{\mathcal{F}}_x(\vec{u})}{\partial x} + \frac{\partial \vec{\mathcal{F}}_y(\vec{u})}{\partial y} + \frac{\partial \vec{\mathcal{F}}_z(\vec{u})}{\partial z} = \overline{\mathcal{G}}(\vec{u}) \tag{15.2b}$$

where $\bar{\bar{\mathcal{F}}}$ is a flux tensor of some variable, and $\mathcal{G}$ is a source term. Now, application of the finite volume method involves integrating this general equation over small volumetric domains (in 3D); the faces and edges of the volumes are known, having been established as a nonuniform, unstructured grid as described in the previous section. We integrate

Equation (15.2a) over a small volume V_i:

$$\int_{V_i} \frac{\partial \vec{u}}{\partial t} dv + \int_{V_i} \nabla \cdot \overline{\overline{\mathcal{F}}}(\vec{u}) dv = \int_{V_i} \mathcal{G}(\vec{u}) dv. \tag{15.3}$$

We next assume the field $\vec{u}$ is constant over the volume V_i and apply the divergence theorem (equation 2.7):

$$V_i \frac{\partial \vec{u}_i}{\partial t} + \oint_{S_i} \mathcal{F}_i^*(\vec{u}) \cdot \hat{n} \, dS = \int_{V_i} \mathcal{G}(\vec{u}) dv \tag{15.4a}$$

$$\rightarrow \qquad \frac{\partial \vec{u}_i}{\partial t} + \frac{1}{V_i} \oint_{S_i} \mathcal{F}_i^*(\vec{u}) \cdot \hat{n} \, dS = \overline{\mathcal{G}}(\vec{u}_i) \tag{15.4b}$$

where $\hat{n}$ is the unit normal to the surface S_i which encloses V_i, and we have assumed the source $\mathcal{G}$ is constant over the volume V_i. The vector $\mathcal{F}_i^*$ is the *numerical flux*, which we have explicitly differentiated from the physical flux $\overline{\overline{\mathcal{F}}}$. This numerical flux is an estimate of the flux leaving or entering the volume V_i through its surface S_i; how we make the estimate of this numerical flux is what defines the particular finite volume method.

Equation (15.4) is the general finite volume algorithm; in short, it states that the time derivative of $\vec{u}$ in the volume V_i is equal to the flux out of the area S_i enclosing V_i. To write this as a time domain algorithm, we can in principle write the time derivative on the left as a centered difference, yielding a usable update equation:

$$\vec{u}_i\big|^{n+1/2} = \vec{u}_i\big|^{n-1/2} - \frac{\Delta t}{V_i} \oint_{S_i} \mathcal{F}_i^*(\vec{u})\big|^n \cdot \hat{n} \, dS + \overline{\mathcal{G}}(\vec{u}_i)\big|^n. \tag{15.5}$$

The problem that remains, of course, is the discretization of the numerical flux term over the surface S_i; the different algorithms for this discretization lead to the many different finite volume methods.

15.2.1 Maxwell's equations in conservative form

The finite volume method can be applied to Maxwell's equations if they are cast in conservative form. We will start with Maxwell's equations in vector form as given in Equations (2.1) and (2.3):

$$\frac{\partial \overline{\mathcal{B}}}{\partial t} = -\nabla \times \overline{\mathcal{E}} \tag{2.1}$$

$$\frac{\partial \overline{\mathcal{D}}}{\partial t} = \nabla \times \overline{\mathcal{H}} - \overline{\mathcal{J}}. \tag{2.3}$$

The component equations can be written in conservative form as follows [8]:

$$\frac{\partial \overline{\mathcal{Q}}}{\partial t} + \frac{\partial \overline{\mathcal{F}}_1}{\partial x} + \frac{\partial \overline{\mathcal{F}}_2}{\partial y} + \frac{\partial \overline{\mathcal{F}}_3}{\partial z} = \mathcal{G} \tag{15.6}$$

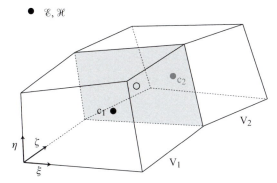

Figure 15.4 Finite volume grid cells. The co-located grid from the method of Shankar et al. [8]. All fields are co-located at the center of each grid cell (c_1, etc.). To evaluate the flux across faces adjoining cells, a spatial average is taken between c_1 and c_2 to find the fields on the adjoining face.

with the composite component vectors defined as:

$$\mathcal{Q} = \left\{ \begin{array}{c} \mathcal{B}_x \\ \mathcal{B}_y \\ \mathcal{B}_z \\ \mathcal{D}_x \\ \mathcal{D}_y \\ \mathcal{D}_z \end{array} \right\} \quad ; \quad \overline{\overline{\mathcal{F}}}_1 = \left\{ \begin{array}{c} 0 \\ -\mathcal{D}_z/\epsilon \\ \mathcal{D}_y/\epsilon \\ 0 \\ \mathcal{B}_z/\mu \\ -\mathcal{B}_y/\mu \end{array} \right\} \quad ; \quad \overline{\overline{\mathcal{F}}}_2 = \left\{ \begin{array}{c} \mathcal{D}_z/\epsilon \\ 0 \\ -\mathcal{D}_x/\epsilon \\ -\mathcal{B}_z/\mu \\ 0 \\ \mathcal{B}_x/\mu \end{array} \right\} \quad ;$$

$$\overline{\overline{\mathcal{F}}}_3 = \left\{ \begin{array}{c} -\mathcal{D}_y/\epsilon \\ \mathcal{D}_x/\epsilon \\ 0 \\ \mathcal{B}_y/\mu \\ -\mathcal{B}_x/\mu \\ 0 \end{array} \right\} \quad ; \quad \overline{\mathcal{G}} = \left\{ \begin{array}{c} 0 \\ 0 \\ 0 \\ -\mathcal{J}_x \\ -\mathcal{J}_y \\ -\mathcal{J}_z \end{array} \right\}. \tag{15.7}$$

The reader can easily verify that this set of equations is equivalent to Maxwell's equations by simply writing out the component equations. Now, in this finite volume method the fields $\overline{\mathcal{E}}$, $\overline{\mathcal{D}}$, $\overline{\mathcal{B}}$, and $\overline{\mathcal{H}}$ are all co-located at the center of each grid cell of volume V_i. In this way, the method differs considerably from the FDTD method, where fields are interleaved in space.

Now, over a regular orthogonal grid such as that used in FDTD, one could quite easily integrate these equations directly, since the field component would then be parallel to cell edges and normal to cell faces. However, the power of the FVTD method comes from its application in conformal, irregular grids.

Consider a general grid cell as shown in Figure 15.4. The geometry of this cell can be defined by a coordinate transformation from (x, y, z) to (ξ, η, ζ). Under this coordinate transformation, Equation (15.6) becomes:

$$\frac{\partial \tilde{\mathscr{D}}}{\partial t} + \frac{\partial \overline{\mathscr{G}}_1}{\partial \xi} + \frac{\partial \overline{\mathscr{G}}_2}{\partial \eta} + \frac{\partial \overline{\mathscr{G}}_3}{\partial \zeta} = \tilde{\mathscr{G}} \tag{15.8}$$

where $\tilde{\mathscr{D}} = \overline{\mathscr{D}}/J$, $\tilde{\mathscr{G}} = \overline{\mathscr{G}}/J$, and the vectors $\overline{\mathscr{G}}$ are given by:

$$\overline{\mathscr{G}}_1 = \frac{1}{J} \left(\frac{\partial \xi}{\partial x} \overline{\mathscr{F}}_1 + \frac{\partial \xi}{\partial y} \overline{\mathscr{F}}_2 + \frac{\partial \xi}{\partial z} \overline{\mathscr{F}}_3 \right) \tag{15.9a}$$

$$\overline{\mathscr{G}}_2 = \frac{1}{J} \left(\frac{\partial \eta}{\partial x} \overline{\mathscr{F}}_1 + \frac{\partial \eta}{\partial y} \overline{\mathscr{F}}_2 + \frac{\partial \eta}{\partial z} \overline{\mathscr{F}}_3 \right) \tag{15.9b}$$

$$\overline{\mathscr{G}}_3 = \frac{1}{J} \left(\frac{\partial \zeta}{\partial x} \overline{\mathscr{F}}_1 + \frac{\partial \zeta}{\partial y} \overline{\mathscr{F}}_2 + \frac{\partial \zeta}{\partial z} \overline{\mathscr{F}}_3 \right) \tag{15.9c}$$

and J is the Jacobian of the coordinate transformation:

$$J = \begin{vmatrix} \partial \xi/\partial x & \partial \xi/\partial y & \partial \xi/\partial z \\ \partial \eta/\partial x & \partial \eta/\partial y & \partial \eta/\partial z \\ \partial \zeta/\partial x & \partial \zeta/\partial y & \partial \zeta/\partial z \end{vmatrix}. \tag{15.10}$$

Now, we arrive at the finite volume algorithm by integrating Equation (15.8) over the cell volume V_i and applying the divergence theorem to the middle term, as in Equations (15.3) and (15.4):

$$\int_{V_i} \frac{\partial \tilde{\mathscr{D}}}{\partial t} dv + \int_{V_i} \left(\frac{\partial \overline{\mathscr{G}}_1}{\partial \xi} + \frac{\partial \overline{\mathscr{G}}_2}{\partial \eta} + \frac{\partial \overline{\mathscr{G}}_3}{\partial \zeta} \right) dv = \int_{V_i} \tilde{\mathscr{G}} dv \tag{15.11a}$$

$$\rightarrow \quad \frac{\partial \tilde{\mathscr{D}}_i}{\partial t} + \frac{1}{V_i} \oint_{S_i} \left(\overline{\mathscr{G}}_1 \cdot \hat{n}_\xi + \overline{\mathscr{G}}_2 \cdot \hat{n}_\eta + \overline{\mathscr{G}}_3 \cdot \hat{n}_\zeta \right) dS = \tilde{\mathscr{G}}_i \tag{15.11b}$$

where $\hat{n}_\xi$, $\hat{n}_\eta$, and $\hat{n}_\zeta$ are the unit normal vectors respectively in the ξ, η, and ζ directions, and we have again assumed $\tilde{\mathscr{D}}$ and $\tilde{\mathscr{G}}$ do not vary over the volume V_i.

Equation (15.11b) constitutes the general finite volume method for electromagnetics; the evaluation of the flux term is what delineates the myriad FVTD methods. The problem lies in the fact that the fields are all defined at the center of each grid cell volume, as shown in Figure 15.4, but the evaluation of the flux terms requires the fields at the cell *faces*. Thanks to the complicated geometries involved, the interpolation of fields at these cell faces is far from trivial, and the specific method used can have significant effects on accuracy and stability. The reader is referred to work by Shankar [8], Remaki [9], and Harmon [10] for details.

15.2.2   Interleaved finite volume method

An alternative form of the finite volume method for electromagnetics was proposed simultaneously with the Shankar method described above. This method, derived by Madsen and Ziolkowski [11], relies on a form of "interleaving" of the fields as shown in Figure 15.5. Because of this interleaving, this method is possibly more intuitive when it follows a thorough background of the FDTD method. In fact, [6] refer to this method

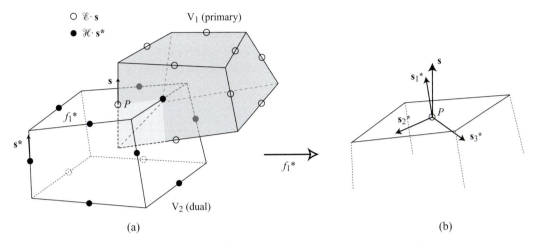

Figure 15.5 Finite volume grid cells, interleaved method. Adapted from [11]; dual and primary grid cells overlap and define the locations of $\overline{\mathscr{E}}$ and $\overline{\mathscr{H}}$ fields respectively. Evaluation of Maxwell's equations relies on computation of fluxes across cell faces, at the centers of which the fields are defined.

as an FDTD method in a locally curvilinear coordinate system, while Yee and Chen [12, 13] refer to it as a hybrid FDTD-FVTD method. For our brief description here, we will follow the notation of Madsen and Ziolkowski [11], and assume an isotropic medium, so that locally $\overline{\mathscr{D}} = \epsilon\overline{\mathscr{E}}$ and $\overline{\mathscr{B}} = \mu\overline{\mathscr{H}}$.

Consider the cells in Figure 15.5a. Cell V_1 is the "primary" cell, and the electric field is defined on the edges of that cell. Cell V_2 is the "dual" cell, on whose edges the $\overline{\mathscr{H}}$ field is defined. At each location, the full 3D vector fields are defined, rather than particular components.

Now, referring to the electric field component at P, we define unit vectors $\mathbf{s}_1^*$, $\mathbf{s}_2^*$, and $\mathbf{s}_3^*$ which form a local orthogonal set defined by the dual cell V_2; $\mathbf{s}_1^*$ is normal to the face f_1^*, and the other two are arbitrary, mutually orthogonal unit vectors. Now, we can decompose $\overline{\mathscr{E}}$ in the direction of these unit vectors:

$$\overline{\mathscr{E}} = (\overline{\mathscr{E}} \cdot \mathbf{s}_1^*)\,\mathbf{s}_1^* + (\overline{\mathscr{E}} \cdot \mathbf{s}_2^*)\,\mathbf{s}_2^* + (\overline{\mathscr{E}} \cdot \mathbf{s}_3^*)\,\mathbf{s}_3^*. \qquad (15.12)$$

On the cell edge defined by $\mathbf{s}$, we are particularly interested in the time derivative of $\overline{\mathscr{E}}$ along $\mathbf{s}$, given by:

$$\frac{\partial(\overline{\mathscr{E}} \cdot \mathbf{s})}{\partial t} = \left(\frac{\partial\overline{\mathscr{E}}}{\partial t} \cdot \mathbf{s}_1^*\right)(\mathbf{s}_1^* \cdot \mathbf{s}) + \left(\frac{\partial\overline{\mathscr{E}}}{\partial t} \cdot \mathbf{s}_2^*\right)(\mathbf{s}_2^* \cdot \mathbf{s}) + \left(\frac{\partial\overline{\mathscr{E}}}{\partial t} \cdot \mathbf{s}_3^*\right)(\mathbf{s}_3^* \cdot \mathbf{s}). \qquad (15.13)$$

The derivation of this finite volume method relies on accurate evaluation of each of the terms on the right in Equation (15.13). The first term is quite simple, since $\mathbf{s}_1^*$ is normal to the face f_1^*. Applying Ampère's law to the face f_1^* and assuming the field $\overline{\mathscr{E}}$ is constant over its area A^*, we have

$$\epsilon A^* \frac{\partial\overline{\mathscr{E}}}{\partial t} \cdot \mathbf{s}_1^* = \int (\nabla \times \overline{\mathscr{H}}) \cdot \mathbf{s}_1^* \, dA^* = \oint \overline{\mathscr{H}} \cdot d\mathbf{l}^* \qquad (15.14)$$

where the line integral is taken over the contour surrounding A^*, which passes through four $\overline{\mathcal{H}}$ fields; the fields are thus assumed to be constant over each edge. The other two terms in Equation (15.13) are evaluated using Ampère's law, integrated over a volume V^* encompassing *two* dual cells: in particular, the two cells that share the face f_1^*. These two terms become:

$$\epsilon V^* \frac{\partial \overline{\mathcal{E}}}{\partial t} \cdot \mathbf{s}_2^* = \int (\nabla \times \overline{\mathcal{H}}) \cdot \mathbf{s}_2^* \, dV^* = \int (\mathbf{n}^* \times \overline{\mathcal{H}}) \cdot \mathbf{s}_2^* \, dA^* \qquad (15.15a)$$

$$\epsilon V^* \frac{\partial \overline{\mathcal{E}}}{\partial t} \cdot \mathbf{s}_3^* = \int (\nabla \times \overline{\mathcal{H}}) \cdot \mathbf{s}_3^* \, dV^* = \int (\mathbf{n}^* \times \overline{\mathcal{H}}) \cdot \mathbf{s}_3^* \, dA^* \qquad (15.15b)$$

where V^* is the volume of the two adjoining cells, and $\mathbf{n}^*$ is a unit normal to the dual face on the surface of V^*. At this point, the complexity of the finite volume method lies in defining the unit normal vectors at each intersection between the primary and dual grids. We refer the reader to Madsen and Ziolkowski [11] and the book by Rao [14] for the details of the evaluation of these equations.

Equations (15.13)–(15.15) make up the finite volume algorithm for the $\overline{\mathcal{E}}$ update. A similar set of equations is necessary for the $\overline{\mathcal{H}}$ field update, which is derived from Faraday's law, and is given by:

$$\frac{\partial (\overline{\mathcal{H}} \cdot \mathbf{s}^*)}{\partial t} = \left(\frac{\partial \overline{\mathcal{H}}}{\partial t} \cdot \mathbf{s}_1 \right) (\mathbf{s}_1 \cdot \mathbf{s}^*) + \left(\frac{\partial \overline{\mathcal{H}}}{\partial t} \cdot \mathbf{s}_2 \right) (\mathbf{s}_2 \cdot \mathbf{s}^*) + \left(\frac{\partial \overline{\mathcal{H}}}{\partial t} \cdot \mathbf{s}_3 \right) (\mathbf{s}_3 \cdot \mathbf{s}^*)$$

$$(15.16a)$$

$$\mu A \frac{\partial \overline{\mathcal{H}}}{\partial t} \cdot \mathbf{s}_1 = - \int (\nabla \times \overline{\mathcal{E}}) \cdot \mathbf{s}_1 \, dA = - \oint \overline{\mathcal{E}} \cdot d\mathbf{l} \qquad (15.16b)$$

$$\mu V \frac{\partial \overline{\mathcal{H}}}{\partial t} \cdot \mathbf{s}_2 = - \int (\nabla \times \overline{\mathcal{E}}) \cdot \mathbf{s}_2 \, dV = - \int (\mathbf{n} \times \overline{\mathcal{E}}) \cdot \mathbf{s}_2 \, dA \qquad (15.16c)$$

$$\mu V \frac{\partial \overline{\mathcal{H}}}{\partial t} \cdot \mathbf{s}_3 = - \int (\nabla \times \overline{\mathcal{E}}) \cdot \mathbf{s}_3 \, dV = - \int (\mathbf{n} \times \overline{\mathcal{E}}) \cdot \mathbf{s}_3 \, dA \qquad (15.16d)$$

where V, A, $\mathbf{s}^*$, and $\mathbf{s}_1$ etc. can be described for the primary face as their counterparts were for the dual face. Finally, note that as usual the time derivative on the left-hand side of Equation (15.13) can be (and is typically) evaluated using the standard leapfrog centered difference in time.

15.2.3   The Yee finite volume method

For the sake of completeness, we briefly introduce here an alternative finite volume method developed by Yee and Chen [12, 13]. The Yee method, as we will refer to it here, is interleaved and uses primary and dual cells as in the Madsen method presented above. However, in the Yee method the fields are located at the *vertices* of the grid cells, as shown in Figure 15.6; this results in a much simpler derivation of update equations, as we will show next.

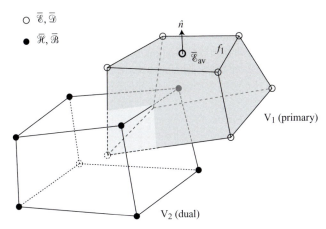

Figure 15.6 Finite volume grid cells, Yee method. Adapted from [12]; dual and primary grid cells overlap and define the locations of $\overline{\mathcal{E}}$ and $\overline{\mathcal{H}}$ fields respectively, but in this case the fields are located at the cell vertices.

The Yee method makes use of the same integral forms of Maxwell's equations:

$$-\int_V \frac{\partial \overline{\mathcal{B}}}{\partial t} dv = \int_A \hat{n} \times \overline{\mathcal{E}} \, dS \tag{15.17a}$$

$$\int_{V*} \frac{\partial \overline{\mathcal{D}}}{\partial t} dv = \int_{A*} \hat{n} \times \overline{\mathcal{H}} \, dS \tag{15.17b}$$

where, again, V and V^* are the primary and dual cells, and A and A^* are the surface areas of those cells. Yee and Chen [13] presents the time-discretized algorithm using the leapfrog algorithm in time, similar to the FDTD algorithm:

$$-\frac{1}{\Delta t} \int_V \left(\overline{\mathcal{B}}\big|^{n+1/2} - \overline{\mathcal{B}}\big|^{n-1/2} \right) dv = \int_A \hat{n} \times \overline{\mathcal{E}}\big|^n \, dS \tag{15.18a}$$

$$\frac{1}{\Delta t} \int_{V*} \left(\overline{\mathcal{D}}\big|^{n+1} - \overline{\mathcal{D}}\big|^n \right) dv = \int_{A*} \hat{n} \times \overline{\mathcal{H}}\big|^{n+1/2} \, dS. \tag{15.18b}$$

Consider the evaluation of Equation (15.18a) on the primary cell in Figure 15.6. The field $\overline{\mathcal{B}}$ is located at the center of the primary cell V_1; we assume the field is constant over this volume, and the volume integral simplifies down to a multiplication by the volume V_1. For the fluxes on the right-hand side of Equation (15.18a), on each face of V_1 we use the average of the $\overline{\mathcal{E}}$ fields at each of four vertices to approximate the $\overline{\mathcal{E}}$ field on the face, as shown for the face f_1. Of course, the vertices and faces need to be well described in the grid formulation, so that the dot product $\overline{\mathcal{E}} \cdot \hat{n}$ can be evaluated on each face.

As a final note on the finite volume methods, it should be recognized that each of the complexities that we have introduced in this book for the FDTD method can also be introduced into the FVTD methods described here. This includes incorporating lossy, dispersive, or anisotropic media; incorporating the total-field / scattered-field method; truncating the simulation space with ABCs or PMLs; and, furthermore, the FVTD methods can be derived as *implicit* schemes. Otherwise, explicit finite volume methods are constrained by stability limitations, which are similar to the CFL stability criterion in FDTD; Madsen and Ziolkowski [11] claim that, while the analytical stability criterion is not known, stability is ensured if the time step Δt times v_p is kept less than half the smallest distance between two grid vertices, which is an intuitively satisfying result.

15.3   The finite element method

In this section we introduce another method that is used frequently for solving electrostatic and electromagnetic problems, namely the finite element method (FEM). This method has been around for many decades, and has long been the method of choice for statics and dynamics problems in mechanical and civil engineering (where it is often referred to as finite element analysis or FEA) since the 1940s. Its use in electromagnetics has a long history as well, beginning in the late 1960s. The dawn of high-speed computing and the development of boundary conditions and near-to-far field transformations pushed the finite difference methods to the forefront of electromagnetic simulations in the 1990s; however, finite element methods are making a comeback in recent years, thanks to novel techniques that enable higher accuracy and efficiency in computations. Furthermore, recent work has developed the Finite Element Method in Time Domain (FE-TD), enabling the simulation of transient behaviors that are so well suited to FDTD.

The two primary reasons for using FEM rather than FDTD for electromagnetic problems are its geometric flexibility and the ability to work in higher orders of accuracy. Geometric flexibility arises because the grid in FEM can use arbitrary polygons or polyhedra (in 2D or 3D, respectively), and these can be designed to match the shapes of objects in the simulation space or the expected field variations. Then, the fields are assumed to take a particular functional form (usually polynomial) over this element; accuracy can be improved by assuming higher-order polynomial basis functions on each element.

Scores of textbooks and papers have been written on the finite element method, and even a full chapter could not cover the method adequately. Hence, here we present a brief introduction to the method and refer students to other resources for further reading.

The primary difference between the finite difference methods discussed so far and the finite element methods is that finite difference methods find an approximation to the *governing differential equations*, and then use these difference equations to solve for the fields at grid locations or within cells. In finite element, on the other hand, one makes an approximation to the *solution* of the differential equation over the domain of the problem, and then tailors that approximation to minimize the difference between it and the ideal, exact solution. This is illustrated in Figure 15.7. On the right, an example 2D finite element mesh is shown, made up of irregular triangles. The mesh can use any type of polygon, but triangles are typical for their simplicity and ease of fitting to structures in the computational space. On the right, example triangular cells are shown with linear (first-order polynomial) and parabolic (second-order) solutions. These solutions are then matched to the adjacent cells to ensure continuity across the cell boundaries.

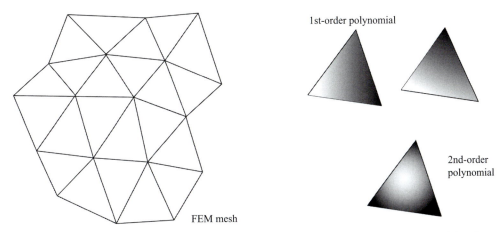

1st-order polynomial

2nd-order polynomial

FEM mesh

Figure 15.7 Finite element mesh in 2D. On the right, individual cells are shown with polynomial solutions assumed across their domains.

The minimization problem can be constructed in a variety of ways; here we present two such methods, corresponding to classical methods for boundary value problems.[1] The first, known as the *Ritz method*, is to minimize some *functional* over the domain whose minimum corresponds to the governing differential equations; for example, in many methods this functional corresponds to the energy or power in the domain, and the physical understanding is that the exact solution will correspond to the field distribution with the minimum loss of power.

The second minimization method, known as *Galerkin's method*, aims to minimize the *residual* between the exact and approximate solutions; for instance, consider the differential equation $\mathscr{L}\phi = f$, where $\mathscr{L}$ is a differential operator, ϕ is the unknown field distribution of interest, and f is some forcing function (i.e., a source). In electromagnetics this equation could correspond to the Poisson's equation (in electrostatics) or the wave equation (in electrodynamics). In Galerkin's method, the approximate solution $\tilde{\phi}$ would have some nonzero residual r in the differential equation:

$$r = \mathscr{L}\tilde{\phi} - f \neq 0.$$

Both the Ritz and Galerkin methods are classical methods that have been used long before the advent of computers, but used to solve problems over the entire domain. The finite element approach to these problems has been to solve boundary value problems using Ritz or Galerkin's methods *over small subdomains*. The complete steps in the methods are as follows:

---

[1] Boundary value problems are those in which a governing differential equation must be solved with constraints on the boundaries of the domain, which could be sources, scatterers, or material structures. The term is used very generally and could even include high-order boundaries with their own governing differential equations. Most, if not all, problems of interest in computational physics fall under the wide umbrella of boundary value problems.

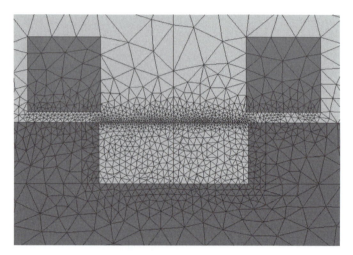

Figure 15.8 Finite element mesh of an integrated circuit transistor. Notice that the mesh is more dense in the channel, where the fields are expected to be complex.

1. *Grid the space.* The domain over which the solution is desired is broken up into small subdomains, known as "elements."
2. *Select basis functions.* The solution is assumed to take a specific functional form over each small subdomain or element.
3. *Formulate a system of equations.* Here is where one chooses either the Ritz, Galerkin, or other methods to set up the system of equations. This system will include information about boundaries and sources, and will have the added constraint of continuity across element boundaries.
4. *Solve the system of equations.* Similar to FDFD, the solution of these systems typically becomes purely a linear algebra problem.

Figure 15.8 shows an example of a 2D simulation using the finite element method, where the simulation space of interest includes some structures resembling an integrated circuit transistor. Notice how the mesh is designed with two purposes in mind: first, it matches the edges of the structures perfectly; if some of these structures were curved, the matching would be much better than our typical FDTD setup. Second, the mesh is more refined in areas where the fields are expected to be more complex, such as along the channel of this transistor. This flexibility allows simulations to be designed efficiently, with greater mesh density only where needed.

15.3.1 Example using Galerkin's method

To illustrate the method, we provide the example of a 1D static field problem in its general form:

$$-\frac{d}{dx}\left(\alpha\frac{d\phi}{dx}\right) + \beta\phi = f. \tag{15.19}$$

Figure 15.9 Nodes and elements in the 1D finite element method.

Here, ϕ is the unknown field quantity, α and β are known parameters or functions which define the physical domain, and f is a source function. This equation, with the correct choices of α, β, and f, can be the Laplace, Poisson, or Helmholtz equations in 1D.

For generality we will assume mixed boundary conditions,

$$\phi\big|_{x=0} = p \tag{15.20a}$$

$$\left[\alpha\frac{d\phi}{dx} + \gamma\phi\right]_{x=L} = q \tag{15.20b}$$

where p, q and γ are known parameters or functions. The first equation above (15.20a) is a boundary condition of the first kind (a constant), known as a *Dirichlet* boundary condition; and the second equation (15.20b) is a boundary condition of the third kind, or *Robin* boundary condition. A boundary condition of the second kind (a *Neumann* boundary condition) results if $\gamma = 0$. Note that the Robin boundary condition is a combination of Dirichlet and Neumann.

Discretization of the space and basis functions

As mentioned above, the first step is to break up the space from $x = 0$ to $x = L$ into smaller elements. Here we must differentiate between *nodes* and *elements*; we solve for the fields on each element, and these elements will each have two or more nodes. In the simplest setup, a 1D problem with linear field variation within each element, each element will have two nodes (one at each end), but adjoining elements will have a common node. In 2D where the elements are set up as triangles, each element has three nodes, and each node is shared with at least two other elements.

In our 1D setup, each of N nodes is located at x_i in space, with $x_1 = 0$ (the left end of the space) and $x_N = L$ (the right end). Each of the 1D elements will have length l^e, where for instance $l^1 = x_2 - x_1$.

For generality with 2D and 3D setups, we must use a *local indexing*, so that each element has nodes x_1^e and x_2^e, where the superscript e refers to the e-th element, up to M elements. In comparison with the i indexing above, this means $x_1^e = x_e$ and $x_2^e = x_{e+1}$. This is described by Figure 15.9.

Next, we must choose basis functions within each element, or in this case, between each pair of nodes. The simplest choice that allows continuity at each node is a linear interpolation, where within each element e the approximate field ϕ takes the form

$$\phi^e(x) = a^e + b^e x.$$

Using this equation at each node of the element x_1^e and x_2^e and solving for a^e and b^e, we find

$$\phi^e(x) = \frac{x_2^e - x}{l^e}\phi_1^e + \frac{x - x_1^e}{l^e}\phi_2^e = N_1^e\phi_1^e + N_2^e\phi_2^e \qquad (15.21)$$

where again $l^e = x_2^e - x_1^e$, and ϕ_1^e and ϕ_2^e are the values of the function at the node points. The notation N_i^e for the linear functions of x will be useful next for keeping our equations manageable.

At this point we have the choice of following the Ritz formulation of minimizing some functional, or using Galerkin's method to minimize the residual; either method is equally valid, but here we will use Galerkin's. Given our approximate solution ϕ, we wish to minimize the residual

$$r = -\frac{d}{dx}\left(\alpha\frac{d\phi}{dx}\right) + \beta\phi - f. \qquad (15.22)$$

The minimization procedure is as follows: at each node $i = 1, 2$, we multiply the residual r above by N_i^e, and integrate over the element from x_1^e to x_2^e to find an integrated residual R_i^e for each node of each element:

$$R_i^e = \int_{x_1^e}^{x_2^e} N_i^e\, r\, dx \qquad ; \qquad i = 1, 2. \qquad (15.23)$$

Plugging in the equation for r in Equation (15.22), with the approximate ϕ given by Equation (15.21), and simplifying,[2] we find

$$R_i^e = \sum_{j=1}^{2}\phi_j^e\underbrace{\int_{x_1^e}^{x_2^e}\left(\alpha\frac{dN_i^e}{dx}\frac{dN_j^e}{dx} + \beta N_i^e N_j^e\right)dx}_{K_{ij}^e} - \underbrace{\int_{x_1^e}^{x_2^e} N_i^e f\,dx}_{b_i^e} - \underbrace{\alpha N_i^e\frac{d\phi}{dx}\Big|_{x_1^e}^{x_2^e}}_{g_i^e}. \qquad (15.24)$$

Following the same procedure for both nodes ($i = 1, 2$), the two equations can be written together in matrix form as

$$\mathbf{R}^e = [\mathbf{K}^e]\bar{\phi}^e - \mathbf{b}^e - \mathbf{g}^e \qquad (15.25)$$

where $\bar{\phi}^e$ denotes that it is a column vector with two entries. $\mathbf{R}^e$, $\mathbf{g}^e$, and $\mathbf{b}^e$ are also two-entry column vectors and $[\mathbf{K}^e]$ is a 2×2 matrix with elements given by

$$K_{i,j}^e = \int_{x_1^e}^{x_2^e}\left(\alpha\frac{dN_i^e}{dx}\frac{dN_j^e}{dx} + \beta N_i^e N_j^e\right)dx.$$

---

[2] Details can be found in Chapter 3 of Jin [15].

Note that if α, β, and f can be approximated as constant over any one element, then $K_{i,j}^e$ and b_i^e can be directly integrated to find

$$K_{1,1}^e = K_{2,2}^e = \frac{\alpha^e}{l^e} + \beta^e \frac{l^e}{3}$$

$$K_{1,2}^e = K_{2,1}^e = -\frac{\alpha^e}{l^e} + \beta^e \frac{l^e}{6}$$

$$b_1^e = b_2^e = f^e \frac{l^e}{2}.$$

Notice also that g_i^e is simply going to be the negative of itself at x_1^e and x_2^e,

$$g_i^e = \alpha N_i^e \frac{d\phi}{dx}\bigg|_{x=x_2^e} - \alpha N_i^e \frac{d\phi}{dx}\bigg|_{x=x_1^e} = \mp\alpha N_i^e \frac{d\phi}{dx}\bigg|_{x=x_i^e} \quad ; \quad i = 1, 2.$$

Now that we have determined the form of the matrix equation and the factors $K_{i,j}^e$, b_i^e, and g_i^e in each element, we must put the elements together. Simply summing the residuals R_i^e for each element, we find:

$$\mathbf{R} = \sum_{e=1}^{M} \mathbf{R}^e = \sum_{e=1}^{M} \left([\mathbf{K}^e]\,\bar{\phi}^e - \mathbf{b}^e - \mathbf{g}^e\right) = 0 \tag{15.26}$$

where it is explicitly noted that we want to set this total residual to zero. In full matrix form,

$$[\mathbf{K}]\,\bar{\phi} = \mathbf{b} + \mathbf{g}. \tag{15.27}$$

Setting up the system of equations and boundary conditions

The next step, which must be done with great care, is to assemble the three elemental vectors $[\mathbf{K}^e]\,\bar{\phi}^e$, $\mathbf{b}^e$, and $\mathbf{g}^e$ into the full vectors in Equation (15.26) that encompass all of the elements. As a simple example, we will show the method for a simple three-element problem (four nodes). We then assemble $\bar{\phi}$ into a four-element vector and $[\mathbf{K}]$ into a 4×4 matrix. The multiplication $[\mathbf{K}^e]\,\bar{\phi}^e$ for the first element $e = 1$ is then

$$[\mathbf{K}^1]\,\bar{\phi}^1 = \begin{bmatrix} K_{1,1}^1 & K_{1,2}^1 & 0 & 0 \\ K_{2,1}^1 & K_{2,2}^1 & 0 & 0 \\ 0 & 0 & 0 & 0 \\ 0 & 0 & 0 & 0 \end{bmatrix} \begin{bmatrix} \phi_1^1 \\ \phi_2^1 \\ 0 \\ 0 \end{bmatrix} = \begin{bmatrix} K_{1,1}^1\phi_1^1 + K_{1,2}^1\phi_2^1 \\ K_{2,1}^1\phi_1^1 + K_{2,2}^1\phi_2^1 \\ 0 \\ 0 \end{bmatrix} \tag{15.28}$$

where the subscripts refer to the node numbers, and the superscripts to the element number. Repeating this process across the other two elements, we find:

$$[\mathbf{K}^2]\,\bar{\phi}^2 = \begin{bmatrix} 0 \\ K_{1,1}^2\phi_1^2 + K_{1,2}^2\phi_2^2 \\ K_{2,1}^2\phi_1^2 + K_{2,2}^2\phi_2^2 \\ 0 \end{bmatrix} \quad ; \quad [\mathbf{K}^3]\,\bar{\phi}^3 = \begin{bmatrix} 0 \\ 0 \\ K_{1,1}^3\phi_1^3 + K_{1,2}^3\phi_2^3 \\ K_{2,1}^3\phi_1^3 + K_{2,2}^3\phi_2^3 \end{bmatrix}. \tag{15.29}$$

Next, we can take the sum of these three vectors, and use the required continuity at the nodes to simplify; i.e., we can write $\phi_1^1 = \phi_1$, $\phi_2^1 = \phi_1^2 = \phi_2$, $\phi_2^2 = \phi_1^3 = \phi_3$, and

$\phi_2^3 = \phi_4$, where now the ϕ's are uniquely stored at each node. Then the sum of the three vectors above becomes

$$\sum_{e=1}^{3} [K^e] \bar{\phi}^e = \begin{bmatrix} K_{1,1}^1 \phi_1 + K_{1,2}^1 \phi_2 \\ K_{2,1}^1 \phi_1 + K_{2,2}^1 \phi_2 + K_{1,1}^2 \phi_2 + K_{1,2}^2 \phi_3 \\ K_{2,1}^2 \phi_2 + K_{2,2}^2 \phi_3 + K_{1,1}^3 \phi_3 + K_{1,2}^3 \phi_4 \\ K_{2,1}^3 \phi_3 + K_{2,2}^3 \phi_4 \end{bmatrix}$$

$$= \begin{bmatrix} K_{1,1}^1 & K_{1,2}^1 & 0 & 0 \\ K_{2,1}^1 & K_{2,2}^1 + K_{1,1}^2 & K_{1,2}^2 & 0 \\ 0 & K_{2,1}^2 & K_{2,2}^2 + K_{1,1}^3 & K_{1,2}^3 \\ 0 & 0 & K_{2,1}^3 & K_{2,2}^3 \end{bmatrix} \begin{bmatrix} \phi_1 \\ \phi_2 \\ \phi_3 \\ \phi_4 \end{bmatrix}. \qquad (15.30)$$

Performing a similar expansion on the $\mathbf{b}^e$ and $\mathbf{g}^e$ terms, we find the vectors

$$\sum_{e=1}^{3} \mathbf{b}^e = \begin{bmatrix} b_1^1 \\ b_2^1 + b_1^2 \\ b_2^2 + b_1^3 \\ b_2^3 \end{bmatrix} \quad ; \quad \sum_{e=1}^{3} \mathbf{g}^e = \begin{bmatrix} g_1^1 \\ g_2^1 + g_1^2 \\ g_2^2 + g_1^3 \\ g_2^3 \end{bmatrix}.$$

At this point, the astute reader will observe that $[K]$ is a tridiagonal matrix, and will remain so for an arbitrary number of elements; similarly, $\mathbf{b}$ has a simple, repeatable form. The last vector $\mathbf{g}$ simplifies even more; note that the middle elements of $\mathbf{g}$ take the form $g_i = g_2^{i-1} - g_1^i$, but comparison with the form above for g_i^e shows that

$$g_i = \alpha \frac{d\phi}{dx}\Big|_{x=x_2^{i-1}} - \alpha \frac{d\phi}{dx}\Big|_{x=x_1^i}$$

$$= \begin{cases} -\alpha \frac{d\phi}{dx}\Big|_{x=0} & i = 1 \\ \alpha \frac{d\phi}{dx}\Big|_{x=L} & i = N \\ 0 & i \neq 1, N. \end{cases}$$

So $\mathbf{g}$ has only two nonzero elements, g_i and g_N. Now we have each of the vectors and matrices for the matrix Equation (15.30) that we wish to solve; all that is left is to impose the boundary conditions.

The simple boundary condition of the first kind (the Dirichlet boundary condition), $\phi(x = 0) = p$, is relatively easy to implement. It requires zeroing out the first row of $[K]$ above, and modification of $\mathbf{b}$ and $\mathbf{g}$, so that

$$\begin{bmatrix} 1 & 0 & 0 & 0 \\ K_{21} & K_{22} & K_{23} & K_{24} \\ K_{31} & K_{32} & K_{33} & K_{34} \\ K_{41} & K_{42} & K_{34} & K_{44} \end{bmatrix} \begin{bmatrix} \phi_1 \\ \phi_2 \\ \phi_3 \\ \phi_4 \end{bmatrix} = \begin{bmatrix} p \\ b_2 \\ b_3 \\ b_4 \end{bmatrix} + \begin{bmatrix} 0 \\ 0 \\ 0 \\ g_4 \end{bmatrix} \qquad (15.31)$$

with $g_4 = g_N$ given above. Here we have changed the entries of the $[K]$ matrix and $\mathbf{b}$ vector to allow for general entries. Notice how evaluating the first row multiplication of this set of equations leaves us with the boundary condition $\phi_1 = p$. However, we have lost the symmetry of the matrix $[K]$, which is problematic for inversion when it becomes large. To get around this, we can move the terms which depend on the known ϕ_1 to the right-hand side to modify **b**; for instance, the second equation has a term $K_{21}\phi_1 = K_{21}p$, which can be moved to the right-hand side; this leaves us with

$$
\begin{bmatrix}
1 & 0 & 0 & 0 \\
0 & K_{22} & K_{23} & K_{24} \\
0 & K_{32} & K_{33} & K_{34} \\
0 & K_{42} & K_{43} & K_{44}
\end{bmatrix}
\begin{bmatrix}
\phi_1 \\
\phi_2 \\
\phi_3 \\
\phi_4
\end{bmatrix}
=
\begin{bmatrix}
p \\
b_2 - K_{21}p \\
b_3 - K_{31}p \\
b_4 - K_{41}p
\end{bmatrix}
+
\begin{bmatrix}
0 \\
0 \\
0 \\
g_4
\end{bmatrix}
\tag{15.32}
$$

or, ignoring the known equation for ϕ_1,

$$
\begin{bmatrix}
K_{22} & K_{23} & K_{24} \\
K_{32} & K_{33} & K_{34} \\
K_{42} & K_{43} & K_{44}
\end{bmatrix}
\begin{bmatrix}
\phi_2 \\
\phi_3 \\
\phi_4
\end{bmatrix}
=
\begin{bmatrix}
b_2 - K_{21}p \\
b_3 - K_{31}p \\
b_4 - K_{41}p
\end{bmatrix}
+
\begin{bmatrix}
0 \\
0 \\
g_4
\end{bmatrix}.
\tag{15.33}
$$

Now the matrix $[\mathbf{K}]$ is symmetric and tridiagonal again, simplifying the inversion process that is still to come. Finally, we must include the third-order Robin boundary condition at $x = L$ given in Equation (15.20b). This is seen to be very simple since:

$$
g_N = \alpha d\phi/dx|_{x=L} = q - \gamma\phi_N.
$$

These can be added into the set of Equations (15.33), to find the final set:

$$
\begin{bmatrix}
K_{22} & K_{23} & K_{24} \\
K_{32} & K_{33} & K_{34} \\
K_{42} & K_{43} & K_{44} + \gamma
\end{bmatrix}
\begin{bmatrix}
\phi_2 \\
\phi_3 \\
\phi_4
\end{bmatrix}
=
\begin{bmatrix}
b_2 - K_{21}p \\
b_3 - K_{31}p \\
b_4 - K_{41}p + q
\end{bmatrix}.
\tag{15.34}
$$

Note how a reversal of the boundary conditions (first-order at $x = L$ and third-order at $x = 0$) would have simply reversed the process from Equations (15.31) to (15.34), and resulted in a matrix equation eliminating ϕ_N rather than ϕ_1.

Solving the system of equations

The final step in the FEM process is to solve the matrix system of Equations (15.34). This particular problem is trivial, since we have only three elements and a 1D problem. The solution is simply found by matrix inversion,

$$
\bar{\phi} = [\mathbf{K}]^{-1}\mathbf{b}'
$$

where $\mathbf{b}'$ is used to differentiate the vector on the right-hand side of Equation (15.34) from the original **b** vector earlier.

15.3.2 TE wave incident on a dielectric boundary

In order to show how this applies to electromagnetic problems, we consider an example involving a TE[3] polarized wave incident on a dielectric slab at an arbitrary angle θ, as shown in Figure 15.10. The slab is placed at $x = L$ and the field is E_z. Behind the dielectric slab, at $x = 0$, is a perfect conductor plate. We wish to measure the reflection

---

[3] TE in the physical sense, with E_z perpendicular to the plane of incidence.

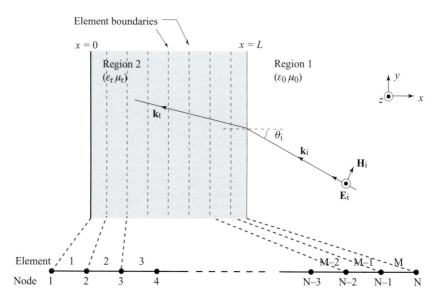

Figure 15.10 Finite element method for a simple electromagnetic problem. A TE wave incident on a dielectric slab, with PEC backing. The FEM method is used to calculate the reflection coefficient.

coefficient as a function of angle θ. Note that this is not a time domain computation; for now, we will work in the frequency domain.

The governing differential equation of interest in the slab is

$$\frac{\partial}{\partial x}\left(\frac{1}{\mu_r}\frac{\partial E_z}{\partial x}\right) + k_0^2\left(\epsilon_r - \frac{1}{\mu_r}\sin^2\theta\right)E_z = 0, \tag{15.35}$$

which has the same form as Equation (15.19) with:

$$\alpha = \frac{1}{\mu_r} \quad ; \quad \beta = -k_0^2\left(\epsilon_r - \frac{1}{\mu_r}\sin^2\theta\right) \quad ; \quad f = 0.$$

It can be shown that the boundary conditions for this problem are just as we had before, a Dirichlet condition on one side and a Robin condition on the other side:

$$E_z\big|_{x=0} = 0 \tag{15.36a}$$

$$\left[\frac{1}{\mu_r}\frac{\partial E_z}{\partial x} + jk_0\cos\theta E_z(x)\right]_{x=L} = 2jk_0\cos\theta E_0 e^{jk_0 L\cos\theta}. \tag{15.36b}$$

The astute reader should see Equation (15.36a) as the first-order boundary condition, and Equation (15.36b) as the third-order boundary condition, exactly as before, having the form:

$$\left[\alpha\frac{d\phi}{dx} + \gamma\phi\right]_{x=L} = q$$

with the following substitutions:

$$\phi = E_z \qquad \alpha = \frac{1}{\mu_r}$$

$$\gamma = jk_0 \cos\theta \qquad q = 2jk_0 \cos\theta \, E_0 e^{jk_0 L \cos\theta}.$$

This third-order boundary condition is equivalent to the electromagnetic boundary conditions in Chapter 2 at the free space–dielectric interface; the first-order boundary condition simply states that the tangential electric field is zero at the PEC wall.

The set of equations to solve can now be transcribed directly from Equation (15.34) with the appropriate substitutions made. This method can be easily applied to a dielectric slab with complex permittivity and permeability, and in addition, these parameters can be made to vary throughout the slab.

Time integration

A note is required about time versus spatial discretization. In the FDTD method, we approximate both the temporal and spatial derivatives with centered differences, as derived in Chapter 3. However, the time and spatial components can easily be integrated separately, using different methods. The finite element method is a method for solving the spatial components of the problem. In electrostatic problems, this is all that is necessary; but in electromagnetic problems, FEM will only solve the spatial derivatives, and a separate method is required for the time integration.

This time integration can be done via explicit or implicit methods, such as those described in Chapter 13. Note, however, that implicit time integration will require matrix solutions for both the spatial components and temporal components.

15.4 Discontinuous Galerkin method

While both finite element and finite volume methods have seen wide use in numerical electromagnetics, both methods suffer from a number of drawbacks in practical use. The finite element method suffers from the *nonlocality* of the scheme. Even if the basis functions are chosen to be local, enforcing continuity of the function or its derivatives at the boundary of each element means that a large (albeit sparse) matrix must be solved in order to solve the system of equations. For 2D or 3D problems, the overhead involved just in solving the system of equations can dominate the overall computational cost.

The finite volume method easily leads to strictly local, explicit schemes for hyperbolic problems and is thus a natural choice for the solution of Maxwell's equations. However, the unique structure of Maxwell's equations, in particular the odd-even coupling of the magnetic and electric fields, means that specialized techniques, such as the dual grids discussed in the previous section, must be used to attain reasonable accuracy. The most severe limitation of the finite volume technique, however, is that achieving high-order accuracy destroys the locality of the scheme, requiring progressively larger and larger stencils for higher order accuracy.

In this section we will briefly discuss a family of numerical methods called discontinuous Galerkin finite element methods (DG-FEM). The discontinuous Galerkin technique combines important ideas from both the finite element method and the finite volume method, producing a scheme that is strictly local, explicit in time, and which can achieve a high order of accuracy.

We begin by considering a time-dependent 1D partial differential equation in flux conservation form:

$$\frac{\partial \phi}{\partial t} + \frac{\partial f(x, \phi)}{\partial x} - s = 0 \qquad (15.37)$$

where ϕ is the unknown quantity of interest, $f(x, \phi)$ is a flux function, and s is some source term. We begin as in the finite element method by forming an approximation to $\phi(x)$ over an element e. Using the same notation as in the previous section:

$$\phi^e(x) = \frac{x_2^e - x}{l^e} \phi_1^e + \frac{x - x_1^e}{l^e} \phi_2^e = N_1^e(x)\phi_1^e + N_2^e(x)\phi_2^e. \qquad (15.38)$$

We will approximate our solution $\phi(x)$ as a weighted sum of functions $N_j^e(x)$. Note that while there are only two terms in this simple example, the method can be easily generalized to use any space of basis functions. We then approximate all the terms in our PDE using the same basis, leading to the following approximations for ϕ, f, and s in the element e:

$$\phi^e(x) \simeq \sum_j \phi_j^e N_j^e(x) \qquad (15.39a)$$

$$f^e(x) \simeq \sum_j f_j^e N_j^e(x) \qquad (15.39b)$$

$$s^e(x) \simeq \sum_j s_j^e N_j^e(x). \qquad (15.39c)$$

Substituting these approximations into our model equation,

$$\sum_j \left[\frac{\partial \phi_j^e}{\partial t} N_j^e(x) + f_j^e \frac{\partial N_j^e}{\partial x} - s_j^e N_j^e(x) \right] = 0. \qquad (15.40)$$

We then apply Galerkin's method, multiplying again by the same set of basis functions N_i, integrating, and setting the integrated residual to zero:

$$\int \sum_j \left[\frac{\partial \phi_j^e}{\partial t} N_j^e(x) + f_j^e \frac{\partial N_j^e}{\partial x} - s_j^e N_j^e(x) \right] N_i^e(x)dx = 0. \qquad (15.41)$$

There is a useful alternative interpretation of this technique so far. The integration *projects* an unknown function onto a finite set of basis functions, yielding a set of coefficients. The sum *reconstructs* that unknown function in terms of its basis. This is exactly analogous to expanding a continuous function in terms of a finite set of Fourier coefficients, then reconstructing that function by a weighted sum of sinusoids.

Next, we interchange the order of integration and summation and integrate the $\partial/\partial x$ term by parts:

$$\sum_j \int \left[\frac{\partial \phi_j^e}{\partial t} N_j^e(x) N_i^e(x) - f_j^e N_j^e(x) \frac{\partial N_i^e}{\partial x} - s_j^e N_j^e(x) N_i^e(x) \right] dx \quad (15.42)$$

$$= -\sum_j N_i^e(x) \left(f_j^e N_j^e(x) \right) \Big|_{x_1}^{x_2}.$$

Note that the right-hand side is a surface integral in 1D. In higher dimensions, we proceed in the same manner but must apply the divergence theorem to the term containing the spatial derivatives. This yields a similar expression, with a volume integral on the left and a surface integral on the right.

The key to the discontinuous Galerkin method is how the right-hand term is handled. Until now, we have only concerned ourselves with a purely element-local expansion of a solution. This cannot by itself, however, possibly lead to a global solution unless we impose additional constraints. In the classic finite element scheme discussed in the previous section, the constraint is *continuity*. This serves to connect each element to the next and ensures a valid global solution, provided the boundary conditions are well-specified.

However, in the discontinuous Galerkin scheme, we relax this restriction and allow the solution to be *discontinuous*: each element has its own independent boundary node, and as such, we have effectively doubled the number of unknowns at the boundaries between elements. At first glance, this might not seem to help matters, but as we will see, we can next borrow an idea from the finite volume technique to connect one element to the next in a physically consistent way.

Identifying the sum on the right-hand side of Equation (15.42) as a numerical flux f^* (as in the finite volume method), we can rewrite the equation as:

$$\sum_j \int \left[\frac{\partial \phi_j^e}{\partial t} N_j^e(x) N_i^e(x) - f_j^e N_j^e(x) \frac{\partial N_i^e}{\partial x} - s_j^e N_j^e(x) N_i^e(x) \right] dx \quad (15.43)$$

$$= -N_i^e(x) f^{*,e}(x) \Big|_{x_1}^{x_2}.$$

Next, we use the fact that the expansion coefficients ϕ_j^e, f_j^e, and s_j^e are independent of x to rewrite Equation (15.43) as:

$$\sum_j \frac{\partial \phi_j^e}{\partial t} M_{ij}^e - f_j^e K_{ij}^e - s_j^e M_{ij}^e = -N_i^e(x) f^{*,e}(x) \Big|_{x_1}^{x_2} \quad (15.44)$$

where:

$$M_{ij}^e = \int N_i^e(x) N_j^e(x) dx \tag{15.45a}$$

$$K_{ij}^e = \int \frac{\partial N_i^e}{\partial x} N_j^e(x) dx. \tag{15.45b}$$

Finally, we observe that this is a matrix equation and write it as:

$$\mathbf{M}^e \frac{\partial \phi^e}{\partial t} - \mathbf{K}^e \mathbf{f}^e - \mathbf{M}^e \mathbf{s}^e = -\mathbf{N}^e(x) f^{*,e}(x) \Big|_{x_1}^{x_2}. \tag{15.46}$$

Thus far, we have not discussed exactly how to calculate the flux function $f^{*,e}(x)$ at the boundaries of the elements. The flux function is what makes the discontinuous Galerkin method feasible: it serves to connect each element to its neighbors and can have dramatic effects on the accuracy and stability of a scheme for a particular problem. The complete description of this flux term is well outside the scope of this book; there are literally decades of finite volume literature devoted to precisely this question. However, a simple (and in practice often very useful) flux function is to simply average the values from neighboring cells. That is:

$$f^{*,e}(x_1) = \frac{1}{2} \left[f^{e-1}(x_2) + f^e(x_1) \right] \tag{15.47a}$$

$$f^{*,e}(x_2) = \frac{1}{2} \left[f^e(x_2) + f^{e+1}(x_1) \right]. \tag{15.47b}$$

The discontinuous Galerkin method has a number of interesting properties. First, the order of accuracy is arbitrary: we can increase the order of accuracy of the entire scheme by simply using a higher-order basis. Second, in contrast to the finite element technique, the mass matrix $\mathbf{M}$ in Equation (15.46) is strictly local, meaning that a complicated inversion is not required, and can even be realistically constructed as a preprocessing step. Finally, this locality can be preserved even for arbitrary orders of accuracy.

Application to Maxwell's equations in 1D

Here we briefly show how the discontinuous Galerkin method can be applied to electromagnetic problems. Maxwell's equations in one dimension in flux conservation form are:

$$\frac{\partial}{\partial t} \begin{bmatrix} \mathcal{E}_y \\ \mathcal{H}_z \end{bmatrix} + \frac{\partial}{\partial x} \begin{bmatrix} 0 & \frac{1}{\epsilon} \\ \frac{1}{\mu} & 0 \end{bmatrix} \begin{bmatrix} \mathcal{E}_y \\ \mathcal{H}_z \end{bmatrix} = 0. \tag{15.48}$$

For simplicity, we will assume that ϵ and μ are constants. As before, we approximate the solutions over each element e as:

$$\mathcal{E}_y(x) \simeq \sum_j \mathcal{E}_{y,j}^e N_j^e(x) \tag{15.49a}$$

$$\mathcal{H}_z(x) \simeq \sum_j \mathcal{H}_{z,j}^e N_j^e(x). \tag{15.49b}$$

As described above, this leads to the following scheme on the element e:

$$\frac{\partial}{\partial t}\mathcal{E}_y^e - (\mathbf{M}^e)^{-1}\mathbf{K}^e\mathbf{f}_{\mathcal{E}_y}^e = -(\mathbf{M}^e)^{-1}\left(\mathbf{N}^e(x_2)f_{\mathcal{E}_y}^{*,e}(x_2) - \mathbf{N}^e(x_1)f_{\mathcal{E}_y}^{*,e}(x_1)\right) \tag{15.50a}$$

$$\frac{\partial}{\partial t}\mathcal{H}_z^e - (\mathbf{M}^e)^{-1}\mathbf{K}^e\mathbf{f}_{\mathcal{H}_z}^e = -(\mathbf{M}^e)^{-1}\left(\mathbf{N}^e(x_2)f_{\mathcal{H}_z}^{*,e}(x_2) - \mathbf{N}^e(x_1)f_{\mathcal{H}_z}^{*,e}(x_1)\right) \tag{15.50b}$$

where we have defined the flux terms as:

$$\mathbf{f}_{\mathcal{E}_y}^e = \frac{1}{\epsilon}\mathcal{H}_z^e \tag{15.51a}$$

$$\mathbf{f}_{\mathcal{H}_z}^e = \frac{1}{\mu}\mathcal{E}_y^e. \tag{15.51b}$$

For the numerical flux terms f^*, as mentioned above, we have some flexibility. The central flux discussed above, for example, is simply:

$$f_{\mathcal{E}_y}^{*,e}(x_1) = \frac{1}{2}\left[f_{\mathcal{E}_y}^{e-1}(x_2) + f_{\mathcal{E}_y}^{e}(x_1)\right] \tag{15.52a}$$

$$f_{\mathcal{E}_y}^{*,e}(x_2) = \frac{1}{2}\left[f_{\mathcal{E}_y}^{e}(x_2) + f_{\mathcal{E}_y}^{e+1}(x_1)\right] \tag{15.52b}$$

$$f_{\mathcal{H}_z}^{*,e}(x_1) = \frac{1}{2}\left[f_{\mathcal{H}_z}^{e-1}(x_2) + f_{\mathcal{H}_z}^{e}(x_1)\right] \tag{15.52c}$$

$$f_{\mathcal{H}_z}^{*,e}(x_2) = \frac{1}{2}\left[f_{\mathcal{H}_z}^{e}(x_2) + f_{\mathcal{H}_z}^{e+1}(x_1)\right] \tag{15.52d}$$

where we have used our basis to reconstruct the value of the flux $f(x)$ at the edges of each element:

$$f_{\mathcal{E}_y}^e(x) = \sum_j f_{\mathcal{E}_y,j}^e N_j^e(x) = \frac{1}{\epsilon}\sum_j \mathcal{H}_{z,j}^e N_j^e(x) \tag{15.53a}$$

$$f_{\mathcal{H}_z}^e(x) = \sum_j f_{\mathcal{H}_z,j}^e N_j^e(x) = \frac{1}{\mu}\sum_j \mathcal{E}_{y,j}^e N_j^e(x). \tag{15.53b}$$

Noting that the right-hand side is a function of the unknowns, we collect all of the terms except the $\partial/\partial t$ term in Equations (15.50) to the right-hand side, yielding the following (where we have dropped the e superscript for clarity):

$$\frac{\partial}{\partial t}\mathcal{E}_y = g(\mathcal{H}_z) \tag{15.54a}$$

$$\frac{\partial}{\partial t}\mathcal{H}_z = h(\mathcal{E}_y) \tag{15.54b}$$

where g and h denote the fact that the right-hand sides here are functions of their arguments. We can now discretize in time using any of the previously discussed techniques. Stability is still a concern, of course, but thankfully much of the intuition gained from working with FDTD carries over into discontinuous Galerkin methods. For instance, using forward Euler with this scheme is unstable. However, for the central flux defined above, the leapfrog technique is stable under specific constraints on Δt:

$$\mathscr{E}_y^{n+1} = \mathscr{E}_y^n + \Delta t \left[g(\mathscr{H}_z^{n+1/2}) \right] \tag{15.55a}$$

$$\mathscr{H}_z^{n+1/2} = \mathscr{H}_z^{n-1/2} + \Delta t \left[h(\mathscr{E}_y^n) \right] . \tag{15.55b}$$

This yields an explicit time-stepping algorithm, similar to the FDTD algorithm, except that the spatial discretization has been applied very differently.

Discontinuous Galerkin methods have become very popular for solutions of Maxwell's equations in complicated geometries, and in hyperbolic problems in general. For further reading in discontinuous Galerkin and DG-FEM methods for electromagnetics, we refer the reader to the book by Hesthaven and Warburton [16], and articles (among many excellent works) by Hu et al. [17], Hesthaven and Warburton [18], and Lu et al. [19], who derive discontinuous Galerkin methods for dispersive materials and a PML boundary condition.

15.5   Summary

In this chapter we have introduced different methods for numerically solving Maxwell's equations in the time domain. We began in Section 15.1 with an introduction to irregular grids. Grids in the FDTD method, or in any numerical method, may be nonuniform, where the spatial grid sizes vary throughout the space; nonorthogonal, where the grid coordinate axes do not form an orthogonal basis; or even unstructured, where the grid is not defined by sequentially indexed locations, but rather is made up of semi-random cells.

Furthermore, grids may be *dynamic*, in that they are redefined at different time steps in the solution; we presented the adaptive mesh refinement algorithm as an example of a dynamic grid.

In Section 15.2 we introduced the finite volume method for electromagnetics. The finite volume method takes a conservative system of equations to relate the density of some parameter inside a small volume to the flux leaving that volume. We presented the finite volume method for a general conservative system, and then for Maxwell's equations specifically, using a number of different formulations. The finite volume method is strictly local, so that an explicit solution can be defined without the need for matrix inversions.

We introduced the finite element method in Section 15.3. The finite element method breaks up the simulation space into small elements, similar to the finite volume method, but adjoins the grid elements by enforcing continuity at the boundaries. We introduced the Galerkin method for finite elements for a general electrostatic problem, and for Maxwell's equations in 1D. The finite element method can achieve arbitrarily high orders of accuracy depending on the basis functions chosen, but at the expense of globally enforced continuity at the nodes of each element.

Finally, in Section 15.4 we introduced the discontinuous Galerkin method, which combines the best of the finite volume and finite element methods. The fields are approximated by basis functions on each element, as in the finite element method, so that high orders of accuracy can be achieved. By joining adjacent cells through the flux leaving each element, the discontinuous Galerkin method takes advantage of the local nature of the finite volume method.

15.6   Problems

15.1.   Robin boundary condition in FEM. Show that the third-order boundary condition, Equation (15.36b), is a statement of the boundary conditions in Chapter 2 at the free space–dielectric interface. Hint: Treat this problem as a dielectric slab with a perfect conductor backing, and use the analytical methods of Chapter 9.

15.2.   Deriving FDTD from FVTD. Show that the finite volume method described by Equations (15.14)–(15.17) collapses to the FDTD algorithm when the grid cells are orthogonal. Hint: on each cell edge, use only the field components parallel to that edge, and note that on the edges, $\mathbf{s} = \mathbf{s}_1^*$, and $\mathbf{s}_2^*$ and $\mathbf{s}_3^*$ are in the plane of the cell face, so that $\overline{\overline{\mathscr{E}}} \cdot \mathbf{s}_2^* = \overline{\overline{\mathscr{E}}} \cdot \mathbf{s}_3^* = 0$.

References

[1] P. Monk and E. Suli, "A convergence analysis of Yee's scheme on non-uniform grids," *SIAM J. Num. Analysis*, vol. 31, pp. 393–412, 1994.

[2] P. Monk, "Error estimates for Yee's method on non-uniform grids," *IEEE Trans. Magnetics*, vol. 30, pp. 3200–3203, 1994.

[3] A. R. Zakharian, M. Brio, and J. V. Moloney, "FDTD based second-order accurate local mesh refinement for Maxwell's equations in two space dimensions," *Comm. Math. Sci.*, vol. 2, pp. 497–513, 2004.

[4] M. J. Berger and J. Oliger, "Adaptive mesh refinement for hyperbolic partial differential equations," *J. Comp. Phys.*, vol. 53, pp. 484–512, 1984.

[5] C. D. Sarris, *Adaptive Mesh Refinement for Time-Domain Numerical Electromagnetics*. Morgan & Claypool, 2007.

[6] A. Taflove and S. Hagness, *Computational Electrodynamics: The Finite-Difference Time-Domain Method*, 3rd edn. Artech House, 2005.

[7] J. F. Thompson, B. K. Soni, and N. P. Weatherill, Eds., *Handbook of Grid Generation*. CRC Press, 1999.

[8] V. Shankar, A. H. Mohammadian, and W. F. Hall, "A time-domain, finite-volume treatment for the Maxwell equations," *Electromagnetics*, vol. 10, pp. 127–145, 1990.

[9] M. Remaki, "A new finite volume scheme for solving Maxwell's system," *COMPEL*, vol. 19, pp. 913–931, 2000.

[10] F. G. Harmon, "Application of a finite-volume time-domain Maxwell equation solver to three-dimensional objects," Master's thesis, Air Force Institute of Technology, 1996.

[11] N. K. Madsen and R. W. Ziolkowski, "A three-dimensional modified finite volume technique for Maxwell's equations," *Electromagnetics*, vol. 10, pp. 147–161, 1990.

[12] K. S. Yee and J. S. Chen, "Conformal hybrid finite difference time domain and finite volume time domain," *IEEE Trans. Ant. Prop.*, vol. 42, pp. 1450–1455, 1994.

[13] K. S. Yee and J. S. Chen, "The finite-difference time-domain (FDTD) and the finite-volume time-domain (FVTD) methods in solving Maxwell's equations," *IEEE Trans. Ant. Prop.*, vol. 45, pp. 354–363, 1997.

[14] S. M. Rao, Ed., *Time Domain Electromagnetics*. Academic Press, 1999.

[15] J. Jin, *The Finite Element Method in Electromagnetics*, 2nd edn. Wiley, 2002.

[16] J. S. Hesthaven and T. Warburton, *Nodal Discontinuous Galerkin Methods: Algorithms, Analysis, and Applications*. Springer, 2008.

[17] F. Q. Hu, M. Y. Hussaini, and P. Rasetarinera, "An analysis of the discontinuous Galerkin method for wave propagation problems," *J. Comp. Phys.*, vol. 151, pp. 921–946, 1999.

[18] J. S. Hesthaven and T. Warburton, "Nodal high-order methods on unstructured grids: I. time-domain solution of Maxwell's equations," *J. Comp. Phys.*, vol. 181, pp. 186–221, 2002.

[19] T. Lu, P. Zhang, and W. Cai, "Discontinuous Galerkin methods for dispersive and lossy Maxwell's equations and PML boundary conditions," *J. Comp. Phys.*, vol. 200, pp. 549–580, 2004.

Index